AF606409

de Gruyter Expositions in Mathematics 2

Combinatorial Homotopy and 4-Dimensional Complexes

by

Hans Joachim Baues

Walter de Gruyter · Berlin · New York 1991

Author
Hans Joachim Baues
Max-Planck-Institut für Mathematik
Gottfried-Claren-Strasse 26
D-5300 Bonn 3, FRG

1980 Mathematics Subject Classification (1985 Revision): Primary: 5502, 5702; 55P05, 55P10, 55P15, 55N25, 55Q05, 55Q15, 55S05, 55S35, 55S37, 55S45, 55S91, 55U10, 55U15, 55U35, 57M05, 57M10, 57M20, 57P10, 57R19.
Secondary: 18D05, 18G35, 18G55, 19D55, 20F14, 20N99.

⊚ Printed on acid-free paper which falls within the guidelines of the ANSI to ensure permanence and durability.

Library of Congress Cataloging-in-Publication Data

Baues, Hans J., 1943–
Combinatorial homotopy and 4-dimensional complexes / by Hans Joachim Baues.
p. cm. – (De Gruyter expositions in mathematics, ISSN 0938-6572 ; 2)
Includes bibliographical references (p.) and index.
ISBN 3-11-012488-2 (acid-free) – ISBN 0-89925-697-X (acid-free)
1. Homotopy theory. 2. CW complexes. 3. Combinatorial topology. I. Title. II. Series.
QA612.7.B386 1991
514′.24–dc20 90-24474

Deutsche Bibliothek Cataloging-in-Publication Data

Baues, Hans J.:
Combinatorial homotopy and 4-dimensional complexes / by Hans Joachim Baues. – Berlin ; New York : de Gruyter, 1991
(De Gruyter expositions in mathematics ; 2)
ISBN 3-11-012488-2
NE: GT

Typesetting: Asco Trade Typesetting Ltd., Hong Kong. Printing: Ratzlow Druck, Berlin. Binding: Dieter Mikolai, Berlin. Cover design: Thomas Bonnie, Hamburg.

Für Barbara
und für unsere Kinder Charis und Sarah

Table of Contents

Chapter III
Crossed modules and homotopy systems of order 3

Chapter IV
Quadratic modules and homotopy systems of order 4

Chapter V
Cohomological invariants

Chapter VI
The cohomology of categories and the calculus of tracks

Preface by Ronald Brown

I believe this book makes important advances for the classical aims of algebraic topology. In order to explain this, I will put the subject and results of this book in the contexts of the history of algebraic topology and of the contribution of algebraic topology to the broad progress of mathematics. In particular, I will explain each of the four main words in the title, as these accurately convey the scope of the book.

1. Homotopy

The process of *classification* is basic in our dealing with the world, and is often used in mathematics. In topology, the idea of *homotopy classification* arises naturally when considering maps between manifolds. It is natural to consider 'nearby' maps as being similar. The equivalence relation generated by this relation gives some kind of classification of maps. Again, the similar idea of 'perturbation' is a common one in mathematics and science: consider a given structure, and then 'perturb' it a bit. However, this idea is not so precise, nor general enough, and a better formulation was found in the notion of *deformation*, or *homotopy*. Such a 'deformation of structure' arises in many areas of mathematics, basically for the same reason, namely to obtain a sensible classification of structure, the standard notion of isomorphism being often too fine to be easy to deal with.

A particular example of classification arose in Poincaré's work on the three-body problem in celestial mechanics. The two body-problem was solved by Newton: two bodies moving under the influence of mutual gravitational forces move in conic sections with the centre of gravity as a focus. The beauty and simplicity of this answer lead people to expect and to hope for a similar kind of solution to the motion of three bodies under gravitational forces. Finally, Poincaré proved that no general solution could be found by a finite number of integrations.

In view of this failure, he sought to ask different questions, of a qualitative kind, by the following method. The position and velocity of a particle in 3-dimensional space are together described by six coordinates. So the positions and velocities of three particles are described by eighteen coordinates. However, if the particles are given initial positions and velocities, then thereafter the total energy of the system remains constant. This gives an equation 'energy = constant' governing the motion, so that the phase space P of the motion is expected to be a seventeen dimensional subspace of the original eighteen dimensional space. A motion of the three bodies with a given energy is then a path in this phase space P. A classification of motions is thus equivalent to a classification of such paths.

The immediate classification to be thought of is that of 'nearby' paths. A more refined idea is that of continuous deformation through paths. That is, the path f_0 from a to b is *deformable* into the path f_1, also supposed from a to b, if there is a family f_t of paths from a to b such that the map $(s,t) \mapsto f_t(s)$ is continuous as a function of two variables. A more modern term for this concept is *homotopy keeping the end points fixed.* The set of these homotopy classes of paths in P is written $\pi_1 P$. The paths which start and finish at the same point a are called *loops* at a. The subset of $\pi_1 P$ of classes of loops at a is written $\pi_1(P,a)$. In fact, Poincaré found that these classes of loops at a could be given a group structure under the obvious composition of paths (first do one, then do another). This group $\pi_1(P,a)$ is called the *fundamental group* of the space P at the point a. The space P is called *simply connected* if it is connected and this fundamental group at any point is trivial.

This is an indication of another application of the homotopy notion. Quite often, convenient algebraic structures can be placed on certain homotopy classes of given structures, but not on the structures themselves. Or again, sometimes an algebraic structure, for example a composition, can be placed on the original structure and this structure is inherited by the homotopy classes, but the structure on the homotopy classes has more convenient and usable properties. In the case of loops, the algebraic structure of the composition of loops is not easy to analyse, whereas the homotopy classes of loops at a have this familiar structure of a group $\pi_1(P,a)$!

Another source for the homotopy notion was that of *simplicial approximation.* It is worth recalling the historical importance of this, because the simplicial theory has some interesting contrasts with that of *combinatorial homotopy.*

The formula $V - E + F = 2$ for a polyhedral decomposition of the surface S of a sphere was known to Euler (1752), and has been said to have been known earlier to Descartes. It is easy to prove that the number

$$\chi(S) = V - E + F$$

is invariant under subdivision of the given polyhedral decomposition, but it is not so clear that this number will be the same whatever decomposition of the sphere is chosen. The same problem arose with the Euler-Poincaré characteristic

$$\chi(K) = \sum_{r=0}^{r=n} (-1)^r \alpha_r$$

of a simplicial complex K, where α_r is the number of r-simplices of K. It is again easy to prove this is invariant under subdivision of the complex, but it is not easy to prove that $\chi(K)$ is dependent only on the space and not on the chosen type of triangulation, that is, decomposition into simplices. The similar problem for the *Betti numbers*, *torsion coefficients*, and the later defined *homology groups* $H_n(K)$, became known as the problem of the *topological invariance of homology*. It was hoped to prove it by proving the *Hauptvermutung* (*fundamental conjecture*), namely that *homeomorphic complexes have isomorphic subdivisions*. This was a hopeless task, since it was proved by J. Milnor in 1962 that the Hauptvermutung is false.

A basic advance in this area was made by J.W. Alexander in 1915. He proved the topological invariance of homology groups by proving a stronger result, namely that homology groups are isomorphic not only for homeomorphic spaces, but also for spaces of the same *homotopy type*. In order to explain this notion, recall that two spaces X and Y are *homeomorphic* if there are continuous maps $f: X \to Y$ and $g: Y \to X$ such that the composites fg and gf are the identity maps on Y and X respectively, i.e. $fg = 1_Y$, $gf = 1_X$. (We do not need to be precise here about the chosen definition of *space* as long as we have a clear notion of *continuity*.)

Now we define a classification of maps $X \to Y$ by saying two maps f_0 and f_1 are *homotopic*, written $f \simeq g$, if there is a family of maps $f_t: X \to Y$, $t \in [0, 1]$, such the map $(x, t) \mapsto f_t(x)$ is continuous in the two variables. It is not hard to prove that this relation is an equivalence relation. There is a corresponding classification of spaces: two spaces X and Y are of the same *homotopy type* if there are maps $f: X \to Y$ and $g: Y \to X$ such that fg and gf are *homotopic* to the identity maps on Y and X respectively; in symbols, $fg \simeq 1_Y$, $gf \simeq 1_X$. Alexander proved that simplicial complexes of the same homotopy type have isomorphic homology groups by showing that if $f: K \to L$ is a continuous map of simplicial complexes, then K has a subdivision K^s such that f is homotopic to a simplicial map $K^s \to L$. This neatly sidestepped the impossible problem of the Hauptvermutung, thus illustrating an important principle of research, that sometimes the interesting question is not 'What is the answer?' but instead 'What is the (right) question?'. Alexander's methods confirmed the importance of the notion of homotopy in algebraic topology.

There is a further advantage of the homotopy classification of maps and of spaces. For finite simplicial complexes K and L, the set of homotopy classes of maps $K \to L$ is countable (this is not so obvious!), whereas in general the set of all such maps is uncountable. Thus there is a greater prospect of describing this set of homotopy classes than there is of describing the set of all maps. Further, the classification of homotopy classes by numerical or algebraic data may actually show us the interesting and significant properties of maps. For example, the Brouwer *degree* of a self map $S^n \to S^n$ of the n-sphere S^n is an integer which is a determination of the homotopy class of the map, and shows the algebraic number of times the map 'wraps the sphere around itself'.

Once the notion of homotopy was established, an early question was to *describe* homotopy types of spaces, or, to be more explicit, of say simplicial complexes. However, the question of deciding by Alexander's methods whether two simplicial complexes X, Y are homotopy equivalent involved arbitrary subdivisions of X and Y. This is a major disadvantage of the simplicial approach. There was a need to answer the question without using subdivisions by the determination of *invariants* or *algebraic models* of the homotopy types.

It was early found that the homology groups were not always sufficient to distinguish homotopy equivalent spaces, since for example the torus and the sphere with two tangential circles have the same homology groups, but are not homotopy equivalent. In fact, these spaces have non isomorphic fundamental groups. The description of all homotopy types of connected 1-dimensional complexes is not hard: such homotopy types are represented by *wedges* (i.e. unions with one point identified) of circles. How then should one describe all homotopy types of say 2-dimensional complexes? This, and its higher dimensional counterparts, is a classical and basic problem which corresponds to the '*main problem of algebraic topology*' in the book of Seifert-Threlfall.

An important advance involved the notion of *homotopy groups*. The fundamental group of a space P with a base point a may be also defined as the set of homotopy classes of maps f from the circle S^1 to P such that f maps the base point e, say, of S^1 to a. In 1932, Čech gave a corresponding definition of *higher homotopy groups*, $\pi_n(P)$, using pointed maps $S^n \to P$. His paper on this was submitted to the International Congress of Mathematicians at Zurich in that year. There now arises a curious story. Alexandroff and Hopf, or maybe someone else, quickly proved that these groups are commutative for $n \geq 2$. On these grounds, it seems, it was felt that the groups had to be the same as the homology groups of the space. Čech was then persuaded to withdraw his paper, so that only a small paragraph appeared in the Proceedings of the Congress. One of the strange features of this story is that the Hopf map $S^3 \to S^2$ was found by Hopf in 1929, so that it was in essence known to Hopf that $\pi_3(S^2)$ was non-trivial and so did not coincide with the (trivial) homology group $H_3(S^2)$.

Čech wrote no more on the subject, but extensive work on the relations between homotopy and homology groups was published by W. Hurewicz in

1935, so that these groups have come to be called the Hurewicz homotopy groups. The calculation of homotopy groups even of such simple spaces as spheres proved a subtle and difficult problem, and there is still no formula, not even an expected formula, for the homotopy group $\pi_r(S^n)$, though it is known that these groups are commutative in all cases and are finite for $r \neq n, 2n - 1$. Because of the difficulty of this problem, a considerable amount of effort has been given to it by algebraic topologists and homotopy theorists, and a large amount of important algebraic techniques have been developed for its analysis.

Nonetheless, the original problem of describing homotopy types remained. It was found that the homotopy groups and the homology groups were not themselves sufficient to determine the homotopy types.

Homotopy groups were used by Postnikov to describe a further presentation of a homotopy type via iterated 'fibrations'. On the other hand, simplicial complexes and more generally Whitehead's *CW*-complexes were built by iterated 'cofibrations'. These two presentations refer to a further basic question, the relation between homology groups and homotopy groups, whose resolution has to be regarded as a part of the homotopy classification problem for spaces. A clear solution for the rational groups $\pi_n(K) \otimes \mathbb{Q}$ and $H_n(K) \otimes \mathbb{Q}$ is available by the work of Quillen and Sullivan, provided $\pi_1(K) = 0$. An integral approach using simplicial groups is due to Kan and Curtis.

2. Complexes

We have several times used the word 'complex'. This is indeed a fundamental idea in algebraic topology, namely that of building a complicated space out of small, standard pieces. This idea can be seen in Euler's formula, where a surface is built out of faces. It was developed further by Poincaré, using the idea of a triangulation, i.e. a breaking of a space into r-simplices for various r, where an r-simplex is the r-dimensional version of the 2-dimensional triangle. He used this method to make precise the ideas of Betti and others on the notion of homology groups. Originally the idea of a *cycle* was imprecise, since it was thought of as some kind of *composition* of small pieces. Dieudonné has suggested that the original idea was moving towards a primitive notion of *bordism.* In any case, the idea of composition was replaced by the idea of working in what we now call the free abelian group on the set of r-simplices, but was earlier simply thought of as *formal sums* of r-simplices. Such formal sums also arose naturally in integration theory, as formal sums of domains of integration. The problem of defining the boundary of such an r-chain was solved initially by the assignment of *orientations* to the r-simplices, for all r.

It was only in the early 1940s that Eilenberg introduced the idea of an *ordered* simplicial complex, so that each simplex σ had a well assigned i-face $\partial_i\sigma$, thus allowing for the famous formula for the boundary of an r-simplex

$$\partial\sigma = \sum_{i=0}^{i=r} (-1)^i \partial_i \sigma,$$

and the proof of the crucial rule

$$\partial^2 = 0.$$

Unfortunately, the computation of homology groups using triangulations was notably tedious. Intuitively, it was much easier to compute the homology of say a Klein bottle by using the standard diagram

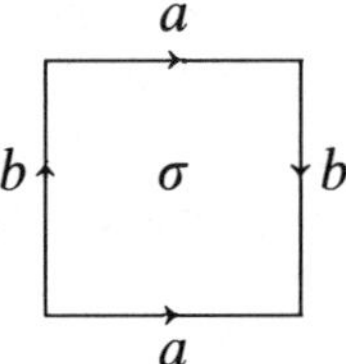

so yielding the formula $\partial\sigma = a + b - a + b = 2b$. The justification of such a formula was somewhat awkward. In a paper published in 1940, J.H.C. Whitehead began to formalise this idea of *conglomerating* simplices into blocks by using an idea of Borsuk of a *membrane complex*. In the late 1940s, this idea was formalised in Whitehead's famous paper defining a CW-complex, which is a decomposition of a space into subspaces homeomorphic to the interior of a cell. The important feature of the definition was that it allowed proofs by induction on the dimension of cells. These CW-complexes are now a basic tool in algebraic topology and its applications.

One useful feature of CW-complexes is that the homology of such a complex X may be computed entirely in terms of the cellular structure, without any subdivisions. Here the (abelian) group of n-dimensional cellular chains of X is defined to be

$$C_n(X) = H_n(X^n, X^{n-1}),$$

which is the free abelian group generated by the n-cells of X. The boundary $\partial: C_n(X) \to C_{n-1}(X)$ is defined to be the boundary operator in the exact homology sequence of the triple (X^n, X^{n-1}, X^{n-2}). It may be shown how to compute this boundary from the cellular structure, and this method leads to the formula for the boundary of the 2-cell of the Klein Bottle given above. It is this idea that may be called *combinatorial homology theory*. What makes it work is the *excision theorem* for relative homology, which is the fact which replaces the classical use of simplicial approximation.

Various geometric notions of *complex*, such as simplicial complex or *CW*-complex, have algebraic counterparts with analogous properties. The simplest of these is that of a *chain complex*. This is simply a sequence of 'boundary' morphisms $\partial_n\colon C_n \to C_{n-1}$ (satisfying $\partial_{n-1}\partial_n = 0$) of abelian groups, or more generally of modules over a ring, or even, in the most abstract and general case, of objects of an abelian category. This last case has the advantage of including the previous ones and also other cases, such as categories of sheaves of modules, which have important applications, for example in complex analysis, and in algebraic geometry.

It is important that chain complexes have not only notions of morphisms but also notions of *homotopy* of morphisms. For a chain complex C, a notion of homology $H_*(C)$ is defined and a fairly easy fact is that homotopic morphisms of chain complexes induce the same map in homology. This is the algebraic residue of Alexander's argument for the homotopical invariance of homology groups.

There are other notions of complex which are important in the literature. However, their status as *algebraic* complexes is more sophisticated. They form sometimes a kind of halfway house between geometry and algebra. This is useful, given the overall role of algebraic topology, to develop algebraic methods for modelling geometric phenomena. Readers should also be aware of the reasons for the importance of this modelling. One is to do calculations, since it is easier to formulate algorithms in algebraic situations. Another is that algebraic proofs are more robust than geometric ones; there are many examples of this in the literature. But the most important reason is for understanding. It is not so much to construct and use algebraic models as to understand the *formal* processes underlying geometric phenomena.

It is in this respect that the 'crossed chain complexes' and 'quadratic chain complexes' are more efficient than for example the large algebraic models given by simplicial sets or simplicial groups. These generalised chain complexes share many properties with the classical cellular chain complex $C_*(\hat{X})$ of the universal cover $\hat{X}$ of the *CW*-complex X. This chain complex of modules over the fundamental group of X was considered by Reidemeister in the 1930s, and was later investigated by Eilenberg-Mac Lane in fundamental papers on the homology of spaces with operators. J.H.C. Whitehead observed that crossed chain complexes have a higher level of realisability than chain complexes with operators. The quadratic chain complexes introduced in this book have a higher level of realisability than crossed chain complexes.

3. Combinatorial

This is a good point to write something explicit about the word 'combinatorial', especially as it has been used several times. It refers to an important notion, but one which is more difficult to make precise. The word is used by Whitehead, but not explained by him.

There is a subject called 'Combinatorial group theory'. What is included is largely a matter of taste, but workers in the field would agree that it includes such matters as free groups, generators and relations, presentations of groups, equivalences of presentations, free products, free products with amalgamation, HNN extentions, graph products, Cayley diagrams.

The term 'combinatorial homotopy' is intended to indicate similar concerns but in a geometric context. For example, it is standard that a presentation of a group determines a 2-dimensional *CW*-complex, whose 1-cells correspond to the generators of the presentation and whose 2-cells correspond to the relators. It is then of interest to compare algebraic operations on presentations of groups with operations on the corresponding complexes. Further, this comparison should be effected without having to resort to subdivision of the cells. It was this 'combinatorial' aim which led Whitehead initially to the notion of membrane complex, and finally to the notion of cell complex and *CW*-complex, since he was deeply versed in the work on combinatorial group theory and combinatorial topology of the 1930s.

In particular, a major aim of combinatorial homotopy would be to describe algebraically the homotopy types of low-dimensional complexes directly in terms of the cell structures. Part of this aim would also be to relate the description of homotopy types clearly to homology. The algebra should also be related to that of group theory on the one hand, particularly in order to deal with non-simply connected spaces, and with the general area of homological algebra on the other.

In this context, it should be remembered that algebraic topology has had an enormous influence on the development of homological algebra. The first example of this was the Hopf formula of 1941, which showed that if X is a connected simplicial complex which has fundamental group G and which is aspherical, i.e. all homotopy groups vanish in dimensions greater than 1, then

$$H_2(X) \cong (R \cap [F,F])/[F,R]$$

for any presentation $1 \to R \to F \to G \to 1$ of G where F is a free group. This result led to the work of Eilenberg-Mac Lane describing *all* the homology groups of such an X directly in terms of the group G. Then Cartan and Eilenberg found that similar methods worked in a wide variety of algebraic situations, and named the resulting subject *homological algebra*. This developed with the work of Serre, Grothendieck, and many others, into a fundamental tool in many algebraic and geometric situations. For example, the solution by Deligne of the Weil conjectures in number theory would have been inconceivable without this kind of machinery. This kind of progress suggests that the development of algebraic tools for specific homotopical problems has the possibility of wide prospects outside its original conception and aims, and must reflect some deep facts about the nature of these problems. Certainly for me, this is one of the purposes of taking an eclectic approach to the development of algebraic tools in homotopy theory.

The development of such tools to include the case of non simply-connected spaces presents special difficulties and problems, and much current work in homotopy theory, particularly stable homotopy theory, tries to avoid these problems by taking the theory as far away from the fundamental group as possible. An argument for this is the intractability of many problems in group theory itself, particularly in combinatorial group theory. On the other hand, in many application of topology, such involvement of the fundamental group is essential. This situation occurs for example in the homology of groups, in algebraic K-theory, in spaces with a group of operators, and in low-dimensional topology, for example in knot theory. Thus the algebraic description of homotopical problems for non-simply connected spaces has wide prospective importance.

Even in the simply connected case, there are still clear calculational problems in low dimensions. For example, one would like to have methods which ensure that one could calculate just the *number* of simply connected homotopy types of 5 dimensional complexes with a given homology. For simply connected 4-dimensional complexes, this kind of problem was solved by J.H.C. Whitehead's famous classification of such complexes. He liked to tell the story that when he announced his results at an international meeting, H. Whitney rose and said he did not believe them. Whitehead felt this to be a tribute to the quality of the results that someone as good as Whitney found them quite unexpected!

The algebraic methods for both the simply-connected and non-simply connected cases should also be instructive, in the sense that they should yield ways of modelling geometric constructions on spaces by corresponding algebraic constructions. To obtain this is a tall order. For example, one cannot expect to be able to calculate all matters related to finite complexes, in view of the algorithmic insolubility of many problems in group theory. But it explains that the aim of combinatorial homotopy is to find for the homotopy theory of complexes general algebraic methods which help ones understanding, and which will in many particular cases allow for calculations which can be carried out in complete detail.

4. Dimension 4

Some explanation should be given of why there is interest enough in the 4-dimensional case to devote such a book to it. After all, if one is interested in general homotopy theory, surely 4 is a long way from say 1000, and if the 4-dimensional case is so complicated then the prospects for the general case look dim.

Part of the answer to such an argument is that in the first place, the low dimensional case is particularly interesting. It is closely related to group theory, which has wide applications in both mathematics and in physics. In some areas of topology, the low dimensional case has proved particularly complicated and fascinating. For example, the Poincaré conjecture, that an n-manifold of the homotopy type of the n-sphere is homeomorphic to the n-sphere, has been proved in every dimension except 3, and the 4-dimensional case presented special problems, which were resolved by Freedman. There is fundamental work by Donaldson on the case of 4-dimensional manifolds which lead for example to the result that the 4-dimensional real vector space has more than one differential structure, in fact uncountably many. This phenomenon occurs only in dimension 4.

Thus there is intrinsic interest in low dimensions and in extending the wide variety of methods of group theory, which itself corresponds to the homotopy theory of 2-dimensional complexes.

A description of the homotopy types of 3-dimensional complexes was found by J.H.C. Whitehead using the notion of *crossed module* and *crossed chain complex*.

The 4-dimensional case needs an extra *quadratic* information, essentially because of the quadratic properties of the Hopf map $S^3 \to S^2$ referred to above. The major work of this book concerns this quadratic information. Some of the methods are due to Whitehead, for example the use of his universal quadratic functor Γ. But this is the first text in which such methods appear. Other methods given here are entirely new. For example, the theory of quadratic modules and quadratic chain complexes yields fundamental new algebraic models of 4-dimensional complexes.

This book thus lays the foundation for the application of such quadratic methods both in homotopy theory and in related areas.

Ronald Brown,
University of Wales,
Bangor, July, 1990

Introduction

The standard algebraic invariants of a topological space depend only on the homotopy type of the space. *Combinatorial homotopy* deals with the converse problem of the determination of the homotopy type by algebraic invariants. In this book we are concerned with the combinatorial homotopy theory of 4-dimensional complexes. Our principal objective is the solution of an old problem of J.H.C. Whitehead who states

> "What has been achieved so far is a purely algebraic description of the homotopy type of any 3-dimensional complex and of any finite, simply connected, 4-dimensional complex", see Whitehead (C),

and later writes

> "Therefore it seems reasonable to hope that these theorems [on the homotopy classification of simply connected 4-dimensional complexes K] in conjunction with the cohomology theory of abstract groups, may lead to similar theorems for $\pi_1(K) \neq 0$", see Whitehead (SB).

We achieve a purely algebraic description of the homotopy types of 4-dimensional complexes with non-trivial fundamental groups by minimal models which we call *quadratic chain complexes.* These models blend with and provide a link between Whitehead's famous classifications of 3-dimensional complexes and simply connected 4-dimensional complexes.

Whitehead's hope above, however, was somewhat optimistic since (as we show) homotopy types of 4-dimensional complexes are not only intimately related with the cohomology theory of abstract groups, but also with the algebraic K-theory of group rings, in particular with $K_3(\mathbb{Z})$. In fact, the exotic element in $K_3(\mathbb{Z})$ corresponds to a universal obstruction for the existence of Pontrjagin squares with local coefficients. Therefore an algebraic description of the homotopy type of a 4-dimensional complex cannot be obtained by the use of Pontrjagin squares as in the simply connected case. Instead the "quadratic structure" of the complex has to be described by a new kind of algebraic object which we call a *quadratic module.* A quadratic module is the 3-dimensional part of a quadratic chain complex; a complete definition of these notions is given at the end of this introduction.

A further task of combinatorial homotopy is the computation of the homology groups $H_n(X)$ and homotopy groups $\pi_n(X)$ of a connected CW-complex X. In addition one would like to know the relations between these groups. This indeed is a very hard problem; for example, Massey in his book on algebraic topology writes:

> "... the problem of determining the structure of the groups $\pi_n(X)$ for a noncontractible space X is difficult or impossible. Most of the higher homotopy groups $\pi_n(X)$ are unknown even in the relatively simple case where X is a k-sphere, $k \geq 2$; ... homology or cohomology groups have many properties similar to those of homotopy groups, and they have one important advantage over the latter: For a wide variety of topological spaces, their algebraic structure is computable. Homology groups were introduced by Poincaré about 1895".

In low dimensions there is, however, a well known combinatorial theory of the fundamental group $\pi_1(X)$, see for example Van Kampen, Seifert-Threlfall or Brown. Moreover the second homotopy group $\pi_2(X)$ can be computed by the Hurewicz homomorphism

$$\hat{h}: \pi_n X \cong \pi_n \hat{X} \to H_n \hat{X}$$

which is an isomorphism for $n = 2$ and surjective for $n = 3$. Here $\hat{X}$ is the universal covering of X. The combinatorial properties of $\pi_2 X$ were already studied by Johansson and Reidemeister and later by J.H.C. Whitehead. The combinatorial nature of $\pi_3(X)$ remained open in the literature. This problem is solved by the quadratic chain complex $\sigma(X)$ of the CW-complex X. In fact $\sigma(X)$ is given by a diagram of σ_1-groups $\sigma_n = \sigma_n(X)$

$$\begin{array}{ccccccc} & & C_2 \otimes C_2 & & & & \\ & & \downarrow{\scriptstyle \omega} & & & & \\ \longrightarrow \sigma_4 & \xrightarrow[d_4]{} & \sigma_3 & \xrightarrow[d_3]{} & \sigma_2 & \xrightarrow[d_2]{} & \sigma_1 \end{array}$$

with additional properties. The row of the diagram is a chain complex of groups with homology groups

$$\pi_n \sigma(X) = \text{kernel}\,(d_n)/\text{image}\,(d_{n+1}).$$

These homology groups of $\sigma(X)$ determine in low dimensions the homotopy groups of X since we show that there is a natural isomorphism of $\pi_1(X)$-modules:

Theorem A.
$$\pi_n\sigma(X) \cong \begin{cases} \pi_n X & \textit{for } n \leq 3, \\ \hat{h}\pi_4 X & \textit{for } n = 4, \\ H_n \hat{X} & \textit{for } n \geq 5. \end{cases}$$

Here $\hat{h}\pi_4 X$ is the image of the Hurewicz homomorphism above.

The definitions of maps, weak equivalences and homotopies in the category of quadratic chain complexes are similar to those in the category of classical chain complexes of abelian groups. A crucial new feature of a quadratic chain complex is the *quadratic map*

$$\omega \colon C_2 \otimes C_2 \to \sigma_3.$$

When we divide the image of the quadratic map in $\sigma(X)$ we obtain the following commutative diagram in which the columns are exact sequences and the maps h_i are isomorphisms for $i \neq 2, 3$.

$$\begin{array}{ccccccccc} & & & & C_2 \otimes C_2 & = & C_2 \otimes C_2 & & \\ & & & & \downarrow \omega & & \downarrow d_3\omega & & \\ \longrightarrow & \sigma_4 & \longrightarrow & & \sigma_3 & \longrightarrow & \sigma_2 & \longrightarrow & \sigma_1 \\ \cdots & \cong \downarrow h_4 & & & \downarrow h_3 & & \downarrow h_2 & & \cong \downarrow h_1 \\ \longrightarrow & \pi_4(X^4, X^3) & \longrightarrow & & \pi_3(X^3, X^2) & \longrightarrow & \pi_2(X^2, X^1) & \longrightarrow & \pi_1 X^1 \end{array}$$

The bottom row of the diagram is given by the relative homotopy groups of the skeletal filtrations of X and

$$C_2 = H_2(\hat{X}^2, \hat{X}^1) = \pi_2(X^2, X^1)^{ab}$$

is the abelianization of the cokernel of $d_3\omega$. The bottom row is the "crossed chain complex" $\rho(X)$ studied in Whitehead (CH). Therefore the quadratic chain complex $\sigma(X)$ can be considered to be the 'quadratic refinement' of the crossed chain complex $\rho(X)$. The homology groups $H_n\hat{X}$ and $H_n X$ can be deduced easily from the crossed chain complex $\rho(X)$. Therefore theorem A shows us the complete algebraic interaction between the low dimensional homotopy groups $\pi_1 X$, $\pi_2 X$, $\pi_3 X$, $\hat{h}\pi_4 X$ and the homology groups $H_*\hat{X}$, $H_* X$.

Theorem A yields indeed an optimal natural algebraic presentation of the $\pi_1(X)$-module $\pi_3(X)$ since the algebraic model $\sigma(X)$ of X is *minimal* with respect to the number of generators (which are the cells of X) and with respect to the 'nilpotency degree' of the groups $\sigma_n(X)$, $n \geq 2$. Moreover the quadratic chain complex $\sigma(X)$ is 'totally free' in the sense that it satisfies a freeness condition in each degree, the basis elements in degree n being the n-cells of X.

For example we get for the 2-sphere S^2 the quadratic chain complex $\sigma(S^2)$ by the simple diagram

$$\begin{array}{ccc} \mathbb{Z} & & \\ \Big\downarrow{\scriptstyle \omega = 1} & & \\ \mathbb{Z} & \xrightarrow[d_3 = 0]{} & \mathbb{Z} \end{array}$$

with $\sigma_i(S^2) = 0$ for $i \neq 2, 3$. For the complex projective plane $\mathbb{C}P_2$ the quadratic chain complex $\sigma(\mathbb{C}P_2)$ is given by the diagram

$$\begin{array}{ccccc} & & \mathbb{Z} & & \\ & & \Big\downarrow{\scriptstyle \omega = 1} & & \\ \mathbb{Z} & \xrightarrow[d_4 = 1]{} & \mathbb{Z} & \xrightarrow[d_3 = 0]{} & \mathbb{Z}. \end{array}$$

These examples demonstrate the minimality of the algebraic model $\sigma(X)$. In general it is enough to describe $\sigma(X)$ by fixing formulas for the boundaries $d(e)$ where e is a cell of X. For example the quadratic chain complex $\sigma(\mathbb{R}P_\infty)$ of the real projective space $\mathbb{R}P_\infty$ is given by the following formulas where we use additive notation for the group structure in σ_n and where e_n is the n-cell of $\mathbb{R}P_\infty$ with $e = e_1$.

$$d(e_n) = \begin{cases} e + e & n = 2 \\ e_2 - e_2^e & n = 3 \\ e_3 + e_3^e + \omega(e_2 \otimes e_2) & n = 4 \\ e_{n-1} + (-1)^n e_{n-1}^e & n \geq 5. \end{cases}$$

Many further examples are discussed in this book.

We now quote several results from the book which describe fundamental properties of the quadratic chain complex $\sigma(X)$.

Theorem B. *The full homotopy category of connected 3-dimensional CW-complexes is equivalent to the homotopy category of 3-dimensional totally free quadratic chain complexes.*

Theorem C. *The homotopy types of connected* 4*-dimensional* CW*-complexes are in* 1-1 *correspondence with the homotopy types of* 4*-dimensional totally free quadratic chain complexes.*

Theorem D. *Each* 5*-dimensional totally free quadratic chain complex* σ *is realizable by a* 5*-dimensional* CW*-complex* X *with* $\sigma(X) \cong \sigma$. *Moreover for* 4*-dimensional connected* CW*-complexes* X, Y *each map* $\sigma(X) \to \sigma(Y)$ *is realizable by a map* $X \to Y$.

Theorem C yields a solution of Whitehead's problem concerning the homotopy classification of 4-dimensional complexes. We deduce from quadratic chain complexes new cohomological invariants with local coefficients which generalize the classical Pontrjagin squares. These invariants are sufficient to classify certain classes of 4-dimensional homotopy types. In general, however, the Pontrjagin square and the Steenrod square

$$\wp\colon \hat{H}^2(X, A) \to \hat{H}^4(X, \Gamma A),$$

$$\mathcal{S}q\colon \hat{H}^n(X, A) \to \hat{H}^{n+2}(X, A \otimes \mathbb{Z}/2)$$

do not exist for arbitrary $\pi_1(X)$-modules A. The universal obstruction for their existence is a non-trivial element of order 2 in the cohomology of the category of free modules, see Chapter V. The third cohomology of this category of modules and the connection with algebraic K-theory is described in Chapter VI.

Also quadratic chain complexes and quadratic modules which are not totally free play an important rôle. For this we recall that an n-type Y is a connected CW-complex with trivial homotopy groups $\pi_j(Y) = 0$ for $j > n$. As pointed out by Whitehead (C), an abstract group is the algebraic equivalent of a 1-type. It is of interest to find appropriate algebraic objects which represent the hierachy of n-types ($n = 1, 2, \ldots$). MacLane-Whitehead observed that 2-types are represented algebraically by crossed modules. The next result shows that a quadratic module is the algebraic equivalent of a 3-type; the specific advantages of quadratic modules in comparison with other algebraic models of 3-types are discussed in the text.

Theorem E. *The full homotopy category of* 3*-types is equivalent to the localization of the category of quadratic modules with respect to weak equivalences.*

In order to give the reader already at this point an impression of the miraculous algebraic nature of 4-dimensional homotopy types we now give the complete definitions of quadratic modules and quadratic chain complexes.

Definition. We write a group N additively and the action in an N-group M is denoted by $x^\alpha, x \in M, \alpha \in N$. The group N is an N-group by $x^\alpha = -\alpha + x + \alpha$ for x, $\alpha \in N$. A *quadratic module* $(\omega, \delta, \partial)$ is a diagram of N-groups and N-equivariant homomorphisms

$$C \otimes C \xrightarrow{\omega} L \xrightarrow{\delta} M \xrightarrow{\partial} N$$

satisfying the equations

$$\begin{cases} \partial\delta = 0, \\ -x - y + x + y^{\partial x} = \delta\omega(\{x\} \otimes \{y\}), \\ -a - b + a + b = \omega(\{\delta a\} \otimes \{\delta b\}), \\ a^{\partial x} = a + \omega(\{\delta a\} \otimes \{x\} + \{x\} \otimes \{\delta a\}) \end{cases}$$

for a, $b \in L$, x, $y \in M$. Here C is the abelianization of the quotient group $M/\langle M, M\rangle$ where $\langle M, M\rangle$ is the subgroup of M generated by all elements $\langle x, y\rangle = -x - y + x + y^{\partial x}$. The element $\{x\} \in C$ is represented by $x \in M$ and the action of $\alpha \in N$ on the $\mathbb{Z}$-tensor product $C \otimes C$ is given by

$$(\{x\} \otimes \{y\})^\alpha = \{x^\alpha\} \otimes \{y^\alpha\}.$$

A *quadratic chain complex* σ is a diagram of σ_1-groups

$$\begin{array}{ccccccccc} & & & & C_2 \otimes C_2 & & & & \\ & & & & \big\downarrow \omega & & & & \\ \cdots \longrightarrow & \sigma_4 & \xrightarrow[d_4]{} & & \sigma_3 & \xrightarrow[d_3]{} & \sigma_2 & \xrightarrow[d_2]{} & \sigma_1 \end{array}$$

where (ω, d_3, d_2) is a quadratic module and where the row is a chain complex of groups with $d_n d_{n+1} = 0$ for $n \geq 2$. For $n \geq 4$ the groups σ_n are abelian and the image $d_2\sigma_2$ acts trivially on σ_n. A map $f: \sigma \to \sigma'$ is a chain map of f_1-equivariant homomorphisms compatible with the quadratic map ω, that is $f_3\omega = \omega(f_* \otimes f_*)$ where f_* is induced by f_2.

The bulk of the book describes the algebraic theory of quadratic modules and their connections with 4-dimensional complexes, Pontrjagin squares, homotopy groups, the cohomology of categories, and algebraic K-theory. The account is essentially self contained and should be accessible to graduate students with some background in algebraic topology and homotopy theory. In the first three chapters we describe all relevant results from the literature needed in the book, in particular the *CW*-tower of categories, the theory of crossed modules and crossed chain complexes. This provides the reader with an introduction to basic combinatorial homotopy theory. Many explicit examples and applications of the theory are described. For a more complete discussion of the contents see the chapter introductions.

The six chapters which comprise the book are subdivided into several sections, §0, §1, etc. and appendices A, B etc.; definitions, propositions, remarks, etc., are consecutively numbered in each section, each number being preceded by the section number, for example (6.1) or (A.5). A reference number such as (II.5.6), resp. (II.B.3), points to (5.6) in chapter II, resp. (B.3) in appendix B of chapter II, while (5.6) and (B.5)(2) point to (5.6) and (B.5)(2) in the chapter at hand. References to the bibliography are given by the author's name, e.g. Steenrod or J.H.C. Whitehead (CH).

I would like to acknowledge the support of the Max-Planck-Institut für Mathematik in Bonn, the Forschungsinstitut für Mathematik ETH Zürich, and the University of Wales Bangor. Moreover, I am grateful to my students in Bonn, in particular to W. Dreckmann, M. Hartl, M. Hennes, A. Hohmann, M. Weick and J. Zobel who read parts of the manuscript and who computed examples. I am very grateful to D. Conduché and R. Brown for comments on an earlier version of this manuscript. Special thanks go also to R. Brown for preparing a preface for this book.

Hans Joachim Baues
Bonn, November, 1990

Chapter I

Homotopy, homology, and Whitehead's classification of simply connected 4-dimensional CW-complexes

In the first part of this chapter we recall some basic definitions and facts of homotopy theory. For example we describe homotopy groups together with the action of the fundamental group. We introduce the cellular chain complex of the universal covering of a CW-complex and we derive from it the homology and cohomology with local coefficients. Moreover the Hurewicz homomorphism is obtained as part of Whitehead's certain exact sequence. The first non trivial Γ-group in the sequence is given by the quadratic functor Γ which plays a major role in the book. In terms of this functor we introduce the classical Pontrjagin square which is a cohomological invariant of a homotopy type.

In the second part of the chapter we discuss and prove classical results of J.H.C. Whitehead concerning the homotopy classification of simply connected 4-dimensional CW-complexes X. For this we associate with X the "Pontrjagin invariant"

$$\mathscr{P}(X) = (C, \xi), \quad \text{with} \quad \xi \in H^4(C, \Gamma H_2 C), \tag{1}$$

where $C = \tilde{C}_*(X)$ is the reduced cellular chain complex of X and where $\xi = \mathfrak{p}_2(X)$ is a cohomological invariant of X given by the Pontrjagin square, see (6.8) and (8.5). A proper map F between pairs (C, ξ) and (C', ξ') as in (1) is a homotopy class of a chain map $F: C \to C'$ satisfying

$$F_*(\xi) = F^*(\xi') \in H^4(C, \Gamma H_2 C'). \tag{2}$$

Such a map is a proper equivalence if F is a homology isomorphism. We prove that any pair (C, ξ), for which C satisfies the general algebraic conditions appropriate to a 1-connected 4-dimensional CW-complex, can be realized geometrically. Moreover, two such CW-complexes X, Y are of the same homotopy type if and only if their Pontrjagin invariants are properly equivalent. This implies that the proper equivalence classes of pairs (C, ξ) as in (2) are in 1-1 correspondence with the homotopy types of 1-connected 4-dimensional polyhedra. In addition we show that any proper map between the Pontrjagin invariants $\mathscr{P}(X) \to \mathscr{P}(Y)$ can be realized geometrically by a cellular map $f: X \to Y$. In fact, the obstruction for the realizability of an arbitrary chain map $F: \tilde{C}_* X \to \tilde{C}_* Y$ is the element

$$\mathcal{O}_{X,Y}(F) = F_* \wp_2(X) - F^* \wp_2(Y) \in H^4(X, \Gamma H_2 Y) \tag{3}$$

which vanishes if and only if F is realizable.

These results serve as an introduction for the homotopy theory of 4-dimensional CW-complexes with non trivial fundamental groups considered in this book. This chapter is of a preparatory nature. Using the simply connected case we hope to familiarize the reader with some of the problems and ideas which will occur later in a more sophisticated context.

§0 Sufficiency, realizability and detecting functors

We introduce some basic notations on categories which will be used frequently. In particular we describe the sufficiency and realizability conditions which define a detecting functor. Such detecting functors arise often in classification problems since they induce a 1-1 correspondence of equivalence classes of objects. For example the Pontrjagin invariant in the introduction above is such a detecting functor on the homotopy category of simply connected 4-dimensional CW-complexes, see §8 below.

A boldface letter like **C** denotes a category, Ob(**C**) and Mor(**C**) are the classes of objects and of maps (morphisms) respectively. The identity of an object A is $1_A = 1 = id$ and $\mathbf{C}(A, B)$ is the set of morphisms $A \to B$. The group of automorphisms of A is $\mathrm{Aut}_{\mathbf{C}}(A)$. An equivalence in **C** is written $f: A \cong B$. An equivalence in **C** is also called an isomorphism. Let $\simeq$ be a natural equivalence relation on a category **A** which is called **homotopy**. Then a **homotopy equivalence** $f: A \simeq B$ is the same as an isomorphism in the quotient category $\mathbf{A}/\simeq$. The **homotopy type** $\{B\}$ of B is the class of all objects A homotopy equivalent to B. **Surjective maps**, resp. **injective maps**, between sets are denoted by

(0.1) $$A \twoheadrightarrow B, \qquad \text{resp. } A \rightarrowtail B.$$

(The arrow $\rightarrowtail$ as well describes a cofibration in a cofibration category, see III §4; yet the meaning of $\rightarrowtail$ will always be clear from the context.) A functor $\lambda: \mathbf{A} \to \mathbf{B}$ is **full**, resp. **faithful** if the induced maps $\lambda: \mathbf{A}(X, Y) \to \mathbf{B}(\lambda X, \lambda Y)$ are surjective, resp. injective for all objects $X, Y \in \mathbf{A}$; we also write $\lambda: \mathbf{A} \twoheadrightarrow \mathbf{B}$, resp. $\lambda: \mathbf{A} \rightarrowtail \mathbf{B}$ in this case. An **equivalence between categories** is denoted by $\mathbf{A} \simeq \mathbf{B}$. For a functor $\lambda: \mathbf{A} \to \mathbf{B}$ let $\lambda\mathbf{A}$ be the **image category** of

λ. Objects in $\lambda\mathbf{A}$ are the same as in $\mathbf{A}$ and morphisms $X \to Y$ in $\lambda\mathbf{A}$ are the maps $f: \lambda X \to \lambda Y$ in the image set $\lambda\mathbf{A}(X, Y)$. Clearly λ induces functors

(0.2) $$\mathbf{A} \xrightarrow{\lambda} \lambda\mathbf{A} \xrightarrow{i} \mathbf{B}$$

where λ is full and where i is faithful. We say that λ is a **quotient functor** if i is an isomorphism of categories. The following notation was introduced by J.H.C. Whitehead, see for example §14 of Whitehead (CE).

(0.3) **Definition.** Let $\lambda: \mathbf{A} \to \mathbf{B}$ be a functor. By the **sufficiency and the realizability conditions**, with respect to λ, we mean the following:

(a) *Sufficiency.* If $\lambda(f)$ is an isomorphism, so is f, where f is a morphism in $\mathbf{A}$. That is, the functor λ "reflects" isomorphisms.

(b) *Realizability.* The functor $i: \lambda\mathbf{A} \to \mathbf{B}$ in (0.2) is an equivalence of categories. This is equivalent to the following two conditions (b1) and (b2).

(b1) *Realizability of objects.* The functor λ is "representative" that is, for each object B in $\mathbf{B}$ there is an object A in $\mathbf{A}$ such that λA is equivalent to B, in this case we say that B is *λ-realizable.*

(b2) *Realizability of morphisms.* The functor λ is full, that is, for objects X, Y in $\mathbf{A}$ and for $f: \lambda X \to \lambda Y$ in $\mathbf{B}$ there is $f_0: X \to Y$ with $\lambda f_0 = f$, in this case we also say that f is *λ-realizable.*

In this book the Whitehead theorem is often used for checking that a functor satisfies the sufficiency condition, compare (2.3) below. The proof of realizability conditions is then the hard part in classification problems. Since the sufficiency and realizability conditions appear frequently it is convenient to condense these conditions in the following definition.

(0.4) **Definition.** We call $\lambda: \mathbf{A} \to \mathbf{B}$ a **detecting functor** if λ satisfies both the sufficiency and the realizability conditions, or equivalently if λ reflects isomorphisms, is representative and full.

Clearly, a faithful detecting functor is the same as an equivalence of categories. By a **1-1 correspondence** we always mean a function which is injective and surjective.

(0.5) **Lemma.** *A detecting functor $\lambda: \mathbf{A} \to \mathbf{B}$ induces a* 1-1 *correspondence between equivalence classes of objects in* $\mathbf{A}$ *and equivalence classes of objects in* $\mathbf{B}$.

Next we consider pairs (A, b) where A is an object in $\mathbf{A}$ and where $b: \lambda A \cong B$ is an equivalence in $\mathbf{B}$. We have an equivalence relation on such pairs by

$(A,b) \sim (A',b')$ if and only if there is an equivalence $g\colon A' \cong A$ in $\mathbf{A}$ with $\lambda(g) = b^{-1}b'$. The equivalence classes form the **class of realizations** of B denoted by

$$\text{(0.6)} \qquad \mathrm{Real}_\lambda(B) = \{(A,b)|b\colon \lambda A \cong B\}/\sim.$$

Let $\{A,b\}$ be the equivalence class of (A,b). In addition to (0.3) we use sometimes the following notation, see (VI.7.1) in Baues (AH). The reader is advised to read through the rest of this section quickly and return to it in case the following notations are actually needed. The statement of the next definition is somewhat long, as it contains three abstract individual points. The value of the definition is in condensing a number of facts which appear frequently in homotopy theory, compare the remark below.

(0.7) **Definition.** Let $(\mathbf{A}, \simeq)$ and $(\mathbf{B}, \simeq)$ be categories together with natural equivalence relations, $\simeq$, which we call homotopy, compare (VI.§3). Let $\lambda\colon \mathbf{A} \to \mathbf{B}$ be a functor which induces the functor $\bar{\lambda}\colon \mathbf{A}/\simeq \to \mathbf{B}/\simeq$ between homotopy categories. We say that $\bar{\lambda}$ satisfies the **strong sufficiency condition** if (a), (b) and (c) hold.

(a) $\bar{\lambda}$ satisfies the sufficiency condition, that is, a map in $\mathbf{A}$ is a homotopy equivalence if and only if the induced map in $\mathbf{B}$ is a homotopy equivalence.

(b) **Existence of models.** for an object A in $\mathbf{A}$ let $\beta\colon B \xrightarrow{\simeq} \lambda(A)$ be a homotopy equivalence in $\mathbf{B}$. Then there exists a λ-realization (M,i) of B together with a map $\alpha\colon M \to A$ in $\mathbf{A}$ such that α realizes β, that is, the diagram

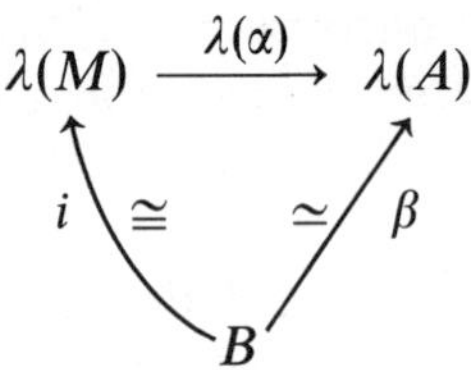

commutes.

(c) For $\xi_0\colon A \to A'$ in $\mathbf{A}$ and $\eta_0 = \lambda\xi_0\colon \lambda A \to \lambda A'$ we have: If $\eta_0 \simeq \eta_1$ then there exists $\xi_0 \simeq \xi_1$ with $\eta_1 = \lambda\xi_1$.

Remark. An old result of J.H.C. Whitehead shows that the functor which carries a CW-complex X to its crossed chain complex $\rho(X)$ satisfies the strong sufficiency condition, see (III.2.9) (a) below. This result can be used for the construction of small models of X, moreover it leads to the definition of "finiteness obstructions" as considered by Wall (F), compare (VI.7.5) and (VI. 7.13) in Baues (AH). One has to have this example in mind for the justification of the notation 'strong sufficiency' in (0.7).

One readily checks that the strong sufficiency condition for $\bar{\lambda}$ implies the following properties (1), (2), and (3).

(1) If $f: \lambda A \to \lambda A'$ is λ-realizable then each element in the homotopy class of f is λ-realizable.

(2) If B is an object in **B** which is λ-realizable, then each object in the homotopy type of B is λ-realizable.

(3) For the functors $\lambda: \mathbf{A} \to \mathbf{B}$, $\bar{\lambda}: \mathbf{A}/\simeq \to \mathbf{B}/\simeq$ we have the equation $\mathrm{Real}_\lambda(B) = \mathrm{Real}_{\bar{\lambda}}(B)$ provided λ satisfies the sufficiency condition.

§1 Homotopy groups

In this section we fix some notation on homotopy groups and we recall some well known facts on the action of the fundamental group.

Let **Top*** be the category of topological spaces with basepoint $*$ and of basepoint preserving maps. A **homotopy** $H: f \simeq g$ in **Top*** is given by a map $H: I_* X \to Y$ with $Hi_0 = f$ and $Hi_1 = g$. Here

(1.1) $$I_* X = (I \times X)/(I \times *)$$

is the (reduced) **cylinder on** X (given by the unit interval $I = [0,1]$) and $i_t: X \to I_* X$ is the inclusion with $i_t(x) = (t, x)$ for $t \in I$. Let

(1.2) $$[X, Y] = \mathbf{Top}^*(X, Y)/\simeq$$

be the set of homotopy classes of maps $X \to Y$ in **Top***. This set is the set of morphisms $\{f\}: X \to Y$ in the quotient category **Top***$/\simeq$. We have the trivial map $0: X \to * \to Y$ which represents $0 \in [X, Y]$. The **cone** of X is $CX = I_* X/i_1 X$ and the **suspension** of X is $\Sigma X = CX/i_0 X$. The n-**sphere** S^n satisfies $\Sigma S^n = S^{n+1}$ and **homotopy groups** are given by

(1.3) $$\begin{cases} \pi_n(X) = [S^n, X] \\ \pi_{n+1}(Y, X) = [(CS^n, S^n), (Y, X)] \end{cases}$$

where (Y, X) is a pair in **Top***. We have the long exact sequence $(n \geq 0)$

(1.4) $$\pi_{n+1}(X) \xrightarrow{i} \pi_{n+1}(Y) \xrightarrow{j} \pi_{n+1}(Y, X) \xrightarrow{\partial} \pi_n(X) \xrightarrow{i} \pi_n(Y)$$

where ∂ is the restriction and where j is induced by the quotient map

$(CS^n, S^n) \to (S^{n+1}, *)$. Clearly i is induced by the inclusion $X \subset Y$. The exact sequence (1.4) is natural with respect to pair maps $F: (Y, X) \to (Y', X')$ in **Top***.

Next we consider the action of the fundamental group $\pi = \pi_1(X)$ on the groups (1.3). To this end we fix the following notation.

(1.5) **Notation.** In this book the group structure $+$, $-$, 0 of a group π is written additively though addition $+$ in π needs not to be abelian. The element 0 denotes the neutral element in π. Let $\mathbb{Z}[\pi]$ be the **group ring** of π, this is the free abelian group generated by elements $[\alpha]$, $\alpha \in \pi$, with the multiplication $[\alpha]\cdot[\beta] = [\alpha + \beta]$. A homomorphism $\varphi: \pi \to \pi'$ induces the ring homomorphism

$$\varphi_{\#}: \mathbb{Z}[\pi] \to \mathbb{Z}[\pi'] \tag{1}$$

with $\varphi_{\#}[\alpha] = [\varphi\alpha]$. A π-**group** M is given by an action of the group π on M denoted by x^α for $x \in M$, $\alpha \in \pi$. We have

$$\left.\begin{array}{ll} (x + y)^\alpha = x^\alpha + y^\alpha, & (-x)^\alpha = -x^\alpha, \\ x^{\alpha+\beta} = (x^\alpha)^\beta, & x^0 = x \end{array}\right\} \tag{2}$$

for $x, y \in M$, $\alpha, \beta \in \pi$. A π-group M is a π-**module** (or a $\mathbb{Z}[\pi]$-**module**) in case M is abelian. A φ-**equivariant** map $F: M \to M'$ from a π-group M to a π'-group M' is a homomorphism between groups satisfying

$$F(x^\alpha) = (Fx)^{\varphi\alpha}. \tag{3}$$

Here F is equivariant or π-equivariant if $\varphi = 1$ is the identity of $\pi = \pi'$.

Let $w: I \to X$ be a path with $w(0) = x_0$ and $w(1) = x_1$. Then w induces the functions $(n \geq 1)$

(1.6)
$$\begin{cases} w^{\#}: \pi_n(X, x_0) \to \pi_n(X, x_1), \\ w^{\#}: \pi_{n+1}(Y, X, x_0) \to \pi_{n+1}(Y, X, x_1), \end{cases}$$

compare (III.§10) below and (II.5.7) is Baues (AH). The 1-sphere S^1 is given by the quotient $S^1 = I/\partial I$. Whence a path w with $w(0) = w(1)$ represents an element $w \in \pi_1(X)$. We now define the **action of** $\pi_1(X)$ on $\pi_n(X)$ and $\pi_{n+1}(Y, X)$ respectively by

$$\xi^w = w^{\#}(\xi). \tag{1}$$

For $\xi \in \pi_1(X)$ we get $\xi^w = -w + \xi + w$. The maps in the exact sequence (1.4) are equivariant with respect to the action of $\pi_1(X)$. Moreover a pair map $F: (Y, X) \to (Y', X')$ with $F(*) = *$ induces a $\pi_1(F)$-equivariant homomorphism $(n \geq 2)$

$$\pi_n(F) = F_*: \pi_n(Y, X) \to \pi_n(Y', X'). \tag{2}$$

This leads to the following notation:

(1.7) **Definition of the category $\mathbf{Mod}_{\mathbb{Z}}^{\wedge}$.** Objects are pairs (π, M) where M is a right π-module and morphisms $(\varphi, f): (\pi, M) \to (\pi', M')$ are φ-equivariant homomorphisms $f: M \to M'$.

For example $(\pi_1 F, \pi_n F)$ in (1.6)(2) is a morphism in $\mathbf{Mod}_{\mathbb{Z}}^{\wedge}$ for $n \geq 3$. The homotopy group π_n yields the functor

(1.8) $$\pi_n: \mathbf{Top}^*/\simeq \; \to \mathbf{Mod}_{\mathbb{Z}}^{\wedge}$$

which carries a space X to the object $(\pi_1 X, \pi_n X)$, $n \geq 2$.

§2 CW-complexes and homology with local coefficients

We describe two different notions of homotopy on the category of path connected CW-complexes and we use the universal covering for the definition of homology and cohomology with local coefficients.

Let X be a CW-complex with skeleta X^n. A map $F: X \to Y$ between CW-complexes is **cellular** if $F(X^n) \subset Y^n$. Let **CW** be the following category. Objects are CW-complexes X with trivial 0-skeleton $* = X^0$ and morphisms are cellular maps $F: X \to Y$. Each path connected CW-complex Y is homotopy equivalent to an object in **CW** by dividing out a maximal tree in the 1-skeleton of Y. The objects in **CW** are also called '0-reduced' CW-complexes. The category **CW** is the special case of the category $\mathbf{CW}_0^D$ in Baues (AH) with $D = *$. The **universal covering**

(2.1) $$p: \hat{X} \to X$$

of X yields a CW-complex $\hat{X}$ with $\pi_1 \hat{X} = 0$ and with skeleta $\hat{X}^n = p^{-1}(X^n)$. We always fix a basepoint $* \in p^{-1}(*) \subset \hat{X}^0$. This determines a functor which carries X to $\hat{X}$ and which carries a map $F: X \to Y$ to the unique basepoint preserving covering map $\hat{F}: \hat{X} \to \hat{Y}$ with $p\hat{F} = Fp$. The fundamental group

$\pi_1(X)$ acts from the right on $\hat{X}$ by covering transformations and $\hat{F}$ is a $\pi_1(F)$-equivariant map with respect to this action. The action of $\pi_1(X)$ on $\hat{X}$ yields an action of $\pi_1(X)$ on the singular **homology group** $H_n(\hat{X})$ so that $\hat{H}_n(X) = H_n(\hat{X})$ is a $\pi_1(X)$-module. Whence we have a functor (see (1.7))

(2.2) $$\hat{H}_n\colon \mathbf{CW}/\simeq \;\to \mathbf{Mod}_{\mathbb{Z}}^{\wedge}$$

which carries X to $(\pi_1 X, H_n(\hat{X}))$ and which carries F to $(\pi_1 F, H_n(\hat{F}))$. The functors π_n in (1.8) and $\hat{H}_n$ in (2.2) have the following well known property:

(2.3) **Whitehead theorem.** *A map $F\colon X \to Y$ in* **CW** *is a homotopy equivalence if and only if* (a) *or equivalently* (b) *is satisfied.*

(a) *The induced map $\pi_n(F)$ is an isomorphism for $n \geq 1$.*

(b) *The induced maps $\pi_1(F)$ and $H_n(\hat{F})$ are isomorphisms for $n \geq 2$.*

The cylinder I_*X of a CW-complex X is a CW-complex in **CW** with skeleta

(2.4) $$(I_*X)^n = X^n \cup I_*X^{n-1} \cup X^n.$$

We call a cellular map $H\colon I_*X \to Y$ a 1-**homotopy** and we write $H\colon H_0 \overset{1}{\simeq} H_1$. Moreover, we call H a 0-**homotopy** if H_t is a cellular map for $t \in I$; in this case we write $H\colon H_0 \overset{0}{\simeq} H_1$. The natural equivalence relations $\overset{0}{\simeq}$ and $\overset{1}{\simeq}$ yield the quotient functors

(2.5) $$\mathbf{CW} \to \mathbf{CW}/\overset{0}{\simeq} \;\to \mathbf{CW}/\overset{1}{\simeq} \;= \mathbf{CW}/\simeq$$

for the corresponding homotopy categories. Here the cellular approximation theorem shows that actually $\mathbf{CW}/\simeq \;= \mathbf{CW}/\overset{1}{\simeq}$. This is a full subcategory of $\mathbf{Top}^*/\simeq$. We now consider the cellular chain complex of the universal covering which is an object in the following category.

(2.6) **Definition of the category $\mathbf{Chain}_{\mathbb{Z}}^{\wedge}$.** Objects are pairs (π, C) where π is a group and where $C = (C_n, d_n; n \in \mathbb{Z})$ is a chain complex of right π-modules. Maps $(\varphi, F)\colon (\pi', C') \to (\pi, C)$ are φ-equivariant chain maps $F\colon C' \to C$ where $\varphi\colon \pi' \to \pi$ is a homomorphism. Two such chain maps are **homotopic**, $(\varphi, F) \simeq (\Psi, G)$, if $\varphi = \Psi$ and if there exists a φ-equivariant map $\alpha\colon C' \to C$ of degree $+1$ with $d\alpha + \alpha d = -F + G$. The chain map (φ, F) is a **weak equivalence** if φ is an isomorphism and if F induces an isomorphism in homology. We say that (π, C) is **free** if each C_n, $n \in \mathbb{Z}$, is a free $\mathbb{Z}[\pi]$-module.

We have the **chain functor**

(2.7) $$\hat{C}_*\colon \mathbf{CW}/\overset{0}{\simeq} \;\to \mathbf{Chain}_{\mathbb{Z}}^{\wedge}$$

which carries X to the cellular chain complex $(\pi_1 X, C_*\hat{X})$. This is a free chain complex given by

$$\hat{C}_n X = C_n \hat{X} = H_n(\hat{X}^n, \hat{X}^{n-1}). \tag{1}$$

The functor $\hat{C}_*$ carries a map F in **CW** to the induced $\pi_1(F)$-equivariant chain map $C_*\hat{F}$. Since cylinders satisfy the equation

$$(I \times X)^\wedge = I \times (\hat{X}) \tag{2}$$

one readily verifies that $\hat{C}_*$ induces a functor

$$\hat{C}_*\colon \mathbf{CW}/\simeq \;\to \mathbf{Chain}_{\mathbb{Z}}^{\wedge}/\simeq \tag{3}$$

between homotopy categories. The Whitehead theorem (2.3) shows that a map F in **CW** is a homotopy equivalence if and only if $(\pi_1 F, C_*\hat{F})$ is a weak equivalence or a homotopy equivalence in $\mathbf{Chain}_{\mathbb{Z}}^{\wedge}/\simeq$. For a subcomplex D of X we obtain the subcomplex $p^{-1}(D)$ of $\hat{X}$. This yields the quotient complex

(2.8) $$\hat{C}_*(X, D) = C_*(\hat{X})/C_*(p^{-1}D).$$

The usual cellular chain complex of the pair (X, D) satisfies

$$C_*(X, D) = \hat{C}_*(X, D) \otimes_{\mathbb{Z}[\pi]} \varepsilon^* \mathbb{Z}$$

where $\mathbb{Z}$ is a $\mathbb{Z}[\pi]$-module via the augmentation $\varepsilon\colon \mathbb{Z}[\pi] \to \mathbb{Z}$. We now define homology and cohomology for chain complexes (π, C) in $\mathbf{Chain}_{\mathbb{Z}}^{\wedge}$.

(2.9) **Definition.** Let Γ be a *left* $\mathbb{Z}[\pi]$-module. Then the tensor product $C \otimes_{\mathbb{Z}[\pi]} \Gamma$ is a chain complex of abelian groups. Its **homology** is denoted by

$$\hat{H}_*(C; \Gamma) = H_*(C \otimes_{\mathbb{Z}[\pi]} \Gamma). \tag{1}$$

Next let Γ be a *right* $\mathbb{Z}[\pi]$-module. Then $\mathrm{Hom}_\pi(C, \Gamma)$ is a cochain complex of abelian groups which yields the **cohomology** groups

$$\hat{H}^*(C; \Gamma) = H^*(\mathrm{Hom}_\pi(C, \Gamma)). \tag{2}$$

A map (φ, F) in $\mathbf{Chain}_{\mathbb{Z}}^{\wedge}$ induces homomorphisms $F_* = (F \otimes 1_\Gamma)_*$ and $F^* = \mathrm{Hom}(F, 1_\Gamma)_*$ between homology groups and cohomology groups respectively. We also use the following special case of (2). Let $\varphi\colon \pi \to \pi'$ be a homomorphism and let Γ' be a right $\mathbb{Z}[\pi']$-module. Then $\Gamma = \varphi^*\Gamma'$ is the induced

π-module and

$$\hat{H}^*(C;\varphi^*\Gamma') = H^*(\mathrm{Hom}_\varphi(C,\Gamma')). \tag{3}$$

Here Hom_φ denotes the cochain complex of φ-equivariant homomorphisms. For X in **CW** and for a subcomplex D of X we obtain by (2.8) the **(co)homology groups with local coefficients** Γ

(2.10)
$$\begin{cases} \hat{H}^n(X,D;\Gamma) = \hat{H}_n(\hat{C}_*(X,D);\Gamma), \\ \hat{H}^n(X,D;\Gamma) = \hat{H}^n(\hat{C}_*(X,D);\Gamma). \end{cases}$$

As usual we omit D in the notation in case $D = \phi$ is the empty space. For $\Gamma = \mathbb{Z}[\pi]$ we get

(2.11)
$$\hat{H}_n(X) = \hat{H}_n(X;\mathbb{Z}[\pi]) = H_n(\hat{X}).$$

This is the functor considered in (2.2). On the other hand we have the 'ordinary' (co)homology with coefficients in an abelian group A

(2.12)
$$\begin{cases} \hat{H}_n(X;A) = H_n(X;A), \\ \hat{H}^n(X;A) = H^n(X;A). \end{cases}$$

On the left hand side A is a **trivial π-module**, with the trivial action $g^\alpha = g$, $\alpha \in \pi$, $g \in A$. The right hand side of (2.12) is the usual singular (co)homology.

§3 Whitehead's certain exact sequence

We consider the Hurewicz homomorphism which is embedded in Whitehead's certain exact sequence and we define the operators in this sequence. Its 4-dimensional part was used by Whitehead (CE) for the homotopy classification of 1-connected 4-dimensional CW-complexes. We shall describe this result in section §8 below where we also discuss the connection of the sequence with the Pontrjagin square. The **Hurewicz homomorphism** h is a natural map ($n \geq 0, X \in$ **Top***)

(3.1)
$$\begin{cases} h\colon \pi_n(X) \to H_n(X,*), \\ h\colon \pi_{n+1}(Y,X) \to H_{n+1}(Y,X) \end{cases}$$

where H_n denotes the homology with coefficients in $\mathbb{Z}$. We define h by $h(\alpha) = \alpha_*(e_n)$ where e_n is an appropriate generator in $H_n(S^n, *) \cong \mathbb{Z}$ or $H_{n+1}(CS^n, S^n) \cong \mathbb{Z}$ such that h is compatible with the exact sequence (1.4). The projection $p: \hat{X} \to X$ of the universal covering induces for an inclusion $i: D \subset X$ in **Top*** the isomorphism

(3.2) $$p_*: \pi_{n+1}(\hat{X}, p^{-1}D) \xrightarrow{\cong} \pi_{n+1}(X, D), \quad n \geq 0.$$

Whence we obtain for $n \geq 0$ the natural map

(3.3) $$\hat{h} = hp_*^{-1}: \pi_{n+1}(X, D) \to H_{n+1}(\hat{X}, p^{-1}D).$$

For $n \geq 1$ this map is $\pi_1(i)$-equivariant with $\pi_1(i): \pi_1(D) \to \pi_1(X)$. As a special case we obtain for $n \geq 2$ the natural Hurewicz map of $\pi_1(X)$-modules

(3.4) $$h_n: \pi_n(X) \xrightarrow{\hat{h}} H_n(\hat{X}, p^{-1}*) \cong H_n(\hat{X}).$$

Now let X be a path connected CW-complex with $* \in X^0$. We define for $n \geq 1$ the $\pi_1(X)$-module

(3.5) $$\Gamma_n(X) = \text{image}(i_*: \pi_n X^{n-1} \to \pi_n X^n)$$

where $i: X^{n-1} \subset X^n$ is the inclusion. Clearly $\Gamma_n(X) = 0$ for $n = 1$, 2 since $\pi_n(X^{n-1}) = 0$ in this case. In the next section we describe $\Gamma_3(X)$. For $n \geq 3$ we obtain the action of $\pi_1(X)$ on $\Gamma_n(X)$ since $\pi_1 X^2 = \pi_1 X$. A cellular map $F: X \to Y$ induces a $\pi_1(F)$-equivariant map $\Gamma_n(F): \Gamma_n(X) \to \Gamma_n(Y)$. The cellular approximation theorem shows that $\Gamma_n(F)$ depends only on the homotopy class of F in **CW**/$\simeq$. Whence we have by (3.5) a well defined functor

(3.6) $$\Gamma_n: \mathbf{CW}/\simeq \;\to \mathbf{Mod}_{\mathbb{Z}}^{\wedge}.$$

The Hurewicz map h_n in (3.4) is embedded in the following long exact sequence of $\pi_1(X)$-modules which is the **certain exact sequence** of Whitehead (CE), $n \geq 2$,

(3.7) $$H_{n+1}\hat{X} \xrightarrow{b_{n+1}} \Gamma_n X \xrightarrow{i_n} \pi_n X \xrightarrow{h_n} H_n\hat{X} \xrightarrow{b_n} \Gamma_{n-1} X.$$

This sequence is natural with respect to maps in **CW**/$\simeq$. The operator i_n is induced by the inclusion $X^n \subset X$. Moreover, the **secondary boundary** b_n is defined by the following commutative diagram where $\hat{h}$ and ∂ are given by (3.3) and (1.4) respectively.

(3.8)

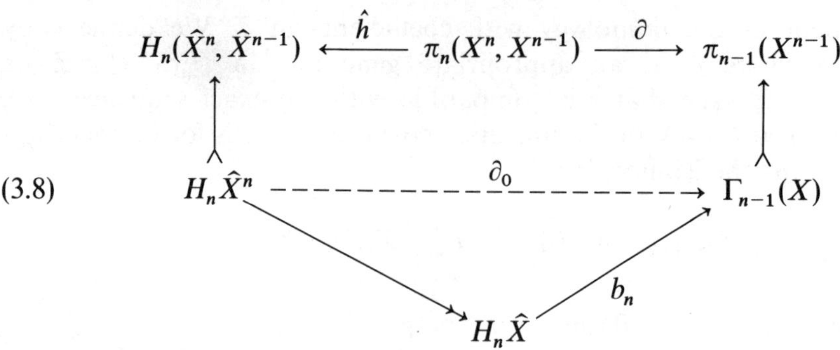

Here the map $\hat{h}$ is an isomorphism for $n \geq 3$. One can check that $\partial(\hat{h})^{-1}$ induces maps ∂_0 and b_n such that the diagram commutes.

The exactness of the sequence (3.7) implies that

$$h_2: \pi_2 X \cong H_2 \hat{X} \tag{3.9}$$

is an isomorphism and that $h_3: \pi_3 X \to H_3 \hat{X}$ is surjective. Here we use $\Gamma_1 = \Gamma_2 = 0$. More generally we have $\Gamma_1 X = \cdots = \Gamma_n X = 0$ provided $\hat{X}$ is $(n-1)$-connected; whence h_n is an isomorphism and h_{n+1} is surjective in this case. This fact corresponds to the **Hurewicz theorem**.

We now describe $\partial(\hat{h})^{-1}$ in (3.8) by use of the **attaching maps** of n-cells in X. Let $Z_n = Z_n(X)$ be the set of n-cells in X. For $n \geq 2$ we can choose a map in **Top***

$$f: \bigvee_{Z_n} S^{n-1} \to X^{n-1} \tag{3.10}$$

and a homotopy equivalence $c: C_f \simeq X^n$ under X^{n-1} where C_f is a mapping cone of f. Let

$$i_e: S^{n-1} \subset \bigvee_{Z_n} S^{n-1}, \quad e \in Z_n,$$

be the inclusion for the one point union $\bigvee S^{n-1}$ of $(n-1)$-spheres in (3.10). The choice of f and c yields a function

$$i: Z_n \xrightarrow{i'} \pi_n(C_f, X^{n-1}) \cong \pi_n(X^n, X^{n-1}) \tag{3.11}$$

where the isomorphism is induced by c and where i' carries $e \in Z_n$ to the element represented by the canonical map $(CS^{n-1}, S^{n-1}) \to (C_f, X^{n-1})$ induced by fi_e. For $n \geq 2$ the map $\hat{h}i$ induces an isomorphism of free $\mathbb{Z}[\pi_1(X)]$-modules

$$\bigoplus_{Z_n} \mathbb{Z}[\pi_1 X] \cong \hat{C}_n X. \tag{3.12}$$

We use this as an identification so that Z_n is a basis of $\hat{C}_n X$. For $n = 1$ this isomorphism is also defined via (3.11) provided $X^0 = *$ is a point. For $n \geq 3$ the map

$$f_n = \partial(\hat{h})^{-1}\colon \hat{C}_n X \to \pi_{n-1} X^{n-1} \tag{3.13}$$

in (3.8) is a $\pi_1(X)$-equivariant homomorphism which is determined on generators $e \in Z_n$ by $f_n(e) = \{fi_e\}$. We call f_n as well the **attaching map of n-cells in X**.

§4 The quadratic functor Γ and central extensions

The first non trivial Γ-group, $\Gamma_3(X)$, in Whitehead's exact sequence (3.7) can be computed in terms of a quadratic functor Γ. This follows from the fact that the Hopf map η satisfies a quadratic distributivity law, see (4.10) (∗). In connection with Γ we introduce the quadratic functors, $\otimes^2$, Λ^2, $\hat{\otimes}^2$ which carry abelian groups to abelian groups (in (II.§7) below a more general approach yields these functors as examples of quadratic tensor products). We then consider central extensions of abelian groups which have a quadratic ingredient by their commutator maps.

A function $f\colon A \to B$ between abelian groups is called a **quadratic map** if $f(-a) = f(a)$ and if the function $A \times A \to B$, given by $(a,b) \mapsto f(a+b) - f(a) - f(b)$, is bilinear. There is a universal quadratic map

$$\gamma\colon A \to \Gamma(A) \tag{4.1}$$

with the property: for all quadratic maps $f\colon A \to B$ there is a unique homomorphism $f^{\square}\colon \Gamma(A) \to B$ with $f = f^{\square}\gamma$. We say that $f^{\square}$ is induced by f. A homomorphism $\varphi\colon A' \to A$ yields the quadratic map $\gamma\varphi$ which induces $\Gamma(\varphi) = (\gamma\varphi)^{\square}\colon \Gamma(A') \to \Gamma(A)$. This shows that Γ is a well defined functor. We associate with Γ the following natural commutative diagram in which the subdiagram 'push' is a push out of abelian groups.

$$\begin{array}{ccccc}
A \otimes A & & & & \\
\downarrow{\scriptstyle [1,1]} & & & & \\
\Gamma A & \xrightarrow{\tau} & A \otimes A & \xrightarrow{q} & A \wedge A \\
\downarrow{\scriptstyle \sigma} & \text{push} & \downarrow{\scriptstyle \bar{\sigma}} & & \| \\
A \otimes \mathbb{Z}/2 & \xrightarrow{\bar{\tau}} & A \hat{\otimes} A & \xrightarrow{q} & A \wedge A
\end{array} \tag{4.2}$$

The column and the rows of the diagram are exact sequences of abelian groups. The homomorphism τ is induced by the quadratic map $A \to A \otimes A$, $a \mapsto a \otimes a$, and the homomorphism σ is induced by the quadratic map $A \to A \otimes \mathbb{Z}/2$ given by $a \mapsto a \otimes 1$. Here $1 \in \mathbb{Z}/2$ is the generator. Since γ in (4.1) is quadratic we can define the homomorphism $[1,1]$ by

(4.3) $$[1,1](a \otimes b) = [a,b] = \gamma(a+b) - \gamma(a) - \gamma(b).$$

We clearly have $[a,b] = [b,a]$, $[a,a] = 2\gamma(a)$, $\Gamma(\varphi)[a,b] = [\varphi a, \varphi b]$, $\sigma[a,b] = 0$ and

(4.4) $$\tau[a,b] = a \otimes b + b \otimes a.$$

We obtain the **exterior product**

(4.5) $$\Lambda^2(A) = A \wedge A = A \otimes A/\sim$$

by the relation $\tau\gamma(a) = a \otimes a \sim 0$ and we obtain

(4.6) $$A \hat{\otimes} A = A \otimes A/\approx$$

by the relation $\tau[a,b] = a \otimes b + b \otimes a \approx 0$. This shows that the rows of (4.2) are exact since the column is exact. The quotient map q carries $a \otimes b$ to $a \wedge b$. If A is free abelian with an ordered basis M, then $A \wedge A$ is free abelian with the basis

(4.7) $$\{m \wedge n : m < n, m, n \in M\}.$$

Moreover, the homomorphisms τ and $\bar{\tau}$ in (4.2) are injective in case A is free abelian. For homomorphism $\varphi\colon A' \to A$, $\psi\colon A'' \to A$ we write $[\varphi, \psi] = [1,1](\varphi \otimes \psi)\colon A' \otimes A'' \to \Gamma(A)$. There is the canonical isomorphism

(4.8) $$\Gamma(A \oplus B) = \Gamma(A) \oplus \Gamma(B) \oplus A \otimes B$$

given by $\Gamma(i_A)$, $\Gamma(i_B)$ and $[i_A, i_B]$ where $i_A\colon A \to A \oplus B$, $i_B\colon B \to A \oplus B$ are the inclusions. Moreover for cyclic groups we have the isomorphism $\tau\colon \Gamma(\mathbb{Z}) = \mathbb{Z}$ and $\Gamma(\mathbb{Z}/n) = \mathbb{Z}/\gcd(n^2, 2n)$. Whence (4.8) yields a computation of $\Gamma(A)$ for any finitely generated abelian group A. We also will use the following crucial property of the functor Γ. An exact sequence of abelian groups $C \xrightarrow{d} D \xrightarrow{q} A \to 0$ induces the exact sequence

(4.9) $$(C \otimes D) \oplus \Gamma C \xrightarrow{\partial} \Gamma D \xrightarrow{\Gamma(q)} \Gamma A \longrightarrow 0$$

where $\partial = ([d,1], \Gamma(d))$. This yields a presentation of ΓA if C and D are free abelian.

We now consider the group $\Gamma_3(X)$ in (3.5). The **Hopf map** $\eta\colon S^3 \to S^2$ yields the commutative diagram

(4.10)
$$\begin{array}{ccc} & \Gamma(\pi_2 X) & \\ {\scriptstyle\gamma}\nearrow & & \searrow{\scriptstyle\eta_0^{\square}} \\ \pi_2 X & \xrightarrow{\quad\eta_0\quad} & \Gamma_3 X \\ \wr\Vert & & \cap \\ \pi_2 X^3 & \xrightarrow[\eta^*]{\qquad} & \pi_3 X^3 \end{array}$$

with $\eta^*(\alpha) = \alpha\eta$. The map η^* is quadratic since we have the left distributivity law

$$\eta^*(\alpha + \beta) = \eta^*\alpha + \eta^*\beta + [\alpha, \beta] \qquad (*)$$

where $[\alpha, \beta]$ is the **Whitehead product** which is bilinear. The map η^* yields a quadratic map η_0, see (4.10), which induces the homomorphism $\eta_0^{\square}$. This homomorphism is a natural isomorphism which we use as an identification

(4.11) $$\Gamma_3(X) = \Gamma(\pi_2 X) = \Gamma(H_2 \hat{X})$$

of $\pi_1(X)$-modules. Here we define the action of $\alpha \in \pi_1(X)$ on $y \in \Gamma(\pi_2 X)$ by $y^\alpha = \Gamma(1^\alpha)(y)$ with $1^\alpha\colon \pi_2 X \to \pi_2 X$ given by $1^\alpha(x) = x^\alpha$. The isomorphism (4.11) can be found in Whitehead (CE), see also (IX.4.5) in Baues (AH). By (4.11) the map $[1, 1]$ in (4.2) corresponds to the Whitehead product in (4.10)(∗). In addition to (4.11) we get the isomorphism of $\pi_1 X$-modules

(4.12) $$\Gamma_{n+1}(X) = (\pi_n X) \otimes \mathbb{Z}/2 = (H_n \hat{X}) \otimes \mathbb{Z}/2$$

provided $\hat{X}$ in $(n-1)$-connected, $n \geq 3$. This isomorphism is induced by $(\Sigma^{n-2}\eta)^*\colon \pi_n X^{n+1} \to \pi_{n+1} X^{n+1}$ similarly as in (4.10). The map σ in (4.2) corresponds to the suspension Σ, namely

$$\sigma\colon \Gamma(\pi_2 X) = \Gamma_3 X \xrightarrow{\Sigma} \Gamma_4(\Sigma X) \cong \pi_2 X \otimes \mathbb{Z}/2$$

provided X is simply connected.

Recall that an **Eilenberg-Mac Lane space** $K(A, n)$ is a CW-complex with a single non vanishing homotopy group in degree n satisfying $\pi_n K(A, n) = A$, $n \geq 0$. A **Moore space** $M(A, n)$ is a simply connected CW-complex with a single non vanishing homology group in degree n satisfying $H_n M(A, n) = A$,

$n \geq 2$. We derive from the exact sequence (3.7) immediately the natural isomorphisms

$$(4.13)\qquad \begin{cases} H_4K(A,2) = \Gamma(A) = \pi_3 M(A,2), \\ H_{n+2}K(A,n) = A \otimes \mathbb{Z}/2 = \pi_{n+1}M(A,n), \quad n \geq 3. \end{cases}$$

We shall use the isomorphism $H_4K(A,2) = \Gamma(A)$ for the definition of the Pontrjagin square in (5.3) below. For a group G and a G-module D we obtain the **cohomology of** G by

$$(4.14)\qquad H^n(G,D) = \hat{H}^n(K(G,1),D)$$

where the right hand side is given by (2.10).

Next we consider some properties of central extensions of abelian groups. Let A, B be abelian groups. The cohomology $H^2(A;B)$ classifies the equivalence classes of **central extensions** $B \overset{i}{\rightarrowtail} E \overset{p}{\twoheadrightarrow} A$ of A by B. One has the exact sequence (see Eilenberg-Mac Lane)

(4.15)
$$0 \longrightarrow \mathrm{Ext}_{\mathbb{Z}}(A,B) \overset{i}{\longrightarrow} H^2(A;B) \overset{K}{\longrightarrow} \mathrm{Hom}_{\mathbb{Z}}(A \otimes A, B) \overset{\tau^*}{\longrightarrow} \mathrm{Hom}_{\mathbb{Z}}(\Gamma A, B).$$

Here $\mathrm{Ext}_{\mathbb{Z}}(A,B)$ classifies the **abelian extensions** of A by B. The homorphism K associates with an extension $\{E\} \in H^2(A;B)$ the **commutator map** $K\{E\} = w_E\colon A \otimes A \to B$. We obtain w_E by the formula $w_E(a \otimes b) = i^{-1}(-\bar{a} - \bar{b} + \bar{a} + \bar{b})$ where $\bar{a}$, $\bar{b} \in E$ are elements with $p\bar{a} = a$, $p\bar{b} = b$. Clearly $w_E(a \otimes a) = 0$ so that w_E induces a map $w_E\colon A \wedge A \to B$. For example let C be a free abelian group, then there is a unique central extension

$$(4.16)\qquad C \wedge C \rightarrowtail G \twoheadrightarrow C$$

for which the quotient map $q\colon C \otimes C \to C \wedge C$ is the commutator map. In fact, G is the free nil(2)-group in this case, see (II.1.19) and (III.1.12) below. Any central extension $B \rightarrowtail E \twoheadrightarrow C$ is up to equivalence determined by its commutator map w_E since $\mathrm{Ext}_{\mathbb{Z}}(C,B) = 0$; we can construct E by the central push out diagram of groups

$$(4.17)\qquad \begin{array}{ccc} C \wedge C & \xrightarrow{w_E} & B \\ {\scriptstyle i}\downarrow & c\text{-push} & \downarrow \\ G & \longrightarrow & E \end{array}$$

Here we use the following notation.

(4.18) **Definition.** We say that a homomorphism between groups, $\alpha: A \to G$, is **central** if A is abelian and if αA lies in the center of G. Now consider the following commutative diagram in the category of groups.

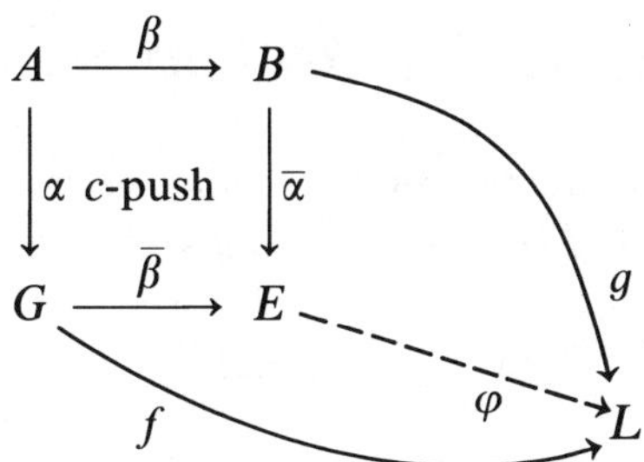

We say that the subdiagram 'c-push' is a **central push out** diagram if α and $\bar{\alpha}$ are central and if the following universal property is satisfied. For all f, g with g central and with $f\alpha = g\beta$ there is a unique φ extending the diagram commutatively.

We can construct the central push out E as follows. The group

$$E = (G \times B)/\sim \tag{4.19}$$

is the quotient of the product group $G \times B$ by the equivalence relation $(x + \alpha(a), y) \sim (x, \beta(a) + y)$ with $x \in G$, $y \in B$, $a \in A$. Here $+$ denotes the group law, see (1.5). The homorphisms $\bar{\alpha}$, $\bar{\beta}$ in (4.18) are induced by the inclusions $B \subset G \times B$ and $G \subset G \times B$ respectively. It is obvious that E in (4.19) has the universal property in (4.18). Moreover, a central push out diagram satisfies

(4.20) **Lemma.** cokernel(α) = cokernel$(\bar{\alpha})$ *and* $\beta\,$kernel(α) = kernel$(\bar{\alpha})$.

Proof. Let $y \in$ kernel$(\bar{\alpha})$. Whence by (4.19) we have $(0, y) \sim (0, 0)$. Equivalently there is $a \in A$ with $(0, y) = (0 + \alpha(a), y) = (0, \beta(a) + y) = (0, 0)$. □

§5 Cup products and Pontrjagin squares

In this section we describe some simple properties of cup products, Pontrjagin squares, and Steenrod squares respectively. We observe that cup products are also defined for cohomology groups with local coefficients while Pontrjagin squares and Steenrod squares are, a priori, only defined for cohomology groups with coefficients in an abelian group G.

For π-modules G, G' let $G \otimes G' = G \otimes_{\mathbb{Z}} G'$ be the π-module with the action $(g \otimes h)^\alpha = g^\alpha \otimes h^\alpha$, $(g \in G, h \in G', \alpha \in \pi)$. The cohomology groups with local coefficients in (2.10) are endowed with the natural **cup product**

$$\cup\colon \hat{H}^n(X, G) \otimes \hat{H}^m(X, G') \to \hat{H}^{n+m}(X, G \otimes G'), \tag{5.1}$$

compare Steenrod. This cup product is associative and the interchange map $T\colon G \otimes G' \to G' \otimes G$ with $T(g \otimes h) = h \otimes g$ yields the commutativity formula

$$T_*(x \cup y) = (-1)^{nm} y \cup x \tag{5.2}$$

for $x \in \hat{H}^n(X, G)$, $y \in \hat{H}^m(X, G)$. Clearly, for trivial $\pi_1(X)$-modules G, G' the cup product (5.1) is the usual one for cohomology groups $H^n(X, G)$ as in (2.12). Whitehead (CE) introduced the **Pontrjagin square** (see also E. Thomas)

$$\mathfrak{p}\colon H^2(X, G) \to H^4(X, \Gamma G). \tag{5.3}$$

This function can be defined as follows. Since $H_3 K(G, 2) = 0$ and since $H_4 K(G, 2) = \Gamma(G)$, see (4.13), we get the isomorphism.

$$\mu\colon H^4(K(G, 2), A) \cong \operatorname{Hom}_{\mathbb{Z}}(\Gamma G, A)$$

by the universal coefficient formula. Whence for $A = \Gamma G$ we can choose the element

$$\mathfrak{p} \in H^4(K(G, 2), \Gamma G) \cong [K(G, 2), K(\Gamma G, 4)]$$

for which $\mu(\mathfrak{p})$ is the identity of ΓG. This element, considered as a map between Eilenberg-Mac Lane spaces, induces the function

$$H^2(X, G) = [X, K(G, 2)] \to [X, K(\Gamma G, 4)] = H^4(X, \Gamma G)$$

which coincides with the Pontrjagin square above. A further characterization of the Pontrjagin square can be derived from (7.6) below. Clearly the function (5.3) is natural with respect to maps $X \to Y$ between spaces and with respect to homomorphisms $G \to G'$ between abelian groups. For $x, y \in H^n(X, G)$ one has the formulas

$$\mathfrak{p}(-x) = \mathfrak{p}(x) \tag{1}$$

$$\mathfrak{p}(x + y) - \mathfrak{p}(x) - \mathfrak{p}(y) = [1, 1]_*(x \cup y) \tag{2}$$

where $[1, 1]\colon G \otimes G \to \Gamma G$ is given by (4.2). Therefore $\mathfrak{p}$ is quadratic and induces the binatural homomorphism (see (4.1))

$$\mathscr{p}^{\square}\colon \Gamma(H^2(X,G)) \to H^4(X,\Gamma G). \tag{3}$$

Moreover, the cup product $x \cup x$ satisfies the formula

$$\tau_* \mathscr{p}(x) = x \cup x \tag{4}$$

where $\tau\colon \Gamma(G) \to G \otimes G$ is defined in (4.2). Now let

$$\mathscr{S}q\colon H^n(X,G) \to H^{n+2}(X, G \otimes \mathbb{Z}/2) \tag{5}$$

be the **Steenrod square**, $n \geq 2$. For $n = 2$ there is a connection between the Pontrjagin square and the Steenrod square, namely

$$\sigma_* \mathscr{p}(x) = \mathscr{S}q(x), \quad x \in H^2(X,G). \tag{6}$$

Here $\sigma\colon \Gamma G \to G \otimes \mathbb{Z}/2$ is the suspension homomorphism in (4.2). Since the function (5) is compatible with the suspension $\Sigma\colon H^n(H,G) \cong H^{n+1}(\Sigma X, G)$, that is $\Sigma \mathscr{S}q = \mathscr{S}q \Sigma$ on $H^n(X,G)$, $n \geq 2$, we see by (6) that the Steenrod square (5) is completely determined by the Pontragin square (5.3).

We have seen in (5.1) that the cup product $x \cup x$ in (4) is also defined for cohomology groups with local coefficients. This leads to the following question.

Problem. Do there exist binatural functions

$$\begin{aligned} &\mathscr{p}\colon \hat{H}^2(X,G) \to \hat{H}^4(X,\Gamma G), \\ &\mathscr{S}q\colon \hat{H}^n(X,G) \to \hat{H}^{n+2}(X, G \otimes \mathbb{Z}/2), \quad n \geq 2, \end{aligned} \tag{7}$$

(where G is a $\pi_1(X)$-module) such that these functions coincide with the cohomology operations above provided G is a trivial $\pi_1(X)$-module? Moreover, these functions should satisfy formulas as above.

We shall solve the problem in chapter V. It turns out that the functions in (7) exist provided an obstruction vanishes. In particular, if $\pi_1 X$ is a finite group with an odd number of elements then this obstruction is trivial. For a complete discussion of the Pontrjagin square with local coefficients see (V.§6) below.

The next example describes a well known and simple space with a non trivial Pontrjagin square for which, however, the Steenrod square is trivial.

(5.4) **Example.** Let $\mathbb{R}P_n$ be the **real projective space** of dimension n. Then the suspension of $\mathbb{R}P_3$ has a non trivial Pontrjagin square

$$0 \neq \wp\colon H^2(\Sigma\mathbb{R}P_3, \mathbb{Z}/2) = \mathbb{Z}/2 \to H^4(\Sigma\mathbb{R}P_3, \mathbb{Z}/4) = \mathbb{Z}/4, \tag{1}$$

where we use $\Gamma(\mathbb{Z}/2) = \mathbb{Z}/4$. The Steenrod square

$$0 = \mathscr{S}q\colon H^2(\Sigma\mathbb{R}P_3, \mathbb{Z}/2) = \mathbb{Z}/2 \to H^4(\Sigma\mathbb{R}P_3, \mathbb{Z}/2) = \mathbb{Z}/2, \tag{2}$$

however, is trivial as follows from (1) by (5.3)(6). A fairly complicated proof of the non triviality of the Pontrjagin square (1) is given in the Appendix to chapter 8 in Hilton. In (IV.A.11) below we shall give a new and simple proof which uses the theory of quadratic chain complexes. There we show that $\Sigma\mathbb{R}P_3$ is homotopy equivalent to a mapping cone C_f of a map

$$f\colon S^3 \vee S^2 \to S^2 \tag{3}$$

for which $f|S^3 = 2\eta$ is twice the Hopf map and for which $f|S^2$ is the map of order 2. In fact, f is an example for the A_n^2-forms described in lemma (7.5) below. Therefore proposition (7.6) shows that the Pontrjagin square (1) is non trivial.

§6 Cohomological invariants

Cup products and Pontrjagin squares are invariants of the homotopy type of the CW-complex X which depend on the choice of the coefficients G, G' above. We now describe 'universal' cohomological invariants, called cup product elements and Pontrjagin elements, which do not depend on coefficient groups and which determine all cup products and Pontrjagin squares respectively. For this we introduce the general notion of a cohomological invariant which is motivated by the examples in this section and by further examples in chapter V where we study Pontrjagin squares with local coefficients. Moreover we describe in this section connections of the Pontrjagin elements with the operators in Whitehead's exact sequence.

(6.1) **Definition.** Let **K** be a full subcategory of $\mathbf{CW}/\simeq$ such that for $X \in \mathbf{K}$ also $I_*X \in \mathbf{K}$. Moreover let

$$A\colon \mathbf{Chain}_{\mathbb{Z}}^{\wedge} \to \mathbf{Mod}_{\mathbb{Z}}^{\wedge} \tag{1}$$

be a functor which carries a chain complex (π, C) to a π-module and which carries φ-equivariant maps to φ-equivariant maps. We call α a **cohomological invariant on K with coefficients in** A if α associates with each object X in **K** a 'natural' element

$$\alpha_X \in \hat{H}^*(X, A\hat{C}_*X). \tag{2}$$

Here we say that α_X is **natural on K** if the equation

$$f_*\alpha_X = f^*\alpha_Y \in \hat{H}^*(X, \varphi^* A\hat{C}_* Y) \tag{3}$$

is satisfied for each cellular map $f: X \to Y$ representing a morphism in **K**, $\varphi = \pi_1(f)$. Clearly the map f_* is induced by the coefficient homomorphism $A\hat{C}_*(f)$.

The map f_* in (3) actually depends on the cellular map f and is not defined by the homotopy class of f. The naturality formula (5.6)(3) implies, however, that for $f \simeq g$

$$f_*\alpha_X = f^*\alpha_X = g^*\alpha_X = g_*\alpha_X.$$

In particular, if $f = i_0$, $g = i_1: X \to I_*X$ are the inclusions of the cylinder we get $(i_0)_*\alpha_X = (i_1)_*\alpha_X$. Therefore α_X is an element in the kernel of

(6.2) $$(i_0)_* - (i_1)_*: \hat{H}^*(X, A\hat{C}_*X) \to \hat{H}^*(X, A\hat{C}_* I_* X).$$

Let $A[X]$ be this kernel. Then clearly $f_*\alpha = g_*\alpha$ for all $\alpha \in A(X)$ since a homotopy $H: f \simeq g$ yields $f_*\alpha = (Hi_0)_*\alpha = H_*(i_0)_*\alpha = H_*(i_1)_*\alpha = (Hi_1)_*\alpha = g_*\alpha$. We now show that cup products as in (5.1) yield a cohomological invariant in the sense of (6.1). For this let $C = \hat{C}_*X$ and let

$$i_n \in \hat{H}^n(X, C_n/dC_{n+1})$$

be the cohomology class represented by the quotient map $C_n \to C_n/dC_{n+1}$ which clearly is a cocycle. The **cup product element**

(6.3) $$\begin{cases} i_n \cup i_m \in \hat{H}^{n+m}(X; AC_*X) \\ A(C) = (C_n/dC_{n+1}) \otimes (C_m/dC_{m+1}) \end{cases}$$

is a cohomological invariant on $\mathbf{CW}/\simeq$. The element $i_n \cup i_m$ determines all cup products in (5.1). In fact, for $\{x\} \in \hat{H}^n(X; G)$, $\{y\} \in \hat{H}^m(X; G')$ where $x: C_n \to G$, $y: C_m \to G'$ are cocycles we get the induced homomorphism $x \otimes y: A(C) \to G \otimes G'$ with

$$\{x\} \cup \{y\} = (x \otimes y)_*(i_n \cup i_m). \tag{1}$$

We now consider the kernel $A[X]$ in (6.2) for A in (6.3). The boundary maps $d = d_n\colon C_n/dC_{n+1} \to C_{n-1}$ and $d = d_m$ yield the homomorphisms

$$\left.\begin{aligned} d_n \otimes 1\colon A(C) &\to C_{n-1} \otimes (C_m/dC_{m+1}), \\ 1 \otimes d_m\colon A(C) &\to (C_n/dC_{n+1}) \otimes C_{m-1}. \end{aligned}\right\} \tag{2}$$

(6.4) **Proposition.** *For A in (6.3) the kernel $A[X]$ is given by the intersection $A[X] = \operatorname{kernel}(1 \otimes d_m)_* \cap \operatorname{kernel}(d_n \otimes 1)_*$. Clearly $i_n \cup i_m \in A[X]$.*

Proof. For the chain complex $C = \hat{C}_*X$ we have the cyclinder I_*C in (III.3.2) with $I_*C = \hat{C}_*(I_*X)$, see (III.3.4). We define an isomorphism of modules

$$\Psi\colon C_n \oplus C_n \oplus C_{n-1} \cong C_n' \oplus C_n'' \oplus sC_{n-1} = (I_*C)_n \tag{1}$$

by $\Psi(x, y, z) = (x', y'' - x'', s(z + dx))$. The inverse of Ψ is φ with $\varphi(a', b'', sc) = (a, a + b, c - da)$. For the boundary $d\colon (I_*C)_{n+1} \to (I_*C)_n$ we get $\varphi d(x') = \varphi(dx)' = (dx, 0, -ddx)$, $\varphi d(x'') = \varphi(dx)'' = (0, dx, 0)$, $\varphi d(sx) = \varphi(-x', x'', -sdx) = (-x, -x + x, +dx - dx)$. This shows that φ induces an isomorphism

$$\bar{\varphi}_n = \bar{\varphi}\colon (I_*C)_n/d(I_*C)_{n+1} \cong (C_n/dC_{n+1}) \oplus C_{n-1}.$$

Moreover, we get the maps

$$\bar{\varphi}i_0, \bar{\varphi}i_1\colon C_n/dC_{n+1} \to (C_n/dC_{n+1}) \oplus C_{n-1}$$

by $\bar{\varphi}i_0 = (1, -d_n)$ and $\bar{\varphi}i_1 = (1, 0)$. Whence $(i_0)_* - (i_1)_*$ is given by

$$(\bar{\varphi}_n i_0 \otimes \bar{\varphi}_m i_0) - (\bar{\varphi}_n i_1 \otimes \bar{\varphi}_m i_1) = -1 \otimes d_m - d_n \otimes 1 + d_n \otimes d_m. \tag{2}$$

This proves the proposition since the kernel of $(d_n \otimes d_m)_*$ contains the intersection in (6.4). □

Next we describe examples of cohomological invariants given by the Pontrjagin square and the Steenrod square respectively. For X in **CW** we have similarly as in (6.3) the cohomology class

$$i_n^{\mathbb{Z}} \in H^n(X; C_n/dC_{n+1})$$

represented by the quotient map $C_n \to C_n/dC_{n+1}$ with $C = C_*X = \hat{C}_*X \otimes_{\mathbb{Z}[\pi]} \mathbb{Z}$.

Now the elements

(6.5) $$\wp(i_2^{\mathbb{Z}}) \in H^4(X; \Gamma(C_2/dC_3)),$$

(6.6) $$\mathcal{S}q(i_n^{\mathbb{Z}}) \in H^{n+2}(X; (C_n/dC_{n+1}) \otimes \mathbb{Z}/2), \quad n \geq 2$$

given by (5.3) and (5.5)(5) are cohomological invariants on **CW**/$\simeq$. Similarly as in (6.3)(1) we see that these elements determine the functions (5.3) and (5.5)(5) completely. In chapter V we consider such elements in case C above is replaced by $\hat{C}_* X$; this leads to a solution of the problem concerning the existence of Pontrjagin squares with local coefficients in (5.3)(7).

Now let X be an $(n-1)$-connected space. Then we have the cohomology class

(6.7) $$i_n \in H^n(X; H_n X) \cong \operatorname{Hom}(H_n X, H_n X)$$

given by the identity on $H_n(X)$. Whence we get the **Pontrjagin-Steenrod** element

(6.8) $$\wp_n(X) = \begin{cases} \wp(i_2) \in H^4(X; \Gamma(H_2 X)) & \text{for } n = 2, \\ \mathcal{S}q(i_n) \in H^{n+2}(X; H_n \otimes \mathbb{Z}/2) & \text{for } n \geq 2, \end{cases}$$

which is a cohomological invariant on the category **K** consisting of all $(n-1)$-connected CW-complexes. These elements determine the cohomology operations $\wp$ and $\mathcal{S}q$ respectively on **K** and whence on **CW**/$\simeq$.

In the next section, see (7.6), we describe the element $\wp_n(X)$ in terms of A_n^2-forms. Moreover, we prove in the next section the following connection of the elements $\wp_n(X)$ with the operators in Whitehead's exact sequence. As above let X be an $(n-1)$-connected space, $n \geq 2$. We define the functor Γ_n^1 by

(6.9) $$\Gamma_n^1(A) = \begin{cases} \Gamma(A) & n = 2, \\ A \otimes \mathbb{Z}/2 & n \geq 3. \end{cases}$$

Using (4.11) and (4.12) we have the natural exact sequence

(6.10) $$H_{n+2} \xrightarrow{b} \Gamma_n^1(H_n) \xrightarrow{i} \pi_{n+1} \xrightarrow{h} H_{n+1}$$

where $H_i = H_i X$, $\pi_{n+1} = \pi_{n+1} X$, see (3.7). Now consider the following commutative diagram where $\Gamma = \Gamma_n^1(H_n)$ and where $q \colon \Gamma \twoheadrightarrow \operatorname{cok}(b)$ is the quotient map for the cokernel of $b = b_{n+2}$ in (6.10).

$$
\begin{array}{ccc}
\operatorname{Ext}(H_{n+1},\Gamma) & \xrightarrow{q_*} & \operatorname{Ext}(H_{n+1},\operatorname{cok}(b)) \\
\downarrow{\scriptstyle\Delta} & & \downarrow{\scriptstyle\Delta} \\
H^{n+2}(X,\Gamma) & \xrightarrow{q_*} & H^{n+2}(X,\operatorname{cok}(b)) \\
\downarrow{\scriptstyle\mu} & & \downarrow{\scriptstyle\mu} \\
\operatorname{Hom}(H_{n+2},\Gamma) & \xrightarrow[q_*]{} & \operatorname{Hom}(H_{n+2},\operatorname{cok}(b))
\end{array} \tag{6.11}
$$

The columns are given by the **universal coefficient sequence**. Using the operators in this diagram we get the following result.

(6.12) **Theorem.** *The element $\wp_n(X)$ in* (6.8) *satisfies the formulas*

$$\mu(\wp_n X) = b_{n+2}, \quad \text{and} \quad -\Delta^{-1}q_*(\wp_n X) = \{\pi_{n+1}\}.$$

Here $\{\pi_{n+1}\}$ represents the extension $\operatorname{cok}(b) \rightarrowtail \pi_{n+1} \twoheadrightarrow H_{n+1}$ given by (6.10).

The second equation is well defined since by the first equation we have $\mu q_*(\wp_n X) = 0$. We prove (6.12) in (7.7) below.

(6.13) **Corollary.** *Let X be an $(n-1)$-connected space, $n \geq 2$. Then the following diagram commutes for any coefficient group G.*

$$
\begin{array}{ccc}
\operatorname{Ker}(b^\#_{n+2}) & \xrightarrow{-\{\pi_{n+1}\}^\#} & \operatorname{Ext}(H_{n+1},\Gamma^1_n G) \\
\cap & & \downarrow{\scriptstyle\Delta} \\
H^n(X,G) & \xrightarrow{\wp_n} & H^{n+2}(X,\Gamma^1_n G) \\
\|{\scriptstyle\mu} & & \downarrow{\scriptstyle\mu} \\
\operatorname{Hom}(H_n,G) & \xrightarrow[b^\#_{n+2}]{} & \operatorname{Hom}(H_{n+2},\Gamma^1_n G)
\end{array} \tag{1}
$$

Here the cohomology operation $\wp_n$ is the Pontrjagin square for $n = 2$ and $\wp_n$ is the Steenrod square for $n \geq 3$. Moreover, we define for $\alpha \in \operatorname{Hom}(H_n, G)$ the element

$$b^\#_{n+2}(\alpha) = \Gamma^1_n(\alpha)b_{n+2}. \tag{2}$$

Therefore $b^\#_{n+2}(\alpha) = 0$ implies that $\Gamma^1_n(\alpha)$ induces a map $\Gamma^1_n(\alpha)\colon \operatorname{cok}(b_{n+2}) \to$

$\Gamma_n^1(G)$. Therefore we can define $\{\pi_{n+1}\}^{\#}$ in (1) by

$$\{\pi_{n+1}\}^{\#}(\alpha) = \Gamma_n^1(\alpha)_*\{\pi_{n+1}\}. \tag{3}$$

The commutativity of the bottom square in (1) was proved by Whitehead (CE) § 17, theorem 20. The equations in (6.12) are also described in V.1.9 of the book of G.W. Whitehead.

§7 A_n^2-polyhedra and A_n^2-forms

In section §5 above the Pontrjagin square was defined as a cohomology operation via a map between Eilenberg-Mac Lane spaces. In this section we describe the Pontrjagin square by the internal structure of a simply connected 4-dimensional CW-complex X, compare (7.6). For this we show that X is homotopy equivalent to the mapping cone C_f of a map f between one point unions of spheres. The homotopy class of f can be described by algebraic data which we call A_2^2-forms. More generally we consider A_n^2-forms, $n \geq 2$, which can be used to describe $(n-1)$-connected $(n+2)$-dimensional homotopy types and which are needed in the proof of theorem (6.12) above.

Let $\mathbf{A}_n^k$ be the full subcategory of $\mathbf{CW}/\simeq$ consisting of $(n-1)$-connected $(n+k)$-dimensional CW-complexes; such complexes are also called A_n^k-**polyhedra**. The suspension Σ gives us the sequence of functors

(7.1) $$\mathbf{A}_2^k \xrightarrow{\Sigma} \mathbf{A}_3^k \xrightarrow{\Sigma} \cdots \xrightarrow{\Sigma} \mathbf{A}_n^k \longrightarrow \mathbf{A}_{n+1}^k \longrightarrow \cdots$$

which we call the k-**stem** of homotopy categories. The Freudenthal suspension theorem shows that for $k+1<n$ the functor $\Sigma\colon \mathbf{A}_n^k \to A_{n+1}^k$ is an equivalence of categories. For $k+1=n$ this functor is full and a 1-1 correspondence of homotopy types. Moreover each object in $\mathbf{A}_{k+1}^k$ has the homotopy type of a suspension. In this section we are only concerned with A_n^2-polyhedra. We describe such polyhedra by the following algebraic data which we call A_n^2-forms.

(7.2) **Definition.** Let $n \geq 2$. An A_n^2-**form** is a triple $f=(C,t,\bar{f})$ where C is a chain complex of free abelian groups $C=\{C_{n+2} \xrightarrow{d} C_{n+1} \xrightarrow{d} C_n\}$ with $C_i=0$ for $i<n$, $i>n+2$. Moreover, the homomorphism $t\colon d(C_{n+1}) \to C_{n+1}$ is a splitting of $d\colon C_{n+1} \twoheadrightarrow d(C_{n+1})$ and $\bar{f}$ is a homomorphism $\bar{f}\colon C_{n+2} \to \Gamma_n^1(C_n)$; for the definition of Γ_n^1 see (6.9).

We associate with each A_n^2-form $f=(C,t,\bar{f})$ the following map (which is also denoted by f). For this we need the following notation. If A is a group in degree n then $MA = M(A, n)$ and $M^-A = M(A, n-1)$ are the corresponding Moore spaces. These are wedges of spheres if A is *free abelian* and in this case we have

(7.3) $$[MA, X] = \operatorname{Hom}(A, \pi_n X).$$

Moreover, we know $\pi_n MA = A$ and $\pi_{n+1} MA = \Gamma_n^1(A)$. Using these identifications the A_n^2-form f gives us the map (up to homotopy)

(7.4) $$f: M^-C_{n+2} \vee M^-C'_{n+1} \to MC''_{n+1} \vee MC_n.$$

Here $C'_{n+1} = tdC_{n+1}$ and $C''_{n+1} = \operatorname{kernel}(d: C_{n+1} \to C_n)$ are groups of degree $n+1$ with

$$C_{n+1} = C'_{n+1} \oplus C''_{n+1}. \tag{1}$$

The map f is given via (7.3) by

$$f|M^-C_{n+2} = i_1 d_{n+2} + i_2 \bar{f}, \tag{2}$$

$$f|M^-C'_{n+1} = i_2 d_{n+1} \tag{3}$$

where i_1 and i_2 denote the inclusions of MC''_{n+1} and MC_n respectively and $d_{n+2}: C_{n+2} \longrightarrow C''_{n+1}$, $d_{n+1}: C'_{n+1} \rightarrowtail C_n$ are given by the boundary maps in C. The following lemma is due to Whitehead (SC) (lemma 8).

(7.5) **Lemma.** *Let $n \geq 2$. For each A_n^2-polyhedron X there is a homotopy equivalence $X \simeq C_f$ where f is given by an A_n^2-form as in* (7.4) *and where C_f is the mapping cone of f.*

We call $X = C_f$ a **reduced A_n^2-polyhedron.** For the convenience of the reader we recall a proof of lemma (7.5), see also the proof of (V.8.5), page 288, in Baues (AH).

Proof of (7.5). We may assume that $X^{n-1} = *$. Let $C = C_*X$ be the cellular chain complex of X. We choose a splitting t as in (7.2). Then we get

$$g = i_2 d_{n+1} : M = M^-C'_{n+1} \to Y = MC''_{n+1} \vee MC_n \tag{1}$$

and a homotopy equivalence $C_g \simeq X^{n+1}$. Below we show that the inclusion $i_g: Y \subset C_g \simeq X^{n+1}$ induces a surjective map

$$(i_g)_*: \pi_{n+1} Y \twoheadrightarrow \pi_{n+1} C_g. \tag{2}$$

Therefore the attaching map of $(n+2)$-cells is homotopic to a map

$$M^- C_{n+2} \xrightarrow{\bar{f}} Y \overset{i}{\subset} C_g \simeq X^{n+1} \tag{3}$$

Here $\bar{f}$ yields $\bar{f}$ in (7.4) since

$$\pi_{n+1}(Y) = C''_{n+1} \oplus \Gamma_n^1(C_n). \tag{4}$$

This completes the proof of (7.5) since clearly $C_{i\bar{f}} = C_f \simeq X^{n+2}$. We obtain the surjectivity of (2) by the exact sequence

$$\begin{array}{ccccccc}
\pi_{n+2}(C_g, Y) & \xrightarrow{\partial} & \pi_{n+1} Y & \xrightarrow{i} & \pi_{n+1} C_g & \xrightarrow{j} & \\
\uparrow & \nearrow_{(g,1)_*} & & & & & \\
\pi_{n+1}(M \vee Y)_2 & & & & & & \\
& & & & & & \\
\pi_{n+1}(C_g, Y) & \xrightarrow{\partial} & \pi_n Y = C_n & & & & \\
\cong \uparrow & \nearrow_{(g,1)_*} & & & & & \\
\pi_n(M \vee Y)_2 = C'_{n+1} & & & & & &
\end{array} \tag{5}$$

Here the vertical arrows are the functional suspensions $\bar{E}_g$, see (II.11.6) and (V.7.6) in Baues (AH). The right hand $(g, 1)_*$ can be identified with the injective map $d: C'_{n+1} \rightarrowtail C_n$ so that $j = 0$. Therefore the map $i = (i_g)_*$ in (2) is surjective. Moreover, for $n = 2$ the left hand $(g, 1)_*$ can be identified with

$$\Gamma C'_3 \oplus C'_3 \otimes C_2 \xrightarrow{\partial} \Gamma C_2 \subset C''_3 \oplus \Gamma(C_2) \tag{6}$$

where ∂ is defined by $d = d_3: C'_3 \rightarrowtail C_2$ as in (4.9). This shows by (4.9) that the following equation holds for $n = 2$.

$$\pi_{n+1}(C_g) = C''_{n+1} \oplus \Gamma_n^1(H_n X). \tag{7}$$

We leave the proof of (7) for $n \geq 3$ as an exercise. □

(7.6) **Proposition.** *Let X be an $(n-1)$-connected CW-complex and let $X^{n+2} \simeq C_f$ be a homotopy equivalence as in (7.5). Then the composition*

$$C_{n+2} \xrightarrow{\bar{f}} \Gamma_n^1(C_n) \xrightarrow{q_*} \Gamma_n^1(H_n X)$$

is a cocycle which represents the Pontrjagin-Steenrod element $\wp_n X$ in (6.8). The map $q_* = \Gamma_n^1(q)$ is induced by the quotient map $q: C_n \twoheadrightarrow H_n C_f \cong H_n X$.

Proof of (7.6). The properties of the quadratic map $\wp$ imply

$$2\wp(x) = [1,1]_*(x \cup x). \tag{1}$$

For $G = \mathbb{Z}$ we have $\Gamma(G) = \mathbb{Z}$ and $[1,1]$ = multiplication by 2. Therefore in this case

$$2\wp(x) = [1,1]_*(x \cup x) = 2(x \cup x). \tag{2}$$

In particular, for $X = \mathbb{C}P_2 = S^2 \cup_\eta e^4$ we get

$$\wp(x) = x \cup x \tag{3}$$

where $x \in H^2(\mathbb{C}P_2) \cong \mathbb{Z}$ and $x \cup x \in H^4(\mathbb{C}P_2) \cong \mathbb{Z}$ are generators. Thus $\wp$ detects the Hopf map η. Now naturality of $\wp$ and the definition of the isomorphism (4.11), see (4.10), show that (7.6) is satisfied for $n = 2$. Using a suspension argument we get (7.6) for $n \geq 3$. □

(7.7) *Proof of* (6.12). We use (7.6). The attaching map of $(n+2)$-cells is given by

$$f_{n+2} = (d_{n+2}, q_*\bar{f})\colon C_{n+2} \to C''_{n+1} \oplus \Gamma_n^1(H_n). \tag{1}$$

Here we use the isomorphism (7.5)(7). This shows by definition of b_{n+2} in (3.8) and by (7.6) that the first equation in (6.12) is satisfied. Clearly π_{n+1} is the cokernel of (1). For the quotient map p of this cokernel we get the exact sequence

$$C_{n+2} \xrightarrow{f_{n+2}} C''_{n+1} \oplus \Gamma_n^1(H_n) \overset{p}{\twoheadrightarrow} \pi_{n+1} \tag{2}$$

where the coordinates of p are $p = (\bar{p}, i)$. Here i is the map in (6.10) and we get by $pf_{n+2} = 0$ the equation

$$\bar{p}d_{n+2} = -iq_*\bar{f}. \tag{3}$$

Moreover, $h\bar{p}\colon C''_{n+1} \to H_{n+1}$ with h as in (6.10) is the quotient map. Therefore (3) yields the second equation in (6.12). □

§8 Whitehead's classification of A_n^2-polyhedra, $n \geq 2$

In this section the homotopy classification of $(n-1)$-connected $(n+2)$-dimensional CW-complexes ($n \geq 2$) is obtained in two different ways by algebraic invariants; on the one hand by the Pontrjagin-Steenrod invariant, on the other hand by Whitehead's certain exact sequence. This corresponds to famous results of Whitehead (CE), (SC), (HT). The invariants actually yield two detecting functors $\mathscr{P}$ and $\mathscr{B}$ respectively on the category $\mathbf{A}_n^2$ of $(n-1)$-connected $(n+2)$-dimensional CW-complexes. The connection of the functors $\mathscr{P}$ and $\mathscr{B}$ is obtained by the classical formulas in theorem (6.12). For $n = 2$ the functor $\mathscr{P}$ was already discussed in the introduction of this chapter. As mentioned there the results in this section form a starting point for the homotopy theory of 4-dimensional CW-complexes with non trivial fundamental group.

First we define algebraic categories $\mathbf{P}_n^2$ and $\mathbf{B}_n^2$ and functors as in the diagram, $n \geq 2$,

(8.1)
$$\begin{array}{ccc} & & \mathbf{P}_n^2 \\ & \overset{\mathscr{P}}{\nearrow} & \\ \mathbf{A}_n^2 & & \downarrow \mu \\ & \underset{\mathscr{B}}{\searrow} & \\ & & \mathbf{B}_n^2 \end{array}$$

(8.2) **Definition of the category $\mathbf{B}_n^2$.** Objects are tuples (H_*, b, β) where $H_* = (H_{n+2}, H_{n+1}, H_n)$ is a sequence of abelian groups with H_{n+2} free abelian. Moreover, b and β are elements

$$b \in \operatorname{Hom}(H_{n+2}, \Gamma_n^1 H_n), \quad \beta \in \operatorname{Ext}(H_{n+1}, \operatorname{cok}(b))$$

where Γ_n^1 is the functor in (6.9). A morphism $f: (H_*, b, \beta) \to (H'_*, b', \beta')$ is a homomorphism $f: H_* \to H'_*$ of degree 0 with $f_* b = f^* b'$, $f_* \beta = f^* \beta'$.

The functor $\mathscr{B}$ in (8.1) carries an $\mathbf{A}_n^2$-polyhedron X to the object

(8.3)
$$\mathscr{B}(X) = (H_* X, b_{n+2}, \{\pi_{n+1}\})$$

where b_{n+2} and $\{\pi_{n+1}\}$ are defined by Whitehead's exact sequence (6.10), see (6.12).

(8.4) **Definition of the category $\mathbf{P}_n^2$.** Objects are paris (C, ξ) where C is an $(n+2)$-dimensional chain complex of free abelian groups with $H_i C = 0$ for $n-1 \geq i \geq 0$ and $C_i = 0$ for $i < 0$. Moreover, ξ is an element $\xi \in H^{n+2}(C, \Gamma_n^1(H_n C))$. Morphisms $F: (C, \xi) \to (C', \xi')$ are homotopy classes of chain maps $F: C \to C'$ which satisfy $F_* \xi = F^* \xi' \in H^{n+2}(C, \Gamma_n^1(H_n C'))$. Here F_* is induced by $\Gamma_n^1 H_n(F)$.

The functor $\mathscr{P}$ in (8.1) carries the A_n^2-polyhedron X to the object

(8.5) $$\mathscr{P}(X) = (C_*(X,*), \wp_n X)$$

where $\wp_n X$ is the Pontrjagin-Steenrod element in (6.8) and where $C_*(X,*)$ is the reduced cellular chain complex of X. Next we define the functor μ in (8.1) by

(8.6) $$\mu(C,\xi) = (H_*C, \mu\xi, -\Delta^{-1}q_*\xi)$$

where the right hand side is defined as in (6.12). Theorem (6.12) shows that there is a natural isomorphism $\mu\mathscr{P} \cong \mathscr{B}$ for the functors in (8.1).

(8.7) **Theorem.** *The functors* $\mathscr{B}, \mathscr{P}$ *and* μ *in* (8.1) *are detecting functors, see* (0.4).

Recall that a detecting functor induces a 1-1 correspondence between equivalence classes of objects, see (0.5). Whence theorem (8.7) implies the following classical result of Whitehead (CE).

(8.8) **Corollary.** *The homotopy types of* A_n^2*-polyhedra,* $n \geq 2$, *are in* 1-1 *correspondence with the isomorphism classes of objects in the category* $\mathbf{B}_n^2$.

The functors in (8.1) are compatible with the suspension functor $\Sigma\colon \mathbf{A}_n^2 \to \mathbf{A}_{n+1}^2$ by (5.3)(6). The following proof of (8.7) uses fairly simple special cases of results obtained in chapter II and chapter III below. This is an entertaining application demonstrating the utility of the methods developed in these chapters.

(8.9) *Proof of* (8.7). For the proof of (8.7) we use the CW-tower of categories in the next chapter (II.§2) and we use the equivalence of categories in (III.2.9). Let X and Y be A_n^2-polyhedra. Then a chain map $F\colon C_*X \to C_*Y$ can be identified with a map in $\mathbf{H}_3^c$. The map F is realizable by a cellular map $f\colon X \to Y$ if and only if the obstruction

$$\mathscr{O}_{X,Y}(F) \in H^{n+2}(X, \Gamma_n^1 H_n Y) \tag{1}$$

is trivial, see (II.3.4). Now the definition of the obstruction (1) and (7.6) show that

$$\mathscr{O}_{X,Y}(F) = F_*(\wp_n X) - F^*(\wp_n Y), \tag{2}$$

see (8.5). This proves that the functor $\mathscr{P}$ satisfies the realizability condition for maps. Moreover, (7.6) implies that each pair (C,ξ) can be realized by an A_n^2-form and whence the functor $\mathscr{P}$ satisfies the realizability condition for objects. We derive from (2) and (6.12) the following formulas for the obstruction

$$\mu \mathcal{O}_{X,Y}(F) = F_*(b_{n+2}X) - F^*(b_{n+2}Y), \tag{3}$$

$$-q_*\Delta^{-1}\mathcal{O}_{X,Y}(F) = F_*\{\pi_{n+1}X\} - F^*\{\pi_{n+1}Y\}. \tag{4}$$

Here (4) is given provided the element (3) is trivial. These formulas hold since X and Y are $(n-1)$-connected. A generalization of these formulas is described in (II.§5).

Next we show that μ in (8.1) is a detecting functor. Let $[C, C']$ be the set of homotopy classes of chain maps with $(C, \xi), (C', \xi') \in \mathbf{P}_n^2$. Then we have the exact sequence

$$\mathrm{Ext}(H_n, H'_{n+1}) \oplus \mathrm{Ext}(H_{n+1}, H'_{n+2}) \rightarrowtail^{i} [C, C'] \xrightarrow{H} \mathrm{Hom}(H_*C, H_*C'), \tag{5}$$

where the surjection is given by the homology functor H. Assume now a map $f: \mu(C, \xi) \to \mu(C', \xi')$ in $\mathbf{B}_n^2$ is given. Then we can choose $F \in [C, C']$ with $H(F) = f$. However, F is not unique by (5). Since the sequence

$$\mathrm{Ext}(H_{n+1}, H'_{n+2}) \xrightarrow[b'_*]{} \mathrm{Ext}(H_{n+1}, \Gamma_n^1 H'_n) \xrightarrow[q_*]{} \mathrm{Ext}(H_{n+1}, \mathrm{cok}\, b') \tag{6}$$

is exact we see that

$$F_*\xi - F^*\xi' = \Delta b'_*(\alpha) \tag{7}$$

for an appropriate $\alpha \in \mathrm{Ext}(H_{n+1}, H'_{n+2})$; here b' is given by $\mu(C', \xi')$. For the proof of (7) we use a diagram as in (6.11) and the assumption that $f = H(F)$ is a map in $\mathbf{B}_n^2$. We now alter F by α via (5), then we obtain $G = F + i\alpha \in [C, C']$ with $H(G) = H(F) = f$ and with

$$\begin{aligned} G_*\xi - G^*\xi' &= F_*\xi - (F + i\alpha)^*\xi' \\ &= F_*\xi - F^*\xi' - \Delta b'_*(\alpha) \\ &= 0. \end{aligned} \tag{8}$$

Therefore G is a μ-realization of f. Moreover, one readily verifies the realizability condition for objects for μ. This completes the proof. □

Chapter II

The CW-tower of categories

A CW-complex X is obtained inductively by constructing the skeleta X^n, $n \geq 0$. However, this skeletal filtration has the disadvantage that the homotopy type of X^n is not well defined by the homotopy type of X. For this reason J.H.C. Whitehead introduced the $(n-1)$-type of X which is also the $(n-1)$-type of X^n and which can be obtained from X^n by "killing" all homotopy groups $\pi_m X^n$, $m \geq n$. The $(n-1)$-type is the $(n-1)$-section of the Postnikov tower of X which we denote by $P_{n-1}(X)$. It is a classical result that the homotopy type of $P_{n-1}(X)$ is well defined by the homotopy type of X. This fact justifies the construction of $P_{n-1}(X)$, though it is to some extent absurd to replace a nice n-dimensional CW-complex X^n by a CW-complex $P_{n-1}(X)$ which in general is infinite dimensional, and the homology of which is very hard to compute. The fact that a homotopy type can be built by constructing inductively Postnikov sections is "Eckmann–Hilton dual" to the fact that a homotopy type can be obtained by constructing inductively CW-skeleta: the Postnikov sections form a sequence of principal fibrations while the CW-skeleta form a sequence of principal cofibrations. Hence Postnikov sections and CW-skeleta are just two ways of representing a homotopy type by a filtered object; this fact does not have a deep impact on the classification of homotopy types, but it marks a kind of starting point for the problem.

In this book we study the skeletal filtration of a CW-complex rather than the Postnikov decomposition. There is, however, an important relationship between $(n-1)$-types and n-dimensional CW-complexes described by the trees of homotopy types, in § 6. We deduce from the skeletal filtration of a CW-complex X the object $(n \geq 2)$

$$r_{n+1}(X) = (C, f_{n+1}, X^n)$$

which is a triple consisting of an algebraic part $C = C_* \hat{X}$ (which is the cellular chain complex of the universal cover of X) and of a topological part X^n (which is the n-skeleton of X). Moreover f_{n+1} is the homotopy class of the attaching map of $(n+1)$-cells in X given by a homomorphism $C_{n+1} \to \pi_n X^n$. We call such a triple a *homotopy system of order* $(n+1)$; the crucial point is the fact that homotopy systems of order $(n+1)$ form a homotopy category $\mathbf{H}^c_{n+1}/\simeq$ and that the homotopy type of the object $r_{n+1}(X)$ in $\mathbf{H}^c_{n+1}/\simeq$ depends only on the homotopy type of X, see (2.6). Whence enriching the n-skeleton X^n by an algebraic part (C, f_{n+1}) yields a new invariant $r_{n+1}(X)$ of the homotopy type of X which has the same kind of naturality as the Postnikov section $P_{n-1}(X)$, $n \geq 2$. We consider the sequence ($n \geq 2$)

$$r_3(X), r_4(X), \ldots, r_{n+1}(X), \ldots$$

of homotopy systems to be the "true" Eckmann-Hilton dual of the sequence

$$P_1(X), P_2(X), \ldots, P_{n-1}(X), \ldots$$

of Postnikov sections of X.

The categories $\mathbf{H}^c_{n+1}/\simeq$ of homotopy systems of order $(n+1)$ yield the *CW-tower* of categories. This is a sequence of functors ($n \geq 3$)

$$\mathbf{CW}/\simeq \xrightarrow{r} \mathbf{H}^c_{n+1} \xrightarrow{\lambda} \mathbf{H}^c_n/\simeq \xrightarrow{\lambda} \cdots \xrightarrow{\lambda} \mathbf{H}^c_3/\simeq$$

which approximates the homotopy category $\mathbf{CW}/\simeq$ of connected CW-complexes. Each functor λ is embedded in an "exact sequence"

$$H^n\Gamma_n + \longrightarrow \mathbf{H}^c_{n+1}/\simeq \xrightarrow{\lambda} \mathbf{H}^c_n/\simeq \xrightarrow{\mathcal{O}} H^{n+1}\Gamma_n \tag{1}$$

where $H^m\Gamma_n$ denotes a natural system of abelian groups on $\mathbf{H}^c_n/\simeq$, (we introduce the useful language concerning natural systems and exact sequences for a functor λ in section § 1). Originally the CW-tower of categories was studied in the book "Algebraic homotopy", see Baues (AH). The tower turned out to be a useful tool for the homotopy classification problems, in particular it yields an obstruction theory for the classification of homotopy types and for the realizability of $\mathbb{Z}[\pi_1]$-chain complexes C by CW-complexes, see (3.13). The obstruction operator $\mathcal{O}$ in the exact sequence above can be described in terms of "Postnikov chain functors" and "cellular boundary invariants" which we deduce from the Postnikov decomposition of a space, § 4. The sequence of cellular boundary invariants determines the homotopy type as does the sequence of k-invariants in a Postnikov decomposition.

We use the cellular boundary invariants for the proof of three formulas which relate the obstruction operator $\mathcal{O}$ with Whitehead's certain exact sequence. We also give an entertaining application of one of these formulas concerning the homotopy classification of certain manifolds, see § 7. Moreover Whitehead's classification of simply connected 4-dimensional homotopy types is a direct consequence of these formulas, see (I.8.9) and (5.7) below.

The algebraic nature of a 4-dimensional homotopy type with non-trivial fundamental group is bizarre in general. In order to probe this nature further we study the two bottom categories $\mathbf{H}^c_3/\simeq$ and $\mathbf{H}^c_4/\simeq$ of the CW-tower: in chapter III we show that the category $\mathbf{H}_3^c/\simeq$ is equivalent to the purely algebraic category $\mathbf{H}/\simeq$ of "totally free crossed chain complexes" and in chapter IV we show that the category $\mathbf{H}_4^c/\simeq$ is equivalent to the purely algebraic category $\mathbf{Q}/\simeq$ of "totally free quadratic chain complexes". Moreover we describe in (IV, § 9) a (purely algebraic) exact sequence

$$H^3\Gamma\pi_2 + \longrightarrow \mathbf{Q}/\simeq \xrightarrow{\lambda} \mathbf{H}/\simeq \xrightarrow{\mathcal{O}} H^4\Gamma\pi_2 \tag{2}$$

which is equivalent to the exact sequence for λ: $\mathbf{H}_4^c/\simeq \rightarrow H_3^c/\simeq$ in (1) above. This equivalence is the main new exploration in this book which dives deeply into the miraculous world of 4-dimensional homotopy types.

The equivalence of categories $\mathbf{H}_3^c/\simeq \xrightarrow{\sim} \mathbf{H}/\simeq$ is essentially due to Whitehead (CH) who introduced 40 years ago crossed chain complexes for the classification of 3-dimensional homotopy types. In this book we introduce quadratic chain complexes for the classification of 4-dimensional homotopy types and we obtain the equivalence $\mathbf{H}_4^c/\simeq \xrightarrow{\sim} \mathbf{Q}/\simeq$. Whence the CW-tower of categories leads to the next challenge of combinatorial homotopy, namely to find a purely algebraic model category for the category $\mathbf{H}_5^c/\simeq$. This task indeed requires great efforts and it may well be some time before a satisfying "minimal" algebraic model of the objects in $\mathbf{H}_5^c/\simeq$ is found. The construction of such models would be necessary in order to understand the connections and algebraic compounds inside a homotopy type.

§ 1 Linear extensions of categories and exact sequences for functors

The concept of exact sequences for groups is fundamental in algebraic topology. We can consider a group to be a category with a single object in which all morphisms are equivalences. Therefore we might try to find a more general notion of an exact sequence for categories and functors. In this section we introduce for a functor λ exact sequences of the form

$$D+ \longrightarrow \mathbf{A} \xrightarrow{\lambda} \mathbf{B} \xrightarrow{\mathcal{O}} H.$$

Here, however, D and H are not categories but 'natural systems' of abelian groups on $\mathbf{B}$. Such natural systems serve as well as coefficients of cohomology groups $H^n(\mathbf{B}, D)$ of the (small) category $\mathbf{B}$. Special exact sequences are the linear extensions of $\mathbf{B}$ by D denoted by

$$D+ \rightarrowtail \mathbf{A} \xrightarrow{\lambda} \mathbf{B}$$

the equivalence classes of which are classified by the cohomology group

$H^2(\mathbf{B}, D)$. This fact generalizes the classical result on extensions of a group B by a B-module D which are classified by the cohomology $H^2(B, D)$.

Exact sequences for a functor λ and linear extensions arise frequently in algebraic topology and in many other fields of mathematics. In fact, once the reader learned about these concepts he shall recognize many examples himself and soon the usefulness and naturality of such notions will become apparent. There are indeed many examples; we urge the reader to consider the example at the end of the section concerning the category **Ext** of group extensions. This example is an illuminating illustration of the abstract notions in the section. The connection with Igusa's cohomology class $\chi(1)$ and hence with the exotic element in the K-group $K_3(\mathbb{Z})$ as well indicates the fundamental importance of these notions. The author introduced the exact sequences for a functor λ in Baues (AH) since they form the natural language for describing the properties of the CW-tower of categories, see section § 3 below. Recall that **Ab** denotes the category of abelian groups.

(1.1) **Definition.** Let **C** be a category. The **category of factorizations** in **C**, denoted by $F\mathbf{C}$, is given as follows. Objects are morphisms $f, g, \ldots$ in **C** and morphisms $f \to g$ are pairs (α, β) for which

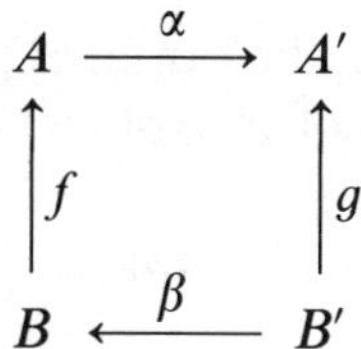

commutes in **C**. Here $\alpha f \beta$ is a factorization of g. Composition is defined by $(\alpha', \beta')(\alpha, \beta) = (\alpha'\alpha, \beta\beta')$. We clearly have $(\alpha, \beta) = (\alpha, 1)(1, \beta) = (1, \beta)(\alpha, 1)$. A **natural system** (of abelian groups) on **C** is a functor $D\colon F\mathbf{C} \to \mathbf{Ab}$. The functor D carries the object f to $D_f = D(f)$ and carries the morphism $(\alpha, \beta)\colon f \to g$ to the induced homomorphism

$$D(\alpha, \beta) = \alpha_* \beta^*\colon D_f \to D_{\alpha f \beta} = D_g$$

Here we set $D(\alpha, 1) = \alpha_*$, $D(1, \beta) = \beta^*$.

We have a canonical forgetful functor $\pi\colon F\mathbf{C} \to \mathbf{C}^{op} \times \mathbf{C}$ so that each *bifunctor* $D\colon \mathbf{C}^{op} \times \mathbf{C} \to \mathbf{Ab}$ yields a natural system $D\pi$, as well denoted by D. Such a bifunctor is also called a **C-bimodule**, it serves as a 'coefficient module' for cohomology groups of **C**; the more general 'natural systems' defined above serve as well as coefficients for cohomology groups of **C**, see Baues (AH).

We now describe bifunctors and natural systems which appear in classical obstruction theory. For $\mathbf{C} = \mathbf{Top}^*/\simeq$ we have the bifunctor D with $(n \geq 0, m \geq 2)$

$$D(X, Y) = H^n(X, \pi_m Y) \tag{1}$$

and the natural system $\hat{D}$ with

$$\hat{D}_f = \hat{D}(f: X \to Y) = \hat{H}^n(X, \pi_1(f)^* \pi_m Y). \tag{2}$$

The bifunctor D is relevant mainly for simply connected spaces X and Y while the natural system $\hat{D}$ is the appropriate functor for the obstruction theory of non simply connected spaces.

The following linear extensions of groups generalize the well known extensions of a group B by a B-module D, compare (IV.6.2) in Baues (AH) and (1.19) below.

(1.2) **Definition.** Let D be a natural system on **C**. We say that

$$D+ \rightarrowtail \mathbf{E} \xrightarrow{p} \mathbf{C}$$

is a **linear extension of the category C** by D if (a), (b) and (c) hold.

(a) **E** and **C** have the same objects and p is a full functor which is the identity on objects.

(b) For each $f: A \to B$ in **C** the abelian group D_f acts transitively and effectively on the subset $p^{-1}(f)$ of morphisms in **E**. We write $f_0 + \alpha$ for the action of $\alpha \in D_f$ on $f_0 \in p^{-1}(f)$.

(c) The action satisfies the *linear distributivity law*:

$$(f_0 + \alpha)(g_0 + \beta) = f_0 g_0 + f_* \beta + g^* \alpha.$$

Two linear extensions **E** and **E**$'$ are **equivalent** if there is an isomorphism of categories $\varepsilon: \mathbf{E} \cong \mathbf{E}'$ with $p'\varepsilon = p$ and with $\varepsilon(f_0 + \alpha) = \varepsilon(f_0) + \alpha$ for $f_0 \in \mathrm{Mor}(\mathbf{E})$, $\alpha \in D_{pf_0}$. The extension **E** is **split** if there is a functor $s: \mathbf{C} \to \mathbf{E}$ with $ps = 1$.

Let **C** be a small category and let $M(\mathbf{C}, D)$ be the set of equivalence classes of linear extensions of **C** by **D**. Then there is a canonical bijection

$$\psi: M(\mathbf{C}, D) \cong H^2(\mathbf{C}, D) \tag{1.3}$$

which maps the split extension to the zero element, see IV § 6 in Baues (AH). Here $H^n(\mathbf{C}, D)$ denotes the *cohomology* of **C** with coefficients in D which is defined by the nerve of **C** as in (VI. § 1) below. We obtain a *representing cocycle* Δ_t of the cohomology class $\{\mathbf{E}\} = \psi\mathbf{E} \in H^2(\mathbf{C}, D)$ as follows. Let t be a "splitting" function for p which associates with each morphism $f: A \to B$ in **C** a morphism $f_0 = t(f)$ in **E** with $pf_0 = f$. Then t yields a cocycle Δ_t by the formula

(1.4) $$t(gf) = t(g)t(f) + \Delta_t(g,f)$$

with $\Delta_t(g,f) \in D(gf)$. The cohomology class $\{\mathbf{E}\} = \{\Delta_t\}$ is trivial if and only if $\mathbf{E}$ is a split extension.

Next we consider exact sequences for functors, see (IV.4.10) in Baues (AH). The reader should compare the following abstract notion of an exact sequence with the particular example discussed at the end of this section which covers classical facts of the theory of extensions of groups, see (1.19).

(1.5) **Definition.** Let $\lambda: \mathbf{A} \to \mathbf{B}$ be a functor and let D and H be natural systems of abelian groups on $\mathbf{B}$. We call the sequence

$$D \xrightarrow{+} \mathbf{A} \xrightarrow{\lambda} \mathbf{B} \xrightarrow{\mathcal{O}} H$$

an **exact sequence** for λ if the following properties are satisfied.

(a) For each morphism $f_0: X \to Y$ in $\mathbf{A}$ the abelian group D_f, $f = \lambda f_0$, acts transitively on the set of morphisms $\lambda^{-1}(f) \subset \mathbf{A}(X, Y)$. Let $I_{f_0} = \{\alpha \in D_f, f_0 + \alpha = f_0\}$ be the isotropy group.

(b) The linear distributivity law (1.2)(c) is satisfied.

(c) For all objects X, Y in $\mathbf{A}$ and for all morphisms $f: \lambda X \to \lambda Y$ in $\mathbf{B}$ an *obstruction element* $\mathcal{O}_{X,Y}(f) \in H(f)$ is given such that $\mathcal{O}_{X,Y}(f) = 0$ if and only if there is a morphism $f_0: X \to Y$ with $\lambda f_0 = f$.

(d) $\mathcal{O}$ is a *derivation*, that is $\mathcal{O}_{X,Z}(gf) = g_* \mathcal{O}_{X,Y}(f) + f^* \mathcal{O}_{Y,Z}(g)$ for $f: \lambda X \to \lambda Y$, $g: \lambda Y \to \lambda Z$.

(e) For all objects X in $\mathbf{A}$ and for all $\alpha \in H(1_{\lambda X})$ there is an object Y in $\mathbf{A}$ with $\lambda Y = \lambda X$ and $\mathcal{O}_{X,Y}(1_{\lambda X}) = \alpha$, we write $X = Y + \alpha$ in this case.

(f) A **tower of categories** is a diagram ($i \in \mathbb{Z}$)

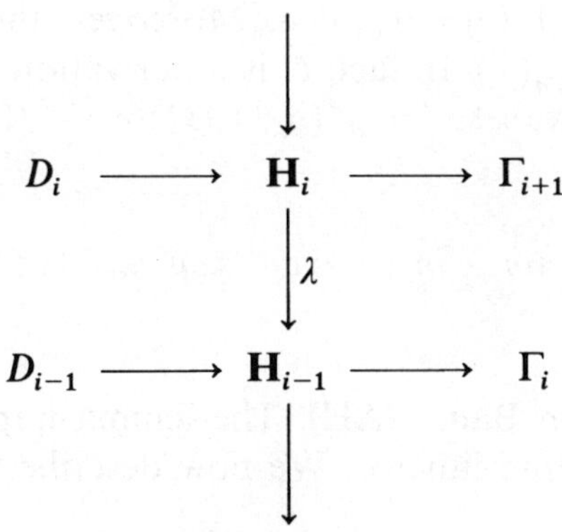

where $D_i \to \mathbf{H}_i \to \mathbf{H}_{i-1} \to \Gamma_i$ is an exact sequence.

We say that D *acts on* λ if (1.5)(a) above is satisfied. Moreover, D *acts linearly* on λ if (a) and (b) are satisfied. We say that D *acts effectively* if all isotropy groups in (a) are trivial. A linear extension as in (1.2) yields an exact sequence

$$D \xrightarrow{+} \mathbf{E} \longrightarrow \mathbf{C} \xrightarrow{\mathcal{O}} 0$$

where 0 is the trivial natural system. On the other hand each exact sequence as in (1.5) yields a linear extension of categories

$$D/I \overset{+}{\rightarrowtail} \mathbf{A} \twoheadrightarrow \lambda\mathbf{A}. \tag{1.6}$$

Here $\lambda\mathbf{A}$ is the image category of $\lambda\colon \mathbf{A} \to \mathbf{B}$, see (I.0.2). The natural system D/I on $\lambda\mathbf{A}$ is given by $(D/I)(f) = D_f/I_{f_0}$, $f_0 \in \lambda^{-1}(f)$, see (1.5)(a).

(1.7) **Definition.** We call $D+ \rightarrowtail \mathbf{A} \overset{\lambda}{\twoheadrightarrow} \mathbf{B}$ a **weak linear extension** if i: $\lambda\mathbf{A} \xrightarrow{\sim} \mathbf{B}$ is an equivalence of categories and if $D+ \rightarrowtail \mathbf{A} \twoheadrightarrow \lambda\mathbf{A}$ is a linear extension where $\mathbf{A} \longrightarrow \lambda\mathbf{A}$ is induced by λ.

For a weak linear extension we obtain the element $\{\mathbf{A}\} \in H^2(\mathbf{B}, D)$ by $\{\mathbf{A}\} = (i^*)^{-1}\psi\mathbf{A}$ where we use ψ in (1.3) and where we use the isomorphism i^* induced by the equivalence i, see (VI.1.4). In particular each exact sequence as above defines by (1.6) the element $\{\mathbf{A}\} \in H^2(\lambda\mathbf{A}, D/I)$.

Next, we consider the **groups of automorphisms** in an exact sequence. Let A be an object in $\mathbf{A}$. Then we obtain by the properties in (1.5) the exact sequence

$$D(1_{\lambda A}) \xrightarrow{1^+} \mathrm{Aut}_{\mathbf{A}}(A) \xrightarrow{\lambda} \mathrm{Aut}_{\mathbf{B}}(\lambda A) \xrightarrow{\bar{\mathcal{O}}} H(1_{\lambda A}) \tag{1.8}$$

Here λ is the homorphism of groups induced by λ and 1^+ is the homorphism of groups given by $1^+(\alpha) = 1_{\lambda A} + \alpha$. Moreover, the function $\bar{\mathcal{O}}$ is defined by $\bar{\mathcal{O}}(f) = (f^{-1})_* \mathcal{O}_{A,A}(f)$. In fact, $\bar{\mathcal{O}}$ is a derivation of groups with $\bar{\mathcal{O}}(fg) = \bar{\mathcal{O}}(f)^g + \bar{\mathcal{O}}(f)$. Here we set $x^g = g^*(g^{-1})_*(x)$ for $x \in H(1_{\lambda A})$. Compare (IV.4.11) Baues (AH).

(1.9) **Lemma.** *A functor λ in an exact sequence* (1.5) *satisfies the sufficiency condition.*

Compare (IV.4.11) in Baues (AH). The lemma implies that a weak linear extension is a detecting functor. We now describe further properties of an exact sequence.

(1.10) **Lemma.** *Let λ be a functor in an exact sequence* (1.5) *and assume*

$\mathrm{Real}_\lambda(B)$ *is not empty. Then* $H(1_B)$ *acts transitively and effectively on* $\mathrm{Real}_\lambda(B)$, *see* (I.0.6). *In particular* $\mathrm{Real}_\lambda(B)$ *is a set.*

Compare (IV.4.12) in Baues (AH). The action in (1.10) is given by $\{A, b\} + \alpha = \{A + \alpha, b\}$, see (1.5)(e). We have the canonical surjection

(1.11) $$\mathrm{Real}_\lambda(B) \twoheadrightarrow \mathrm{types}_\lambda(B)$$

where $\mathrm{types}_\lambda(B)$ is the set of all equivalence classes $\{A\}$ of objects A in $\mathbf{A}$ for which there exists an equivalence $\lambda(A) \cong B$ in $\mathbf{B}$. The function (1.11) carries $\{A, b\}$ to the equivalence class $\{A\}$. For $\{A, b\} \in \mathrm{Real}_\lambda(B)$ we obtain the derivation

(1.12) $$\bar{\mathcal{O}} = \bar{\mathcal{O}}\{A, b\} \colon \mathrm{Aut}_{\mathbf{B}}(B) \to H(1_B)$$

similarly as in (1.8). Here $H(1_B\}$ is a right $\mathrm{Aut}_{\mathbf{B}}(B)$-module by $x^g = g^*(g^{-1})_*(x)$, $x \in H(1_B)$, $g \in \mathrm{Aut}_{\mathbf{B}}(B)$. The derivation $\bar{\mathcal{O}}$ is defined by the obstruction operator $\mathcal{O}$ in the exact sequence (1.5), namely

$$\bar{\mathcal{O}}(f) = (f^{-1}b)_*(b^{-1})^*\mathcal{O}_{A,A}(b^{-1}fb).$$

Let

(1.13) $$\Delta_B = \{\bar{\mathcal{O}}\} \in H^1(\mathrm{Aut}_{\mathbf{B}}(B), H(1_B))$$

be the cohomology class represented by $\bar{\mathcal{O}}$. This class does not depend on the choice of $\{A, b\} \in \mathrm{Real}_\lambda(B)$, in fact, the set $\{\bar{\mathcal{O}}\{A, b\}; \{A, b\} \in \mathrm{Real}_\lambda(B)\}$ coincides with the full cohomology class Δ_B. This follows from (1.5)(e).

(1.14) **Proposition.** *The cohomology class* Δ_B *determines the number of elements in the set* $\mathrm{types}_\lambda(B)$. *In fact, let* Δ *be a derivation which represents* Δ_B. *Then there is a bijection*

$$\mathrm{types}_\lambda(B) \approx H(1_B)/\sim$$

where the equivalence relation $\sim$ *on* $H(1_B)$ *is defined by* Δ, *that is* $\alpha \sim \beta$ *if and only if there exists* $f \in \mathrm{Aut}_{\mathbf{B}}(B)$ *with* $\Delta(f) = f^*(\beta) - f_*(\alpha)$.

Proof. We can choose $\{A, b\} \in \mathrm{Real}_\lambda(B)$ with $\Delta = \bar{\mathcal{O}}\{A, b\}$; for simplicity we assume $b = 1$. We define the bijection (1.14) by the surjection $\alpha \mapsto q(\{A, b\} + \alpha)$ where q is the surjection in (1.11) and where we use the action in (1.10). Now we have an equivalence $A + \alpha \sim A + \beta$ in $\mathbf{A}$ if and only if there is $f \in \mathrm{Aut}_{\mathbf{B}}(B)$ with

$$0 = \mathcal{O}_{A+\alpha, A+\beta}(f) = \mathcal{O}_{A,A}(f) + f_*\alpha - f^*\beta = \Delta(f) + f_*\alpha - f^*\beta. \qquad \square$$

(1.15) **Definition.** We say that an obstruction operator $\mathcal{O}$ with the properties in (1.5) is determined by **invariants** I if I associates with each object X in $\mathbf{A}$ an element

$$I(X) \in H(1_{\lambda X}) \tag{1}$$

such that

$$\mathcal{O}_{X,Y}(f) = f_* I(X) - f^* I(Y) \in H(f) \tag{2}$$

for $f\colon \lambda X \to \lambda Y \in \mathbf{B}$. Here f_* and f^* are defined by the natural system H as in (1.1). Formula (2) shows that the obstruction is an **inner derivation** in this case, see (IV.3.3) Baues (AH). We define the **category of invariants** $(\mathbf{B}, I)$ as follows. Objects are pairs (B, I_B) where B is an object in $\mathbf{B}$ and where I_B is an element in $H(1_B)$. Morphisms $f_0\colon (B, I_B) \to (B', I_{B'})$ are maps $f_0\colon B \to B'$ with $(f_0)_* I_B = (f_0)^* I_{B'}$. By (2) and (1.5)(c) we have the full functor

$$\lambda^I\colon \mathbf{A} \to (\mathbf{B}, I) \tag{3}$$

which carries the object A to the object $(\lambda(A), I(A))$.

(1.16) **Lemma.** *Assume the functor $\lambda\colon \mathbf{A} \to \mathbf{B}$ in the exact sequence (1.5) satisfies the realizability condition for objects and assume the obstruction operator $\mathcal{O}$ is determined by invariants I. Then the functor λ^I above induces an equivalence of categories $\lambda\mathbf{A} \xrightarrow{\sim} (\mathbf{B}, I)$. This implies that*

$$D/I + \rightarrowtail \mathbf{A} \xrightarrow{\lambda^I} (\mathbf{B}, I)$$

is a weak linear extension of categories and that λ^I is a detecting functor.

Proof. It is enough to show that $\lambda\mathbf{A} \to (\mathbf{B}, I)$ is full and faithful and satisfies the realizability condition for objects. In fact, the realizability condition for objects follows from (1.10). Moreover we use (1.6) and (1.9) for the second part of the lemma. $\square$

(1.17) **Remark.** Assume the obstruction $\mathcal{O}$ is determined by invariants as in (1.15). Then the cohomology class Δ_B in (1.13) is trivial. For this compare also (IV.§ 7) in Baues (AH).

(1.18) **Example.** Consider the CW-tower in (3.3) below. The restriction of the exact sequence

$$H^3\Gamma_3+ \longrightarrow \mathbf{H}_4^c/\simeq \xrightarrow{\lambda} \mathbf{H}_3^c/\simeq \xrightarrow{\mathcal{O}} H^4\Gamma_3$$

to simply connected objects has an obstruction operator which is determined by invariants $I = \wp_2$. Here the invariant $\wp_2(X)$ is the Pontrjagin element of X defined as in (I.6.8), compare also formula (I.8.9)(2). More generally the obstruction operator $\mathcal{O}$ is as well determined by invariants on non simply connected objects X for which Pontrjagin elements exist, that is for special objects X, see (V.6.7) below. In general, however, the obstruction operators in the CW-tower are not determined by invariants.

(1.19) **Example.** We describe an example of an exact sequence which summarizes some well known facts on group extensions. Recall that the category $\mathbf{Mod}_{\mathbb{Z}}^{\wedge}$ in (I.1.7) consists of objects (π, M) where π is a group and where M is a π-module, morphisms are pairs $(\varphi, f)\colon (\pi, M) \to (\pi', M')$ where f is a φ-equivariant homomorphism. A **group extension** $\mathscr{E}$ of π by M is a short exact sequence of groups

$$\mathscr{E}\colon M \rightarrowtail^{i} G \overset{p}{\twoheadrightarrow} \pi \tag{1}$$

satisfying $i(x^\alpha) = -a + i(x) + a$ for $p(a) = \alpha$, $x \in M$, $\alpha \in \pi$, $a \in G$; again we use additive notation for the group structure of G and π though these groups need not to be commutative. We now obtain the **category Ext of group extensions**; the objects are extensions as in (1) and the morphisms are triple $(\varphi, f, F)\colon \mathscr{E} \to \mathscr{E}'$ of homomorphisms for which the diagram

$$\begin{array}{cccccc} \mathscr{E}\colon & M & \rightarrowtail & G & \twoheadrightarrow & \pi \\ & \downarrow f & & \downarrow F & & \downarrow \varphi \\ \mathscr{E}'\colon & M' & \rightarrowtail & G & \longrightarrow & \pi \end{array} \tag{2}$$

commutes. There is an obvious functor

$$\lambda\colon \mathbf{Ext} \to \mathbf{Mod}_{\mathbb{Z}}^{\wedge} \tag{3}$$

which carries the extension $\mathscr{E}$ to the pair (π, M). This functor is embedded in the exact sequence

$$\mathrm{Der}+ \longrightarrow \mathbf{Ext} \xrightarrow{\lambda} \mathbf{Mod}_{\mathbb{Z}}^{\wedge} \xrightarrow{\mathcal{O}} H^2 \tag{4}$$

Here Der is the natural system of derivation groups on $\mathbf{Mod}_{\mathbb{Z}}^{\wedge}$; for the morphism $(\varphi, f)\colon (\pi, M) \to (\pi', M')$ we define

$$\mathrm{Der}(\varphi, f) = \mathrm{Der}_\varphi(\pi, M') \tag{5}$$

to be the group of **φ-equivariant derivations** $d: \pi \to M'$. Recall that such a derivation is a function d satisfying $d(\alpha + \beta) = d(\alpha)^{\varphi(\beta)} + d(\beta)$. The action Der+ in (4) is defined by $(\varphi, f, F) + d = (\varphi, f, F_d)$ where F_d is the homorphism given by $F_d(\alpha) = F(\alpha) + i\,dp(\alpha)$ for $\alpha \in G$. One readily checks that F_d is a well defined homomorphism.

Next we define the obstruction operator $\mathcal{O}$ in (4) which actually is determined by invariants. The natural system H^2 on $\mathbf{Mod}_{\mathbb{Z}}^{\wedge}$ is given by the cohomology groups, see (III.4.16),

$$H^2(\varphi, f) = H^2(\pi, \varphi^* M'). \tag{6}$$

This natural system is similar to the examples following the definition (1.1). There is a classical bijection (see (I.0.6))

$$\psi: \mathrm{Real}_\lambda(\pi, M) \cong H^2(\pi, M) \tag{7}$$

which can be considered to be a special case of (1.3), compare (VI.1.7). The invariants I for the obstruction operator $\mathcal{O}$ in (4) are simply given by $I(\mathscr{E}) = \psi\{\mathscr{E}\} \in H^2(\pi, M)$ so that we have, see (1.15)(2),

$$\mathcal{O}_{\mathscr{E}, \mathscr{E}'}(\varphi, f) = f_* \psi\{\mathscr{E}\} - \varphi^* \psi(\mathscr{E}'). \tag{8}$$

We leave it to the reader to show that all the properties of an exact sequence listed in (1.5) are actually satisfied by the example in (4) above. Moreover, since there is always the split extension of (π, M) in (1) we see that λ satisfies the realizability condition for objects. Therefore we obtain by (1.16) the weak linear extension of categories

$$\mathrm{Der}+ \rightarrowtail \mathbf{Ext} \twoheadrightarrow (\mathbf{Mod}_{\mathbb{Z}}^{\wedge}, I) \tag{9}$$

where we use the category of invariants described in (1.15). In (9) we also use the fact that the isotropy groups of the action Der+ are obviously trivial. Whence the extension (9) yields a canonical element

$$\{\mathbf{Ext}\} \in H^2(\mathbf{C}, \mathrm{Der}) \tag{10}$$

in the cohomology of the category $\mathbf{C} = (\mathbf{Mod}_{\mathbb{Z}}^{\wedge}, I)$. This element is actually non trivial. We now discuss a particular restriction of the class $\{\mathbf{Ext}\}$ which is of deep significance for the theory of 4-dimensional homotopy types. Let $\mathbf{Ab}_c$ be the full subcategory of $\mathbf{Ab}$ consisting of free abelian groups. For a group π in $\mathbf{Ab}_c$ and for a trivial π-module M we have the natural isomorphism

$$H^2(\pi, M) \cong \mathrm{Hom}(\Lambda^2\pi, M) \tag{11}$$

where $\Lambda^2\pi = \pi \wedge \pi$ is the exterior product in (I.4.5), see (I.4.14). For $M = \Lambda^2\pi$ there is the canonical element $I_\pi \in H^2(\pi, \Lambda^2\pi)$ given by the identity of $\Lambda^2\pi$. We thus obtain the full inclusion

$$j\colon \mathbf{Ab}_c \rightarrowtail (\mathbf{Mod}_{\mathbb{Z}}^{\wedge}, I) \tag{12}$$

which carries π to the object $(\pi, \Lambda^2\pi, I_\pi)$ where π acts trivialy on $\Lambda^2\pi$. The restriction $j^*\,\mathrm{Der}$ of the natural system Der in (5) is the $\mathbf{Ab}_c$-bimodule $\mathrm{Hom}(-, \Lambda^2-)$ defined by the group of homomorphisms $\mathrm{Hom}_{\mathbb{Z}}(\pi, \Lambda^2\pi)$. Moreover the restriction $j^*\,\mathbf{Ext}$ is equivalent to the full category $\mathbf{E} = \mathbf{nil}$ consisting of free nil(2)-groups; whence the restriction of (9) to the subcategory $\mathbf{Ab}_c$ corresponds to the linear extension

$$\mathrm{Hom}(-, \Lambda^2-) \rightarrowtail \mathbf{nil} \twoheadrightarrow \mathbf{Ab}_c \tag{13}$$

discussed in (VI.§ 2) below. This extension represents the element

$$\{\mathbf{nil}\} = j^*\{\mathbf{Ext}\} \in H^2(\mathbf{Ab}'_c, \mathrm{Hom}(-, \Lambda^2-)) = \mathbb{Z}/2 \tag{14}$$

which is actually the generator. For this we restrict to the category $\mathbf{Ab}'_c$ which is the full and small subcategory of *finitely generated* free abelian groups. We refer the reader to Baues-Dreckmann where one also finds a proof that the cohomology group (14) is a group of order 2. Moreover there is the following connection with **Igusa's associativity class** $\chi(1)$ which is used in Igusa to detect the exotic element in the algebraic K-group $K_3(\mathbb{Z})$ of the ring $\mathbb{Z}$. Let $\pi = \mathbb{Z}^N$ and let $\mathbf{End}(\mathbb{Z}^N)$ with

$$i\colon \mathbf{End}(\mathbb{Z}^N) \subset \mathbf{Ab}_c \tag{15}$$

be the ring of integral $N \times N$-matrices considered as the full subcategory of $\mathbf{Ab}_c$ with the single object π. Then the restriction $i^*\{\mathbf{nil}\}$ to $\mathbf{End}(\mathbb{Z}^N)$ satisfies the formula

$$\beta_0 i^*\{\mathbf{nil}\} = \chi(1) \tag{16}$$

where β_0 is the Bockstein operator induced by the natural exact sequence

$$\pi \otimes \mathbb{Z}/2 \rightarrowtail \hat{\otimes}^2\pi \longrightarrow \Lambda^2\pi \tag{17}$$

which is given by the bottom row of (I.4.2). The deep connection of the elements $\{\mathbf{nil}\}$ and $\chi(1)$ with homotopy theory is discussed in chapter VI.

§2 Homotopy systems of order $(n + 1)$

As motivation for the notion of a "homotopy system of order $(n + 1)$" we consider the following question: Given a chain complex C of free π-modules; is there a CW-complex which realizes (π, C)? That is, is there a CW-complex X together with an isomorphism $(\pi, C) \cong (\pi_1 X, \hat{C}_* X)$ where $\hat{C}_* X$ is the cellular chain complex of the universal covering of X? There is an obvious procedure to find such a realization by constructing inductively realizations of the n-dimensional part of C by an n-dimensional CW-complex X^n. This leads to partial realizations of C; "homotopy systems of order $(n + 1)$" are essentially such partial realizations of a chain complex. Recall that **CW** is the category of CW-complexes with trivial 0-skeleton $*$ and of cellular maps and that $\mathbf{CW}^n$ is the full subcategory consisting of n-dimensional CW-complexes. Moreover morphisms in $\mathbf{CW}/\overset{0}{\simeq}$ are 0-homotopy classes of cellular maps, where a 0-homotopy is a homotopy running through cellular maps.

(2.1) **Definition.** Let $n \geq 2$. A **homotopy system of order** $(n + 1)$ is a triple (C, f_{n+1}, X^n) where X^n is an object in $\mathbf{CW}^n$ and where $(\pi_1 X^n, C)$ is a chain-complex of free $\pi_1(X^n)$-modules which coincides with $\hat{C}_* X^n$ in degree $\leq n$. Moreover, f_{n+1} is a homorphism of right $\pi_1(X^n)$-modules for which the following diagram commutes

$$\begin{array}{ccc} C_{n+1} & \xrightarrow{f_{n+1}} & \pi_n(X^n) \\ {\scriptstyle d}\downarrow & & \downarrow{\scriptstyle j} \\ C_n & \xleftarrow{h_n} & \pi_n(X^n, X^{n-1}) \end{array} \tag{1}$$

Here d is the boundary in C and

$$h_n\colon \pi_n(X^n, X^{n-1}) \underset{\cong}{\xrightarrow{p_*^{-1}}} \pi_n(\hat{X}^n, \hat{X}^{n-1}) \underset{\cong}{\xrightarrow{h}} H_n(\hat{X}^n, \hat{X}^{n-1}) \tag{2}$$

is given by the Hurewicz isomorphism h, see (I.3.3) and (I.3.8). In addition f_{n+1} satisfies the **cocycle condition**

$$f_{n+1} d(C_{n+2}) = 0. \tag{3}$$

A **map** between homotopy systems of order $(n + 1)$ is a pair (ξ, η),

$$(\xi, \eta)\colon (C, f_{n+1}, X^n) \to (C', g_{n+1}, Y^n), \tag{4}$$

with the following properties. The map $\eta\colon X^n \to Y^n$ is a morphism in $\mathbf{CW}/\overset{0}{\simeq}$ and $\xi\colon C \to C'$ is a $\pi_1(\eta)$-equivariant chain map which coincides with $\hat{C}_*(\eta)$ in degree $\leq n$ and for which the following diagram commutes:

$$\begin{array}{ccc} C_{n+1} & \xrightarrow{\xi_{n+1}} & C'_{n+1} \\ \Big\downarrow f_{n+1} & & \Big\downarrow g_{n+1} \\ \pi_n X^n & \xrightarrow[\eta_*]{} & \pi_n Y^n \end{array} \tag{5}$$

Let $\mathbf{H}_{n+1}^c$ be the **category of homotopy systems of order** $(\boldsymbol{n+1})$ and of such maps. Clearly composition is defined by $(\xi,\eta)(\bar{\xi},\bar{\eta}) = (\xi\bar{\xi},\eta\bar{\eta})$. We have obvious functors, $n \geq 3$,

(2.2) $$\mathbf{CW} \xrightarrow[r_{n+1}]{} \mathbf{H}_{n+1}^c \xrightarrow[\lambda]{} \mathbf{H}_n^c$$

with $\lambda r_{n+1} = r_n$. Clearly the functor r_{n+1} which as well is denoted by $r = r_{n+1}$ carries the CW-complex X to $r_{n+1}X = (C, f_{n+1}, X^n)$ where $C = \hat{C}_* X$ and where f_{n+1} is the composition, see (I.3.13), $f_{n+1}\colon C_{n+1} \cong \pi_{n+1}(X^{n+1}, X^n) \xrightarrow{\partial} \pi_n(X^n)$. Here we use the isomorphism h_{n+1} in (2.1)(2).

For the definition of the homotopy relation on the category $\mathbf{H}_{n+1}^c$ we need the coaction $(n \geq 2)$

(2.3) $$\mu\colon X^n \to X^n \vee \bigvee_{Z_n} S^n \qquad \text{under } X^{n-1}$$

where Z_n denotes the set of n-cells of X. The coaction μ is obtained by the coaction μ_f on C_f in (I.3.10), that is $\mu = (c \vee 1)\mu_f c'$ where c' is a homotopy inverse of the homotopy equivalence c in (I.3.10). The coaction μ induces the action

(2.4) $$[X^n, Y] \times E(X^n, Y) \xrightarrow{+} [X^n, Y]$$

with $F + \alpha = \mu^*(F, \alpha)$ where we set

$$E(X^n, Y) = \left[\bigvee_{Z_n} S^n, Y\right] = \bigtimes_{Z_n} \pi_n Y \tag{1}$$

$$= \mathrm{Hom}_\varphi(\hat{C}_n X^n, \pi_n Y) \tag{2}$$

Here $\varphi\colon \pi_1 X \to \pi_1 Y$ is any homomorphism of groups, the isomorphism (2) is induced by (I.3.12). The action (2.4) induces an action

$$[X^n, Y]_\varphi \times \hat{H}^n(X^n, \varphi^*\pi_n Y) \xrightarrow{+} [X^n, Y]_\varphi \tag{3}$$

by $F + \{\alpha\} = F + \alpha$. Here $[X^n, Y]_\varphi$ is the set of elements in $[X^n, Y]$ which induce φ on fundamental groups. For $n = 2$ the action (3) is transitive and effective. For $n > 2$ the isotropy groups of the action (3) can be computed by the spectral sequence (III.5.10) in Baues (AH).

(2.5) **Definition.** Let

$$(\xi, \eta), (\xi', \eta'): (C, f_{n+1}, X^n) \to (C', g_{n+1}, Y^n)$$

be maps in $\mathbf{H}^c_{n+1}$. We set $(\xi, \eta) \simeq (\xi', \eta')$ if $\pi_1(\eta) = \pi_1(\eta') = \varphi$ and if there exist φ-equivariant homomorphisms $\alpha_{j+1}: C_j \to C'_{j+1}$, $j \geq n$, such that

(a) $$\{\eta\} + g_{n+1}\alpha_{n+1} = \{\eta'\} \quad \text{and}$$

(b) $$\xi'_k - \xi_k = \alpha_k d + d\alpha_{k+1}, \qquad k \geq n+1.$$

The action $+$ in (a) is defined in (2.4) above; $\{\eta\}$ denotes the homotopy class of η in $[X^n, Y^n]$. We call $\alpha: (\xi, \eta) \simeq (\xi', \eta')$ a **homotopy** in $\mathbf{H}^c_{n+1}$.

One can check that this homotopy relation is a natural equivalence relation and that the functors in (2.2) induce functors between homotopy categories ($n \geq 3$)

(2.6) $$\mathbf{CW}/\simeq \xrightarrow[r_{n+1}]{} \mathbf{H}^c_{n+1}/\simeq \xrightarrow[\lambda]{} \mathbf{H}^c_n/\simeq.$$

We refer the reader to Baues (AH) where we actually study homotopy systems in any "cofibration category", homotopy systems as above are discussed in (VI, § 5) of Baues (AH) where, however, we deal with a slightly more general notion of homotopy system under a space D; for the purposes here we simply set $D =$ the basepoint.

The functor r_{n+1} is compatible with products $X \times Y$ of CW-complexes X and Y in the sense that there is a natural isomorphism

(2.7) $$r_{n+1}(X \times Y) \cong r_{n+1}(X) \otimes r_{n+1}(Y)$$

in the category $\mathbf{H}^c_{n+1}$. For this we introduce the **tensor product of homotopy systems** which is a bifunctor

$$\otimes: \mathbf{H}^c_{n+1} \times \mathbf{H}^c_{n+1} \to \mathbf{H}^c_{n+1} \tag{1}$$

For homotopy systems $\bar{X} = (C, f_{n+1}, X^n)$ and $\bar{Y} = (C', g_{n+1}, Y^n)$ we get the homotopy system

$$\bar{X} \otimes \bar{Y} = (C \otimes_{\mathbb{Z}} C', h_{n+1}, (X^n \times Y^n)^n) \tag{2}$$

as follows. We choose CW-complexes X^{n+1} and Y^{n+1} which have attaching maps given by f_{n+1} and g_{n+1} respectively. Then h_{n+1} in (2) is given by the attaching maps of $(n+1)$-cells in the product $X^{n+1} \times Y^{n+1}$. The following argument shows that (2) is a well defined homotopy system. It is clear that the universal covering of a product $X \times Y$ satisfies

$$(X \times Y)^\wedge = \hat{X} \times \hat{Y} \tag{3}$$

so that the cellular chain complex $\hat{C}_*(X \times Y)$ is given by

$$\hat{C}_*(X \times Y) = C_*(X \times Y)^\wedge = C_*\hat{X} \otimes_{\mathbb{Z}} C_*\hat{Y}. \tag{4}$$

This shows that $C \otimes_{\mathbb{Z}} C'$ in (2) coincides with $\hat{C}_*(X^n \times Y^n)^n$ in degree $\leq n$. We obtain the cocycle condition (2.1)(3) for $\bar{X} \otimes \bar{Y}$ as follows. Let $\bar{X}^{n+2}$ and $\bar{Y}^{n+2}$ be the $(n+2)$-skeleton of $\bar{X}$ and $\bar{Y}$ respectively. Then an argument as in (3.13)(2) below (where we replace n by $n+1$) shows that there exist CW-complexes X^{n+2}, Y^{n+2} with $r_{n+1}X^{n+2} = \bar{X}^{n+2}$ and $r_{n+1}Y^{n+2} = \bar{Y}^{n+2}$. Now the $(n+2)$-skeleton of $\bar{X} \otimes \bar{Y}$ is exactly the $(n+2)$-skeleton of $r_{n+1}(X^{n+2} \times Y^{n+2})$ and whence the cocycle condition is satisfied. We leave it as an exercise to show that the functor (1) given by (2) is well defined and is compatible with homotopies (2.5), that is, the functor (1) induces a functor

$$\otimes\colon \mathbf{H}^c_{n+1}/\simeq \times \mathbf{H}^c_{n+1}/\simeq \to \mathbf{H}^c_{n+1}/\simeq. \tag{5}$$

§ 3 The CW-tower of categories

The categories of homotopy systems introduced in section § 2 above form a sequence of categories and functors $(n \geq 2)$

$$\mathbf{CW}/\simeq \xrightarrow{r} \mathbf{H}^c_{n+1}/\simeq \xrightarrow{\lambda} \mathbf{H}^c_n/\simeq \xrightarrow{\lambda} \cdots \xrightarrow{\lambda} \mathbf{H}^c_3/\simeq$$

which establishes an inductive approximation of the homotopy category $\mathbf{CW}/\simeq$ of path connected CW-complexes. The main purpose of this section is to describe the fundamental properties of this sequence of categories, namely we observe that each functor λ is embedded in an exact sequence as discussed in section § 1. We call the collection of these exact sequences the *CW-tower of categories.*

We first observe that Whitehead's functor Γ_n in (I.3.6) factors through the functor $r_n\colon \mathbf{CW}/\simeq \to \mathbf{H}^c_n/\simeq$, $n \geq 3$; that is, there is a functor

(3.1) $$\Gamma_n\colon \mathbf{H}^c_n/\simeq \to \mathbf{Mod}^{\wedge}_{\mathbb{Z}}$$

for which $\Gamma_n r_n(X) = \Gamma_n(X)$. We define $\Gamma_n(Y)$ for an object $Y = (C', g_{n+1}, Y^n)$ in $\mathbf{H}^c_n$ as follows. Let Y^{n+1} be a CW-complex such that g_{n+1} is the attaching map of $(n+1)$-cells in Y^{n+1}, see (I.3.13). The CW-complex Y^{n+1} is well defined by (g_{n+1}, Y^n) up to homotopy equivalence under Y^n. Whence $\Gamma_n(Y) = \Gamma_n(Y^{n+1})$ is well defined by Γ_n in (I.3.6). A map $(\xi, \eta)\colon X \to Y$ in $\mathbf{H}_n^c$ admits an extension $\bar{\eta}\colon X^{n+1} \to Y^{n+1}$ of η so that induced maps $\Gamma_n(\xi, \eta) = \Gamma_n(\bar{\eta})$ are given by Γ_n in (I.3.6).

We use Γ_n in (3.1) for the definition of the natural system of abelian groups

(3.2) $$H^m\Gamma_n\colon F(\mathbf{H}_n^c/\simeq) \to \mathbf{Ab}$$

on the category of homotopy systems $\mathbf{H}_n^c/\simeq$, see (1.1). The functor $H^m\Gamma_n$ carries the map $f\colon X \to Y$ in $\mathbf{H}_n^c/\simeq$ to the abelian group

$$H^m\Gamma_n(f) = \hat{H}^m(X, \varphi^*\Gamma_n Y) \tag{1}$$

Here $f = \{(\xi, \eta)\}$ is a homotopy class in $\mathbf{H}_n^c/\simeq$ given by (ξ, η) as in (2.1)(4). In (1) we set $\varphi = \pi_1(f) = \pi_1(\eta)$ and we set

$$\hat{H}^m(X, -) = \hat{H}^m(C, -) \tag{2}$$

for $X = (C, f_{n+1}, X^n)$. Thus (3.2) is a well defined natural system which is of a similar nature as the example (1.1)(2) of obstruction theory. Using the natural systems (3.2) we can state the following theorem on the **CW-tower of categories**; the main result of this book describes an "algebraic model" for the exact sequence and for the diagram of the theorem in case $n = 3$, see chapter III and IV.

(3.3) **Theorem.** *The functors* λ *in* (2.2) *and* (2.6) *are part of the following commutative diagram in which the rows are exact sequences in the sense of* (1.5), $n \geq 3$.

$$\begin{array}{ccccccc}
H^n\Gamma_n + & \longrightarrow & \mathbf{H}_{n+1}^c & \xrightarrow{\lambda} & \mathbf{H}_n^c & \xrightarrow{\mathcal{O}} & H^{n+1}\Gamma_n \\
\downarrow 1 & & \downarrow q_{n+1} & & \downarrow q_n & & \downarrow 1 \\
H^n\Gamma_n + & \longrightarrow & \mathbf{H}_{n+1}^c/\simeq & \longrightarrow & \mathbf{H}_n^c/\simeq & \longrightarrow & H^{n+1}\Gamma_n
\end{array}$$

Here q_n is the quotient functor and 1 denotes the identity. The functor q_{n+1} is equivariant with respect to the action of $H^n\Gamma_n$. The theorem is proved in (VI.5.11) Baues (AH); (where we observe that the coboundaries in $Z^n\Gamma_n$ are actually part of the isotropy group so that the action in the top row above is well defined too). For the convenience of the reader we describe the obstruction operator $\mathcal{O}$ and the action + explicitely below. With the notation in (1.5)(f) the exact sequences in (3.3) yield towers of categories which approximate $\mathbf{CW}/\overset{0}{\simeq}$ and $\mathbf{CW}/\simeq$ respectively. In particular we obtain the tower of homotopy categories

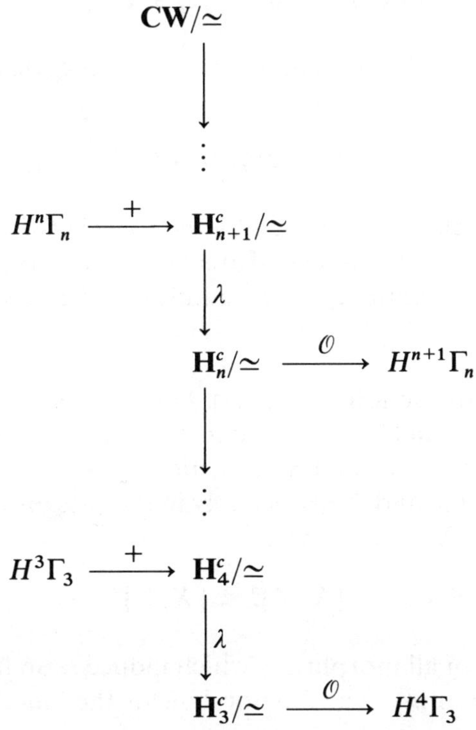

which somehow resembles the Postnikov tower of a space since we have obstructions and actions as we are used to in Postnikov towers.

We now describe the obstruction operator in (3.3). Let X, Y be objects in $\mathbf{H}^c_{n+1}$ and let $(\xi, \eta)\colon \lambda X \to \lambda X$ be a map in $\mathbf{H}^c_n$ with $\varphi = \pi_1(\eta)$. Then an element

(3.4) $$\mathcal{O}_{X,Y}(\xi, \eta) \in \hat{H}^{n+1}(X, \varphi^*\Gamma_n Y)$$

is defined such that $\mathcal{O}_{X,Y}(\xi, \eta) = 0$ if and only if there exists a map $(\xi, \bar{\eta})\colon X \to Y$ in $\mathbf{H}^c_{n+1}$ with $\lambda(\xi, \bar{\eta}) = (\xi, \eta)$.

We define the **obstruction** $\mathcal{O}_{X,Y}(\xi, \eta)$ in (3.4) as follows. Since (ξ, η) is a map in $\mathbf{H}^c_n$ we can choose a map $F\colon X^n \to Y^n$ in $\mathbf{CW}/\overset{0}{\simeq}$ which extends η and for which $\hat{C}_* F$ coincides with ξ in degree $\leq n$. The diagram

$$\begin{array}{ccc} C_{n+1} & \longrightarrow & C'_{n+1} \\ {\scriptstyle f_{n+1}}\downarrow & & \downarrow{\scriptstyle g_{n+1}} \\ \pi_n X^n & \xrightarrow[F_*]{} & \pi_n Y^n \end{array} \tag{1}$$

needs not to be commutative. The difference

$$\mathcal{O}(F) = -g_{n+1}\xi_{n+1} + F_* f_{n+1}, \tag{2}$$

maps C_{n+1} to $\Gamma_n Y \subset \pi_n Y^n$ and this difference is a cocycle in $\mathrm{Hom}_\varphi(C_{n+1}, \Gamma_n Y)$. The obstruction

$$\mathcal{O}_{X,Y}(\xi,\eta) = \{\mathcal{O}(F)\} \tag{3}$$

is the cohomology class represented by the cocycle $\mathcal{O}(F)$. This class does not depend on the choice of F in (1). Moreover, $\mathcal{O}_{X,Y}(\xi,\eta)$ depends only on the homotopy class of (ξ,η) in $\mathbf{H}_n^c/\simeq$. In addition the properties (1.5)(c), (d), (e) are satisfied.

Next we consider the action $+$ in (3.3). Let X, Y be CW-complexes in **CW** or let X, Y be objects in $\mathbf{H}_m^c$ and let $\varphi: \pi_1 X \to \pi_1 Y$ be a homomorphism. We denote by $[X, Y]^n$ the set of all morphisms $X_0 \to Y_0$ in $\mathbf{H}_n^c/\simeq$ where X_0 and Y_0 are the images of X and Y respectively in the category $\mathbf{H}_n^c$, $n \leq m$. Here we use the functors in (2.2). Moreover,

(3.5) $$[X, Y]_\varphi^n \subset [X, Y]^n$$

denotes the subset of all morphisms which induce φ on fundamental groups. This subset can be empty. The functor λ yields the function

(3.6) $$\lambda: [X, Y]_\varphi^{n+1} \to [X, Y]_\varphi^n.$$

The action $+$ in the bottom row of (3.3) is given by

(3.7) $$[X, Y]_\varphi^{n+1} \times \hat{H}^n(X, \varphi^*\Gamma_n Y) \xrightarrow{+} [X, Y]_\varphi^{n+1}.$$

We have $\lambda f = \lambda g$ if and only if $g = f + \alpha$ for an appropriate α. We define the action as follows. Let $(\xi,\eta): X \to Y$ be a map in $\mathbf{H}_{n+1}^c$ which induces $\varphi = \pi_1(\eta)$. Moreover, let $\{\alpha\} \in \hat{H}^n(X, \varphi^*\Gamma_n Y)$ be a class represented by the cocycle

$$\alpha \in \mathrm{Hom}_\varphi(C_n, \Gamma_n(Y)). \tag{1}$$

Then we obtain by $i: \Gamma_n(Y) \subset \pi_n Y^n$ the composition $i\alpha$ such that $\eta + i\alpha = \mu^*(\eta, i\alpha)$ is defined by μ in (2.3). Here $\eta + i\alpha$ is a well defined map in $\mathbf{CW}/\overset{0}{\simeq}$. We now set

$$\{(\xi,\eta)\} + \{\alpha\} = \{(\xi, \eta + i\alpha)\}. \tag{2}$$

Here $\{(\xi,\eta)\}$ denotes the homotopy class of (ξ,η) in $\mathbf{H}_{n+1}^c/\simeq$. The action (2) is well defined. The action + in the top row of (3.3) is given similarly by

$$(\xi, \eta) + \{\alpha\} = (\xi, \eta + i\alpha). \tag{3}$$

Here $\eta + i\alpha$ depends only on the cohomology class $\{\alpha\}$. The actions in (2) and (3) satisfy the properties (1.5)(a), (b).

(3.8) **Remark.** The **isotropy groups** of the action (3.7) are described in (VI.5.16) Baues (AH). For the computation we use the spectral sequence (VI.5.9)(3) Baues (AH). Such a spectral sequence is also available for the computation of the isotropy groups of the action (3.7)(3); for this we work in the cofibration category of filtered objects and we apply (III.4.18) of Baues (AH) in this category.

We now list some properties of the CW-tower of categories in (3.3) which are immediate consequences of theorem (3.3). Let X, Y be CW-complexes in **CW** and let $\varphi: \pi_1 X \to \pi_1 Y$ be a homomorphism. As in (3.5) we have the subset $[X, Y]_\varphi \subset [X, Y]$ consisting of all maps $\{\eta\}$ with $\pi_1(\eta) = \varphi$. The CW-tower yields the following diagram of exact sequences of sets.

(3.9)
$$\begin{array}{ccccc}
 & & [X,Y]_\varphi & & \\
 & & \downarrow & & \\
 & & \vdots & & \\
\hat{H}^n(X, \varphi^*\Gamma_n Y) & \xrightarrow{+} & [X,Y]^{n+1}_\varphi & & \\
 & & \downarrow \lambda & & \\
 & & [X,Y]^n_\varphi & \xrightarrow{\mathcal{O}} & \hat{H}^{n+1}(X, \varphi^*\Gamma_n Y) \\
 & & \downarrow & & \\
 & & \vdots & & \\
H^3(X, \varphi^*\Gamma_3 Y) & \xrightarrow{+} & [X,Y]^4_\varphi & & \\
 & & \downarrow \lambda & & \\
 & & [X,Y]^3_\varphi & \xrightarrow{\mathcal{O}} & \hat{H}^4(X, \varphi^*\Gamma_3 Y)
\end{array}$$

We have kernel($\mathcal{O}$) = image(λ) and we have $\lambda(f) = \lambda(g)$ if and only if there is

α with $g = f + \alpha$. Moreover, for an N-dimensional CW-complex $X = X^N$ the map

$$r_n\colon [X, Y]_\varphi \to [X, Y]^n_\varphi \tag{3.10}$$

is bijective for $n = N + 1$ and is surjective for $n = N$. This follows from the definition of $\mathbf{H}^c_n/\simeq$.

Next we derive from the CW-tower a structure theorem for the group of homotopy equivalences. For a CW-complex X in **CW** let $\operatorname{Aut}(X)^* \subset [X, X]$ be the **group of homotopy equivalences** of X in $\mathbf{CW}/\simeq$. Moreover, let $E_n(X) \subset [X, X]^n$, $n \geq 3$, be the group of equivalences of $r_n X$ in $\mathbf{H}^c_n/\simeq$. Then the CW-tower yields the following tower of groups where the arrows $\bar{\mathcal{O}}$ denote derivations and where all the other arrows are homomorphisms between groups.

(3.11)

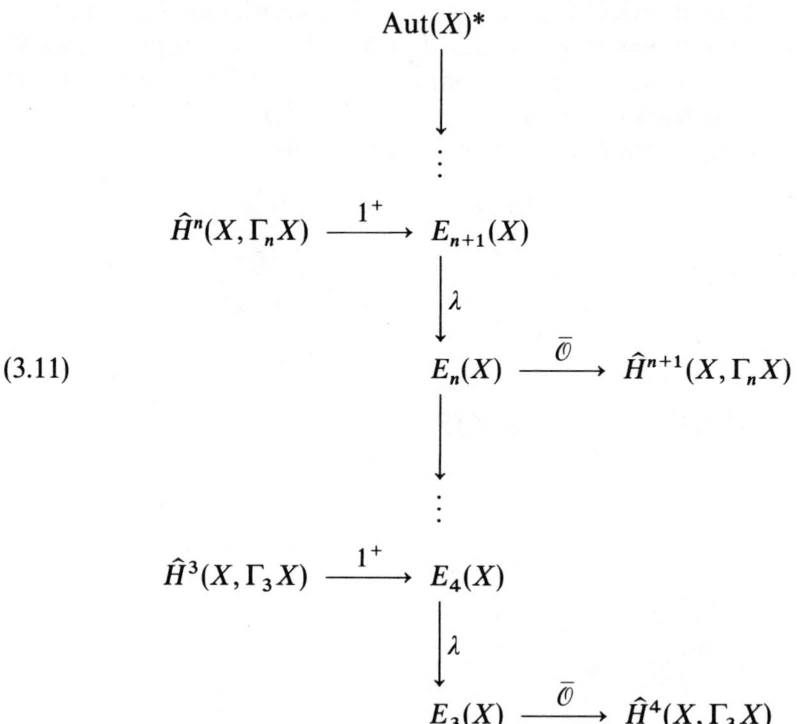

Here we define the derivation $\bar{\mathcal{O}}$ by the obstruction $\mathcal{O}$ as in (1.8) and we set $1^+(\alpha) = 1 + \alpha$ where 1 is the identity and where $1 + \alpha$ is given by (3.7). We have exactness (see (1.6))

$$\operatorname{image}(1^+) = \operatorname{kernel}(\lambda), \qquad \operatorname{image}(\lambda) = \operatorname{kernel}(\bar{\mathcal{O}}).$$

Moreover, as in (3.10) we see that for $X = X^N$ the homomorphism

(3.12) $$r_n\colon \operatorname{Aut}(X)^* \to E_n(X)$$

is an isomorphism for $n = N + 1$ and is an epimorphism for $n = N$. The kernel of 1^+ can be computed by a spectral sequence, see (3.8). Next, we obtain by (1.12) the following result which establishes an **obstruction theory for the realizability of a chain complex** (π, C) by a CW-complex X. In fact, given a "partial" realization of (π, C) in $\mathbf{H}_n^c$ we obtain an obstruction for the realizability of (π, C) by an object in $\mathbf{H}_{n+1}^c$.

(3.13) **Proposition.** *Let X be an object in $\mathbf{H}_n^c/\simeq$ and let $\lambda\colon \mathbf{H}_{n+1}^c/\simeq \to \mathbf{H}_n^c/\simeq$ be the functor in (2.6). Then the group $\hat{H}^{n+1}(X, \Gamma_n X) = \hat{H}^{n+1}(\Gamma_n)$ acts transitively and effectively on the set $\operatorname{Real}_\lambda(X)$ provided this set is non empty. Moreover, the set $\operatorname{Real}_\lambda(X)$ is non empty if and only if an obstruction $\mathcal{O}(X) \in \hat{H}^{n+2}(X, \Gamma_n(X))$ vanishes.*

The action of $\hat{H}^{n+1}(\Gamma_n)$ on $\operatorname{Real}_\lambda(X)$ can be defined by the obstruction (3.4) as in (1.10). Equivalently this action is given as follows. Let $\{\alpha\} \in \hat{H}^{n+1}(X, \Gamma_n X)$ and let $\{\bar{X}\} \in \operatorname{Real}_\lambda(X)$, that is $\lambda\bar{X} = X$. For $\bar{X} = (C, f_{n+1}, X^n)$ the cocycle $\alpha\colon C_{n+1} \to \Gamma_n X = \Gamma_n X^n$ yields the object

$$\bar{X} + \alpha = (C, f_{n+1} + i\alpha, X^n) \tag{1}$$

where $i\colon \Gamma_n X^n \subset \pi_n X^n$ is the inclusion. Now the **action** is given by $\{\bar{X}\} + \{\alpha\} = \{\bar{X} + \alpha\}$. Next we define the obstruction $\mathcal{O}(X)$ in (3.13) as follows. The object $X = (C, f_n, X^{n-1})$ yields the following commutative diagram for which the column is an exact sequence, ($X^n = C_f$ with f representing f_n)

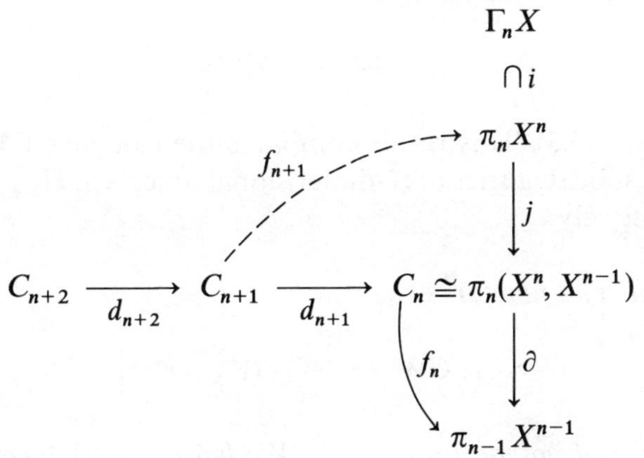

$$\begin{array}{ccccc}
 & & & & \Gamma_n X \\
 & & & & \cap\, i \\
 & & & \overset{f_{n+1}}{\dashrightarrow} & \pi_n X^n \\
 & & & & \downarrow j \\
C_{n+2} & \xrightarrow[d_{n+2}]{} & C_{n+1} & \xrightarrow[d_{n+1}]{} & C_n \cong \pi_n(X^n, X^{n-1}) \\
 & & & & f_n \searrow \quad \downarrow \partial \\
 & & & & \pi_{n-1} X^{n-1}
\end{array} \tag{2}$$

The cocycle condition $f_n d_{n+1} = 0$ shows that there is f_{n+1} with $jf_{n+1} = hd_{n+1}$. Moreover $jf_{n+1}d_{n+2} = 0$ since $d_{n+1}d_{n+2} = 0$. Whence $i^{-1}f_{n+1}d_{n+2}\colon C_{n+2} \to \Gamma_n X$

is a cocycle which represents

$$\mathcal{O}(X) = \{i^{-1} f_{n+1} d_{n+2}\} \in H^{n+2}(X, \Gamma_n X). \tag{3}$$

This is the **obstruction for the λ-realizability** of $X \in \mathbf{H}_n^c$. In fact, $\mathcal{O}(X) = 0$ if and only if there is a map $\bar{f}_{n+1}$ satisfying the cocycle condition $\bar{f}_{n+1} d_{n+2} = 0$. In this case $\bar{X} = (C, \bar{f}_{n+1}, X^n)$ satisfies $\lambda \bar{X} = X$. This proves proposition (3.13).

The obstruction $\mathcal{O}(X)$ in (3) as well can be described by use of the secondary boundary operator b_{n+1} in Whitehead's exact sequence, see (I.3.7). For this we choose an $(n+1)$-dimensional CW-complex X^{n+1} which realizes the $(n+1)$-skeleton (C^{n+1}, f_n, X^{n-1}) of X. The attaching map of $(n+1)$-cells in X^{n+1} is given by f_{n+1} in (2). We therefore obtain the commutative diagram

$$\begin{array}{ccccc} & \overset{d}{\dashrightarrow} & \hat{H}_{n+1} X^{n+1} & \overset{b_{n+1}}{\dashrightarrow} & \Gamma_n X \\ & & \cap & & \cap\, i \\ C_{n+2} & \xrightarrow[d_{n+2}]{} & C_{n+1} & \xrightarrow[f_{n+1}]{} & \pi_n X^n \end{array}$$

so that $b_{n+1} d$ is a cocycle which represents the cohomology class

$$\mathcal{O}(X) = \{b_{n+1} d\} \in H^{n+2}(C, \Gamma_n X). \tag{4}$$

Now let $\mathbf{CW}^N$ be the full subcategory of $\mathbf{CW}$ consisting of N-dimensional CW-complexes and let $(\mathbf{H}_n^c)^N$ be the full subcategory of $\mathbf{H}_n^c$ consisting of homotopy systems $X = (C, f_n, X^{n-1})$ with $C_i = 0$ for $i > N$. One gets the following result on the functors

$$r_n \colon \mathbf{CW}/\simeq \;\to \mathbf{H}_n^c/\simeq$$

in the CW-tower (3.3). This result compares the category $\mathbf{CW}^N/\simeq$ with the corresponding subcategories of N-dimensional objects in $\mathbf{H}_{N+1}^c/\simeq$, $\mathbf{H}_N^c/\simeq$ and $\mathbf{H}_{N-1}^c/\simeq$ respectively.

(3.14) **Theorem.** *The functor*

$$r_{N+1} \colon \mathbf{CW}^N/\simeq \;\xrightarrow{\sim} (\mathbf{H}_{N+1}^c)^N/\simeq \tag{1}$$

is an equivalence of categories ($N \geq 2$). We have a weak linear extension of categories ($N \geq 3$)

$$H^N \Gamma_N / I \xrightarrow{+} \mathbf{CW}^N/\simeq \;\xrightarrow{r_N} (\mathbf{H}_N^c)^N/\simeq. \tag{2}$$

In particular, the functor r_N in (2) *is a detecting functor which induces a* $1-1$ *correspondence on equivalence classes of objects, see* (1.9). *Moreover, the functor*

$$r_{N-1}\colon \mathbf{CW}^N/\simeq \;\to (\mathbf{H}^c_{N-1})^N/\simeq \tag{3}$$

satisfies the realizability condition for objects ($N \geq 4$), *see* (I.0.3)(b1), *and for each object X in $(\mathbf{H}^c_{N-1})^N$ the group $\hat{H}^N(X, \Gamma_{N-1}X)$ acts transitively and effectively on the set* $\mathrm{Real}_r(X)$ *where $r = r_{N-1}$ is the functor in* (3), *see* (1.10).

The proposition in (3.14)(3) follows from (3.13). The weak linear extension (3.14)(2) is obtained as in (1.6) by the corresponding exact sequence in (3.3). Finally we recall the following theorem, see (IV.7.4) Baues (AH).

(3.15) **Theorem.** *The functors $r_n\colon \mathbf{CW}/\simeq \;\to \mathbf{H}^c_n/\simeq$ and $\lambda = \bar{\lambda}\colon \mathbf{H}^c_{n+1}/\simeq \;\to \mathbf{H}^c_n/\simeq$ satisfy the strong sufficiency condition, see* (I.0.7).

Moreover, the functor $\lambda\colon \mathbf{H}^c_{n+1} \to \mathbf{H}^c_n$ satisfies the sufficiency condition as follows from (1.9) and (3.3), so that we can apply (I.0.7)(1), (2) and (3).

§ 4 The Postnikov chain functor

The computation of the obstruction operator $\mathcal{O}$ in the CW-tower of categories is of high importance for the classification of homotopy types. This operator associates with a triple (X, Y, f) an element ($n \geq 3$)

$$\mathcal{O}_{X,Y}(f) \in H^{n+1}(X, \varphi^*\Gamma_n Y) \tag{1}$$

where X and Y are objects in $\mathbf{H}^c_{n+1}$ and where $f\colon \lambda X \to \lambda Y$ is a map in $\mathbf{H}^c_n$ which induces $\varphi = \pi_1(f)$. The question arises whether the obstruction operator is determined by invariants I as described in (1.15) and as in the example concerning the category **Ext** of extensions of groups in (1.19). Such invariants

$$I(X) \in H^{n+1}(X, \Gamma_n X) \tag{2}$$

would be most convenient for the computation of $\mathcal{O}_{X,Y}(f)$ since they would satisfy the formula

$$\mathcal{O}_{X,Y}(f) = f_* I(X) - f^* I(Y). \tag{3}$$

Unfortunately such invariants do not exist in general though there are special objects for which they exist. For example in case $X = (C, f_{n+1}, X^n)$ is $(n-2)$-

connected (that is, in case X^n is $(n-2)$-connected) there is the invariant

$$I(X) = \begin{cases} \wp(i_2) & \text{for } n = 3, \\ \mathscr{S}q(i_{n-1}) & \text{for } n \geq 4 \end{cases} \tag{4}$$

which is defined in the same way as the Pontrjagin-Steenrod element, see (I.6.7). These invariants satisfy the formula in (3). Hence we are led to the question: what is the largest class of objects X in $\mathbf{H}_{n+1}^c$ which admit invariants $I(X)$ as above? This question (for $n = 3$) is studied in (V.§4) below.

The purpose of this section is to deduce from an object X in $\mathbf{H}_{n+1}^c$ the algebraic ingredient which may serve for the computation of the obstruction element (1). For this we introduce the "cellular boundary invariant" β_X which arises in connection with the $(n-1)$-type $P_{n-1}X^n$ of the Postnikov decomposition of X^n. In fact, β_X is determined by a chain map

$$\beta: C \to P\lambda X \tag{5}$$

where $P\lambda X$ is the "$(n+1)$-type" of the chain complex $\hat{C}_* P_{n-1} X^n$. The main result of the section shows that an appropriate homotopy class β_X of β is a natural invariant of X and that $\mathcal{O}_{X,Y}(f)$ can be computed by the triple (β_X, β_Y, f), see (4.21). Moreover, we show that there is an isomorphism

$$H_{n+1} C_\beta \cong \pi_n X \tag{6}$$

of $\pi_1(X)$ modules where C_β is the algebraic mapping cone of the chain map β, see (4.12), and we describe the connection of β_X with the k-invariants of a Postnikov decomposition. The constructions in this section are somewhat technical. The reader is advised to read through the section quickly and return to it when he finds the methods are needed. We only make use of the cellular boundary invariants in the next section §5 and in chapter V where we compute algebraically the cellular boundary invariants of objects in $\mathbf{H}_4^c$. This leads to the homotopy classification of 4-dimensional CW-complexes in terms of pairs (C, β) where C is a chain complex and where β is an appropriate chain map; for this the final proposition (4.31) below is significant.

Let n-**types** be the full subcategory of $\mathbf{Top}^*/\simeq$ consisting of connected CW-spaces Y with $\pi_i(Y) = 0$ for $i > n$. We introduce the n-th **Postnikov functor**

(4.1) $$P_n: \mathbf{CW}/\simeq \; \to n\text{-}\mathbf{types}$$

as follows. For X in $\mathbf{CW}$ we obtain $P_n X$ by 'killing homotopy groups'; that is, we choose a CW-complex $P_n X$ with $(n+1)$-skeleton

$$(P_n X)^{n+1} = X^{n+1} \tag{1}$$

and with $\pi_i P_n X = 0$ for $i > n$. For a cellular map $F\colon X \to Y$ in **CW** we choose a map $PF^{n+1}\colon P_n X \to P_n Y$ which extends the restriction $F^{n+1}\colon X^{n+1} \to Y^{n+1}$ of F. This is possible since $\pi_i P_n Y = 0$ for $i > n$. The functor P_n carries X to $P_n X$ and carries F to the homotopy class of PF^{n+1}. Different choices for $P_n X$ yield canonically isomorphic functors P_n. The space

$$P_1(X) = K(\pi_1(X), 1) \tag{2}$$

is an Eilenberg-Mac Lane space and P_1, as a functor, is equivalent to the functor π_1 of fundamental groups. A map

$$p_n\colon X \to P_n X \tag{3}$$

which extends the inclusion $X^{n+1} \subset P_n X$ in (1) is called the ***n*-type** or the ***n*-th Postnikov section** of X. Clearly p_n induces isomorphisms of homotopy groups $(p_n)_*\colon \pi_i X \cong \pi_i P_n X$ for $i \leq n$. The map p_n is natural with respect to the functor in (4.1). The **Postnikov tower** $\{q_n\}$ is given by maps

$$q_n\colon P_n X \to P_{n-1} X \tag{4}$$

with $q_n p_n \simeq p_{n-1}$. For further properties we refer the reader to section (VIII.§ 1) in Baues (AH) where we derive from the Postnikov tower a tower of categories.

(4.2) **Definition.** For a chain complex P and $n \in \mathbb{Z}$ we consider the commutative diagram

$$\begin{array}{ccccccccc} \cdots \longrightarrow & P_{n+1} & \longrightarrow & P_n & \longrightarrow & P_{n-1} & \longrightarrow \cdots \\ & \downarrow & & \downarrow{\scriptstyle q_n} & & \downarrow{\scriptstyle q_{n-1}} & \\ \cdots \longrightarrow & 0 & \longrightarrow & P_n/dP_{n+1} & \longrightarrow & P_{n-1} & \longrightarrow \cdots \end{array}$$

The bottom row is a chain complex $P^{(n)}$ with

$$P_i^{(n)} = \begin{cases} P_i, & i < n \\ P_n/dP_{n+1}, & i = n \\ 0, & i > n. \end{cases}$$

The map q_i is the identity for $i < n$ and is the projection for $i = n$. We call $P^{(n)}$ the ***n*-type** of P. Clearly $q\colon P \to P^{(n)}$ induces isomorphisms $q_*\colon H_i P \to H_i P^{(n)}$ for $i \leq n$. Moreover $q\colon P \to P^{(n)}$ is natural with respect to chain maps and chain homotopies.

(4.3) **Definition.** Let $F, G\colon C \to P$ be φ-equivariant chain maps in $\mathbf{Chain}_{\mathbb{Z}}^{\wedge}$ and let $\alpha\colon F \simeq G$ be a homotopy as in (I.2.6) given by

$$\alpha = \alpha_i\colon C_i \to P_{i+1}, \quad i \in \mathbb{Z}.$$

We say that α is a **homotopy relative the $(n-1)$-skeleton** and we write $\alpha\colon F \simeq G \operatorname{rel} C^{n-1}$ if $\alpha_i = 0$ for $i < n$. Moreover, we say that α is a $(\simeq n)$-**homotopy** and we write $\alpha\colon F \simeq_n G$ if $\alpha_i = 0$ for $i < n$ and if $d\alpha_n = 0$. One readily checks that $\simeq_n$ is a natural equivalence relation on $\mathbf{Chain}_{\mathbb{Z}}^{\wedge}$. Let $\mathbf{Chain}_{\mathbb{Z}}^{\wedge}/\simeq_n$ be the quotient category.

Recall that C^n denotes the n-**skeleton** of C; this is the chain subcomplex $C^n \subset C$ given by $\{C_i, i \leq n\}$. A chain map $F\colon C \to P$ induces the restriction $F^n\colon C^n \to P^n$. Clearly a homotopy $F \simeq_n G$ implies $F^n = G^n$. Thus we have the functor

$$\mathbf{Chain}_{\mathbb{Z}}^{\wedge}/\simeq_n \to \mathbf{Chain}_{\mathbb{Z}}^{\wedge} \tag{4.4}$$

which carries C to the skeleton C^n and which carries a morphism $\{F\}$ to the restriction F^n. We now are ready for the construction of a functor $(n \geq 3)$:

$$P\colon \mathbf{H}_n^c \to \mathbf{Chain}_{\mathbb{Z}}^{\wedge}/\simeq_n \tag{4.5}$$

which we call the **Postnikov chain functor**. Here $\mathbf{H}_n^c$ is the category of homotopy systems of order n in § 2. The functor P carries an object $X = (C, f_n, X^{n-1})$ in $\mathbf{H}_n^c$ to the $(n+1)$-type of the chain complex $\hat{C}_* P_{n-1} X^n$, that is

$$PX = (\hat{C}_* P_{n-1} X^n)^{(n+1)}. \tag{1}$$

Here X^n is a CW-complex with attaching map f_n and with $(n-1)$-skeleton X^{n-1} and $P_{n-1}X^n$ is the $(n-1)$-type of X^n. By (4.1)(1) the n-skeleton of PX is

$$(PX)^n = (\hat{C}_* P_{n-1} X^n)^n = \hat{C}_* X^n = C^n. \tag{2}$$

A map $(\xi, \eta)\colon X \to Y$ in $\mathbf{H}_n^c$ with $Y = (C', g_n, Y^{n-1})$ induces a map

$$P(\xi, \eta)\colon PX \to PY \tag{3}$$

in $\mathbf{Chain}_{\mathbb{Z}}^{\wedge}/\simeq_n$ as follows. Let $F^{n-1}\colon X^{n-1} \to Y^{n-1}$ be a cellular map in $\mathbf{CW}$ which represents η. This map has an extension $F\colon X^n \to Y^n$ which induces $\xi^n = \hat{C}_* F$, see (3.4)(1). We choose $P_{n-1}F$ as in (4.1) and we define $P(\xi, \eta)$ by the $(\simeq_n)$-class of the chain map

$$(\hat{C}_* P_{n-1} F)^{(n+1)}\colon PX \to PY. \tag{4}$$

We now check that $P(\xi, \eta)$ is well defined. Let $G^{n-1}: X^{n-1} \to Y^{n-1}$ be a further map which represents η and let $G: X^n \to Y^n$ be an extension with $\xi^n = \hat{C}_* G$. Then there is a 0-homotopy $H: F^{n-1} \overset{0}{\simeq} G^{n-1}$ which has an extension $\bar{H}: P_{n-1}F \simeq P_{n-1}G$. Here $\bar{H}$ exists since $\pi_i P_{n-1} Y^n = 0$ for $i \geq n$. We may assume that $\bar{H}$ is a 1-homotopy. Now $\hat{C}_* \bar{H} = \alpha$ yields a homotopy

$$\alpha: \hat{C}_* P_{n-1} F \simeq \hat{C}_* P_{n-1} G$$

which is a homotopy relative C^{n-1} since $\bar{H}$ is an extension of H. Since, however,

$$(\hat{C}_* P_{n-1} F)^n = \xi^n = (\hat{C}_* P_{n-1} G)^n$$

the homotopy α is actually *a* $(\simeq_n)$-homotopy. Thus we obtain the $(\simeq)_n$-homotopy

$$\alpha^{(n+1)}: (\hat{C}_* P_{n-1} F)^{(n+1)} \simeq_n (\hat{C}_* P_{n-1} G)^{(n+1)}$$

This proves that the functor P in (4.5) is well defined.

Now let $X = (C, f_{n+1}, X^n)$ be an object in $\mathbf{H}^c_{n+1}$ with $\lambda X = (C, f_n, X^{n-1})$, see (2.2). Then we can construct a map

(4.6) $$\beta_X: C \to P\lambda X \qquad \text{in } \mathbf{Chain}^{\wedge}_{\mathbb{Z}}/\simeq_n$$

which we call the **cellular boundary invariant of** X. We obtain β_X as follows. By (3.14)(3) we can find a CW-complex X^{n+2} which realizes the $(n+2)$-skeleton of X. Moreover we can choose a cellular map, see (4.1)(3),

$$p_{n-1}: X^{n+2} \to P_{n-1} X^n \tag{1}$$

which extends the inclusion $X^n \subset P_{n-1} X^n$. This map induces a chain map

$$\hat{C}_* p_{n-1}: C^{n+2} = \hat{C}_* X^{n+2} \to \hat{C}_* P_{n-1} X^n \tag{2}$$

the $(n+1)$-type of which represents β_X. We point out that the restriction of β_X to the n-skeleton is the identity

$$1 = \beta_X^n: C^n \to (P\lambda X)^n = C^n, \tag{3}$$

see (4.5)(2). It is easy to see that β_X is well defined. In fact, a different choice p'_{n-1} in (1) admits a homotopy $p_{n-1} \simeq p'_{n-1}$ relative X^n since $\pi_i P_{n-1} X = 0$ for $i \geq n$. This homotopy can be assumed to be a 1-homotopy and whence we get a homotopy

$$\alpha\colon \hat{C}_* p_{n-1} \simeq \hat{C}_* p'_{n-1} \quad \text{relative } C^n.$$

Clearly α as well is a $(\simeq_n)$-homotopy.

(4.7) **Lemma.** *The cellular boundary invariant is natural.*

This means that a map $(\xi, \eta)\colon X \to Y$ in $\mathbf{H}^c_{n+1}$ induces a commutative diagram

$$\begin{array}{ccc} C & \xrightarrow{\beta_X} & P\lambda X \\ {\scriptstyle\xi}\downarrow & & \downarrow{\scriptstyle P\lambda(\xi,\eta)} \\ C' & \xrightarrow{\beta_Y} & P\lambda Y \end{array}$$

in the category $\mathbf{Chain}^{\wedge}_{\mathbb{Z}}/\simeq_n$. Here we use the functor $P\lambda$ given by (4.5) and (2.6).

Proof of (4.7). We consider a diagram of cellular maps

$$\begin{array}{ccc} X^{n+1} \subset X^{n+2} & \longrightarrow & P_{n-1}X^n \\ {\scriptstyle F^{n+1}}\downarrow & & \downarrow{\scriptstyle P_{n-1}F^n} \\ Y^{n+1} \subset Y^{n+2} & \longrightarrow & P_{n-1}Y^n \end{array}$$

where F^n represents η and where F^{n+1} extends F^n such that $\hat{C}_* F^{n+1} = \xi^{n+1}$. The horizontal arrows are given by (4.6)(1). Clearly the diagram homotopy commutes relative the n-skeleton X^n. Thus the induced chain maps are homotopic relative C^n. □

Next we apply Whitehead's certain exact sequence (I.3.7) to the Postnikov section in (4.1)(3). This yields the following commutative diagram of $\pi_1(X)$-equivariant homomorphisms.

(4.8)
$$\begin{array}{ccccccc} 0 \longrightarrow & \hat{H}_{n+1}P_{n-1}X & \xrightarrow{\cong} & \Gamma_n P_{n-1}X & \longrightarrow & 0 \\ & \uparrow & & \uparrow\cong & & \uparrow \\ & \hat{H}_{n+1}X & \xrightarrow[b_{n+1}]{} & \Gamma_n X & \xrightarrow[i_n]{} & \pi_n X \\ \longrightarrow & \hat{H}_n P_{n-1}X & \rightarrowtail & \Gamma_{n-1}P_{n-1}X & \longrightarrow & \pi_{n-1}P_{n-1}X \\ & \uparrow & & \uparrow\cong & & \uparrow\cong \\ \xrightarrow[h_n]{} & \hat{H}_n X & \xrightarrow[b_n]{} & \Gamma_{n-1}X & \xrightarrow[i_{n-1}]{} & \pi_{n-1}X \end{array}$$

The vertical arrows are induced by p_{n-1} in (4.1)(3). Since $(P_{n-1}X)^n = X^n$ the map p_{n-1} induces isomorphisms $\Gamma_i(p_{n-1})$ for $i \leq n$, see (I.3.5). Whence we have the natural identifications

$$(4.9) \qquad \begin{cases} \Theta\colon \hat{H}_{n+1}P_{n-1}X \cong \Gamma_n X \\ \Theta\colon \hat{H}_n P_{n-1}X \cong \operatorname{kernel}(i_{n-1}) \end{cases}$$

and the maps $\hat{H}_{n+1}(p_{n-1})$ and $\hat{H}_n(p_n)$ can be identified with the secondary boundary operators b_{n+1} and b_n respectively. Now let X be an object in $\mathbf{H}^c_{n+1}$. We show that the cellular boundary invariant β_X determines the bottom sequence in (4.8). For this we choose a chain map β which represents β_X and we choose a commutative diagram

$$(4.10) \qquad \begin{array}{ccccc} & & P\lambda X & & \\ & \overset{\beta}{\nearrow} & \uparrow{\scriptstyle q}\ {\scriptstyle\sim} & & \\ C & \rightarrowtail & Z_\beta & \overset{r}{\twoheadrightarrow} & C_\beta \end{array}$$

where Z_β is a mapping cylinder in the cofibration category of $\mathbb{Z}[\pi_1 X]$-chain complexes. Moreover, $C_\beta = Z_\beta / C$ is the mapping cone of β. The short exact sequence of chain complexes $C \rightarrowtail Z_\beta \twoheadrightarrow C_\beta$ induces the long exact homology sequence in the top row of the following diagram.

$$(4.11) \qquad \begin{array}{ccccccccccc} H_{n+1}C & \xrightarrow{\beta_*} & H_{n+1}P\lambda X & \xrightarrow{r'} & H_{n+1}C_\beta & \xrightarrow{\partial} & H_n C & \xrightarrow{\beta_*} & H_n P\lambda X & \longrightarrow & 0 \\ \| & & \cong\downarrow\Theta & & \cong\downarrow\Theta' & & \| & & \cong\downarrow\Theta & & \\ \hat{H}_{n+1}X & \xrightarrow[b_{n+1}]{} & \Gamma_n X & \xrightarrow[i_n]{} & \pi_n X & \xrightarrow[h_n]{} & \hat{H}_n X & \xrightarrow[b_n]{} & \ker(i_{n-1}) & & \end{array}$$

The map $r' = r_*(q_*)^{-1}$ is given by the chain maps in (4.10) and ∂ is the boundary operator. The isomorphisms Θ are obtained in the same way as in (4.9). The bottom row of (4.11) is Whitehead's exact sequence of an $(n+2)$-dimensional CW-complex which realizes the $(n+2)$-skeleton of the object X in $\mathbf{H}^c_{n+1}$. Clearly the bottom row is natural with respect to maps in $\mathbf{H}^c_{n+1}/\simeq$. The top row is determined up to isomorphism by the homotopy class of β_X in $\mathbf{Chain}^{\wedge}_{\mathbb{Z}}/\simeq$.

(4.12) **Proposition.** *There is an isomorphism Θ' of $\pi_1(X)$-modules such that diagram* (4.11) *commutes.*

Proof. Since $P = P\lambda X$ is an $(n+1)$-type, $P = P^{(n+1)}$, and since β is the identity on n-skeleta we see that

$$H_{n+1}(C_\beta) = \text{cokernel}(\beta_{n+1}). \tag{1}$$

We now define Θ'. The attaching map of $(n+1)$-cells in $P_{n-1}X$ induces a map of $\pi_1(X)$-modules

$$f^0_{n+1}: P_{n+1} \longrightarrow \pi_n X^n \tag{2}$$

since $(P_{n-1}X)^n = X^n$. Moreover we have the following commutative diagram with $Z_i = \text{kernel}(d: C_i \to C_{i-1})$.

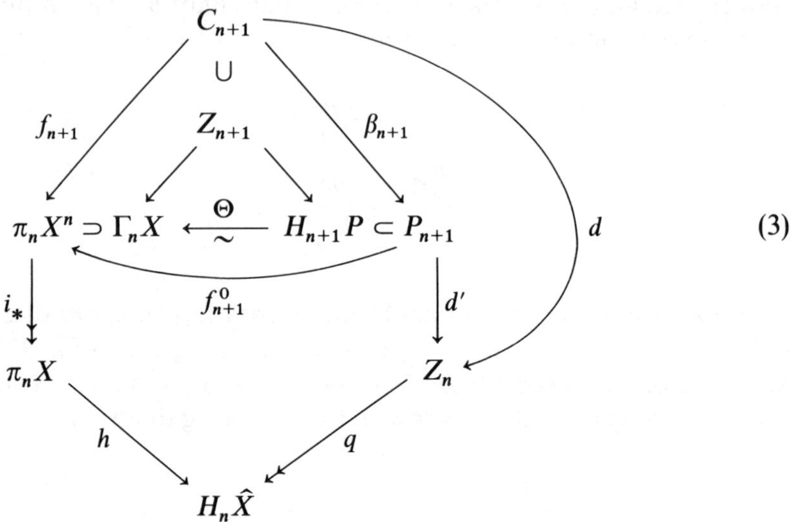

(3)

Here d' is the boundary of P moreover $f_{n+1}|Z_{n+1}$ induces b_{n+1} and $\beta|Z_{n+1}$ induces $\beta_* = H_{n+1}\beta$ in (4.11). The isomorphism

$$\Theta': \text{cokernel}(\beta_{n+1}) \xrightarrow{\cong} \pi_n X \tag{4}$$

is induced by $i_* f^0_{n+1}$ where i_* is given by $i: X^n \subset X$ with kernel $(i_*) = \text{im}(f_{n+1})$. Moreover ∂ in (4.11) is induced by d' since $\text{kernel}(q) = dC_{n+1}$. □

The object X in $\mathbf{H}^c_{n+1}$ determines the k-**invariant**

(4.13) $$k^X_n \in \hat{H}^{n+1}(P\lambda X, \pi_n X)$$

as follows. We choose a CW-complex X^{n+1} which realizes the $(n+1)$-skeleton of X. The n-th k-**invariant** k_n of X^{n+1} is an element

$$k_n \in \hat{H}^{n+1}(P_{n-1}X^{n+1}, \pi_n X) \overset{\psi}{\cong} \hat{H}^{n+1}(P\lambda X, \pi_n X)$$

which yields $k_n^X = \psi k_n$. The isomorphism ψ is obtained by $P_{n-1}X^{n+1} = P_{n-1}X^n$ and by definition of the functor P in (4.5). Clearly k_n^X in (4.13) is natural with respect to maps in $\mathbf{H}_{n+1}^c/\simeq$. We now show that the cellular boundary invariant β_X determines the k-invariant k_n^X. With the notation in the proof of (4.12) we get

(4.14) **Proposition.** *Assume the chain map $\beta: C \to P\lambda X$ represents β_X. Then we have the composition $\Theta' q: P_{n+1} \twoheadrightarrow \operatorname{cokernel}(\beta_{n+1}) \cong \pi_n X$ where q is the quotient map. The map $\Theta' q$ is a cocycle which represents*

$$k_n^X = \{\Theta' q\} \in H^{n+1}(P\lambda X, \pi_n X).$$

This follows easily from the description of the k-invariant of a space in (VI.8.13) (1) Baues (AH).

The following two lemmas are purely algebraic properties of sets of homotopy classes in $\mathbf{Chain}_{\mathbb{Z}}^{\wedge}/\simeq_n$. Let C and P be chain complexes in $\mathbf{Chain}_{\mathbb{Z}}^{\wedge}$ and let $F^n: C^n \to P^n$ be a chain map. Then we obtain the subset

$$[C, F^n, P] \subset \mathbf{Chain}_{\mathbb{Z}}^{\wedge}(C, P)/\simeq_n \tag{4.15}$$

which consists of all ($\simeq_n$)-homotopy classes $G: C \to P$ which have the restriction $G^n = F^n$, see (4.4).

(4.16) **Lemma.** *Let C be free and let $T: P \to Q$ be a weak equivalence which induces an isomorphism $T^n: P^n \cong Q^n$. Then T induces a bijection*

$$T_*: [C, F^n, P] \xrightarrow{\approx} [C, T^n F^n, Q]$$

Proof of (4.16). Let $u = F^{n-1}$ and let $[C, P]^u$ be the set of homotopy classes relative u. Moreover, let $E(u)$ be the subset of all $\{F\}: C \to P$ in $\mathbf{Chain}_{\mathbb{Z}}^{\wedge}/\simeq_n$ which are extensions of u. Then we have the short exact sequence of sets.

$$\operatorname{Hom}_\varphi(C_n, dP_{n+1}) \rightarrowtail^{+} E(u) \twoheadrightarrow [C, P]^u$$

Here the left hand group acts on $E(u)$ as follows. For $\alpha: C_n \to dP_{n+1}$ we choose $\bar{\alpha}: C_n \to P_{n+1}$ with $d\bar{\alpha} = \alpha$. Then we define $\{F\} + \alpha = \{G\}$ by

$$G_i = \begin{cases} F_n + d\bar{\alpha}, & i = n, \\ F_{n+1} + \bar{\alpha} d, & i = n+1, \\ F_i, & \text{otherwise.} \end{cases}$$

One readily checks that the above sequence is exact. Now naturality of the sequence yields the proposition since a weak equivalence T induces a bijection $T_*: [C,P]^u \to [C,Q]^{Tu}$, compare (II.2.11) and (I.6.4) in Baues (AH). □

(4.17) **Lemma.** *Let C be free and assume that $H_iP = 0$ for $i > n+1$. Moreover let $F^n: C^n \to P^n$ be a φ-equivariant chain map. Then the group $H^{n+1}(C, \varphi^* H_{n+1}P)$ acts transitively and effectively on the set $[C, F^n, P]$.*

Proof. The map $q: P \to P^{(n+1)}$ induces a bijection

$$q_*: [C, F^n, P] \xrightarrow{\approx} [C, qF^n, P^{(n+1)}] \tag{1}$$

by (4.16). Whence it is enough to prove (4.17) in case $P_i = 0$ for $i > n+1$. Let elements

$$\left.\begin{array}{ll} \{\xi\} \in H^{n+1}(C, \varphi^* H_{n+1}P), & \xi \in \operatorname{Hom}_\varphi(C_{n+1}, H_{n+1}P) \\ \{F\} \in [C, F^n, P] & \end{array}\right\} \tag{2}$$

be given. We define the *action* + in the lemma by $\{F\} + \{\xi\} = \{G\}$ where $G_i = F_i$ for $i \le n$ and where

$$G_{n+1} = F_{n+1} + \xi. \tag{3}$$

Here we use the inclusion

$$H_{n+1}P = \operatorname{kernel}(d: P_{n+1} \to P_n) \subset P_{n+1}.$$

One readily checks that the class $\{G\}$ is well defined by $\{F\}$ and $\{\xi\}$.

Assume $\{F\} + \{\xi\} = \{F\}$ or equivalently assume $\alpha: F \simeq_n G$ is given. Then we get

$$-F_{n+1} + (F_{n+1} + \xi) = \alpha_n d \tag{4}$$

where $\alpha_n: C_n \to H_{n+1}P \subset P_{n+1}$. Thus $\xi = \alpha_n d$ is a coboundary and whence $\{\xi\} = 0$. This proves that the action is effective.

Next assume $\{F\}, \{G\} \in [C, F^n, P]$ are given with $F^n = G^n$. Then we obtain $\xi: C_{n+1} \to H_{n+1}P$ by

$$-F_{n+1} + G_{n+1} = \xi. \tag{5}$$

This shows $\{G\} = \{F\} + \{\xi\}$ and thus $\{G\}$ and $\{F\}$ are in the same orbit. □

We apply lemma (4.17) in case $P = PY$ is given by the functor P in (4.5). Let $F, G \in [C, \xi^n, PY]$ where Y is an object in $\mathbf{H}_n^c$. Then we have the difference

(4.18) $$-F + G = \delta \in H^{n+1}(C, \varphi^* H_{n+1} PY)$$

by the action in (4.17), namely δ is the unique element with $F + \delta = G$. We now are ready for the computation of the obstruction operator $\mathcal{O}$ in (3.4) in terms of the cellular boundary invariants. As in (3.4) let X, Y be objects in $\mathbf{H}^c_{n+1}$ and let $(\xi, \eta)\colon \lambda X \to \lambda Y$ be a map in $\mathbf{H}^c_n$. Then the following diagram in $\mathbf{Chain}^{\wedge}_{\mathbb{Z}}/\simeq_n$ is defined.

(4.19) $$\begin{array}{ccc} C & \xrightarrow{\beta_X} & P\lambda X \\ \Big\downarrow \xi & & \Big\downarrow P(\xi,\eta) \\ C' & \xrightarrow{\beta_Y} & P\lambda Y \end{array}$$

This diagram needs not to commute, however, it commutes by (4.7) provided (ξ, η) is λ-realizable. We observe that diagram (4.19) satisfies $(P(\xi, \eta)\beta_X)^n = \xi^n = (\beta_Y \xi)^n$ so that by (4.18) the difference

(4.20) $$\Theta_*(-\beta_Y \xi + P(\xi, \eta)\beta_X) \in H^{n+1}(X, \varphi^* \Gamma_n Y)$$

is defined with $\varphi = \eta_*\colon \pi_1 X^{n-1} \to \pi_1 Y^{n-1}$. Here we use the isomorphism $\Theta\colon H_{n+1} P\lambda Y \cong \Gamma_n Y$ in (4.11). The element (4.20) coincides with the obstruction in (3.4), that is:

(4.21) **Proposition.** $\mathcal{O}_{X,Y}(\xi, \eta) = \Theta_*(-\beta_Y \xi + P(\xi, \eta)\beta_X)$

Proof. Let $F\colon X^n \to Y^n$ and $G = P_{n-1}F$ be given as in (4.5)(4). We apply the functor r_{n+1} in (2.2) to the cellular map G. This yields the following commutative diagram of unbroken arrows

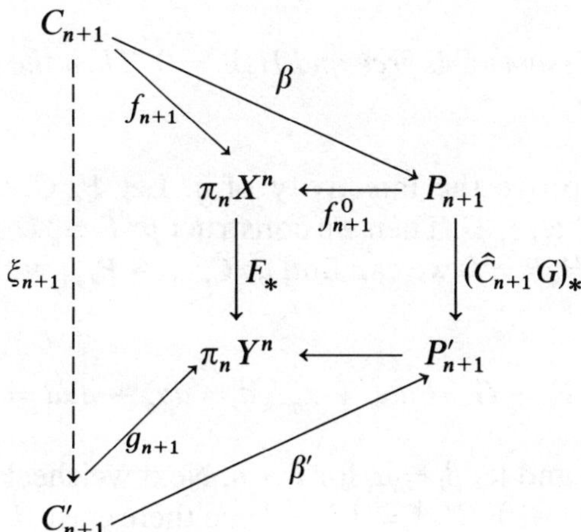

Here $P = P\lambda X$, $(P' = P\lambda Y)$ are given by the functor P and $\beta \in \beta_X$, $(\beta' \in \beta_Y)$ are chosen as in (4.12)(3). Since g_{n+1}^0 induces Θ as in (4.12)(3) we get the result. □

(4.22) **Corollary.** *A map* $(\xi, \eta)\colon \lambda X \to \lambda Y$ *in* $\mathbf{H}_n^c$ *is* λ*-realizable if and only if diagram* (4.19) *commutes in* $\mathbf{Chain}_{\mathbb{Z}}^{\wedge}/{\simeq_n}$.

Next let X be an object in $\mathbf{H}_n^c$ with $\mathcal{O}(X) = 0$ and consider the set $\mathrm{Real}_\lambda(X)$ as in (3.13) and (I.O.6).

(4.23) **Corollary.** *The cellular boundary invariant yields a bijection of sets*

$$\beta\colon \mathrm{Real}_\lambda(X) \xrightarrow{\approx} [C, I^n, PX]$$

where $X = (C, f_n, X^{n-1})$ *and where* I^n *is the identity of* C^n. *The bijection* β *carries* $\{Y\}$ *to* β_Y. *Moreover* β *is equivariant with respect to the actions in* (3.13) *and* (4.17) *respectively, that is* $\beta(Y + \alpha) = \beta(Y) + \Theta_*^{-1}\alpha$.

There is the obvious quotient functor

$$q\colon \mathbf{Chain}_{\mathbb{Z}}^{\wedge}/{\simeq_n} \to \mathbf{Chain}_{\mathbb{Z}}^{\wedge}/{\simeq_{n-1}} \tag{4.24}$$

induced by the identity on $\mathbf{Chain}_{\mathbb{Z}}^{\wedge}$. Let C and P be chain complexes and let $F^n\colon C^n \to P^n$ be a chain map with restriction $F^{n-1}\colon C^{n-1} \to P^{n-1}$. We consider the maps between homotopy sets

$$[C, F^n, P] \xrightarrow{q} [C, F^{n-1}, P] \xleftarrow{p} [C, P]^{F^{n-1}} \tag{4.25}$$

Here q is given by the functor p above and p carries the homotopy class of F relative F^{n-1} to the $(\simeq_{n-1})$-homotopy class of F.

(4.26) **Lemma.** *Assume* C *is free and* $H_n P = 0$. *Then the maps* p *and* q *in* (4.25) *both are bijective.*

Proof. We first prove the injectivity of q. Let F, $G\colon C \to P$ be given with $F^n = G^n$ and $\alpha\colon F \simeq_{n-1} G$. Then we construct $\beta\colon F \simeq_n G$ as follows. Since C_{n-1} is free and since $H_n P = 0$ we can find $\bar{\alpha}\colon C_{n-1} \to P_{n+1}$ with $d\bar{\alpha} = d_{n-1}$. Whence we get

$$0 = -F_n + G_n = d\alpha_n + \alpha_{n-1} d = d\alpha_n + d\bar{\alpha} d = d(\alpha_n + \bar{\alpha} d)$$

Let $\beta_n = \alpha_n + \bar{\alpha} d$ and let $\beta_i = \alpha_i$ for $i > n$. Next we check surjectivity of q. Let $G\colon C \to P$ be given with $G^{n-1} = F^{n-1}$. Then there exist $H\colon C \to P$ with $H^n = F^n$

and $\alpha: H \simeq_{n-1} G$. In fact, since C_n is free and since $H_n P = 0$ we can find $\alpha_n: C_n \to P_{n+1}$ with $d\alpha_n = -F_n + G_n$. Let $H_{n+1} = G_{n+1} - \alpha_n d$ and let $H_i = G_i$ for $i > n+1$. Moreover, let $\alpha_i = 0$ for $i \neq n$.

Finally we consider p. Clearly p is surjective. Moreover p is injective. In fact, let $\alpha: F \simeq_{n-1} G$ be given. Then we obtain $\beta: F \simeq G \operatorname{rel} C^{n-1}$ as follows. We choose $\bar{\alpha}: C_{n-1} \to P_{n+1}$ with $d\bar{\alpha} = \alpha_{n-1}$ and we set $\beta_n = \alpha_n + \bar{\alpha}d$, $\beta_i = \alpha_i$ for $i > n$. □

Again let X be an object in $\mathbf{H}^c_{n+1}$. Then we see by (4.11) that $H_n(P\lambda X) = 0$ if and only if the secondary boundary $b^X_n = b_n: \hat{H}_n X \to \Gamma_{n-1} X$ is trivial. Whence we derive from (4.26):

(4.27) **Proposition.** *Let X be an object in $\mathbf{H}^c_{n+1}$ with $b^X_n = 0$. Then the cellular boundary invariant β_X is determined by the element*

$$q\beta_X \in [C, I^{n-1}, P\lambda X] \cong [C, I^n, P\lambda X].$$

For example we have $\Gamma_2 X = 0$ for all X in $\mathbf{H}^c_4$ so that the assumption in (4.27) is satisfied for all X if $n = 3$. This shows that for $n = 3$ the functor P in (4.5) can be replaced by the functor

$$qP: \mathbf{H}^c_3 \to \mathbf{Chain}^{\wedge}_{\mathbb{Z}}/\simeq_2 \tag{4.28}$$

Using the functors $P = q^0 P$ or $qP = q^1 P$ we obtain the following category which is completely determined by $q^i P$, $i \in \{0, 1\}$.

(4.29) **Definition.** Let $(\mathbf{H}^c_n, q^i P)$ be the **category of boundary invariants**: Objects are pairs (X, β) where $X = (C, f_n, X^{n-1})$ is an object in $\mathbf{H}^c_n$ and where β is an element $\beta \in [C, I^{n-i}, PX]$. Here I^{n-i} is the identity of C^{n-i}. A morphism $(X, \beta) \to (Y, \beta')$ is a map $(\xi, \eta): X \to Y$ in $\mathbf{H}^c_n$ which satisfies $\beta'\xi = P(\xi, \eta)\beta$ in $\mathbf{Chain}^{\wedge}_{\mathbb{Z}}/\simeq_{n-i}$. Morphisms are homotopic in $(\mathbf{H}^c_n, q^i P)$ if they are homotopic in $\mathbf{H}^c_n$. We have obvious functors

$$\begin{cases} \beta: \mathbf{H}^c_{n+1} \to (\mathbf{H}^c_n, q^i P) \\ \beta: \mathbf{H}^c_{n+1}/\simeq \;\to (\mathbf{H}^c_n, q^i P)/\simeq \end{cases} \tag{4.30}$$

which carry the object X to $(\lambda X, q^i \beta_X)$.

(4.31) **Proposition.** *The functors* (4.30) *are detecting functors for* $(n \geq 3, i = 0)$ *and* $(n = 3, i = 1)$.

This is a consequence of (4.21) and (4.27). The proposition shows that the classification of objects in $\mathbf{H}^c_{n+1}/\simeq$ can be completely obtained by the computation of the functor P on the category $\mathbf{H}^c_n$, see (I.0.5). We can use the functor qP for the classification of objects X in $\mathbf{H}^c_{n+1}$ with $b_n^X = 0$; this follows from (4.27). In (V.§ 1) we shall give a purely algebraic characterization of the functor qP in (4.28).

§ 5 Three formulas for the obstruction

In this section we describe the connection of the obstruction operator in the CW-tower with the operators in Whitehead's certain exact sequence, see (I.3.7). A simple application of one of these formulas (see (5.4)) yields a homotopy classification of certain $2n$-dimensional manifolds, see § 7 below. For a complex X in **CW** we obtain the natural invariants

(5.1) $$\begin{cases} b_{n+1}X \in \mathrm{Hom}_\pi(H_{n+1}\hat{X}, \Gamma_n X) \\ \{\pi_n X\} \in \mathrm{Ext}_\pi(\ker b_n, \operatorname{cok} b_{n+1}) \end{cases}$$

by the exact sequence (I.3.7). Here $\pi = \pi_1 X$ is the fundamental group and the element $\{\pi_n X\}$ is represented by the extension of π-modules

$$\operatorname{cok}(b_{n+1}) \rightarrowtail^{i} \pi_n \xrightarrow{h} \ker(b_n)$$

given by i_n and h_n in (I.3.7). In (I.6.12) we have seen that the elements b_{n+2} and $\{\pi_n\}$ are completely determined by the cohomological invariant $\mathscr{k}_n$ provided X is $(n-1)$-connected. In general, however, we do not have such a cohomological invariant, but we still have the following connection of the obstruction in the CW-tower with the elements (5.1). This generalizes the corresponding formulas in (I.8.9)(3), (4). To this end we need the following general form of the universal coefficient sequence.

(5.2) **Lemma.** *Let C be a chain complex of free π-modules and let Γ be a π-module. Then we have the exact sequence*

$$\mathrm{Ext}_\Lambda(C_n/B_n, \Gamma) \rightarrowtail^{\Delta} \hat{H}^{n+1}(C, \Gamma) \xrightarrow{\mu} \mathrm{Hom}_\Lambda(H_{n+1}C, \Gamma) \xrightarrow{k} \mathrm{Ext}_\Lambda(B_n, \Gamma)$$

where $B_n = dC_{n+1}$ is the module of n-boundaries and where $\Lambda = \mathbb{Z}[\pi]$.

The map μ in (5.2) carries the cohomology class $\{x\}$ of the cocycle x to the composition

$$\mu\{x\}: H_{n+1}C \subset C_{n+1}/B_{n+1} \xrightarrow{x} \Gamma. \tag{1}$$

Moreover the extension $B_n \overset{i}{\rightarrowtail} C_n \twoheadrightarrow C_n/B_n$ yields the exact sequence

$$\mathrm{Hom}_\Lambda(C_n, \Gamma) \xrightarrow{i^*} \mathrm{Hom}_\Lambda(B_n, \Gamma) \overset{\delta}{\twoheadrightarrow} \mathrm{Ext}_\Lambda(C_n/B_n, \Gamma). \tag{2}$$

Here we use the assumption that C_n is a free Λ-module. We use δ for the identification

$$\mathrm{Hom}_\Lambda(B_n, \Gamma)/i^* \mathrm{Hom}_\Lambda(C_n, \Gamma) = \mathrm{Ext}_\Lambda(C_n/B_n, \Gamma) \tag{3}$$

Then $\Delta\{\beta\}$ with $\beta \in \mathrm{Hom}_\Lambda(B_n, \Gamma)$ is defined by the composition

$$\Delta\{\beta\} = \{\beta d\} \quad \text{with } \beta d\colon C_{n+1} \twoheadrightarrow B_n \longrightarrow \Gamma. \tag{4}$$

Finally we obtain the map k in (5.2) as follows. Let

$$k_{n+1}(C) \in \mathrm{Ext}_\Lambda(B_n, H_{n+1}) \tag{5}$$

by the element represented by the extension $H_{n+1} \rightarrowtail C_{n+1}/B_{n+1} \twoheadrightarrow B_n$. We call this element a ***k*-invariant** of the chain complex C. Now k in (5.2) is simply given by the formula

$$k(\varphi) = \varphi_*(k_{n+1} C) \tag{6}$$

We leave the proof of the exactness of the sequence in (5.2) as an exercise. In addition to (5.2) the extension $H_n \overset{i}{\rightarrowtail} C_n/B_n \twoheadrightarrow B_{n-1}$ yields the long exact sequence

$$\mathrm{Ext}_\Lambda(B_{n-1}, \Gamma) \longrightarrow \mathrm{Ext}_\Lambda(C_n/B_n, \Gamma) \xrightarrow{i^*} \mathrm{Ext}_\Lambda(H_n, \Gamma) \longrightarrow \mathrm{Ext}^2_\Lambda(B_{n-1}, \Gamma) \tag{7}$$

Therefore we obtain the usual universal coefficient formula by (7) and (5.2) provided B_n and B_{n-1} are projective Λ-modules; in this case i^* in (7) is an isomorphism and μ in (5.2) is surjective.

We now consider the obstruction.

(5.3) $$\mathcal{O}_{X,Y}(\xi, \eta) \in \hat{H}^{n+1}(X, \varphi^*\Gamma_n Y)$$

where $(\xi, \eta)\colon \lambda X \to \lambda Y$ is a map in the category $\mathbf{H}^c_n$, see (3.4). For the objects X and Y in $\mathbf{H}^c_{n+1}$ we have as well the natural invariants in (5.1).

(5.4) **Proposition.** *The obstruction* (5.3) *satisfies the following formulas* (1) *and* (2).

$$\mu \mathcal{O}_{X,Y}(\xi,\eta) = \eta_*(b_{n+1}X) - \xi^*(b_{n+1}Y) \tag{1}$$

$$-j^*i^*q_*\Delta^{-1}\mathcal{O}_{X,Y}(\xi,\eta) = \eta_*\{\pi_n X\} - \xi^*\{\pi_n Y\} \tag{2}$$

Here equation (2) is given provided the element (1) is trivial. The map i^* is defined as in (5.2)(7) and j^* and q_* are induced by the inclusion j: $\ker(b_n X) \rightarrowtail \hat{H}_n X$ and by the projection q: $\Gamma_n Y \twoheadrightarrow \operatorname{cok}(b_{n+1} Y)$ respectively.

Proof of (5.4). Equation (1) is an immediate consequence of the definition in (3.4) and of the definition in (I.3.8). Equation (1) as well is obtained by applying the homology functor H_{n+1} to the chain maps in (4.19); here we use (4.21). Now assume that the element (1) is trivial. Then both sides of equation (2) are well defined. We now use (4.14) for the presentation of the group π_n. The map $\beta = \beta_{n+1}$ is embedded in the following commutative diagram

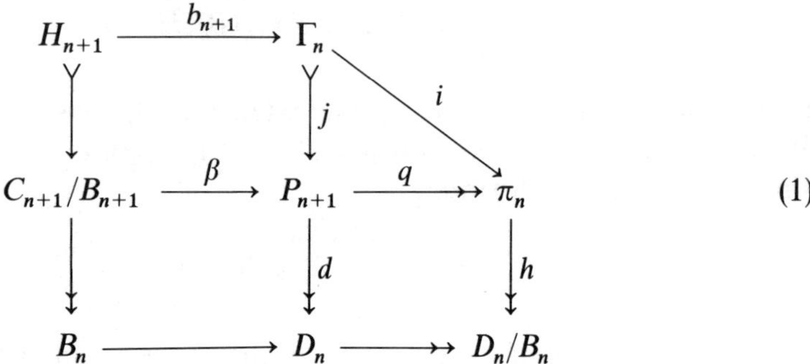

Here we set $B_n = dC_{n+1}$ and $D_n = dP_{n+1}$; these are both submodules of C_n. Moreover, the map q is the surjection $\Theta' q$ in (4.14). The columns and the rows of diagram (1) are exact sequences. The maps b_{n+1}, i, h are the operators in the certain exact sequence where we identify

$$\ker(b_n) = D_n/B_n \tag{2}$$

by use of diagram (4.8). In fact, D_n/B_n is canonically the kernel of the surjection $\hat{H}_n X \twoheadrightarrow \hat{H}_n P_{n-1} X$ and this kernel is $\ker(b_n)$ by (4.8). Now let $\eta_{n+1}\colon P_{n+1} \to P'_{n+1}$ be given by the map $P(\xi,\eta)$ in (4.19). Then the difference

$$j^{-1}(\eta_{n+1}\beta - \beta'\xi_{n+1})\colon C_{n+1}/B_{n+1} \to \Gamma'_n \tag{3}$$

represents the obstruction (5.3), see (4.21). If the element (5.4)(1) is trivial then diagram (1) shows that the map (3) factors over $C_{n+1}/B_{n+1} \twoheadrightarrow B_n$. Therefore we consider the next diagram which we obtain by dividing out the group H_{n+1} in (1).

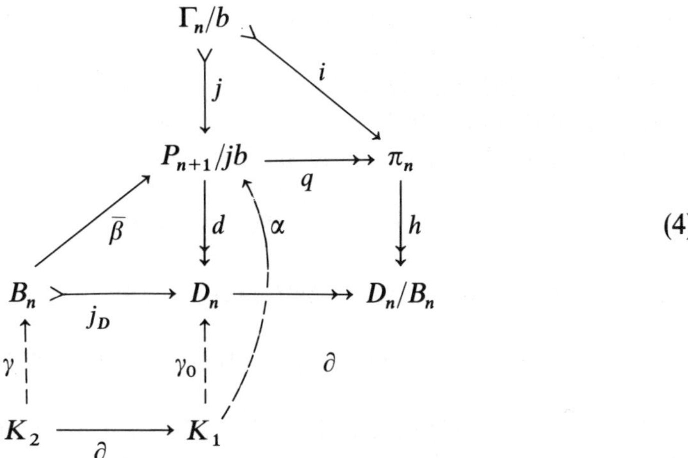

(4)

Here $\bar{\beta}$ is induced by β in (1); clearly P_{n+1}/jb denotes the cokernel of jb_{n+1}. We choose free π-modules K_1 and K_2 such that the sequence (∂, ∂) in the diagram is exact. Moreover, we choose maps γ_0, γ and α such that the diagram commutes, in particular $d\alpha = \gamma_0$. Now we get

$$d(\bar{\beta}\gamma - \alpha\partial) = \gamma_0\partial - d\alpha = 0 \tag{5}$$

so that

$$\pi = j^{-1}(\bar{\beta}\gamma - \alpha\partial)\colon K_2 \to \Gamma_n/b \tag{6}$$

is defined. In fact π represents the element

$$-\{\pi_n\} = \{\pi\} \in \operatorname{Ext}(D_n/B_n, \Gamma_n/b) = \operatorname{cok} \partial^* \tag{7}$$

where

$$\partial^*\colon \operatorname{Hom}(K_1, \Gamma_n/b) \to \operatorname{Hom}(K_2, \Gamma_n/b). \tag{8}$$

We obtain (7) since

$$i\pi = q(\bar{\beta}\gamma - \alpha\partial) = -q\alpha\partial \tag{9}$$

where $h(q\alpha) = \partial$. The diagram

$$\begin{array}{ccc} \operatorname{Ext}(C_n/B_n, \varphi^*\Gamma'_n/b') & \xrightarrow{j_0^* i_1^*} & \operatorname{Ext}(\ker b_n, \varphi^*\Gamma'_n/b') \\ \| & & \| \\ \operatorname{cok} i_0^* & \xrightarrow[\gamma^*]{} & \operatorname{coker} \partial^* \end{array} \tag{10}$$

commutes. Here $i_0: B_n \subset C_n$ is the inclusion and the identification on the left hand side is the one in (5.2)(3). Moreover, the identification on the right hand side of (10) is the one in (7) above. The map $j_0^* i_1^*$ coincides with the map $j^* i^*$ used in equation (5.4)(2). We now are ready for the proof of this equation. Using (3) and (10) the left hand side of (5.4)(2) is represented by

$$-\gamma^* j^{-1}(\eta_{n+1}\bar{\beta} - \bar{\beta}'\xi_n): K_2 \to \Gamma_n'/b'. \tag{11}$$

The right hand side of (5.4)(2) is represented by

$$\eta_*(-\{\pi\}) - \xi^*(-\{\pi'\}) = \eta_*(j^{-1}(\alpha\partial - \bar{\beta}\gamma)) - \xi^*(j^{-1}(\alpha'\partial' - \bar{\beta}'\gamma')). \tag{12}$$

Here we use π in (7) and π' is defined for Y in the same way as π for X. Moreover the map $\xi^* = k_2^*$ is induced by a lift k_2 in the commutative diagram

$$\begin{array}{ccccc} K_2 & \xrightarrow{\partial} & K_1 & \longrightarrow & D_n/B_n \\ \downarrow{\scriptstyle k_2} & & \downarrow{\scriptstyle k_1} & & \downarrow{\scriptstyle \xi_n^D} \\ K_2' & \longrightarrow & K_1' & \longrightarrow & D_n'/B_n' \end{array} \tag{13}$$

Here ξ_n^D is induced by ξ_n. We now can find $\rho: K_1 \to B_n'$ such that

$$\gamma' k_2 - \xi_n \gamma = \rho\partial. \tag{14}$$

Here $\xi_n: B_n \to B_n'$ is the restriction of ξ_n. Now we get the following formula for the difference $\Delta = (12) - (11)$ of the elements (12) and (11).

$$\begin{aligned} j\Delta &= (\eta_{n+1}\alpha\partial - \eta_{n+1}\bar{\beta}\gamma - \alpha'\partial' k_2 + \bar{\beta}'\gamma' k_2) + (\eta_{n+1}\bar{\beta}\gamma - \bar{\beta}'\xi_n\gamma) \\ &= \eta_{n+1}\alpha\partial - \alpha' k_1\partial + \bar{\beta}'\rho\partial \\ &= (\eta_{n+1}\alpha - \alpha' k_1 + \bar{\beta}'\rho)\partial \end{aligned} \tag{15}$$

In addition we get

$$d'(\eta_{n+1}\alpha - \alpha' k_1 + \bar{\beta}'\rho) = \xi_n\gamma_0 - \gamma_0' k_1 + j_D'\rho = 0 \tag{16}$$

Here the second equation follows from (14). By (16) and (15) we get

$$\Delta = (j^{-1}(\eta_{n+1}\alpha - \alpha' k_1 + \bar{\beta}'\rho))\partial \tag{17}$$

This shows that (11) and (12) represent the same elements and therefore the proof of equation (5.4)(2) is complete. □

In the next formula we alter the φ-equivariant chain map $\xi: C \to C'$ in (5.3), $\varphi = \pi_1(\eta)$. For this we choose a φ-equivariant map α as in the following diagram

$$\begin{array}{ccc} C_{n+1} & \xrightarrow{\ \xi_{n+1}\ } & C'_{n+1} \\ {\scriptstyle d}\downarrow & & \uparrow{\scriptstyle i} \\ dC_{n+1} = B_n & \xrightarrow{\ \alpha\ } & \ker(d'_{n+1}) \\ & & \downarrow{\scriptstyle q} \\ & & H_{n+1}C' \end{array} \tag{5.5}$$

Here d is given by d_{n+1}. Moreover, i is the inclusion and q is the quotient map. One readily checks that the map α yields a well defined chain map

$$\xi + \alpha: C \to C' \tag{1}$$

where we set $(\xi + \alpha)_{n+1} = \xi_{n+1} + i\alpha d$ and $(\xi + \alpha)_i = \xi_i$ otherwise. Using (5.2) (3) the map α represents an element

$$\{q\alpha\} \in \operatorname{Ext}(C_n/B_n, \varphi^* H_{n+1} C'). \tag{2}$$

Since $(\xi + \alpha, \eta): \lambda X \to \lambda Y$ is a well defined map in $\mathbf{H}_n^c$ we have an obstruction as in (5.3). This obstruction satisfies the formula

(5.6) **Proposition.** $\mathcal{O}_{X,Y}(\xi + \alpha, \eta) = \mathcal{O}_{X,Y}(\xi, \eta) - \Delta(b'_{n+1})_* \{q\alpha\}$.

Here $b'_{n+1} = b_{n+1} Y$ is the secondary boundary operator of Y and Δ is the map in (5.2). The proposition is an immediate consequence of the definitions for Δ in (5.2)(4), for the obstruction in (3.4), and for the secondary boundary in (I.3.8).

(5.7) **Remark.** Consider the functor $\mathcal{B}: \mathbf{A}_n^2 \to \mathbf{B}_n^2$ in (I.8.1). Using the formulas (5.4) and (5.6) above one can find a direct proof that $\mathcal{B}$ is a detecting functor. This proof is described in (IX.4.13) Baues (AH).

In general the formulas (5.4), (5.6) do not suffice to compute the obstruction (5.3). There are, however, important special cases for which μ in (5.4) is actually injective. In this simple case the obstruction is determined by the secondary boundary; actually this case can be used for the homotopy classification of certain $2n$-dimensional manifolds, compare (7.12) and (7.14) below.

§ 6 Trees of homotopy types

Theorem 14 of Whitehead (CH) shows that the homotopy types of finite n-dimensional connected CW-complexes which have the same $(n-1)$-type form a connected tree. Whitehead actually proved that as well 'simple' homotopy types of this kind form a tree. In this section we give only a proof for homotopy types. I am grateful to J. Zobel who explained to me the nice proofs of theorem (6.1) and (6.6) below which essentially rely on the Hurewicz theorem and on some algebraic facts.

Recall that two n-dimensional connected CW-complexes X^n and Y^n have the same $(n-1)$-type iff one of the following conditions (A) or (B) is satisfied:
(A) There is a map $F: X^n \to Y^n$ which induces isomorphisms $\pi_i(F)$ for $i \leq n-1$.
(B) There is a homotopy equivalence $P_{n-1}(X^n) \simeq P_{n-1}(Y^n)$, compare (4.1).

(6.1) **Theorem.** *Let X^n, Y^n be two finite n-dimensional* CW*-complexes which have the same $(n-1)$-type. Then there exist natural numbers A, B such that the one point unions*

$$X^n \vee \bigvee_A S^n \simeq Y^n \vee \bigvee_B S^n$$

are homotopy equivalent.

The theorem shows that each $(n-1)$-type Q determines a connected tree $\mathrm{HT}(Q,n)$ which we call the **tree of homotopy types** of (Q,n). The vertices of this tree are the homotopy types $\{X^n\}$ of finite n-dimensional CW-complexes with $P_{n-1}X^n \simeq Q$. The vertex $\{X^n\}$ is connected by an edge to the vertex $\{Y^n\}$ if Y^n has the homotopy type of $X^n \vee S^n$. The **roots** of this tree are the homotopy types $\{Y^n\}$ which do not admit a homotopy equivalence $Y^n \simeq X^n \vee S^n$. Theorem (6.1) shows that the tree $\mathrm{HT}(Q,n)$ is connected.

(6.2) **Remark.** There are various results on the trees $\mathrm{HT}(Q,n)$ in case $Q = K(\pi,1)$ is an Eilenberg-Mac Lane space of degree 1. In this case the tree $\mathrm{HT}(K(\pi,1),n) = \mathrm{HT}(\pi,n)$ is determined by the group π. Results of Metzler, Sieradski and Sieradski-Dyer show that for $n \geq 2$ there exist trees $\mathrm{HT}(\pi,n)$ with at least two roots; (this is as well true for such trees of simple homotopy types). These examples are the crucial ingredients of examples in Kreck-Schafer concerning the "stable" classification of manifolds.

We say that two R-modules M, N are **stably isomorphic** if there exist finitely generated free R-modules A, B and an isomorphism $M \oplus A \cong N \oplus B$. We say that M is **stably free** if there exists an isomorphism $M \oplus A \cong B$, that is, if

M is stably isomorphic to the trivial module $N = 0$. We derive from (6.1) immediately the

(6.3) **Corollary.** *Let X^n, Y^n be two finite n-dimensional* CW*-complexes which have the same $(n-1)$-type, $n \geq 2$. Then the π_1-modules $\pi_n X^n$, $\pi_n Y^n$ are stably isomorphic and also the π_1-modules $H_n \hat{X}^n$, $H_n \hat{Y}^n$ are stably isomorphic.*

Proof. We use the canonical isomorphism of π_1-modules $\pi_n(X \vee S^n) \cong \pi_n(X) \oplus \mathbb{Z}[\pi_1]$ and $\hat{H}_n(X \vee S^n) = \hat{H}_n(X) \oplus \mathbb{Z}[\pi_1]$. □

For the proof of theorem (6.1) we use the following purely algebraic lemma.

(6.4) **Lemma.** *Let C be a chain complex of finitely generated free R-modules with $C_i = 0$ for $i < 0$ and $i > n$. If $H_i(C) = 0$ for $i < n$ the R-module $H_n(C)$ is stably free.*

Proof. The result is true for an $n = 1$. Now assume the lemma is true for $n = m - 1$, $m \geq 2$. Then we obtain for $n = m$ the short exact sequence

$$H_n(C) \rightarrowtail C_n \overset{p}{\twoheadrightarrow} H_{n-1}(C^{n-1})$$

where $H_{n-1}(C^{n-1})$ is stably free by the inductive assumption. Whence there is a finitely generated free module A and an exact sequence

$$H_n(C) \rightarrowtail C_n \oplus A \overset{p \oplus 1}{\twoheadrightarrow} H_{n-1}(C^{n-1}) \oplus A$$

where $H_{n-1}(C^{n-1}) \oplus A$ is a finitely generated free module. Now this sequence is split so that $H_n(C)$ is stably free since $C_n \oplus A$ is free and finitely generated.

□

(6.5) *Proof of* (6.1). Let $n \geq 2$ and choose a cellular map $F\colon X^n = X \to Y^n = Y$ which is an $(n-1)$-equivalence. Then F induces also a morphism $\hat{H}_i(F)$ for $i \leq n - 1$. The mapping cylinder $Z_F = Y^n \cup_F (X^n \times I)$ is a CW-complex with the obvious cell decomposition. We shall use the n-skeleton of Z_F which is given by

$$Z_F^n = Y \cup X^{n-1} \times I \cup X. \tag{1}$$

The Hurewicz theorem yields isomorphisms $(1 < n)$

$$\hat{H}_i(Z_F^n, X) \cong \pi_i(Z_F^n, X) = 0, \tag{2}$$

$$\hat{H}_n(Z_F^n, X) \cong \pi_n(Z_F^n, X). \tag{3}$$

Here the chain complex $\hat{C}_*(Z_F^n, X)$, see (I.2.8), satisfies by (2) the assumption in (6.4) so that (3) is stably free. Therefore we can find a natural number α such that

$$\hat{H}_n(Z_F^n \vee S_\alpha^n, X) \cong \pi_n(Z_F^n \vee S_\alpha^n, X) \tag{4}$$

is a finitely generated free π_1-module where $S_\alpha^n = \bigvee_\alpha S^n$. We choose a π_1-Basis $a_1, \ldots, a_A$ of (4) and we consider the exact sequence

$$\pi_n(Z_F^n \vee S_\alpha^n) \xrightarrow{j} \pi_n(Z_F^n \vee S_\alpha^n, X) \xrightarrow{\partial} \pi_{n-1} X \xrightarrow{i_*} \pi_{n-1}(Z_F^n \vee S_\alpha^n)$$

where i_* is an isomorphism since $i: X \subset Z_F$ is a homotopy equivalence. Whence j is surjective and we can find elements $b_1, \ldots, b_A \in \pi_n(Z_F^n \vee S_\alpha^n)$ with $j(b_r) = a_r$, $r = 1, \ldots, A$. Using the elements b_r we obtain a map

$$\varphi: X \vee S_A^n \to Z_F^n \vee S_\alpha^n \tag{5}$$

where $\varphi|X$ is given by the inclusion i and where $\varphi|S_A^n$ is given by the elements $b_1, \ldots, b_A$. We now consider the following exact sequence of relative homotopy groups

$$\pi_n(X \vee S_A^n, X) \underset{\cong}{\xrightarrow{\varphi_*}} \pi_n(Z_F^n \vee S_\alpha^n, X) \longrightarrow \pi_n(\varphi) \longrightarrow 0$$

where we use the notation $\pi_n(\psi) = \pi_n(Z_\psi, U)$ for a map $\psi: U \to V$. The map φ_* is an isomorphism which carries the basis elements i_r given by the inclusions $S^n \subset S_A^n$ to the basis elements a_r. Since φ_* is an isomorphism we observe that $\pi_n(\varphi) = 0$ and that

$$\pi_{n+1}(Z_F^n \vee S_\alpha^n, X) \twoheadrightarrow \pi_{n+1}(\varphi) \tag{6}$$

is surjective. Now the Hurewicz theorem shows that $\hat{h}: \pi_i(\varphi) \to \hat{H}_i(\varphi)$ is an isomorphism for $i \leq n + 1$. This implies that $\hat{H}_i(\varphi) = 0$ for $i \leq n$ and that indeed also $\hat{H}_{n+1}(\varphi) = 0$ since (6) is surjective. Whence the Whitehead theorem shows that φ is a homotopy equivalence.

We now consider the pair (Z_F^n, Y). For $i < n$ we have the isomorphisms

$$j_*: \pi_i(Y) \xrightarrow{F_*} \pi_i(X) \xrightarrow{i_*} \pi_i Z_F^n$$

with $j_* = i_* F_*$ for the inclusion $j: Y \subset Z_F^n$. Whence the same argument as above yields numbers B, β and a homotopy equivalence $\varphi': Y \vee S_B^n \to Z_F^n \vee S_\beta^n$. This completes the proof of (6.1) by an appropriate alteration of A and B above. □

Next we describe an algebraic criterion which detects roots in the trees above. Let M be an R-module. We say that an element $a \in M$ is **unimodular** if the submodule Ra generated by a is isomorphic to R and if Ra is a direct summand of M.

(6.6) **Theorem.** *Let X be an n-dimensional finite* CW*-complex for which $\pi_1(X)$ is a finite group. Then $\{X\}$ is a root if and only if the image of the Hurewicz map $h\colon \pi_n X \to H_n \hat{X}$ contains no elements which are unimodular in the π_1-module $H_n \hat{X}$.*

Proof. It is clear that for $X \simeq Y \vee S^n$ the image of h contains an unimodular element. Now assume that the image of h in (6.6) contains an unimodular element a. Then we construct a finite- CW-complex $Y = Y^n$ and a homotopy equivalence $X \simeq Y \vee S^n$ as follows. We choose $\bar{a} \in \pi_n X$ with $h(\bar{a}) = a$. Then we have the commutative diagram of R-modules, $R = \mathbb{Z}[\pi_1 X]$,

$$\begin{array}{ccccccc}
R\bar{a} & \rightarrowtail & \pi_n X & \xrightarrow{\bar{j}} & \pi_n(X, X^{n-1}) & & \\
\downarrow \cong & & \downarrow h & & \downarrow \cong & & \\
Ra & \xrightarrow{i} & \hat{H}_n X & \xrightarrow{j} & \hat{H}_n(X, X^{n-1}) & \xrightarrow{\partial} & \hat{H}_{n-1} X^{n-1}
\end{array}$$

Since $\hat{H}_{n-1} X^{n-1}$ is $\mathbb{Z}$-free also cokernel (j) is $\mathbb{Z}$-free so that j is an admissible map in the sense of K.S. Brown page 129. Since a is unimodular in $\hat{H}_n X$ also the composition ji is admissible. Now (VI.2.3) of K.S. Brown shows that $Ra \cong R$ is relatively injective (here we use the assumption that π_1 is a finite group), so that by definition of 'relative injectivity' the map ji admits a retraction (since ji is admissible). Whence Ra is as well a direct summand of the free R-module $\hat{H}_n(X, X^{n-1}) = \hat{C}_n(X)$. Therefore one has an isomorphism

$$\hat{C}_n X \cong \pi_n(X, X^{n-1}) \cong R\bar{j}\bar{a} \oplus S(X) \tag{1}$$

where $S(X)$ is stably free since $R \oplus S(X)$ is free. We now replace X by the CW-complex $\bar{X} = X \vee D^n$ where D^n is the n-disk with $D^n = S^{n-1} \cup e^n$. Then we can choose $S(\bar{X})$ to be the free module $S(\bar{X}) \cong S(X) \oplus R$. Therefore we may assume that $S(X)$ above is actually free. Let $\alpha_i\colon (D^n, S^{n-1}) \to (X, X^{n-1})$ $(i = 1, \ldots, A)$ be maps which yield a basis of $S(X)$ and let $\beta_i = \partial\alpha_i\colon S^{n-1} \to X^{n-1}$. Then we define Y by

$$Y = X^{n-1} \cup_\beta \bigvee_{i=1}^{A} D_i^n, \quad D_i^n = D^n, \tag{2}$$

where we attach cells via the β_i. We now define

$$\varphi: Y \vee S^n \to X \tag{3}$$

on X^{n-1} by the inclusion, on D_i^n by α_i and on S^n by $\bar{a}$ above. The induced homorphism φ_* in the exact sequence

$$\longrightarrow \pi_n(Y \vee S^n, X^{n-1}) \xrightarrow[\cong]{\varphi_*} \pi_n(X, X^{n-1}) \longrightarrow \pi_n(\varphi) \longrightarrow 0$$

is an isomorphism which carries the basis elements D_i and S^n to the basis elements α_i and $j\bar{\alpha}$ respectively. As in the proof (6.5) we now see that φ is a homotopy equivalence. □

The next example shows that the knowledge of roots is of great help for the classification of homotopy types.

(6.7) **Example.** Let $\pi = \mathbb{Z}/N\mathbb{Z} = \mathbb{Z}/N$ be the cyclic group of order N. Then the tree $\mathrm{HT}(\pi, 2)$ has exactly one root given by the **pseudo projective plane**

$$P_N = S^1 \cup_N e^2 \tag{1}$$

which is obtained by attaching a 2-cell to S^1 via a map $S^1 \to S^1$ of degree N; for $N = 2$ this is the real projective plane $P_2 = \mathbb{R}P_2$. Whence any finite 2-dimensional CW-complex X^2 with $\pi_1 X^2 \cong \mathbb{Z}/N$ admits a homotopy equivalence

$$\varphi: X^2 \xrightarrow{\simeq} P_N \vee S_A^2 \tag{2}$$

where $S_A^2 = S^2 \vee \cdots \vee S^2$ is an A-fold one point union of 2-spheres. This as well implies that any finite CW-complex X with $\pi_1 X \cong \mathbb{Z}/N$ is homotopy equivalent to a finite CW-complex Y for which the 2-skeleton Y^2 has the **normal form** $Y^2 = P_n \vee S_A^2$. Indeed we obtain Y by the push out diagram

$$\begin{array}{ccc} X & \xrightarrow{\simeq} & Y \\ \cup & & \cup \\ X^2 & \xrightarrow{\simeq} & P_N \vee S_A^2 \end{array} \tag{3}$$

where we use φ in (2) for the 2-skeleton of X. The normal form of the 2-skeleton of Y simplifies considerably the algebraic models of Y given by crossed chain complexes and quadratic chain complexes respectively, compare chapter III and chapter IV.

§7 On the homotopy classification of manifolds with finite fundamental group

In this section we consider the homotopy types of $2n$-dimensional closed manifolds or more generally of $2n$-dimensional Poincaré complexes X for which the universal covering is $(n-1)$-connected. We derive from the properties of the CW-tower nicely a result on the homotopy classification of such complexes which generalizes a result of Hambleton-Kreck for the case $n = 2$. For the computation of the Γ-groups $\Gamma_{2n-1}X$ we introduce quadratic $\mathbb{Z}$-modules and the quadratic tensor product; for further background on such quadratic constructions we refer the reader to Baues (QF).

The Poincaré duality theorem (in a strong form) asserts that a closed oriented topological manifold has the homotopy type of an oriented Poincaré complex. To this end we recall the following definition and results, compare Wall (P) and Wall (S).

(7.1) **Definition.** A finite connected n-dimensional CW-complex X is an **oriented Poincaré complex of dimension** n if a class $[X] \in H_n(X, \mathbb{Z})$ is given such that the cap product with $[X]$ induces an isomorphism

$$[X]\cap\colon \hat{H}^r(X,\Lambda) \xrightarrow{\cong} \hat{H}_{n-r}(X,\Lambda)$$

where $\Lambda = \mathbb{Z}[\pi_1 X]$, $r \geq 0$.

The condition in (7.1) implies that for any (right) Λ-module B the cap product

$$[X]\cap\colon \hat{H}^r(X,B) \cong \hat{H}_{n-r}(X,B) \tag{1}$$

is an isomorphism. Here the left π_1-module structure of B is given by ${}^{\alpha}b = b^{-\alpha}$, $b \in B$, $\alpha \in \pi_1$. Moreover one has a homotopy equivalence

$$\varphi\colon C^* \to C_* \tag{2}$$

of degree n which induces (1). Here $C_* = \hat{C}_* X$ is the cellular chain complex of the universal covering of X and $C^* = \mathrm{Hom}_\Lambda(C_*, \Lambda)$ is the Λ-cochain complex with $C^n = \mathrm{Hom}_\Lambda(C_n, \Lambda)$. As usual we identify a cochain complex V^* with a (negative) chain complex V_* by setting $V_{-n} = V^n$. Hence a map φ of degree n as in (2) carries C^r to C_{n-r}.

Now assume that X has a finite fundamental group $\pi_1(X)$. Then the universal covering $\hat{X}$ of X is a finite CW-complex which again is a Poincaré-complex with fundamental class $[\hat{X}] = \mathrm{tr}[X]$ where $\mathrm{tr}\colon \mathbb{Z} = H_n(X) \cong H_n(\hat{X})$ is the isomorphism induced by the **transfer** or **norm map**

$$\left.\begin{aligned} &\mathrm{tr}\colon C_* \otimes_\Lambda \varepsilon^*\mathbb{Z} \to C_* \\ &\mathrm{tr}(x \otimes 1) = \sum_{\alpha\in\pi} x^\alpha = x\cdot\left(\sum_{\alpha\in\pi}[\alpha]\right) \end{aligned}\right\} \tag{3}$$

In particular π_1 acts trivially on $H_n(\hat{X})$. It is convenient to introduce the following notation.

(7.2) **Definition.** Let P_n^k be the class of all $(2n+k)$-dimensional Poincaré-complexes which are $(n-1)$-connected, $n \geq 2$. Moreover, for a group π let $P_n^k(\pi)$ be the class of all $(2n+k)$-dimensional Poincaré-complexes X for which there is an isomorphism $\pi \cong \pi_1 X$ and for which the universal covering is $(n-1)$-connected. If π is finite we have the function

$$P_n^k(\pi) \to P_n^k \tag{7.3}$$

which carries X to $\hat{X}$.

From now on we suppose in this section that π is a *finite group*. We want to study the Poincaré complexes in $P_n^0(\pi)$; for $n = 2$ the class $P_2^0(\pi)$ contains all 4-dimensional Poincaré complexes with fundamental group π. It is well known that each Y in P_n^0 has the homotopy type of a mapping cone

$$Y \simeq C_f,\ f\colon S^{2n-1} \to \bigvee_N S^n = S_N^n \tag{7.4}$$

where N is the rank of the free abelian group $H_n(Y) = \mathbb{Z}^N$. This shows that also $\pi_n X$ for $X \in P_n^0(\pi)$ is free as a $\mathbb{Z}$-module. Moreover we get by use of (7.4) applied to $\hat{X}$ the next result on Whitehead's Γ-groups.

(7.5) **Proposition.** *For $X \in P_n^0(\pi)$ there are natural isomorphisms of $\pi_1(X)$-modules for $m \leq n$*

$$\Gamma_m(X) = \begin{cases} 0 & \text{for } m \leq n, \\ \pi_n(X) \otimes_{\mathbb{Z}} \pi_m(S^n) & \text{for } n < m < 2n-1, \\ \pi_n(X) \otimes_{\mathbb{Z}} \pi_{2n-1}\{S^n\} & \text{for } m = 2n-1. \end{cases}$$

Here we use for $m < 2n-1$ the tensor product of abelian groups; for $m = 2n-1$, however, we use the following quadratic tensor product.

(7.6) **Definition. A quadratic $\mathbb{Z}$-module**

$$M = (M_e \xrightarrow{H} M_{ee} \xrightarrow{P} M_e) \tag{1}$$

is a pair of abelian groups M_e, M_{ee} together with homorphisms H, P satisfying $PHP = 2P$ and $HPH = 2H$. For an abelian group A the **quadratic tensor product** $A \otimes_{\mathbb{Z}} M$ is the abelian group generated by the symbols $a \otimes m$ and $[a,b] \otimes n$ with a, $b \in A$, $m \in M_e$, $n \in M_{ee}$ satisfying the following relations where $T = HP - 1$.

$$(a+b) \otimes m = a \otimes m + b \otimes m + [a,b] \otimes H(m),$$

$$a \otimes (m + m') = a \otimes m + a \otimes m',$$

$$[a,a] \otimes n = a \otimes P(n),$$

$$[a,b] \otimes n = [b,a] \otimes T(n),$$

$[a,b] \otimes n$ is linear in each variable a, b, and n

There is a natural homorphism

$$H\colon A \otimes_{\mathbb{Z}} M \to A \otimes_{\mathbb{Z}} A \otimes_{\mathbb{Z}} M_{ee} \tag{2}$$

with $H(a \otimes m) = a \otimes a \otimes H(m)$ and with $H([a,b] \otimes n) = a \otimes b \otimes n + b \otimes a \otimes T(n)$. A homorphism $f\colon A \to A'$ induces the homomorphism $f \otimes 1$: $A \otimes_{\mathbb{Z}} M \to A' \otimes_{\mathbb{Z}} M$ with $f(a \otimes m) = f(a) \otimes m$ and $f([a,b] \otimes n) = [fa, fb] \otimes n$. If A is a π-module then also $A \otimes_{\mathbb{Z}} M$ is one by $(a \otimes m)^\alpha = (a^\alpha) \otimes m$ and $([a,b] \otimes n)^\alpha = [a^\alpha, b^\alpha] \otimes n$, $\alpha \in \pi$.

The following examples show that quadratic $\mathbb{Z}$-modules and quadratic tensor products appear naturally in algebra and topology. The quadratic functors Γ, Λ^2, $\hat{\otimes}^2$ and $\otimes^2$ in (I.4.2) satisfy the natural equations

$$\Gamma(A) = A \otimes_{\mathbb{Z}} \mathbb{Z}^\Gamma \quad \text{with } \mathbb{Z}^\Gamma = (\mathbb{Z} \xrightarrow{1} \mathbb{Z} \xrightarrow{2} \mathbb{Z}),$$

$$\Lambda^2(A) = A \otimes_{\mathbb{Z}} \mathbb{Z}^\Lambda \quad \text{with } \mathbb{Z}^\Lambda = (0 \longrightarrow \mathbb{Z} \longrightarrow 0),$$

$$\hat{\otimes}^2(A) = A \otimes_{\mathbb{Z}} P(1) \quad \text{with } P(1) = (\mathbb{Z}/2 \xrightarrow{0} \mathbb{Z} \xrightarrow{1} \mathbb{Z}/2),$$

$$\otimes^2(A) = A \otimes_{\mathbb{Z}} \mathbb{Z}^\otimes \quad \text{with } \mathbb{Z}^\otimes = (\mathbb{Z} \xrightarrow{(1,1)} \mathbb{Z} \oplus \mathbb{Z} \xrightarrow{(1,1)} \mathbb{Z}).$$

Moreover we use in (7.5) above the following topological example in case $m = 2n - 1$. For $m < 3n - 2$ one has the quadratic $\mathbb{Z}$-module

$$\pi_m\{S^n\} = (\pi_m(S^n) \xrightarrow{H} \pi_m(S^{2n-1}) \xrightarrow{P} \pi_m(S^n)) \tag{3}$$

which is given by homotopy groups of spheres, by the Hopf invariant H and by the Whitehead product square $[i_n, i_n]: S^{2n-1} \to S^n$ which induces $P = [i_n, i_n]_*$. Using results in Toda one gets for example the following list of quadratic $\mathbb{Z}$-modules $\pi_{2n-1}\{S^n\}$, $n \leq 6$.

$$\pi_3\{S^2\} = \mathbb{Z}^\Gamma$$

$$\pi_5\{S^3\} = \mathbb{Z}^\Lambda \oplus \mathbb{Z}/2$$

$$\pi_7\{S^4\} = E^\Gamma \oplus \mathbb{Z}/3, \quad E^\Gamma = (\mathbb{Z} \oplus \mathbb{Z}/4 \xrightarrow{(1,0)} \mathbb{Z} \xrightarrow{(2,-1)} \mathbb{Z} \oplus \mathbb{Z}/4)$$

$$\pi_9\{S^5\} = P(1)$$

$$\pi_{11}\{S^6\} = \mathbb{Z}^S = (\mathbb{Z} \xrightarrow{2} \mathbb{Z} \xrightarrow{1} \mathbb{Z})$$

Here we use the fact that an abelian group C, like $C = \mathbb{Z}/2$ or $C = \mathbb{Z}/3$, yields the quadratic $\mathbb{Z}$-module $(C \to 0 \to C)$ as well denoted by C for which the quadratic tensor product is just the tensor product with C. Moreover we observe that direct sums of quadratic $\mathbb{Z}$-modules, M_1, M_2 are defined in an obvious fashion satisfying the natural isomorphism

$$A \otimes_{\mathbb{Z}} (M_1 \oplus M_2) = A \otimes_{\mathbb{Z}} M_1 \oplus A \otimes_{\mathbb{Z}} M_2.$$

Whence we get by the list above for example the natural isomorphisms

$$A \otimes_{\mathbb{Z}} \pi_3\{S^2\} = A \otimes \mathbb{Z}^\Gamma = \Gamma(A),$$

$$A \otimes_{\mathbb{Z}} \pi_5\{S^4\} = A \otimes (\mathbb{Z}^\Lambda \oplus \mathbb{Z}/2) = \Lambda^2(A) \oplus A \otimes \mathbb{Z}/2,$$

Such functors appear in proposition (7.5) since we have for one point unions of n-spheres the natural isomorphism

$$\text{(7.7)} \qquad \pi_m\Big(\bigvee_\alpha S^n\Big) = \pi_n\Big(\bigvee_\alpha S^n\Big) \otimes_{\mathbb{Z}} \pi_m\{S^n\}, \qquad m < 3n - 2,$$

where $\pi_n(\bigvee_\alpha S^n) = \bigoplus_\alpha \mathbb{Z}$ is a free abelian group, compare Baues (QF). For $X \in P_n^0(\pi)$ the secondary boundary $b: H_{2n}\hat{X} \to \Gamma_{2n-1}X$ in Whiteheads exact sequence gives us (by use of (7.5)) the **boundary invariant**

$$\text{(7.8)} \qquad b[X] \in \pi_n(X) \otimes_{\mathbb{Z}} \pi_{2n-1}\{S^n\}.$$

Via (7.7) the element $b[X]$ corresponds to the attaching map f in (7.4) where we set $Y = \hat{X}$. Clearly the element $b[X]$ is fixed under the action of $\pi_1(X)$

since $\pi_1(X)$ acts trivially on $H_{2n}\hat{X} = \mathbb{Z}$. Moreover the element

$$(7.9) \qquad Hb[X] \in \pi_n(X) \otimes \pi_n(X)$$

corresponds to the **intersection form** on $\hat{X}$ via $\pi_n(X) = H_n\hat{X}$. Here we use H in (7.5)(3) and $\pi_{2n-1}(S^{2n-1}) = \mathbb{Z}$.

For $Y \in P_n^0$ the element $b[Y]$ is a complete homotopy invariant, that is, for $Y, Y' \in P_n^0$ there is an orientation preserving homotopy equivalence $Y \simeq Y'$ if and only if there is an isomorphism $\psi\colon H_nY \cong H_nY'$ with $\psi b[Y] = b[Y']$. The intersection form $Hb[Y]$ needs not to be a complete invariant since H might not be injective, (though H, for example, is injective for $n = 2$).

In case the fundamental group π is non trivial we get, in addition to the boundary invariant $b[X]$ for X in $P_n^0(\pi)$, the **first k-invariant** of X which is an element

$$(7.10) \qquad k(X) \in \hat{H}^{n+1}(\pi_1(X), \pi_n(X)).$$

Now the question arises whether the quadruple

$$Q(X) = (\pi_1(X), \pi_n(X), k(X), b[X])$$

is a complete homotopy invariant of X. By results of Hambleton-Kreck and Bauer this is actually true for $n = 2$ and for special groups π:

(7.11) **Theorem.** *Suppose the 2-Sylow subgroup of the finite group π has 4-periodic cohomology. Then the homotopy type of $X \in P_2^0(\pi)$ is determined by the quadruple $Q(X)$ above where $b[X]$ is given by the intersection form of X.*

Here 'determined' means that complexes $X, Y \in P_2^0(\pi)$ admit an orientation preserving homotopy equivalence $X \simeq Y$ iff there is an isomorphism (φ, f): $Q(X) \cong Q(Y)$; this is an isomorphism $(\varphi, f)\colon (\pi_1 X, \pi_n X) \cong (\pi_1 Y, \pi_n Y)$ in $\mathbf{Mod}_{\mathbb{Z}}^{\wedge}$ which is compatible with the invariants: $\varphi_* k(X) = f^* k(Y)$ and $f_* b[X] = b[Y]$.

The theorem corresponds to the following more general result on the homotopy classification of complexes in $P_n^0(\pi)$, $n \geq 2$. We say that a module (π, π_n) in $\mathbf{Mod}_{\mathbb{Z}}^{\wedge}$ is ***n*-tame** if the homology groups

$$\hat{H}_{k-1}(\pi, \pi_n \otimes \pi_{2n-k}(S^n))$$

are trivial for $k = 2, \ldots, n-1$. For example if the order $|\pi|$ of the group π has no prime factor p with $2p \leq n+1$ then (π, π_n) is n-tame for all π_n. This follows since in this case the greatest common divisor $(|\pi|, |\pi_{2n-k}(S^n)|)$ is trivial, compare for example Toda. We point out that all modules (π, π_2) are 2-tame since there is no prime p with $2p \leq 3$. Moreover for $n \in \{3, 4\}$ all modules (π, π_n) are n-tame for $|\pi|$ odd. The next result is an easy application of the properties of the CW-tower in (3.3).

(7.12) **Theorem.** *Let $X \in P_n^0(\pi)$ where π is finite and suppose $(\pi_1 X, \pi_n X)$ is n-tame. Moreover suppose that the $\pi_1(X)$-module $M = \pi_n(X) \otimes_{\mathbb{Z}} \pi_{2n-1}\{S^n\}$ has an injective transfer map tr: $M \otimes_{\pi_1(x)} \varepsilon^*\mathbb{Z} \to M$. Then the homotopy type of X is determined by the pair $(\hat{C}_* X, b[X])$.*

Here 'determined' means that two complexes X, $Y \in P_n^0(\pi)$ which satisfy the assumption in (7.12) admit an orientation preserving homotopy equivalence $X \simeq Y$ iff there is an isomorphism $\varphi\colon \pi_1 X \cong \pi_1 Y$ and a φ-equivariant chain map $\xi\colon \hat{C}_* X \to \hat{C}_* Y$ which induces an isomorphism $H_*(\xi)$ in homology with

$$(7.13) \qquad \begin{cases} H_{2n}(\xi)[\hat{X}] = [\hat{Y}] \quad \text{and} \\ (H_n(\xi) \otimes 1) b[X] = b[Y]. \end{cases}$$

Here we use the identification $H_n \hat{C}_* X = H_n \hat{X} = \pi_n X$.

Proof of (7.12). For ξ above we can choose a map $\{(\xi, \eta)\} \in [X, Y]_\varphi^3$ in $\mathbf{H}_3^c/\simeq$; this follows from (III.2.9) and (III.2.12) below. Now the assumption that (π_1, π_n) is n-tame implies that certain obstruction groups in (3.9) vanish. Here we use the Poincaré duality isomorphism

$$\hat{H}^{2n-k+1}(X, \varphi^*\Gamma_{2n-k} Y) \cong \hat{H}_{k-1}(X, \varphi^*\Gamma_{2n-k} Y) \tag{1}$$

and the isomorphism $(k < n)$

$$\begin{aligned} \hat{H}_{k-1}(X, \varphi^*\Gamma_{2n-k} Y) &\cong \hat{H}_{k-1}(\pi_1(X), \varphi^*\Gamma_{2n-k} Y) \\ &\cong \hat{H}_{k-1}(\pi_1(Y), \Gamma_{2n-k}(Y)). \end{aligned} \tag{2}$$

We obtain (2) by the map $X \to K(\pi_1 X, 1)$ which induces the identity on π_1 and by φ^*. By (7.5) we see that the obstruction groups (1) vanish for $k \geq 2$. Whence the CW-tower shows that there is a map $\{(\xi, \bar{\eta})\} \in [X, Y]_\varphi^{2n-1}$. Now we get the obstruction

$$\mathcal{O}_{X,Y}(\xi, \bar{\eta}) \in \hat{H}^{2n}(X, \varphi^*\Gamma_{2n-1}(Y)) \tag{3}$$

which satisfies the formula (5.4)(1) with respect to the map μ. This map is part of the following commutative diagram

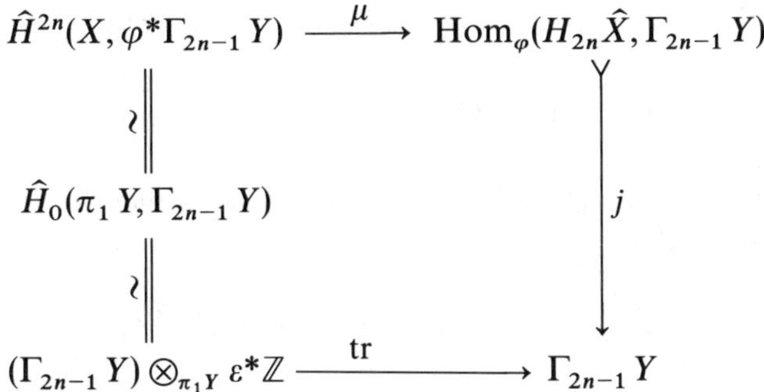

where we use the isomorphisms in (1) and (2). Clearly the inclusion j carries α to $\alpha[\hat{X}]$. The assumption on the transfer in (7.12) shows by (7.5) that the transfer in the diagram is injective and whence μ is injective. Now (7.13) and (5.4)(1) imply that the obstruction element (3) is actually trivial. Thus ξ is realizable by an orientation preserving homotopy equivalence $X \simeq Y$ and the proof of (7.12) is complete. □

(7.14) **Remark.** The assumption on π in (7.11) is used for the proof that for $n = 2$ the transfer map in (7.12) is injective. Moreover in this case the chain complex $\hat{C}_* X$ is up to homotopy equivalence determined by the k-invariant $k(X)$, so that (7.11) can be deduced from (7.12).

If we want to apply theorem (7.12) it is useful to have some knowledge on the possible modules $\pi_n X$ for $X \in P_n^0(\pi)$; to this end the following notation is convenient.

(7.15) **Definition.** Let $\Lambda = \mathbb{Z}[\pi]$ be the group ring of a finite group and let M be a Λ-module which is free as a $\mathbb{Z}$-module, that is a Λ-lattice. We write $\underset{\sim}{\Omega} M$, $\underset{\sim}{S} M$ for any Λ-modules which fit into a short exact sequence

$$\underset{\sim}{\Omega} M \rightarrowtail F \twoheadrightarrow M, \quad \text{and}$$

$$M \rightarrowtail F' \twoheadrightarrow \underset{\sim}{S} M$$

respectively where F and F' are finitely generated free Λ-modules. Up to stable isomorphism the modules $\underset{\sim}{\Omega} M$ and $\underset{\sim}{S} M$ are well defined. In particular the module $\underset{\sim}{S} M$ can be taken to be $\underset{\sim}{S} M = \mathrm{Hom}_\Lambda(\underset{\sim}{\Omega} M, \Lambda) = (\underset{\sim}{\Omega} M)^*$. Using iteration we get the modules $\underset{\sim}{\Omega}^k M$ and $\underset{\sim}{S}^k M$ which are again well defined up to stable isomorphism.

(7.16) **Example.** Let U be a finite n-dimensional CW-complex in the tree $\mathrm{HT}(\pi, n)$, see (6.2). Then the π_1-module $\pi_n(U) \cong H_n(\hat{U})$ depends up to stable isomorphism only on π, see (6.3). From the exact sequence

$$H_n\hat{U} \rightarrowtail \hat{C}_n U \longrightarrow \hat{C}_{n-1} U \longrightarrow \cdots \longrightarrow \hat{C}_0 U \twoheadrightarrow \mathbb{Z}$$

we deduce that actually $\pi_n U = \underset{\sim}{\Omega}^{n+1}(\mathbb{Z})$.

(7.17) **Lemma.** *Let $X \in P_n^0(\pi)$. Then there exists an integer r and a short exact sequence of Λ-modules*

$$\underset{\sim}{\Omega}^{n+1}\mathbb{Z} \rightarrowtail \pi_n X \oplus \Lambda^r \twoheadrightarrow \underset{\sim}{S}^{n+1}\mathbb{Z}.$$

Proof. Let $C_* = \hat{C}_* X$. Then we have exact sequences

$$\underset{\sim}{\Omega}^{n+1}\mathbb{Z} \overset{i}{\rightarrowtail} C_n \longrightarrow \cdots \longrightarrow C_0 \twoheadrightarrow \mathbb{Z},$$

$$\mathbb{Z} \rightarrowtail C_{2n} \longrightarrow \cdots \longrightarrow C_n \overset{p}{\twoheadrightarrow} \underset{\sim}{S}^{n+1}\mathbb{Z}$$

as follows by Poincaré duality, see also (7.4). Moreover we have the commutative diagram of exact sequences ($\pi_n = \pi_n X$)

$$\begin{array}{ccccc} C_{n+1} & \longrightarrow & C_n & \overset{p}{\twoheadrightarrow} & \underset{\sim}{S}^{n+1}\mathbb{Z} \\ \| & & \uparrow i & & \uparrow \bar{i} \\ C_{n+1} & \longrightarrow & \underset{\sim}{\Omega}^{n+1}\mathbb{Z} & \overset{\bar{p}}{\twoheadrightarrow} & \pi_n \end{array}$$

Since this corresponds to a pull back we see that

$$\underset{\sim}{\Omega}^{n+1}\mathbb{Z} \overset{\alpha}{\rightarrowtail} \pi_n \oplus C_n \overset{\beta}{\twoheadrightarrow} \underset{\sim}{S}^{n+1}\mathbb{Z}$$

is exact where $\alpha = (\bar{p}, i)$, $\beta = (\bar{i}, -p)$, compare (III.1.1) in Hilton Stammbach. We found the case $n = 2$ of this lemma in Hambleton-Kreck. □

Chapter III

Crossed modules and homotopy systems of order 3

The category $\mathbf{H}_3^c$ of homotopy systems of order 3 is the bottom category of the CW-tower in chapter II. Recall that a homotopy system of order 3 is a triple (C, f_3, X^2) consisting of a topological part X^2 and an algebraic part (C, f_3). It is well known that the 2-dimensional CW-complex X^2 corresponds to a presentation of the fundamental group $\pi_1 X^2$ and that this presentation can be described by the boundary map

$$\partial\colon \pi_2(X^2, X^1) = \rho_2 \to \pi_1 X^1 = \rho_1 \tag{1}$$

which satisfies $j\colon \pi_2 X^2 \cong \text{kernel}(\partial)$.

Whitehead (CH) called homomorphisms $\partial\colon M \to N$ between groups with the algebraic properties of the boundary map (1) *crossed modules*. Moreover (1) is a free crossed module and ρ_1 is a free group. We now consider the function

$$(C, f_3, X^2) \mapsto (C, jf_3, \partial) \tag{2}$$

which carries a homotopy system of order 3 to the triple (C, jf_3, ∂) given by $\partial\colon \rho_2 \to \rho_1$ in (1). This triple encapsulates the algebraic data of a *crossed chain complex* ρ:

$$\cdots \xrightarrow{d_5} C_4 \xrightarrow{d_4} C_3 \xrightarrow{d_3} \rho_2 \xrightarrow{d_2} \rho_1$$

with $d_3 = jf_3$ and $d_2 = \partial$. Here ρ is "totally free" in the sense that ρ satisfies a freeness condition in each degree. We write $\rho = \rho X$ in case the homotopy system $(C, f_3, X^2) = r_3 X$ is given by a CW-complex X. Then the "homotopy groups"

$$\pi_n(\rho X) = \text{kernel}\, d_n/\text{image}\, d_{n+1}$$

of ρX satisfy the natural equation

$$\pi_n(\rho X) = \begin{cases} \pi_n X & \text{for } n = 1, 2, \\ \hat{h}\pi_3 X & \text{for } n = 3, \\ H_n \hat{X} & \text{for } n \geq 4 \end{cases}$$

where $\hat{h}\colon \pi_3 X \to H_3 \hat{X}$ is the (surjective) Hurewicz homomorphism.

The crossed chain complexes may be thought of as chain complexes with operators from a group but with non abelian features in degree one and two. The somewhat more intricate non abelian part corresponds to the topological part X^2 of a homotopy system of order 3. The main observation in this chapter states that the function (2) yields equivalences of categories, see (2.9),

$$\rho\colon \mathbf{H}_3^c \xrightarrow{\sim} \mathbf{H}, \qquad \rho\colon \mathbf{H}_3^c/\simeq \xrightarrow{\sim} \mathbf{H}/\simeq \tag{3}$$

where $\mathbf{H}$ is the category of totally free crossed chain complexes. This shows that the bottom category $\mathbf{H}_3^c$ in the CW-tower can be replaced by the algebraic category $\mathbf{H}$.

Totally free crossed chain complexes are the same as the "homotopy systems" of Whitehead (CH); by the equivalence (3) they correspond to homotopy systems of order 3. In this sense the homotopy systems of order n in chapter II are generalizations of the homotopy systems of Whitehead which he introduced in his classical paper "Combinatorial homotopy II".

The functor

$$\rho\colon \mathbf{CW} \xrightarrow{r_3} \mathbf{H}_3^c \xrightarrow{\sim} \mathbf{H} \tag{4}$$

which carries a CW-complex X to its crossed chain complex ρX is part of the CW-tower of categories so that the category $\mathbf{H}/\simeq$ can be considered as the "first algebraic approximation" of the homotopy category $\mathbf{CW}/\simeq$ of connected CW-complexes.

We show that the functor ρ in (4) has further nice properties. In particular ρ carries homotopy push outs in $\mathbf{CW}$ to homotopy push outs in $\mathbf{H}$, see § 6. This result generalizes the well known Van Kampen theorem for fundamental groups; it allows the computation of the homotopy group π_2 of a union of spaces. For the definition of homotopy push outs in $\mathbf{H}$ we observe that $\mathbf{H}$ has in a canonical way the structure of a cofibration category. Therefore all basic notions of homotopy theory make sense in the category $\mathbf{H}$, see § 4. Moreover the functor ρ carries a cylinder in $\mathbf{CW}$ to an algebraic cylinder in $\mathbf{H}$. In addition ρ carries a product $X \times Y$ of CW-complexes to the tensor product $\rho X \otimes \rho Y$ of crossed chain complexes, see § 9. These very useful facts are due to Brown Higgins (TP), (CC).

Using the functor ρ in (4) we give an elementary proof of the equivalence of categories (see § 8)

$$\textbf{2-types} \xrightarrow{\sim} \mathrm{Ho}(\textbf{cross}) \tag{5}$$

where 2-**types** is the full homotopy category of 2-types and where Ho(**cross**) is the localization of the category of crossed modules with respect to weak equivalences. The equivalence (5) essentially goes back to the work of J.H.C. Whitehead who pointed out in Whitehead (C) that one has to consider the hierachy of categories

$$\textbf{1-types}, \qquad \textbf{2-types}, \qquad \textbf{3-types}, \ldots \tag{6}$$

Here a 1-type can be identified with an abstract group. The equivalence (5) shows that a 2-type is represented algebraically by a crossed module. In the next chapter we shall see that a 3-type is represented by a new kind of algebraic object which we call a "quadratic module".

We now summarize the basic properties of crossed modules which indeed show that they are an *optimal algebraic system* representing a 2-type.

(a) A crossed module is an algebraic model of a 2-type by an equivalence of categories as above, see (8.2).

(b) A crossed module is an algebraic model of a 2-type with minimal Peiffer nilpotency degree, see (1.10).

(c) A totally free crossed module is an algebraic model of a 2-dimensional CW-complex by an equivalence of categories, see (7.1).

(d) Crossed modules form the degree 2 part of crossed chain complexes which are algebraic models of homotopy systems of order 3 by an equivalence of categories, see (4.12) and (2.9).

(e) The category of crossed chain complexes satisfies the axioms of a cofibration structure and whence a homotopy theory is available in this category, see (4.10).

(f) The functor from spaces to crossed chain complexes carries weak equivalences to weak equivalences and homotopy push outs to homotopy push outs, see (6.12). Moreover this functor carries products to tensor products, see (9.3).

(g) Homotopy types of 3-dimensional totally free crossed chain complexes are in 1-1 correspondence with homotopy types of 3-dimensional connected CW-complexes. See (II. 3.14) (2) with $N = 3$ and (2.9). Moreover the cells of the CW-complex are the generators of the corresponding totally free crossed chain complex.

(h) Each 4-dimensional totally free crossed chain complex ρ can be realized, that is, there is a 4-dimensional CW-complex X^4 together with an isomorphism $\rho \cong \rho(X^4)$ of crossed chain complexes.

Most of the results in this chapter are essentially known in the literature, in particular we refer the reader to chapter VI in Baues (AH) where one finds a further account of crossed chain complexes and homotopy systems. In order to make this book self contained we recall all the basic definitions and facts from the literature which we shall use. The concepts and results in this chapter will serve as a guide in the next chapter where we study in an analogous fashion homotopy systems of order 4 and 3-types.

§ 1 Nilpotent groups and Peiffer nilpotent pre-crossed modules

The algebraic properties of the boundary homorphism of homotopy groups,

$$\partial\colon \pi_2(X, Y) \to \pi_1(Y),$$

lead to the notion of a crossed module. In § 8 below we shall see that each 2-type can be algebraically represented by such a crossed module. Moreover it is a classical result of J.H.C. Whitehead that the boundary map ∂ above is actually a free crossed module if X is obtained from Y by attaching 2-cells. Such a free crossed module can be defined in a usual fashion by a universal property. In this section we also describe an explicit construction of a free crossed module by first constructing a free pre-crossed module and then by dividing out the subgroup of Peiffer commutators.

From our point of view pre-crossed modules are generalizations of groups and crossed modules are generalizations of abelian groups. Moreover the notion of a Peiffer commutator in a pre-crossed module corresponds to the notion of a commutator in a group. Therefore one is as well led to consider the Peiffer central series of a pre-crossed module which generalizes the central series of a group. Using this central series one defines a "nilpotency class" of a pre-crossed module similarly as the nilpotency class of a group. The abelian groups and crossed modules respectively have just nilpotency class 1. For the purposes of this chapter we only need crossed modules; in the next chapter IV, however, pre-crossed modules of nilpotency class 2 play an essential role.

A **pre-crossed module** $\partial\colon M \to N$ is a group homorphism together with an action of N on M (see (I.1.5)) such that

(1.1) $$\partial(x^\alpha) = -\alpha + \partial(x) + \alpha, \quad \alpha \in N.$$

This is a **crossed module** if in addition

(1.2) $$x^{\partial y} = -y + x + y, \quad x, y \in M.$$

For example the boundary homomorphism

(1.3) $$\partial\colon \pi_2(Y, X) \to \pi_1(X)$$

in (I. 1.4) is a crossed module with respect to the action of $\pi_1(X)$ on $\pi_2(Y, X)$. A map between pre-crossed modules, $(m, n)\colon \partial \to \partial'$, is given by a commutative diagram

(1.4) $$\begin{array}{ccc} M & \xrightarrow{m} & M' \\ {\scriptstyle\partial}\big\downarrow & & \big\downarrow{\scriptstyle\partial'} \\ N & \xrightarrow[n]{} & N' \end{array}$$

in the category of groups where m is n-equivariant. Let **precross** be the category of pre-crossed modules and of such maps. Moreover, let **cross** be the full subcategory consisting of crossed modules. For the category **Gr** of groups we have the full inclusion

(1.5) $$\mathbf{Gr} \subset \mathbf{precross}.$$

Here a group G in **Gr** is identified with the trivial pre-crossed module $G \to 0$. In this sense pre-crossed modules form a generalization of groups. Moreover $G \to 0$ is a crossed module if and only if G is an abelian group, see (1.2). Therefore crossed modules generalize abelian groups.

We define the **commutator** in a group G by

(1.6) $$(x, y) = -x - y + x + y \quad (x, y \in G)$$

and we define the **Peiffer commutator** in a pre-crossed module $\partial\colon M \to N$ by

(1.7) $$\langle x, y\rangle = -x - y + x + y^{\partial x} \quad (x, y \in M).$$

Thus ∂ is a crossed module if and only if $\langle x, y\rangle = 0$ for all x, $y \in M$.

In a group G we have the **lower central series**

(1.8) $$\Gamma_{n+1} \subset \Gamma_n \subset \cdots \subset \Gamma_1 = G$$

where $\Gamma_n = \Gamma_n(G)$ is the subgroup of G generated by all iterated commutators $(x_1, \ldots, x_n)_c$ of length n. Here $\Gamma_2(G)$ is the **commutator subgroup** of G. Similarly we obtain the **lower Peiffer central series**

(1.9) $$P_{n+1} \subset P_n \subset \cdots \subset P_1 = M$$

in a pre-crossed module $\partial\colon M \to N$. Here $P_n = P_n(\partial)$ is the subgroup of M generated by all iterated Peiffer commutators $\langle x_1, \ldots, x_n \rangle_c$ of length n. The group $P_2(\partial)$ is the **Peiffer subgroup** of M, this generalizes the commutator subgroup in a group. Clearly (1.8) is a special case of (1.9) when we set $N = 0$. One readily checks that the groups Γ_n and P_n are normal subgroups of G and M respectively. We refer the reader to Baues-Conduché for further properties of the Peiffer central series.

(1.10) **Definition.** (1) A group G is **nilpotent of class** n if $\Gamma_{n+1}(G) = 0$, in this case we call G a **nil(n)-group**. Let **Ab**(n) be the full subcategory of **Gr** consisting of nil(n)-groups.

(2) A pre-crossed module $\partial\colon M \to N$ is **Peiffer nilpotent of class** n if $P_{n+1}(\partial) = 0$, in this case we call ∂ a **nil(n)-module**. Let **cross** (n) be the full subcategory of **precross** consisting of nil(n)-modules.

Thus we have full inclusion of categories

$$\mathbf{Ab} = \mathbf{Ab}(1) \subset \cdots \subset \mathbf{Ab}(n) \subset \mathbf{Ab}(\infty) = \mathbf{Gr},$$

$$\mathbf{cross} = \mathbf{cross}(1) \subset \cdots \subset \mathbf{cross}(n) \subset \mathbf{cross}(\infty) = \mathbf{precross}$$

where $n \geq 1$. These inclusions admit the retractions

$$(1.11) \qquad \begin{cases} r_n\colon \mathbf{Gr} \to \mathbf{Ab}(n), \\ r_n\colon \mathbf{precross} \to \mathbf{cross}(n) \end{cases}$$

where $r_n(G) = G/\Gamma_{n+1}(G)$ and where $r_n(\partial)$ is given by the natural commutative diagram

$$\begin{array}{ccc} M & \xrightarrow{q} & M/P_{n+1}(\partial) \\ \partial \downarrow & \swarrow r_n(\partial) & \\ N & & \end{array} \qquad (1)$$

Here the quotient map q of groups is equivariant with respect to the action of N. The group

$$r_2(G) = G^{ab} = G/\Gamma_2(G) \qquad (2)$$

is the **abelianization of** G and

$$(3) \qquad r_2(\partial) = \partial^{cr} : M^{cr} = M/P_2(\partial) \to N$$

is the **crossed module associated to** ∂.

Next we consider free objects. Let $F = \langle Z \rangle$ be the free group generated by the set Z. The **free nil(n)-group** generated by Z is given by

(1.12) $$r_n(F) = \langle Z \rangle / \Gamma_{n+1}(\langle Z \rangle).$$

This is a free object in the category **Ab**(n), in particular, we obtain for $n = 1$ the free abelian group

(1.13) $$\mathbb{Z}[Z] = \bigoplus_Z \mathbb{Z} = \langle Z \rangle / \Gamma_2(\langle Z \rangle)$$

generated by Z. Next we define free objects in the category **cross**(n) as follows.

(1.14) **Definition.** Let F be a free group and let $f: F \to N$ be a homomorphism between groups. We say that a nil(n)-module $\partial: M \to N$ is a **free nil(n)-module with basis** f if a homomorphism $i: F \to M$ is given such that $\partial i = f$ and such that the following universal property is satisfied. For each nil(n)-module $\partial': M' \to N'$ and for each commutative diagram of unbroken arrows

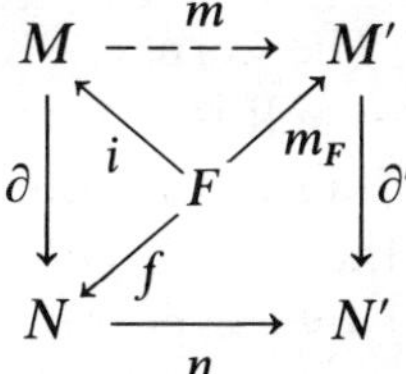

in the category of groups there is a unique map $(m, n): \partial \to \partial'$ in **cross**(n) with $mi = m_F$. Clearly ∂ is a **free crossed module** for $n = 1$ and ∂ is a free **pre-crossed module** for $n = \infty$. For $F = \langle Z \rangle$ the homomorphism f is determined by the function $f|Z: Z \to N$. In this case we call $f|Z$ as well a basis of ∂. We say that a nil(n)-module $\partial: M \to N$ is **totally free** if ∂ is free as above and if N is a free group.

A free nil(n)-module with basis $f: F = \langle Z \rangle \to N$ can be constructed as follows. The **free N-group** generated by Z is the free group $\langle Z \times N \rangle$ where $Z \times N$ is the product set. The action of $\beta \in N$ on $\langle Z \times N \rangle$ is given on generators $(x, \alpha) \in Z \times N$ by

(1.15) $$(x, \alpha)^\beta = (x, \alpha + \beta).$$

A **free pre-crossed module** $\partial = \partial_f$ with basis $f: F \to N$ is given by the commutative diagram of groups

(1.16)
$$\begin{array}{ccc} M = \langle Z \times N \rangle & \xleftarrow{\ i\ } & F = \langle Z \rangle \\ \downarrow \partial_f & \swarrow f & \\ N & & \end{array}$$

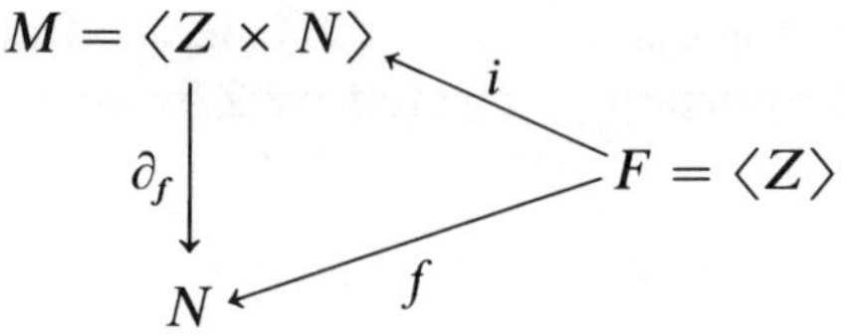

where i is the identity on Z and where

$$\partial_f(x, \alpha) = -\alpha + f(x) + \alpha \tag{1}$$

on generators $(x, \alpha) \in Z \times N$. We associate with ∂_f the following commutative diagram of groups in which the row is a short exact sequence

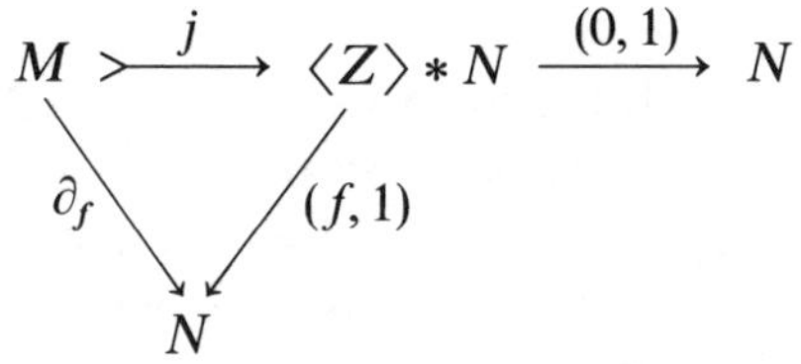

$$\begin{array}{ccccc} M & \overset{j}{\rightarrowtail} & \langle Z \rangle * N & \xrightarrow{(0,1)} & N \\ & \partial_f \searrow & \downarrow (f,1) & & \\ & & N & & \end{array} \tag{2}$$

The group $\langle Z \rangle * N$ is the free product of groups, $0\colon \langle Z \rangle \to 0 \in N$ is the trivial map, and 1 is the identity of N. The kernel of $(0, 1)$ is the normal subgroup of $\langle Z \rangle * N$ generated by Z; this as well is the image of the injection j given by $j(x, \alpha) = -\alpha + x + \alpha$. Clearly diagram (2) commutes by (1). Let $P_{n+1}(\partial_f) \subset \langle Z \times N \rangle$ be the n-th term of the Peiffer central series. This yields the **free nil(n)-module** $r_n(\partial_f)$ with basis f by the commutative diagram

(1.17)
$$\begin{array}{ccccc} \langle Z \times N \rangle / P_{n+1}(\partial_f) & \xleftarrow{\ q\ } & \langle Z \times N \rangle & & \\ & \searrow r_n(\partial_f) & \downarrow \partial_f & \nwarrow & \langle Z \rangle \\ & & N & \xleftarrow{\ f\ } & \end{array}$$

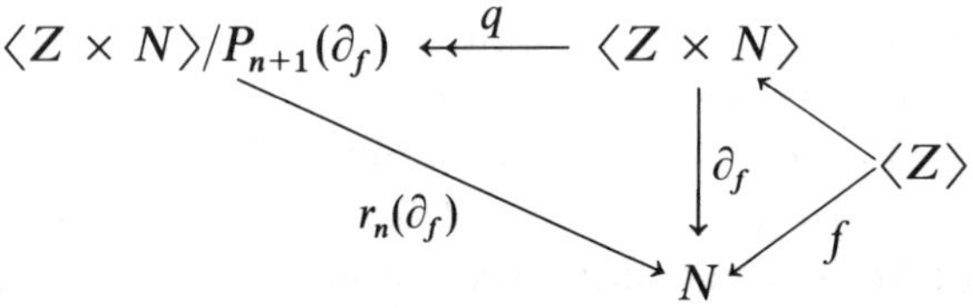

Here q is the quotient map as in (1.11) (1). In particular, we obtain for $n = 1$ the **free crossed module**

(1.18)
$$r_1(\partial_f)\colon \langle Z \times N \rangle / P_2(\partial_f) \to N$$

where $P_2(\partial_f)$ is the Peiffer subgroup of $\langle Z \times N \rangle$. We point out that the Peiffer group $P_2(\partial_f)$ is the normal subgroup generated by the Peiffer commutators

(1.19)
$$\langle x^\alpha, y^\beta \rangle = -x^\alpha - y^\beta + x^\alpha + y^{\beta - \alpha + f(x) + \alpha}$$

of generators $x^\alpha = (x, \alpha)$ and $y^\beta = (y, \beta)$ in $Z \times N$. The following example of a free crossed module is a classical result of Whitehead (CH).

(1.20) **Theorem.** *Let X be path connected and assume Y is obtained from X by attaching 2-cell. Then $\partial: \pi_2(Y, X) \to \pi_1(X)$ is a free crossed module, see* (1.3).

A basis of ∂ in (1.20) is obtained as follows. Let Z_2 be the set of 2-cells in $Y - X$ and let

$$i: Z_2 \to \pi_2(Y, X) \tag{1.21}$$

be chosen as in (I.3.11). Then $\partial i = f: Z_2 \to \pi_1(X)$ is the attaching map of 2-cells which is a basis of the free crossed module ∂ in (1.20).

§ 2 Crossed chain complexes and homotopy systems of order 3

In this section we introduce the category **cross chain** of crossed chain complexes and its subcategory **H** of totally free crossed chain complexes. Each filtered space $* \in X_1 \subset X_2 \subset \cdots \subset X$ yields canonically a crossed chain complex

$$\cdots \xrightarrow{\partial} \pi_3(X_3, X_2) \xrightarrow{\partial} \pi_2(X_2, X_1) \xrightarrow{\partial} \pi_1 X_1 \tag{1}$$

where ∂ denotes the corresponding boundary maps of homotopy groups. A crossed chain complex may be thought of as a chain complex with operators from a group but with a crossed module in degree 1 and 2. Of course the individual rules of crossed chain complexes are commonly used in homotopy theory without necessarily considering the total structure.

Remark. An extensive study of crossed chain complexes from a slightly wider point of view can be found in the work of Brown-Higgins; the crossed chain complexes here are exactly what they call "reduced crossed complexes". Brown-Higgins prefer the base point free approach, or the one which allows a set of base points; in this case the fundamental group $\pi_1 X_1$ in (1) is replaced by the fundamental groupoid of X_1. This creates some extra difficulty of formulation which is not needed in this book. Moreover, the reader is better prepared to follow the wider and definitive approach of Brown-Higgins if he first studies the important special case of crossed chain complexes.

The standard example (1) of a crossed chain complex is totally free (that is, it satisfies a freeness condition in each degree) in case the spaces X_i are the skeleta $X_i = X^i$ of a CW-complex X. Therefore we obtain a functor

$$\rho\colon \mathbf{CW} \to \mathbf{H} \tag{2}$$

which carries a CW-complex X to the crossed chain complex (1) of X. This functor was originally studied by Whitehead (CH). In this section we describe an equivalence of categories

$$\rho\colon \mathbf{H}_3^c \xrightarrow{\sim} \mathbf{H}$$

where $\mathbf{H}_3^c$ is the category of homotopy systems of order 3 in chapter II. Moreover we show that the functor ρ in (2) is naturally isomorphic to the composition of functors

$$\mathbf{CW} \xrightarrow{r_3} \mathbf{H}_3^c \xrightarrow{\sim} \mathbf{H} \tag{4}$$

where r_3 is part of the CW-tower of categories in chapter II. This shows that the bottom category of the CW-tower can be replaced by the algebraic category **H**. We also describe in this section the connection between totally free crossed chain complexes and chain complexes which shows that for many computational purposes it is enough to consider chain complexes, see theorem (2.12).

We use the following notations on crossed homomorphisms. Let π be a group and let M be a π-group. Moreover, let $\eta\colon N \to \pi$ be a homomorphism between groups. An **η-crossed homomorphism** $A\colon N \to M$ is a function which satisfies

(2.1) $$A(x + y) = (Ax)^{\eta(y)} + (Ay).$$

We call A a **crossed homomorphism** if η is the identity; this is a **derivation** if M is a π-module.

Let X be a path connected CW-complex with basepoint $* \in X^0$. The Hurewicz homomorphism $\hat{h}$ in (I.3.3) yields the following commutative diagram

(2.2) $$\begin{array}{ccccccccc} \cdots \xrightarrow{d_4} & \rho_3(X) & \xrightarrow{d_3} & \rho_2(X) & \xrightarrow{d_2} & \rho_1(X) & \xrightarrow{i} & \pi_1(X) \\ \cong & \downarrow h_3 & & \downarrow h_2 & & \downarrow h_1 & & \\ \longrightarrow & \hat{C}_3X & \longrightarrow & \hat{C}_2X & \xrightarrow[d_2]{} & \hat{C}_1X & \longrightarrow & \hat{C}_0X \end{array}$$

where for $n \geq 2$

$$(2.3) \qquad h_n = \hat{h}: \rho_n(X) = \pi_n(X^n, X^{n-1}) \to \hat{C}_n X = H_n(\hat{X}^n, \hat{X}^{n-1})$$

and where $d_n = j\partial$ is given by the operators in the exact sequence (I.1.4). For $n = 1$ let h_1 be given by the composition

$$h_1 = j_0\hat{h}: \rho_1(X) = \pi_1(X^1) \to H_1(\hat{X}^1, p^{-1}*) \to \hat{C}_1 X$$

where j_0 is induced by the inclusion $* \subset X^0$. Clearly $d_2 = \partial$ in (2.2) is the boundary map and the surjective homomorphism i in (2.2) is induced by the inclusion $X^1 \subset X$. The bottom sequence of (2.2) is the cellular chain complex $\hat{C}_*(X)$ of the universal covering, see (I.2.7). The Hurewicz Theorem yields the next result on the functions h_n in (2.3).

(2.4) **Proposition.** *h_n is an isomorphism of $\pi_1(X)$-modules for $n \geq 3$ and h_2 is a surjective i-equivariant homomorphism for which* kernel$(h_2) = \Gamma_2(\rho_2(X))$ *is the commutator subgroup. The map h_2 induces an isomorphism of $\pi_1(X)$-modules*

$$h_2: \pi_2(X^2) = kernel(\partial) \xrightarrow{\cong} H_2(\hat{X}^2) = kernel(d_2).$$

Moreover, h_1 is an i-crossed homomorphism.

Compare Whitehead (CH). The groups in (2.2) satisfy the following freeness properties, see (1.20).

(2.5) **Proposition.** *$\rho_n(X)$, $n \geq 3$, and $\hat{C}_n(X)$, $n \geq 0$, are free $\pi_1(X)$-modules, $d_2: \rho_2(X) \to \rho_1(X)$ is a free crossed module and $\rho_1(X)$ is a free group.*

We have seen in (I.3.11) that h_n carries a basis Z_n to a basis, $n \geq 2$. This is also true for $n = 1$ provided $X^0 = *$ is a point.

The top row of (2.2) motivates the following definition.

(2.6) **Definition. A crossed chain complex** $\rho = \{\rho_n, d_n\}$ is a sequence of homomorphisms between groups

$$\cdots \longrightarrow \rho_3 \xrightarrow{d_3} \rho_2 \xrightarrow{d_2} \rho_1 \overset{i}{\twoheadrightarrow} \pi \qquad (1)$$

such that $d_{n-1}d_n = 0$ for $n \geq 3$ and such that the following properties are satisfied. The homomorphism d_2 is a crossed module with $\pi = $ cokernel(d_2); hence kernel(d_2) is a π-module. Moreover, ρ_n is a π-module $(n \geq 3)$ and d_n $(n \geq 4)$ and $d_3: \rho_3 \to$ kernel(d_2) are homomorphisms between π-modules. A **crossed chain map** $f: \rho \to \rho'$ between crossed chain complexes is a family of homomorphisms between groups $(n \geq 1)$

$$f_n\colon \rho_n \to \rho_n' \quad \text{with } f_{n-1}d_n = d_n' f_n, \tag{2}$$

such that f_2 is f_1-equivariant and such that f_n is $\bar{f}_1$-equivariant, where $\bar{f}_1\colon \pi \to \pi'$ is induced by f_1, $n \geq 3$. Hence (f_2, f_1) is a map between crossed modules. Let **crosschain** be the category of crossed chain complexes and of crossed chain maps.

For morphisms $f, g\colon \rho \to \rho'$ in **crosschain** we define a **homotopy** $\alpha\colon f \simeq g$ as follows. The homotopy α is given by a sequence of functions

$$\alpha = \alpha_n : \rho_n \to \rho_{n+1}' \quad (n \geq 1) \tag{3}$$

satisfying

$$\left.\begin{aligned} -f_1 + g_1 &= d_2'\alpha_1, \\ -f_n + g_n &= d_{n+1}'\alpha_n + \alpha_{n-1}d_n \quad (n \geq 2). \end{aligned}\right\} \tag{4}$$

Here α_n is a homomorphism between groups for $n \geq 2$ which is f_1-equivariant for $n = 2$ and which is $\bar{f}_1$-equivariant for $n > 2$. Moreover, α_1 is an f_1-crossed homomorphism. Next we define for a crossed chain complex ρ the **homotopy groups**

$$\left.\begin{aligned} \pi_1(\rho) &= \pi = \operatorname{cokernel}(d_2) \\ \pi_n(\rho) &= \operatorname{kernel}(d_n)/\operatorname{image}(d_{n+1}), \quad n \geq 2. \end{aligned}\right\} \tag{5}$$

Clearly, $\pi_n(\rho)$ is a $\pi_1(\rho)$-module for $n \geq 2$. A crossed chain map $f\colon \rho \to \rho'$ is a **weak equivalence** if f induces isomorphisms

$$\pi_n(f)\colon \pi_n(\rho) \cong \pi_n(\rho') \quad \text{for } n \geq 1. \tag{6}$$

A crossed chain complex ρ is n-**dimensional** if $\rho_i = 0$ for $i > n$. The n-**skeleton** of ρ, denoted by ρ^n, is the n-dimensional crossed chain complex given by $\{d_i\colon \rho_i \to \rho_{i-1}, i \leq n\}$. Clearly the category **cross** of crossed modules is the full subcategory of **crosschain** consisting of 2-dimensional objects.

(2.7) **Definition.** We call a crossed chain complex ρ **totally free** if ρ_1 is a free group, if $d_2\colon \rho_2 \to \rho_1$ is a free crossed module, see (1.18), and if ρ_n is a free π-module for $n \geq 3$. Let $\mathbf{H} \subset$ **cross chain** be the full subcategory of totally free crossed chain complexes and let $\mathbf{H}/\simeq$ be the homotopy category of $\mathbf{H}$, see (2.6).

We now consider the functors

$$\begin{array}{ccc} & \mathbf{CW}/\overset{0}{\simeq} & \\ {}^{r_3}\swarrow & & \searrow^{\rho} \\ \mathbf{H}_3^c & \xrightarrow[\rho]{} & \mathbf{H} \end{array} \tag{2.8}$$

Here r_3 is defined in (II.2.2) and ρ carries a CW-complex X to $\rho(X)$ in (2.2). Moreover, the functor ρ on $\mathbf{H}_3^c$ carries the homotopy system (C, f_3, X^2) of order 3 to the crossed chain complex

$$\cdots \longrightarrow C_4 \xrightarrow{d_4} C_3 \xrightarrow{jf_3=d_3} \pi_2(X^2, X^1) \xrightarrow{\partial=d_2} \pi_1(X^1) \xrightarrow{i} \pi_1(X^2).$$

Whence we have a natural isomorphism $\rho r_3(X) \cong \rho(X)$ since h_n, $n \geq 3$, in (2.2) is an isomorphism. The functors in (2.8) induce functors between homotopy categories, see (3.4) below. The next result is proved in (IV.S.13) Bauses (AH).

(2.9) **Theorem.** *The functors ρ: $\mathbf{H}_3^c \to \mathbf{H}$ and ρ: $\mathbf{H}_3^c/\simeq \to \mathbf{H}/\simeq$ are equivalences of categories.*

Whence we can replace the bottom category $\mathbf{H}_3^c$ in the CW-tower by the algebraic category $\mathbf{H}$.

(a) **Remark.** The functor ρ: $\mathbf{CW}/\simeq \to \mathbf{H}/\simeq$ also satisfies the strong sufficiency condition, compare the notation in (I.0.7). This fact corresponds to theorem 17 in Whitehead (SH), a proof can be found in (VI.7.5) Baues (AH).

Next we want to compare crossed chain complexes with chain complexes. For this we need the following notations.

(2.10) **Definition.** A chain complex (π, C) is **reduced** if $C_0 = \mathbb{Z}[\pi]$ is the group ring with the generator $*$ and if $C_i = 0$ for $i < 0$. A chain map (φ, f): $C \to C'$ in $\mathbf{Chain}_{\mathbb{Z}}^{\wedge}$ is reduced if $f_0 = \varphi_\#$ is induced by φ, see (I.1.5)(1). Moreover a chain homotopy α as in (2.6) is reduced if $\alpha_0 = 0$. Let $\mathbf{H}_1$ be **the subcategory of reduced free chain complexes** in $\mathbf{Chain}_{\mathbb{Z}}^{\wedge}$. Morphisms in $\mathbf{H}_1$ are reduced chain maps and homotopies in $\mathbf{H}_1$ are reduced homotopies.

For example $\hat{C}_* X$ is reduced if $X^0 = *$ and if a basepoint $* \in \hat{X}$ with $p* = *$ is fixed. We now define a functor

$$C\colon \mathbf{H} \to \mathbf{H}_1. \tag{2.11}$$

We obtain this functor by choosing a basis Z_n, $Z_n \subset \rho_n$, for $n \geq 1$ in each object ρ of **H**. Let

$$C_n(\rho) = C_n = \bigoplus_{Z_n} \mathbb{Z}[\pi] \tag{1}$$

be the free π-module generated by Z_n. We get the boundaries $C_n \to C_{n-1}$ by those π-equivariant maps which make the following diagram commutative

$$\begin{array}{ccccccccc} \longrightarrow & \rho_3 & \longrightarrow & \rho_2 & \xrightarrow{\partial} & \rho_1 & \overset{i}{\twoheadrightarrow} & \pi = \text{cokernel}(\partial) \\ \cdots & \cong\downarrow h_3 & & \downarrow h_2 & & \downarrow h_1 & & \\ \longrightarrow & C_3 & \xrightarrow[d_3]{} & C_2 & \xrightarrow[d_2]{} & C_1 & \xrightarrow[d_1]{} & C_0 = \mathbb{Z}[\pi]. \end{array} \tag{2}$$

Here h_n, $h \geq 1$, is the unique function which has properties as in (2.4) and which satisfies $h_n(e) = e$ for $e \in Z_n$. In particular $C_2 = \rho_2^{ab}$ is the abelianization of ρ_2 and h_2 is the quotient map which induces an isomorphism

$$h_2\colon \text{kernel}(\partial) \xrightarrow{\cong} \text{kernel}(d_2).$$

Moreover, we set

$$d_1(e) = 1 - [i(e)] = * - *^e \tag{3}$$

for $e \in Z_1 \subset \rho_1$. Here $* = 1$ is the generator of $C_0 = \mathbb{Z}[\pi]$. For a map $f\colon \rho \to \rho'$ we obtain the unique map

$$\begin{aligned} C(f)\colon C(\rho) \to C(\rho') \quad &\text{in } \mathbf{H}_1 \text{ by the condition } C(f)_0 = \varphi_\# \quad \text{and} \\ &C(f)_n h_n = h_n f_n, \quad n \geq 1. \end{aligned} \tag{4}$$

(5) **Remark.** There is a natural isomorphism $C\rho(X) \cong \hat{C}_*(X)$ provided X is a CW-complex with trivial 0-skeleton $X^0 = *$. In this case we choose a basis $Z_n \subset \rho_n(X)$ as in (2.5). Then diagram (2.11) (2), with $\rho = \rho(X)$, is naturally isomorphic to diagram (2.2).

(2.12) **Theorem.** *The functor* $C\colon \mathbf{H} \to \mathbf{H}_1$ *is full and the induced functor* $C\colon \mathbf{H}/{\simeq} \to \mathbf{H}_1/{\simeq}$ *is full and faithful.*

This result is originally proved in Whitehead (CH); in (VI.4.9) of Baues (AH) we prove the theorem as a special case of a result on relative CW-complexes.

We now describe chain complexes which are weakly equivalent to C-realizable chain complexes where C is the functor (2.11).

(2.13) **Proposition.** *Let (π, C) be a chain complex in* $\mathbf{Chain}_{\mathbb{Z}}^{\wedge}$ *with the properties*

$$\begin{cases} C_i = 0 \quad \text{for } i < 0, \\ H_0 C = \mathbb{Z}, \quad \text{and} \\ H_1 C = 0. \end{cases}$$

Then there exists a weak equivalence $(\pi, C') \xrightarrow{\sim} (\pi, C)$ where (π, C') is a C-realizable chain complex in $\mathbf{H}_1$.

Proof. The chain complex (π, C) gives us the extension of π-modules

$$H_2 C \rightarrowtail C_2/d_3 C_3 \longrightarrow C_1 \longrightarrow C_0 \twoheadrightarrow \mathbb{Z} \tag{1}$$

which represents an element (see Hilton Stammbach)

$$k_2 \in \operatorname{Ext}^3_\pi(\mathbb{Z}, H_2 C) = H^3(\pi, H_2 C). \tag{2}$$

Let X be a CW-complex for which the first k-invariant is k_2, $X^0 = *$. Then there is a chain map

$$\psi\colon (\pi, \hat{C}_* X^3) \to (\pi, C) \tag{3}$$

which induces isomorphisms of homology groups H_i, $i \le 2$. We obtain the chain map since $\hat{C}_* X^3$ is free and since the extension (1) represents the Postnikov invariant $k_2 = k_2(X)$ if we set $C = \hat{C}_* X^3$. Now we construct inductively a free chain complex C' together with a weak equivalence $\varphi\colon C' \xrightarrow{\sim} C$, see (I.6.12) (2) in Baues (AH). Here the 2-skeleton of C' is $\hat{C}_* X^2$ and φ extends the restriction $\psi | \hat{C}_* X^2$ of ψ in (3). By (2.2) we see that (π, C') is a C-realizable chain complex in $\mathbf{H}_1$. □

The next result is an immediate consequence of (2.9), (2.12) and the CW-tower (II.3.3).

(2.14) **Theorem.** *Let X, Y be CW-complexes with $X^0 = * = Y^0$ and let X be finite dimensional. Assume that*

$$\hat{H}^p(X, \varphi^* \Gamma_n Y) = 0 \tag{$*$}$$

for $p = n, n+1, n \geq 3, \varphi \in \operatorname{Hom}(\pi_1 X, \pi_1 Y)$. *Then the functor* $\hat{C}_* = C\rho$ *yields the bijection of sets*

$$\hat{C}_*\colon [X, Y] \xrightarrow{\approx} [\hat{C}_* X, \hat{C}_* Y] \subset \operatorname{Mor} \mathbf{H}_1/\simeq.$$

Assume $(*)$ *holds for* $X = Y$ *and* $\varphi = 1$ *then*

$$\hat{C}_*\colon \operatorname{Aut}(X)^* \xrightarrow{\cong} \operatorname{Aut}(\hat{C}_* X) \subset \operatorname{Mor} \mathbf{H}_1/\simeq$$

is an isomorphism of groups.

This theorem is a special case of (VI.6.15) in Baues (AH). A result like (2.14) was used in Whitehead (CH) for the homotopy classification of lens spaces.

§ 3 Cylinders of CW-complexes and of crossed chain complexes

Cylinders are used to define homotopies; if a functor carries cylinders to cylinders we know that this functor as well carries homotopies to homotopies so that it induces a functor between homotopy categories. In this section we apply this principle to the functors

$$\hat{C}_*\colon \mathbf{CW} \xrightarrow{\rho} \mathbf{H} \xrightarrow{C} \mathbf{H}_1 \subset \mathbf{Chain}_{\hat{\mathbb{Z}}}$$

described in section § 2 above. We consider the reduced cylinder $I_* X$ of a CW-complex since $I_* X$ again has a canonical base point; homotopies $H\colon I_* X \to X$ are base point preserving homotopies. In section § 10 below we study as well homotopies which are not base point preserving.

Let X be a CW-complex with $X^0 = *$. Then the reduced cylinder $I_* X$ is a CW-complex with skeleta described in (I.2.4). We have the **structure maps of the cylinder**.

(3.1) $$X \vee X \overset{i}{\rightarrowtail} I_* X \xrightarrow{p} X,$$

where $I_* X = (I \times X)/(I \times *)$ and where $X \vee X$ is the one point union. Clearly $i = (i_0, i_1)$ is the inclusion of a subcomplex and the projection p with $p(t, x) = x$ is a homotopy equivalence. A homotopy $H\colon f \simeq g$ is a map $H\colon I_* X \to Y$ with $Hi = (f, g)$. We urge the reader to look at I.§ 3 in Baues (AH) where the main properties of cylinders are listed.

For $n \geq 1$ let $Z_n = Z_n(X)$ be the set of n-cells of X and let $Z_0 = \phi$ be the empty set. The set of n-cells, $n \geq 1$, of $I_* X$ is

$$Z_n(I_* X) = Z'_n \cup sZ_{n-1} \cup Z''_n. \tag{1}$$

Here $\cup$ denotes the disjoint union of sets and $Z_n = Z'_n = Z''_n = sZ_n$ are copies of Z_n. We denote the identity maps $Z_n = Z'_n$, $Z_n = Z''_n$ and $Z_n = sZ_n$ by $x \mapsto x'$, $x \mapsto x''$ and $x \mapsto sx$ respectively. For a cell $x \in Z_n$ we obtain associated cells in $Z_n(I_* X)$ by

$$\left.\begin{aligned} i_0 x &= x', \\ i_1 x &= x'', \\ e \times x &= sx \end{aligned}\right\} \tag{2}$$

where $e = (0, 1) \subset I$ is the open interval.

Let $C = (\pi, C)$ be a reduced free chain complex, that is $C \in \mathbf{H}_1$, see (2.10). We obtain a cylinder $I_* C \in \mathbf{H}_1$ with structure maps

(3.2) $$C \vee C \overset{i}{\rightarrowtail} I_* C \xrightarrow{p} C,$$

as follows. Let $C \vee C$ be the sum of objects under $R = \mathbb{Z}[\pi]$ in the category $\mathbf{H}_1$, as a graded module one has $C \vee C = R \oplus \tilde{C} \oplus \tilde{C}$ with $\tilde{C} = C/R$. Let $Z_n = Z_n(C)$ be a basis of the free π-module C_n, $n \geq 1$, and let $Z_0 = \phi$. Then

$$Z_n(I_* C) = Z'_n \cup sZ_{n-1} \cup Z''_n \tag{1}$$

is a basis of the free π-module $(I_* C)_n$. Here we use the same notation as in (3.1)(1) above. The maps $i = (i_0, i_1)$ and p in (3.2) are given by $(x \in Z_n)$

$$\left.\begin{aligned} i_0 x &= x' \\ i_1 x &= x'' \\ px' &= px'' = x \\ psx &= 0 \end{aligned}\right\} \tag{2}$$

We now define a homorphisms of π-modules

$$S\colon C_n \to (I_* C)_{n+1} \tag{3}$$

by setting $SC_0 = 0$ and $S(x) = sx$ for $x \in Z_n$. The π-equivariant differential d of the chain complex $I_* C$ is defined on generators by

$$\left.\begin{aligned} dx' &= i_0(dx) \\ dx'' &= i_1(dx) \\ d(sx) &= -x' + x'' - S\,dx. \end{aligned}\right\} \tag{4}$$

This shows that i in (3.2) is an inclusion of chain complexes and that p in (3.2) is a weak equivalence. Moreover S in (3) is a homotopy, $S\colon i_0 \simeq i_1$ with

$$-i_0 + i_1 = dS + Sd, \quad SC_0 = 0. \tag{5}$$

A chain map $H\colon I_*C \to C'$ in $\mathbf{Chain}_{\mathbb{Z}}^{\wedge}$ with $Hi = (f, g)$ yields a homotopy $\alpha\colon f \simeq g$ by $\alpha = HS$. Vice versa a homotopy α with $\alpha_0 = 0$ gives us a chain map H by setting

$$\left.\begin{aligned} H(x') &= f(x) \\ H(x'') &= g(x) \\ H(sx) &= \alpha(x). \end{aligned}\right\} \tag{6}$$

These are exactly the reduced homotopies used in the subcategory $\mathbf{H}_1$ of $\mathbf{Chain}_{\mathbb{Z}}^{\wedge}$, see (2.10).

Let ρ be a crossed chain complex which is totally free, see (2.7). For example $\rho = \rho(X)$ in (2.8) is such a crossed chain complex. We obtain a cylinder $I\rho$ with structure maps

(3.3) $$\rho \vee \rho \overset{i}{\rightarrowtail} I\rho \overset{p}{\longrightarrow} \rho$$

as follows. The object $\rho \vee \rho$ is the sum in the category **cross chain**. Let $Z_n = Z_n(\rho) \subset \rho_n$ be a basis of ρ and let $Z_0 = \phi$. Then

$$Z_n(I\rho) = Z_n' \cup sZ_{n-1} \cup Z_n'' \tag{1}$$

is a basis of the totally free crossed chain complex $I\rho$ in (3.3). The crossed chain maps $i = (i_0, i_1)$ and p in (3.3) are defined in terms of the basis (1) by the same formulas as in (3.2). Moreover the boundary d of $I\rho$ is defined on the basis (1) by

$$\left.\begin{aligned} dx' &= i_0(dx) \\ dx'' &= i_1(dx) \\ dsx &= -x' + x'' - S_{n-1}\,dx \end{aligned}\right\} \tag{2}$$

for $x \in Z_n$. Here we set $S_0\,dx = 0$ and

$$S_1\colon \rho_1 \to (I\rho)_2 \tag{3}$$

is the i_0-crossed homomorphism with $S_1(x) = sx$ for $x \in Z_1$. For $n \geq 2$ the function

$$S_n\colon \rho_n \to (I\rho)_{n+1} \tag{4}$$

is the i_0-equivariant homomorphism with $S_n x = sx$ for $x \in Z_n$. One can check that (2) satisfies $dd = 0$. It is also clear that i and p in (3.3) are well defined crossed chain maps and that p is a weak equivalence. Moreover, $S\colon i_0 \simeq i_1$ is a homotopy in the sense of (2.6). A crossed chain map $H\colon I\rho \to \rho'$ with $Hi = (f, g)$ yields a homotopy $\alpha\colon f \simeq g$ by $\alpha = HS$. Vice versa a homotopy $\alpha\colon f \simeq g$ as in (2.6) yields a crossed chain map H by defining H in the same way as in (3.2)(6).

(3.4) **Proposition.** *The cylinder objects above are compatible with the functors $\hat{C}_*$, ρ and C. That is $\hat{C}_*(I_* X) = I_*(\hat{C}_* X)$, $\rho(I_* X) = I(\rho X)$ and $C(I\rho) = I_*(C\rho)$. The structure maps of the cylinders are compatible with these equations. In the first equation we assume $X^0 = *$.*

The proposition implies that the functors $\rho\colon \mathbf{CW} \to \mathbf{H}$ and $C\colon \mathbf{H} \to \mathbf{H}_1$ induce functors between homotopy categories.

§ 4 Cofibrations in the category of crossed chain complexes

In this section we consider the basic properties of weak equivalences and cofibrations in the category **cross chain** of crossed chain complexes. These properties imply a homotopy theory which for example we shall need for the homotopy push outs of crossed chain complexes in section § 6 below. In order to make this book self contained we here first recall the basic definitions of "algebraic homotopy" concerning cofibration categories, cofibrant models, localized categories Ho(**C**), lifting lemmas and cylinders. To this end the reader is urged to study the rich homotopy theory implied by the axioms of a cofibration category in Baues (AH). In this book we show that such axioms are satisfied in two algebraic categories, namely in the category **cross chain** of crossed chain complexes and similarly in the next chapter in the category **quad chain** of quadratic chain complexes.

(4.1) **Definition.** A **cofibration category** is a category **C** with an additional structure (**C**, *cof*, *we*) subject to axioms C1, C2, C3, C4. Here *cof* and *we* are classes of morphisms in **C**, called **cofibrations** and **weak equivalences** respectively. Morphisms in **C** are also called *maps* in **C**. We write $i\colon B \rightarrowtail A$ for a cofibration and we write $X \xrightarrow{\sim} Y$ for a weak equivalence. A map in **C** is a **trivial cofibration** if it is both a weak equivalence and a cofibration. An object R in a cofibration category is **fibrant** if each trivial cofibration $i\colon R \overset{\sim}{\rightarrowtail} Q$ admits a retraction $r\colon Q \to R$, $re = 1$. The axioms in question are:

(C1) **Composition axiom.** The isomorphisms in **C** are weak equivalences and also cofibrations. For two maps $A \xrightarrow{f} B \xrightarrow{g} C$ if any two of f, g, and gf are weak equivalences, then so is the third. The composite of cofibrations is a cofibration.

(C2) **Push out axiom.** For a cofibration $i\colon B \rightarrowtail A$ and a map $f\colon B \to Y$ there exists the push out in **C**

$$\begin{array}{ccc} A & \xrightarrow{\bar{f}} & A \cup_B Y \\ {\scriptstyle i}\uparrow & & \uparrow{\scriptstyle \bar{i}} \\ B & \xrightarrow[f]{} & Y \end{array}$$

and $\bar{i}$ is a cofibration. Moreover: (a) if f is a weak equivalence, so is $\bar{f}$, (b) if i is a weak equivalence, so is $\bar{i}$.

(C3) **Factorization axiom.** For a map $f\colon B \to Y$ in **C** there exists a commutative diagram

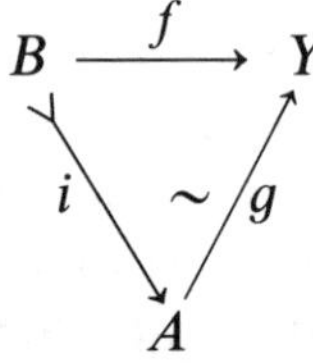

where i is a cofibration and where g is a weak equivalence.

(C4) **Axiom on fibrant models.** For each object X in **C** there is a trivial cofibration $X \overset{\sim}{\rightarrowtail} RX$ where RX is fibrant in **C**. We call $X \overset{\sim}{\rightarrowtail} RX$ a fibrant model of X; if X is fibrant we take $RX = X$.

(4.2) **Definition.** We call $(\mathbf{C}, cof, we)$ as in (4.1) a **cofibration structure** if all axioms of a cofibration category are satisfied except possibly the properness axiom C2(a).

Hence a cofibration structure which satisfies (C2)(a) is a cofibration category. For example let $(\mathbf{C}, cof, fib, we)$ be a model category in the sense of Quillen, then $(\mathbf{C}, cof, we)$ is a cofibration structure.

(4.3) **Definition.** Assume $\mathbf{C}$ has an initial object $*$. Then an object X in $\mathbf{C}$ is **cofibrant** if $* \to X$ is a cofibration. Let $\mathbf{C}_c$ be the full subcategory of $\mathbf{C}$ consisting of cofibrant objects, see (I.1.3) in Baues (AH). □

In (I.1.4) Baues (AH) we show:

(4.4) **Lemma.** *Let* $\mathbf{C}$ *be a cofibration structure. Then* $\mathbf{C}_c$ *with cofibrations and weak equivalences as in* $\mathbf{C}$ *is a cofibration category.*

Actually many definitions and results for a cofibration category described in Baues (AH) are as well available for a cofibration structure. In particular, we make use of the following facts.

(4.5) **Lemma.** *Let* $\mathbf{C}$ *be a cofibration structure in which all objects are fibrant. Then we have an equivalence of localized categories*

$$M\colon \mathrm{Ho}(\mathbf{C}) \xrightarrow{\sim} \mathrm{Ho}(\mathbf{C}_c) = \mathbf{C}_c/\simeq.$$

For the definition of $\mathrm{Ho}(\mathbf{C})$ we consider an arbitrary category $\mathbf{C}$ and a subclass S of the class of morphisms of $\mathbf{C}$. By the **localization** of $\mathbf{C}$ with respect to S we mean the category $S^{-1}\mathbf{C}$ together with a functor $q\colon \mathbf{C} \to S^{-1}\mathbf{C}$ having the following universal property: For every $s \in S$, $q(s)$ is an isomorphism; given any functor $F\colon \mathbf{C} \to \mathbf{B}$ with $F(s)$ an isomorphism for all $s \in S$, there is a unique functor $\theta\colon S^{-1}\mathbf{C} \to \mathbf{B}$ such that $\theta q = F$. Except for set-theoretic difficulties the category $S^{-1}\mathbf{C}$ exists, see Gabriel-Zisman. Let $\mathrm{Ho}(\mathbf{C})$ be the localization of $\mathbf{C}$ with respect to the given class of weak equivalences in $\mathbf{C}$.

For the definition of the functor M in (4.5) we use the next lemma. Consider the commutative diagram of unbroken arrows in $\mathbf{C}$.

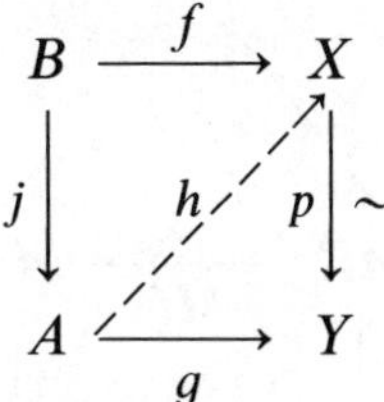

We say that h is a **lifting** of the diagram if $hj = f$ and $ph \simeq g$ rel B, see (4.6)(1) below.

(4.6) **Lemma.** *Let* $\mathbf{C}$ *be a cofibration structure in which all objects are fibrant. Then a lifting as above exists and two liftings are homotopic* rel B.

For the proof we use the same arguments as in (II.1.11) of Baues (AH). Recall that homotopies relative B or under B are defined as follows. For the cofibration $j: B \rightarrowtail A$ we choose a factorization

$$(1,1): A \cup_B A \overset{i}{\rightarrowtail} I_B A \overset{p}{\longrightarrow} A \tag{1}$$

of the folding map $(1,1)$. Such a factorization is a **cylinder** on $B \rightarrowtail A$. A **homotopy** $H: a \simeq b$ **rel** B is a map $H: I_B A \to X$ with $Hi = (a,b)$. Here (a,b) is defined provided a and b coincide on B, that is $aj = bj = f$. The set

$$[A,X]^B = [A,X]^f = \{h \in \mathbf{C}(A,X); hj = f\}/\simeq \text{rel}\, B \tag{2}$$

denotes the set of all homotopy classes $\{h\}$ rel B provided X is fibrant. Moreover we write $a \simeq b$ if $a \simeq b$ rel $*$ and we set

$$[A,X] = [A,X]^* = \mathbf{C}(A,X)/\simeq \text{rel}\, *$$

where we assume that A is cofibrant and that X is fibrant. Lemma (4.6) shows that a weak equivalence p induces a bijection

$$p_*: [A,X]^f \overset{\approx}{\longrightarrow} [A,Y]^{pf}.$$

Next we use (4.6) for the definition of the functor M in (4.5). For each object X in $\mathbf{C}$ we choose a factorization

$$* \rightarrowtail MX \overset{\sim}{\longrightarrow} X \tag{3}$$

of $* \to X$ by (C3). We call MX a **cofibrant model** of X. For a map $f: X \to Y$ we have the diagram

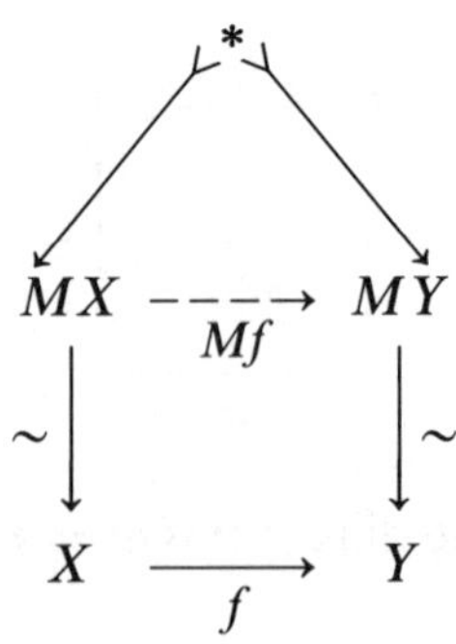

where Mf is a lifting by (4.6). Since Mf is unique up to homotopy we obtain a well defined functor $M: \mathbf{C} \to \mathbf{C}_c/\simeq$. This functor induces the equivalence of categories in (4.5), see (II.3.6) in Baues (AH).

We now describe a cofibration structure on the category **cross chain** of crossed chain complexes.

(4.7) **Definition.** A map $f: \lambda \to \rho$ in **cross chain** is a **cofibration** if f is a free extension in each degree n, $n \geq 1$. Here we say that f is a **free extension** in degree n with basis $\partial_n: Z_n \to \rho_{n-1}$ if a commutative diagram of sets,

(1)

is given such that the following universal property is satisfied (where we set $\rho_{-1} = 0$ for $n = 1$ and where ρ^n and λ^n denote the n-dimensional part of ρ and λ respectively). Let ρ' be any crossed chain complex and let $b: \lambda^n \to \rho'$ and $a^{n-1}: \rho^{n-1} \to \rho'$ be crossed chain maps with $a^{n-1}f^{n-1} = b^{n-1}: \lambda^{n-1} \to \rho'$ and assume a function $a_Z: Z_n \to \rho'_n$ is chosen such that the following diagram of unbroken arrows commutes.

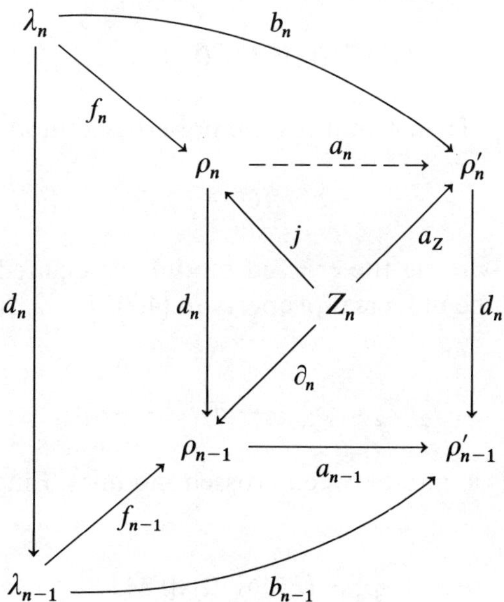

Then there is a unique crossed chain map $a: \rho^n \to \rho'$ for which a_n extends the diagram commutatively and for which $a|\rho^{n-1} = a^{n-1}$. This implies $af^n = b$.

It is clear that **cofibrant** objects in **cross chain** are exactly the totally free crossed chain complexes; whence we have with the notation in (4.3), (2.7):

(4.8) $$\mathbf{H} = (\textbf{cross chain})_c.$$

The next lemma shows that free extensions in degree n exist.

(4.9) **Lemma.** *Let $\lambda = \lambda^n$ be an n-skeleton and assume $f^{n-1}: \lambda^{n-1} \to \rho^{n-1}$ and a function $\partial_n: Z_n \to \rho_{n-1}$ are given. Then a free extension $f: \lambda \to \rho = \rho^n$ as in (4.7) with basis ∂_n exists provided $d_{n-1}\partial_n = 0$. In this case we write $\rho_n = \lambda_n(Z_n)$.*

Proof. In degree 1 we take the free product of groups

$$\rho_1 = \lambda_1 * \langle Z_1 \rangle. \tag{1}$$

In degree 2 we first consider the free pre-crossed module

$$\bar{\partial}: \bar{\rho}_2 = \langle (\lambda_2 \cup Z_2) \times \rho_1 \rangle \to \rho_1 \tag{2}$$

with basis $(f_1 d_2, \partial_2): \lambda_2 \cup Z_2 \to \rho_1$ where $\lambda_2 \cup Z_2$ is the disjoint union. The inclusion $i: \lambda_2 \subset \bar{\rho}_2$, however, is not a map between crossed modules. Let U be the normal ρ_1-subgroup of $\bar{\rho}_2$ generated by the relations

$$\left.\begin{aligned} &i(x) + i(y) - i(x+y) \sim 0 \\ &i(x^\alpha) - (ix)^{f_1\alpha} \sim 0 \end{aligned}\right\} \tag{3}$$

for $x, y \in \lambda_2$, $\alpha \in \lambda_1$. Then $\bar{\partial}$ induces the pre-crossed module

$$\partial: \bar{\rho}_2/U \to \rho_1. \tag{4}$$

Let $d_2 = r_2(\partial): \rho_2 \to \rho_1$ be the crossed module associated to ∂. One readily checks that d_2 has the universal property in (4.7), $n = 2$. In particular (f_2, f_1) with

$$f_2: \lambda_2 \subset \bar{\rho}_2 \twoheadrightarrow \bar{\rho}_2/U \twoheadrightarrow \rho_2,$$

given by i above, is a map between crossed modules. Finally let ρ_n in degree $n \geq 3$ be given by

$$\rho_n = (\lambda_n \otimes_\Lambda R) \oplus M. \tag{5}$$

Here Λ and R are the group rings of $\pi_1\lambda$ and $\pi_1\rho^{n-1}$ respectively and the tensor product is defined via $\pi_1(f^{n-1})$. Moreover, M is the free R-module generated by Z_n. □

(4.10) **Theorem.** *The category* **cross chain** *with cofibrations as in* (4.7) *and with weak equivalences as in* (2.6) *is a cofibration structure for which all objects are fibrant.*

We prove this result in (4.17) below. The lemmas (4.4), (4.5) and (4.8) imply the following corollaries of (4.10).

(4.11) **Corollary.** *The category* **H** *is a cofibration category. Moreover, the cylinder* (3.3) *is a cylinder in the cofibration category* **H**, *see* (I.1.5) *Baues* (AH).

(4.12) **Corollary.** *There is an equivalence of categories*

$$M\colon \mathrm{Ho}(\textbf{cross chain}) \xrightarrow{\sim} \mathrm{Ho}(\mathbf{H}) = \mathbf{H}/\simeq$$

where the homotopy relation on **H** *is defned as in* (2.6)(3).

(4.13) **Corollary.** *A weak equivalence in* **H** *is a homotopy equivalence in* **H**.

(4.14) **Remark.** The cylinder $I\rho$ in (3.3) is natural. In fact, a map $f\colon \rho \to \rho'$ in **H** induces a unique map If: $I\rho \to I\rho'$ in **H** with the properties (1) $i_t f = (If)i_t$ for $t = 0, 1$ and (2) $Sf = (If)S$. Now (4.11) can be used for the proof that **H** with the cylinder (3.3) and with cofibrations (4.7) is an **I-category** in the sense of (I.3.1) Baues (AH).

Let $D(n)$, $n \geq 2$, be the totally free crossed chain complex generated by elements c_n and $d_n c_n$ in degree n and $n-1$ respectively. Whence $D(n)$ is in degree n and $n-1$ an infinite cyclic group and is otherwise trivial. Let $S(n-1)$ be the $(n-1)$-skeleton of $D(n)$, with inclusion $S(n-1) \rightarrowtail D(n)$. If D^n and S^{n-1} denote the skeletal filtrations of the standard **n-ball** $D^n = e^0 \cup e^{n-1} \cup e^n$ and **(*n* − 1)-sphere** $S^{n-1} = e^0 \cup e^{n-1}$, then it is clear that $D(n) = \rho(D^n)$ and $S(n-1) = \rho(S^n)$ where ρ is the functor in (2.8). We say that a cofibration $\lambda \rightarrowtail \rho$ in **cross chain** is given by **attaching an *n*-cell** to λ if for $n \geq 2$ there exists a push out diagram

(4.15)
$$\begin{array}{ccc} D(n) & \longrightarrow & \rho \\ \uparrow & \text{push} & \uparrow \\ S(n-1) & \longrightarrow & \lambda \end{array}$$

in the category **cross chain** and if for $n = 1$ there is an isomorphism $\rho \cong \lambda \vee S(1)$ where $\lambda \vee S(1)$ is the sum in **cross chain.** Now one readily checks that a cofibration as in (4.7) is always given by inductively attaching cells to λ.

Remark. Brown-Golasinski show that the category of crossed complexes is actually a (closed) Quillen model category. This implies as well theorem (4.10) since they show that the cofibrations above are special cofibrations in their model category.

The classical co(homology) of groups can be defined by cofibrant models in the category **cross chain** as follows. Let G be a group considered as a crossed chain complex concentrated in degree 0. Then we obtain a cofibrant model $* \rightarrowtail MG \xrightarrow{\sim} G$ in **cross chain** which is unique up to homotopy equivalence. Since for a $K(G, 1)$ also $\rho K(G, 1) \xrightarrow{\sim} G$ is a cofibrant model we see that there is a homotopy equivalence of chain complexes of free G-modules $C(MG) \simeq C\rho K(G, 1) = \hat{C}_* K(G, 1)$, see (I.4.14). Whence the **homology** and **cohomology of the group** G can be defined by

(4.16) $$H_*(G, \Gamma) = H_* C(MG) \otimes_G \Gamma, \qquad H^*(G, \Gamma) = H^* \mathrm{Hom}_G(C(MG), \Gamma).$$

Also a crossed module $A = (\partial: A_2 \to A_1)$ is an object in **cross chain** concentrated in degree 1 and 2. Whence we obtain as well a cofibrant model $* \rightarrowtail MA \xrightarrow{\sim} A$ which is unique up to homotopy equivalence. Similarly as in (4.16) we therefore can define the **algebraic (co)homology** of A by

$$H_*(A, \Gamma) = H_* C(MA) \otimes_G \Gamma, \qquad H^*(A, \Gamma) = H^* \mathrm{Hom}_G(C(MA), \Gamma) \quad (1)$$

where $G = \pi_1 A$ and where Γ is a left, resp. right G-module. This is a weak notion of (co)homology which coincides with the (co)homology of groups if $A_2 = 0$. We obtain a strong notion of (co)homology of A by use of the equivalence of categories in (8.2) below. Using this equivalence we obtain a 2-type $K(A)$ corresponding to A and we define the **topological (co)homology** of A by

$$\hat{H}_*(A, \Gamma) = \hat{H}_*(K(A), \Gamma), \qquad \hat{H}^*(A, \Gamma) = \hat{H}^*(K(A), \Gamma). \quad (2)$$

If $A_2 = 0$ the groups in (4.16), (1) and (2) respectively coincide. For $A_2 \neq 0$, however, the groups (1) and (2) in general do not coincide. This is easily seen for $A_1 = 0$ since then $K(A) = K(A_2, 2)$ is an Eilenberg-MacLane space of degree 2. The connection of the groups (1) and (2) is given by the fact that only the 2-types of the chain complexes $C(MA)$ and $\hat{C}_* K(A)$ coincide, see (II.4.2).

(4.17) *Proof of* (4.10). We first check (C3). We obtain a factorization

$$f = qi: \lambda \rightarrowtail \rho \xrightarrow{\sim} \tau \tag{1}$$

of $f: \lambda \to \tau$ by choosing "enough" generators for the cofibration $\lambda \rightarrowtail \rho$. We start the induction by

$$f_1: \lambda_1 \rightarrowtail \rho_1 = \lambda_1(Z_1) \to \tau_1$$

where we choose Z_1 and q_1 such that $\rho_1 \to \tau_1 \to \pi_1(\tau)$ is surjective, see (4.9).

Next we consider

$$\begin{array}{ccccc} \lambda_2 & \xrightarrow{i_2'} & \lambda_2(Z_2') & \xrightarrow{q_2'} & \tau_2 \\ \downarrow & & \downarrow{\scriptstyle \partial_2} & & \downarrow{\scriptstyle d_2} \\ \lambda_1 & \rightarrowtail & \rho_1 & \xrightarrow[q_1]{} & \tau_1 \end{array}$$

Here we choose a basis Z_2' and ∂_2 such that $\partial_2(\lambda_2(Z_2'))$ is the kernel of $\rho_1 \to \tau_1 \to \pi_1(\tau)$. Whence $q_1\partial_2$ maps onto the image of d_2 so that we can find q_2' such that the diagram commutes and such that $q_2'i_2' = f_2$. The map

$$q_2': \text{kernel}(\partial_2) \to \text{kernel}(d_2)$$

needs not yet to be surjective. Therefore we choose $q_2: Z_2'' \to \text{kernel}(d_2)$ such that the extension of q_2';

$$q_2: \rho_2 = \lambda_2(Z_2')(Z_2'') \to \tau_2,$$

carries kernel$(d_2: \rho_2 \to \rho_1)$ surjectively to kernel$(d_2: \tau_2 \to \tau_1)$. Here $d_2: \rho_2 \to \rho_1$ is given by ∂_2 and by $d_2 Z_2'' = 0$. In a similar way we obtain $\rho_n = \lambda_n(Z_n')(Z_n'')$. This completes the construction of the factorization.

We next check that all objects are fibrant, this implies (C4). Let $i: \overset{\sim}{\rightarrowtail} \rho$ be given. We construct inductively a retraction $r: \rho \to \lambda$ with $ri = 1$ and a homotopy $\alpha: ir \simeq 1$ relative λ. This shows that i is actually a **strong deformation retract** morphism. Let $\rho_n = \lambda_n(Z_n)$, see (4.9), and let

$$\begin{array}{ccccc} i: \lambda & \rightarrowtail & \rho_n & \xrightarrow[g_{(n)}]{} & \rho \\ & \underset{r_{(n)}}{\nwarrow\!\!-\!\!-\!\!-\!\!-\!\!\swarrow} & & & \end{array} \tag{2}$$

be given by the subcomplex ρ^n of ρ with $(\rho^n)_k = \rho_k$ for $k \le n$ and $(\rho^n)_k = \lambda_k$ for $k > n$. The map $g_{(n)}$ in (2) is the inclusion. We choose inductively a retraction $r_{(n)}$ and a homotopy $\alpha_{(n)}: ir_{(n)} \simeq g_{(n)}$ relative λ. Assume $r_{(n)}$ and $\alpha_{(n)}$

are defined. By

$$-ir_{(n)} + g_{(n)} = d\alpha_{(n)} + \alpha_{(n)}d \tag{3}$$

we have

$$ir_{(n)}d = g_{(n)}d - d\alpha_{(n)}d = d(g_{(n)} - \alpha_{(n)}d). \tag{4}$$

Since i is a weak equivalence we can choose by (4) a map $x: Z_{n+1} \to \lambda_{n+1}$ with $dx = r_n d$. Moreover $(-ix + g_{n+1} - \alpha_n d)$ carries Z_{n+1} to the cycles of ρ by (3). Again since i is a weak equivalence we can choose maps $z: Z_{n+1} \to \lambda_{n+1}$, $y: Z_{n+1} \to \rho_{n+2}$ such that

$$iz + dy = -ix + g_{n+1} - \alpha_n d. \tag{5}$$

We now define the extension $r_{(n+1)}$ of $r_{(n)}$ by $r_{n+1} = x + z$ on Z_{n+1} and we define the extension $\alpha_{(n+1)}$ of $\alpha_{(n)}$ by $\alpha_{n+1} = y$ on Z_{n+1}. This completes the induction.

We now consider push outs

$$\begin{array}{ccc} \rho & \xrightarrow{\bar{f}} & \rho' \\ {\scriptstyle i}\uparrow & \text{push} & \uparrow{\scriptstyle \bar{i}} \\ \lambda & \xrightarrow[f]{} & \lambda' \end{array} \tag{6}$$

is **cross chain** for which i is a cofibration. Such push outs exist and $\bar{i}$ is a cofibration. In fact, let $\rho_n = \lambda_n(Z_n)$ as in (4.9) then we set $\rho'_n = \lambda'_n(Z_n)$. The basis of ρ' is given in degree n by $f_{n-1}d|_{Z_n}: Z_n \to \lambda_{n-1} \to \lambda'_{n-1} = \rho'_{n-1}, n \geq 1$. The map $\bar{f}$ is the identity on Z_n, compare also (4.15).

Finally we prove (C2)(b). If i in (6) is a weak equivalence, then i is a strong deformation retract by (2). This implies that also $\bar{i}$ is a strong deformation retract. In fact we define the retraction $\bar{r}$ of $\bar{i}$ by fr in (2) and we define the homotopy $\bar{\alpha}: \bar{i}\bar{r} \simeq 1$ relative λ' by $\bar{f}\alpha$ on generators. This shows that $\bar{i}$ is a weak equivalence and whence (C2)(b) is satisfied. □

§5 The homotopy addition lemma

The standard n-simplex Δ^n is a CW-complex with the usual cell decomposition given by the faces of Δ^n. In this section we describe explicitely the crossed chain complex $\rho(\Delta^n)$ associated to Δ^n, see (2.2). The classical homotopy addition lemma determines exactly a formula for the boundaries in $\rho(\Delta^n)$. In the next section this formula will be used for the computation of the crossed chain complex of any (reduced) simplicial set. We also shall use $\rho(\Delta^n)$ for the computation of the quadratic chain complex of Δ^n in the next chapter.

Let Δ^n be the standard n-simplex with ordered set of vertices $\Delta_0^n = \{v_0, \ldots, v_n\}$, and let Δ^n have its filtration by skeletons Δ_r^n, $r \geq 0$. The base point of Δ^n is the vertex v_0. The top cell in Δ^n yields a generator

$$\Delta = \Delta_{0\ldots n} \in \pi_n(\Delta^n, \Delta_{n-1}^n, v_0) \cong \mathbb{Z}. \tag{5.1}$$

For a subset $a \subset \{0, \ldots, n\}$ with $a = \{a(0) < \cdots < a(r)\}$ let

$$\Delta_a = (i_a)_* \Delta \in \pi_r(\Delta_r^n, \Delta_{r-1}^n, v_{a(0)}) \tag{1}$$

be given by the inclusion $i_a\colon \Delta^r \subset \Delta^n$. Here i_a carries v_i to the vertex $v_{a(i)}$. For $a = \{i < k\}$ we identify $\Delta_a = \Delta_{ik}$ with the corresponding **path from v_i to v_k**. Moreover, for $i \in \{0, \ldots, n\} = \bar{n}$ let

$$\partial_i \Delta = \Delta_{\bar{n} - \{i\}} \tag{2}$$

be the ***i*-th face** of Δ. We now consider the boundary ($n \geq 3$)

$$\left.\begin{aligned} d_n&\colon \pi_n(\Delta^n, \Delta_{n-1}^n, v_0) \to \pi_{n-1}(\Delta_{n-1}^n, \Delta_{n-2}^n, v_0) \\ d_2&\colon \pi_2(\Delta_2^n, \Delta_1^n, v_0) \to \pi_1(\Delta_1^n, v_0) \end{aligned}\right\} \tag{3}$$

in the crossed chain complex $\rho(\Delta^n)$. Since $\pi_1(\Delta^n) = 0$ we know that

$$C\rho(\Delta^n) \cong \hat{C}_*(\Delta^n/T) = C_*(\Delta^n/T) \tag{4}$$

is the cellular chain complex of the CW-complex Δ^n/T. Here T is a maximal tree in the 1-skeleton Δ_1^n and C is the functor in (2.11). The isomorphism (4) is given by (2.11).

(5.2) **Homotopy addition lemma.** *The object $\rho = \rho(\Delta^n)$ in* **H** *is determined up to isomorphism by $d_2 = \partial\colon \rho_2 \to \rho_1$ and by the isomorphism* (5.1)(4). *The generators* (5.1) *can be chosen such that*

$$d_n(\Delta) = \begin{cases} \Delta_{01} + \Delta_{12} - \Delta_{02}, & n = 2 \\ -\partial_1\Delta - \partial_3\Delta + (-\Delta_{01})^{\#}\partial_0 + \partial_2\Delta, & n = 3 \\ (-\Delta_{01})^{\#}\partial_0\Delta + \sum\limits_{i=1}^{n} (-1)^i \partial_i \Delta, & n \geq 4. \end{cases}$$

For $n = 2$ the right hand side is defined by addition of paths.

The following proof of (5.2) uses the Hurewicz theorem as in (2.4).

Proof. Consider diagram (2.2) where we set $X = \Delta^n$. The map h_2: kernel$(\partial) \to$ kernel(d_2) is an isomorphism by (2.4). Therefore $d_n: \rho_n \to \rho_{n-1}$, $n \geq 3$, is determined by $C_*\Delta^n$. This shows that $\rho\Delta^n$ is determined by d_2 and (5.1)(4). It is well known that the boundary d in $C_*\Delta^n$ satisfies the formula

$$d\bar{\Delta} = \sum_{i=0}^{n} (-1)^i \partial_i \bar{\Delta} \tag{1}$$

where $\bar{\Delta} = h_n\Delta$. The operator h_n carries the formula for $d_n(\Delta)$ in (5.2) to the boundary formula (1). This proves the formula for $d_n(\Delta)$ in case $n \geq 4$ and also in case $n = 3$ provided we show $\partial d_3(\Delta) = 0$ where $\partial: \rho_2 \to \rho_1$ is defined via $d_2(\Delta)$ in (5.2). We get $\partial d_3(\Delta) = 0$ by the following addition of paths.

$$\begin{aligned} \partial d_3(\Delta) &= \partial(-\Delta_{023} - \Delta_{012} + (-\Delta_{01})^{\#}\Delta_{123} + \Delta_{013}) \\ &= -(\Delta_{02} + \Delta_{23} - \Delta_{03}) - (\Delta_{01} + \Delta_{12} - \Delta_{02}) \\ &\quad + \Delta_{01} + (\Delta_{12} + \Delta_{23} - \Delta_{13}) - \Delta_{01} + (\Delta_{01} + \Delta_{13} - \Delta_{03}). \end{aligned} \tag{2}$$

On the right hand side all summands cancel. □

(5.3) **Remark.** The formulas in (5.2) are not unique. For example they depend on the basepoint chosen in Δ^n. In particular, we can choose v_n as a basepoint of Δ^n. We also can alter the formula in (2), say $d_3\Delta = \xi$, by $d_3(\Delta) = -\alpha + \xi + \alpha$ where α is any element, for example $\alpha = -\Delta_{023}$. This way we get the various formulas for the homotopy addition lemma used in the literature, see for example Brown-Higgins (AC).

§ 6 A model functor from spaces to crossed chain complexes

The functor ρ: **CW** $\to$ **H** which carries a CW-complex X to its crossed chain complex $\rho(X)$ is only defined on cellular maps between CW-complexes, see § 2. We here introduce a functor ρ_S which carries any path connected pointed CW-space X to a crossed chain complex $\rho_S(X)$ in **H** and which is well defined on all base point preserving maps. The functor ρ_S is simply given by applying ρ to the (reduced) singular set of X. We prove that ρ_S is a model functor between cofibration categories. Such model functors (which carry weak equivalences to weak equivalences and homotopy push outs to homotopy push outs) have the nice feature that they are compatible with most homotopy theoretic constructions in a cofibration category, compare Baues (AH). The fact that ρ_s is a model functor generalizes the well known Van Kampen theorem for fundamental groups.

(6.1) **Definition.** Let $\mathbf{CW\text{-}spaces}_0^*$ be the following cofibration category. Objects are path connected CW-spaces X with basepoint $* \in X$ for which $* \to X$ is a cofibration in **Top**. Morphisms are basepoint preserving maps $f: X \to Y$. The map f is a **cofibration** if f is a cofibration in **Top**. Moreover, f is a **weak equivalence** if f induces isomorphisms $\pi_n f: \pi_n X \to \pi_n Y$ for $n \geq 1$, equivalently the map f is a homotopy equivalence in **Top*** by the Whitehead theorem.

Recall that a commutative diagram

$$\begin{array}{ccc} A & \longrightarrow & C \\ {\scriptstyle f}\uparrow & & \uparrow \\ B & \underset{g}{\longrightarrow} & D \end{array} \tag{6.2}$$

in a cofibration category **C** is a **homotopy push out** if for some factorization $B \rightarrowtail W \xrightarrow{\sim} A$ of f the induced map $W \cup_B D \xrightarrow{\sim} C$ is a weak equivalence, compare (I.1.9) in Baues (AH). A functor α: $\mathbf{C} \to \mathbf{K}$ between cofibration categories is a **model functor** if α carries weak equivalences to weak equivalences and if α carries homotopy push outs to homotopy push outs.

We will show that there is a model functor

$$\rho_S: \mathbf{CW\text{-}spaces}_0^* \to \mathbf{H} \tag{6.3}$$

which corresponds to the functor ρ in (2.8). In fact, we obtain a diagram of functors

$$\begin{array}{ccc} \mathbf{CW}/\simeq & \underset{i}{\xrightarrow{\sim}} & \mathbf{CW\text{-}spaces}_0^*/\simeq \\ & {\scriptstyle\rho}\searrow \quad \swarrow{\scriptstyle\rho_S} & \\ & \mathbf{H}/\simeq & \end{array} \tag{6.4}$$

where ρ is induced by ρ in (2.8) and where ρ_S is induced by (6.3). The equivalence of categories i is induced by the inclusion $\mathbf{CW} \subset \mathbf{CW\text{-}spaces}_0^*$. Moreover the construction of the functor ρ_S in (6.10) below yields a natural homotopy equivalence

$$\rho_S(X) \simeq \rho(X) \quad \text{in } \mathbf{H}/\simeq$$

for X in **CW**, see (6.11).

For the definition of the functor ρ_S we consider the crossed chain complex of a simplicial set. Let V be a **simplicial set** with boundaries

(6.5) $$\partial_i\colon V_n \to V_{n-1}, \quad (i = 0, \ldots, n)$$

and degeneracies $s_i\colon V_{n-1} \to V_n$, $(i = 0, \ldots, n-1)$. The functions ∂_i, s_i, $n \geq 1$, satisfy the usual simplicial identities. Elements in the images of s_i are said to be degenerate. Let $Z_n \subset V_n$ be the subset of non degenerate elements in V_n. For $\xi \in V_n$ we define

(6.6) $$|\xi| = \begin{cases} \xi & \text{if } \xi \in Z_n, \\ 0 & \text{otherwise.} \end{cases}$$

The **realization** $|V|$ of the simplicial set V is a CW-complex and Z_n is the set of n-cells of $|V|$.

We say that the simplicial set V is **0-reduced** if $V_0 = *$ consist only of a point. In this case the 0-skeleton $|V|^0 = *$ is a point. For example let X be a space with basepoint $*$ and let $(SX)_n$ be the set of all maps $\varphi\colon \Delta^n \to X$ with $\varphi(\Delta_0^n) = *$. Then SX is a 0-reduced simplicial set with the usual boundaries and degeneracies. Moreover we have a canonical map

(6.7) $$p_X\colon |SX| \to X$$

which is a homotopy equivalence provided X is a path connected CW-space. We call SX the **0-reduced singular set** of the pointed space X. A map $f\colon Y \to X$ induces the simplicial map $Sf\colon SY \to SX$ by $(Sf)(\varphi) = f\varphi$. The next definition is derived from the homotopy addition lemma (5.2).

(6.8) **Definition.** For a 0-reduced simplicial set V let $\rho(V) = \rho$ be the following totally free crossed chain complex in **H**. The basis of ρ_n is Z_n, see (6.6), and the boundary $d_n\colon \rho_n \to \rho_{n-1}$ is defined for $\xi \in Z_n$ by the formula

$$d_n(\xi) = \begin{cases} |\partial_2\xi| + |\partial_0\xi| - |\partial_1\xi|, & n = 2 \\[2ex] -|\partial_1\xi| - |\partial_3\xi| + |\partial_0\xi|^{u(\xi)} + |\partial_2\xi|, & n = 3 \\[2ex] |\partial_0\xi|^{u(\xi)} + \sum\limits_{i=1}^{n} (-1)^i |\partial_i\xi|, & n \geq 4. \end{cases}$$

Here we set $u(\xi) = -|\partial_2\partial_3 \ldots \partial_n\xi| \in \langle Z_1 \rangle = \rho_1$. A simplicial map $f\colon V \to W$ yields a map $\rho(f)\colon \rho(V) \to \rho(W)$ in **H** by $(\rho f)(\xi) = |f\xi|$ for $\xi \in Z_n$, $n \geq 1$.

Using the simplicial identities one can check that ρ in (6.8) is a well defined functor from the category of 0-reduced simplicial sets to the category **H**. The next result shows the significance of this functor.

(6.9) **Proposition.** *Let V be a 0-reduced simplicial set and let $\rho(|V|)$ be the crossed chain complex of the CW-complex $|V|$ given by (2.8). Then there is a natural isomorphism $\rho(V) = \rho(|V|)$ which is the identity on the basis Z_n.*

Proof. Let $\tilde{\Delta}^n$ be the 0-reduced simplicial set with $|\tilde{\Delta}^n| = \Delta^n/\Delta^n_0$. Then the homotopy addition lemma shows that $\rho(\tilde{\Delta}^n) = \rho(|\tilde{\Delta}^n|)$, see § 5. Now let $\xi \in Z_n$. Then we obtain a unique simplicial map $i_\xi\colon \tilde{\Delta}^n \to V$ such that $|i_\xi|\colon |\tilde{\Delta}^n| \to |V|$ is the characteristic map of the cell ξ. This implies that d_n in $\rho(|V|)$ satisfies the same formula as in (6.8). Clearly $\partial_2\partial_3 \ldots \partial_n \xi$ corresponds to Δ_{01} in (5.2). □

Using (6.8) we define the functor ρ_S in (6.3) by

(6.10) $$\rho_S(X) = \rho(SX).$$

Then (6.7) and (6.9) yield the canonical homotopy equivalence in **H**

(6.11) $$(p_X)_*\colon \rho_S(X) = \rho(SX) = \rho(|SX|) \xrightarrow{\simeq} \rho(X)$$

provided X is an object in **CW**. Here $(p_X)_*$ is defined by the functor $\rho\colon \mathbf{CW}/\simeq \to \mathbf{H}/\simeq$, see (2.8).

(6.12) **Theorem.** *ρ_S is a model functor.*

Proof. Consider a push out diagram (6.2) where A is a CW-complex in **CW** and where B is a subcomplex such that $A - B$ consists of a single n-cell. Then it is clear that $C - D$ consists of a single n-cell too and that

$$\rho(C) = \rho A \cup_{\rho B} \rho D, \tag{1}$$

compare the definition of the push out in (4.15). Now an induction and limit argument shows that (1) also holds for any subcomplex B of A. This implies by (6.11) that ρ_S carries homotopy push outs to homotopy push outs; a slightly more general result is proved in Brown-Higgins (*AC*). □

The proposition in (6.12) is a kind of a "generalized" **Van Kampen theorem.** In fact, (6.12) implies that

(6.13) $$\pi_1(A \cup_B D) = \pi_1 A *_{\pi_1 B} \pi_1 D$$

is a push out of groups, provided $A \cup_B D$ is a homotopy push out of path connected spaces in **Top***. For this we simply observe that a similar formula as in (6.13) is satisfied for homotopy push outs in **H**.

§ 7 The homotopy category of 2-dimensional CW-complexes

In this section we consider the homotopy classification of maps in low dimensions. The functor $\rho\colon \mathbf{CW}/\simeq \to \mathbf{H}/\simeq$ which carries a CW-complex X to its crossed chain complex yields actually a bijection of sets of homotopy classes

$$\rho\colon [X, Y] \approx [\rho X, \rho Y]$$

if $\dim(X) \le 2$. This fact is very useful for computations since ρX and ρY are often computable. In the next chapter we use the more sophisticated functor σ of quadratic chain complexes which in fact yields a similar bijection for the homotopy set $[X, Y]$ if $\dim(X) \le 3$, see (IV.§ 8).

Recall that $\mathbf{CW}^n$ and $\mathbf{H}^n$ denote the full subcategories of **CW** and **H** respectively consisting of n-dimensional objects. For $n = 2$ we get the following result where we use 0-homotopies, $\overset{0}{\simeq}$, running through cellular maps and ordinary (base point preserving) homotopies, $\simeq$. Clearly $\mathbf{H}^2$ is just the full subcategory of the category **cross** consisting of totally free crossed modules.

(7.1) **Theorem.** *The functor ρ in* (2.9) *induces equivalences of categories* $\rho\colon \mathbf{CW}^2/\overset{0}{\simeq} \xrightarrow{\sim} \mathbf{H}^2$ *and* $\rho\colon \mathbf{CW}^2/\simeq \xrightarrow{\sim} \mathbf{H}^2/\simeq$.

Proof. This follows from (II.3.3) and (2.9), compare also (VI.3.5) in Baues (AH) where we prove a more general result for 2-dimensional CW-complexes under a space D. □

In addition to (7.1) the functors

$$\mathbf{CW} \xrightarrow{\rho} \mathbf{H} \xrightarrow{C} \mathbf{H}_1$$

in (2.9) and (2.11) give us the following bijection of homotopy sets.

(7.2) **Theorem.** *Let X, Y be* CW*-complexes in* **CW** *with* $\dim(X) \le 2$. *Then one has bijections* $[X, Y] \approx [\rho X, \rho Y] \approx [C\rho X, C\rho Y] = [\hat{C}_* X, \hat{C}_* Y]$ *where* $C\rho X = \hat{C}_* X$ *since* $X^0 = *$. *More generally let X, Y be* CW*-complexes in* **CW** *and let A be a subcomplex of X with* $\dim(X/A) \le 2$, $* \in A$. *Moreover suppose* *that a cellular map $a: A \to Y$ is given. Then one has bijections*

$$[X,Y]^A \approx [\rho X, \rho Y]^{\rho A} \approx [\hat{C}_* X, \hat{C}_* Y]^{\hat{C}_* A}$$

of relative homotopy sets, see (4.6)(2).

Proof. The first part is a consequence of (II.3.3) and (2.9). The second part follows from the exact sequence (II.13.10) in Baues (AH) which is compatible with the model functor ρ_S in (6.12). □

Now we consider the action in (II.2.4)(3) for $n = 2$ which yields the top row of the following commutative diagram where $\dim(X) \leq 2$.

(7.3)
$$\begin{array}{ccc}
[X,Y]_\varphi \times \hat{H}^2(X, \varphi^* \pi_2 Y) & \xrightarrow{+} & [X,Y]_\varphi \\
\approx \downarrow \rho \times \hat{h}_* & & \approx \downarrow \rho \\
[\rho,\rho']_\varphi \times \hat{H}^2(C, \varphi^* H_2 C') & \xrightarrow{+} & [\rho,\rho']_\varphi \\
\approx \downarrow C \times 1 & & \approx \downarrow C \\
[C,C']_\varphi \times \hat{H}^2(C, \varphi^* H_2 C') & \xrightarrow{+} & [C,C']_\varphi .
\end{array}$$

Here we set $\rho = \rho X$, $\rho' = \rho Y$, $C = C\rho$, $C' = C\rho'$. Recall that $[A,B]_\varphi$ denotes the subset of $[A,B]$ consisting of all maps which induce φ on fundamental groups. We now describe formulas for the actions in (7.3). Let $f: \rho \to \rho'$ be a map in **H** which induces φ. We consider the diagram.

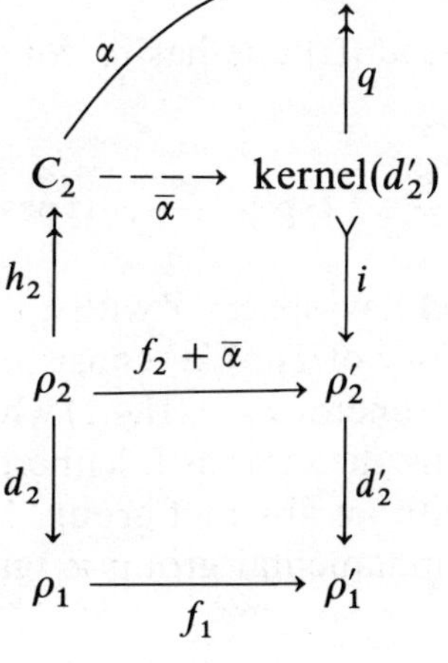

$$\begin{array}{ccc}
 & \overset{\alpha}{\nearrow} & H_2(C') \\
 & & \uparrow q \\
C_2 & \overset{\bar{\alpha}}{\dashrightarrow} & \operatorname{kernel}(d_2') \\
h_2 \uparrow & & \downarrow i \\
\rho_2 & \xrightarrow{f_2 + \bar{\alpha}} & \rho_2' \\
d_2 \downarrow & & \downarrow d_2' \\
\rho_1 & \xrightarrow[f_1]{} & \rho_1'
\end{array} \qquad (1)$$

The surjections q and h_2 are given by (2.11)(2). Let α be a φ-equivariant homomorphism which represents the cohomology class $\{\alpha\} \in H^2(C, \varphi^* H_2 C')$. Then we can choose a φ-equivariant homomorphism $\bar{\alpha}$ with $q\bar{\alpha} = \alpha$. We define the f_1-equivariant homomorphism $f_2 + \bar{\alpha}$ in (1) by the formula

$$(f_2 + \bar{\alpha})(x) = f_2(x) + i\bar{\alpha}h_2(x), \quad x \in \rho_2. \tag{2}$$

One readily checks that the map

$$f + \bar{\alpha} = (f_2 + \bar{\alpha}, f_1): \rho^2 \to \rho' \tag{3}$$

is a well defined map in **H** which induces φ. Now we define the action on $\{f\} \in [\rho, \rho']_\varphi$ by

$$\{f\} + \{\alpha\} = \{f + \bar{\alpha}\}. \tag{4}$$

Similarly we obtain the action in the bottom row of (7.3).

(7.4) **Proposition.** *Diagram* (7.3) *commutes.*

Proof. This is a consequence of (II.8.27)(3) in Baues (AH), we leave it to the reader to give a more direct proof. □

(7.5) **Remark.** Using the definitions of $\mathbf{H}_3^c$ and $\mathbf{H}_3^c/\simeq$ we see that theorem (2.9) is actually a consequence of theorem (7.1) and of (7.4). We leave such a proof of (2.9) as an exercise.

§8 The homotopy category of 2-types

As pointed out by Whitehead (C) one has to consider the hierachy of categories and functors

$$\textbf{1-types} \xleftarrow{P} \textbf{2-types} \xleftarrow{P} \textbf{3-types} \longleftarrow \cdots$$

where n-types are connected CW-spaces Y with $\pi_i(Y) = 0$ for $i > n$ and where **n-types** is the full subcategory of $\mathbf{Top}^*/\simeq$ consisting of n-types. The functor P is given by the Postnikov functor P_n in (II.4.1) which carries an $(n+1)$-type to its n-type. Since 1-types are the same as Eilenberg-Mac Lane spaces $K(\pi, 1)$ we can identify a 1-type with an abstract group. In fact, let **Gr** be the category of groups then the fundamental group π_1 gives us the equivalence of categories

(8.1) $$\pi_1\colon \textbf{1-types} \xrightarrow{\sim} \textbf{Gr}.$$

From this point of view n-types are natural objects of higher complexity extending abstract groups. Following up this idea Whitehead looked for a purely algebraic equivalent of an n-type, $n \geq 2$. An important requirement for such an algebraic system is "realizability", in two senses. In the first instance this means that there is an n-type which is in the appropriate relation to a given one of these algebraic systems, just as there is a 1-type whose fundamental group is isomorphic to a given group. The second kind is the "realizability" of "homorphisms" between such algebraic systems by maps of the corresponding n-types.

We now describe a purely algebraic equivalent of a 2-type. Recall that **cross** is the category of crossed modules $\partial\colon M \to N$ with $\pi_1(\partial) = \text{cokern}\ \partial$ and $\pi_2(\partial) = \text{kernel}\ \partial$ and that weak equivalences in **cross** are morphisms F which induce isomorphisms $\pi_1(F)$ and $\pi_2(F)$, see (2.6). Let Ho(**cross**) be the localization of the category **cross** with respect to weak equivalences.

(8.2) **Theorem.** *There is an equivalence of categories* $\bar{\rho}\colon \textbf{2-types} \xrightarrow{\sim} \text{Ho}(\textbf{cross})$.

Crossed modules form indeed an *optimal algebraic system* representing a 2-type since we have the properties (a)...(h) listed in the introduction of this chapter. In the next chapter we describe as well such an optimal algebraic system representing a 3-type which we call a quadratic module; each of the basic properties (a)...(h) in the introduction of this chapter corresponds to a similar property of quadratic modules.

Remark. Theorem (8.2) goes back to the work of Whitehead (CH) and Mac-Lane Whitehead though they do not formulate the result as an equivalence of categories. In Loday one finds a generalization of the equivalence (8.2) for the category **n-types**, $n \geq 2$. The algebraic systems of Loday (called cat^n-groups), however, do not satisfy for $n \geq 3$ optimality conditions like (b)...(h) in the introduction. Moreover there are various other generalizations of crossed modules in the literature which are algebraic models of 3-types, see Conduché, Brown-Gilbert and unpublished work of Joyal-Tierney. These models again do not satisfy such optimality conditions.

We give a new proof of (8.2) which only uses elementary arguments of homotopy theory and which is essentially based on the equivalence of categories in (7.1). We write down the proof in such a way that the same arguments can be used (almost literally) for the corresponding result (IV.10.1) below concerning models of 3-types.

(8.3) **Definition of $\bar{\rho}$.** Let X be a 2-type which is a CW-complex in **CW**. Then we get the commutative diagram

$$\begin{array}{ccccc} \pi_2(X^2) & \overset{j}{\rightarrowtail} & \rho_2(X) & \overset{d_2}{\longrightarrow} & \rho_1(X) \\ {\scriptstyle i}\downarrow & \text{push} & \downarrow & & \| \\ \pi_2(X) & \rightarrowtail & \bar{\rho}_2 & \overset{\partial}{\longrightarrow} & \bar{\rho}_1 \end{array} \tag{1}$$

where $\rho(X)$ is the crossed chain complex of X and where "push" is a push out diagram of groups. One readily checks that the bottom row is a crossed module with homotopy groups

$$\pi_n(\partial) = \pi_n(X) \quad \text{for } n = 1, 2. \tag{2}$$

The functor $\bar{\rho}$ carries X to $\bar{\rho}(X) = \partial$. Clearly diagram (1) is natural with respect to cellular maps F. This yields the induced map $\bar{\rho}(F)$. The class of $\bar{\rho}(F)$ in Ho(**cross**) depends only on the homotopy class of F in $\mathbf{CW}/\simeq$. Whence the definition of $\bar{\rho}$ is complete. The following notations are only used in proof of (8.2).

(8.4) **Definition.** We define a natural equivalence relation $\sim$ on the category $\mathbf{CW}^{n+1}$. For maps $F, G\colon Y^{n+1} \to X^{n+1}$ we set $F \sim G$ if $F|Y^n \simeq G|Y^n$ are homotopic in **Top***. Here $F|Y^n, G|Y^n\colon Y^n \to X^{n+1}$ are the restrictions of F and G respectively. Let $\mathbf{CW}^{n+1}/\sim$ be the quotient category. We have a quotient functor $\mathbf{CW}^{n+1}/\simeq \to \mathbf{CW}^{n+1}/\sim$

(8.5) **Proposition.** *The Postnikov functor P_n in* (II.4.1) *yields an equivalence of categories* $\mathbf{CW}^{n+1}/\sim \xrightarrow{\sim}$ *n*-**types**.

This result is due to Whitehead (*C*) who called an equivalence class of objects in $\mathbf{CW}^{n+1}/\sim$ an $(n+1)$-type which is now called an n-type since it corresponds to a space with only n non trivial homotopy groups.

(8.6) **Definition.** Let $\mathbf{PCW}^n$ be the following category, $n \geq 2$. Objects are pairs $(X^n, p\colon \pi_n(X^n) \twoheadrightarrow \pi_n)$ where X^n is a CW-complex in $\mathbf{CW}^n$ and where p is a surjective map of $\pi_1(X^n)$-modules. Morphisms

$$(\eta, \varphi_n)\colon (X^n, p) \to (Y^n, p') \tag{1}$$

are given by a homotopy class $\eta \in [X^n, Y^n]$ and by a $\pi_1(\eta)$-equivariant homomorphism $\varphi_n\colon \pi_n \to \pi_n'$ for which the diagram

$$\begin{array}{ccc} \pi_n(X^n) & \xrightarrow{\eta *} & \pi_n(Y^n) \\ {\scriptstyle p'}\big\downarrow & & \big\downarrow{\scriptstyle p'} \\ \pi_n & \xrightarrow[\varphi_n]{} & \pi'_n \end{array} \tag{2}$$

commutes. We obtain a natural equivalence relation $\sim$ on $\mathbf{PCW}^n$ by setting $(\eta, \varphi_n) \sim (\eta', \varphi'_n)$ if there exists a $\pi_1(\eta)$-equivariant homomorphism

$$\beta\colon \hat{C}_n(X^n) \to \ker(p') \subset \pi_n(Y^n) \tag{3}$$

with $\eta + \beta = \eta'$, see (II.2.4). We have the canonical functor

(8.7) $$\mathbf{CW}^{n+1}/\overset{0}{\simeq} \to \mathbf{PCW}^n$$

which carries the object X^{n+1} to the pair (X^n, p) where X^n is the n-skeleton of X^{n+1} and where $p\colon \pi_n X^n \twoheadrightarrow \pi_n X^{n+1}$ is induced by the inclusions $X^n \subset X^{n+1}$.

(8.8) **Proposition.** *The functor* (8.7) *induces an equivalences of categories*

$$\mathbf{CW}^{n+1}/\sim \xrightarrow{\sim} \mathbf{PCW}^n/\sim.$$

Proof. The functor (8.7) has the factorization

$$\mathbf{CW}^{n+1}/\overset{0}{\simeq} \xrightarrow{r} (\mathbf{H}^c_{n+1})^{n+1} \xrightarrow{r'} \mathbf{PCW}^n.$$

Here r is defined by (II.2.6) and r' carries the object $(C, f_{n+1}, X^n) = r(X^{n+1})$ to (X^n, p) where

$$p\colon \pi_n X^n \twoheadrightarrow \operatorname{cokernel}(f_{n+1}) = \pi_n(X^{n+1})$$

is the projection. A map in $(\mathbf{H}^c_{n+1})^{n+1}$ yields a map in $\mathbf{PCW}^n$ by (II.2.1)(5). We define a natural equivalence relation $\sim$ on $(\mathbf{H}^c_{n+1})^{n+1}$ similarly as in (8.6). For this we need (II.2.5)(a). Then we obtain induced functors

$$\mathbf{CW}^{n+1}/\sim \xrightarrow{r} (\mathbf{H}^c_{n+1})^{n+1}/\sim \xrightarrow{r'} \mathbf{PCW}^n/\sim.$$

One readily checks that these functors are equivalences of categories. □

(8.9) *Proof of* (8.2). For each object $\partial\colon M \to N$ in **cross** we choose a cofibrant model

$$* \rightarrowtail \rho \xrightarrow{\sim} \partial \tag{1}$$

in **cross chain**, see (4.13) and (4.17). By (1) we obtain the diagram

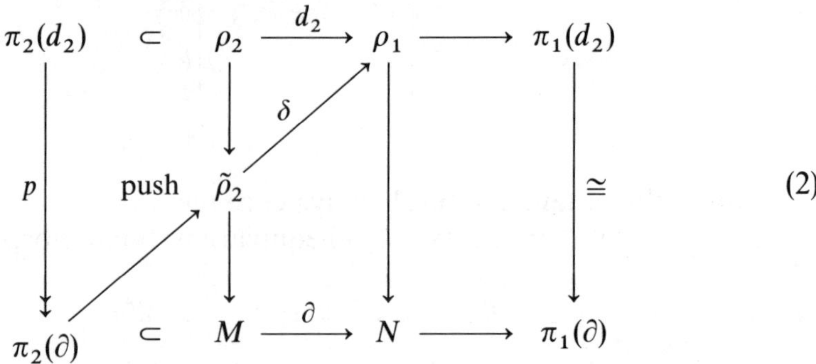

(2)

where "push" is a push out diagram of groups. The map $\delta \to \partial$ given by (2) is a weak equivalence in the category **cross** and d_2 is a totally free crossed module since ρ in (1) is cofibrant. Let **Pcross** be the following category. Objects are pairs (d_2, p) as in (2) and morphisms are defined as in $\mathbf{PCW}^2$, see (8.6). We obtain a natural equivalence relation $\sim$ on **Pcross** in the same way as in (8.6)(3), see (7.3) for the action. Now theorem (7.1) shows that

$$\rho\colon \mathbf{PWC}^2/{\sim} \xrightarrow{\sim} \mathbf{Pcross}/{\sim} \tag{3}$$

is an equivalence of categories. Moreover, we obtain by M in (4.12) the functor

$$M\colon \mathrm{Ho}(\mathbf{cross}) \xrightarrow{\sim} \mathbf{Pcross}/{\sim} \tag{4}$$

which is an equivalence of categories. The functor M carries the object ∂ to the object (d_2, p) which is given by the choice in (1). □

§9 The crossed chain complex of a product

It is well known that the cellular chain complex of a product $X \times Y$ of CW-complexes in **CW** is given by the formula

$$C_*(X \times Y) = C_*(X) \otimes C_*(Y)$$

where the right hand side is the tensor product of chain complexes with the differential

$$d(x \otimes y) = (dx) \otimes y + (-1)^{|x|} x \otimes dy$$

where $|x|$ is the degree of x, that is $|x| = n$ iff $x \in C_n(X)$. In this section we compute the crossed chain complex $\rho(X \times Y)$ by a similar formula

$$\rho(X \times Y) = \rho(X) \otimes \rho(Y).$$

Here the right hand side denotes the tensor product of crossed chain complexes which is originally due to Brown Higgins (TP).

Recall that **cross chain** denotes the category of crossed chain complexes. We define the **tensor product functor**.

(9.1) $\otimes$: **cross chain** $\times$ **cross chain** $\rightarrow$ **cross chain**

as follows. Let A, B be crossed chain complexes. As in the case of chain complexes we define the **degree** $|a|$ of an element $a \in A$ by $|a| = n$ iff $a \in A_n$. The sum $a + a'$ of elements $a, a' \in A$ is defined only on case $|a| = |a'|$. The crossed chain complex $A \otimes B$ is generated by elements $a \otimes b$, $a \otimes *$, $* \otimes b$ where $a \in A$, $b \in B$ with the following defining relation (plus, of course, the laws for crossed chain complexes):

$$|a \otimes b| = |a| + |b|, \qquad |a \otimes *| = |a|, \qquad |* \otimes b| = |b|. \tag{1}$$

$$\left.\begin{aligned}
&(a \otimes b)^{*\otimes t} = a \otimes (b^t) \quad \text{for } |t| = 1, \qquad |b| \geq 2 \quad \text{and} \\
&(* \otimes b)^{*\otimes t} = * \otimes (b^t) \quad \text{for } |t| = 1, \\
&(a \otimes b)^{s\otimes *} = (a^s) \otimes b \quad \text{for } |s| = 1, \qquad |a| \geq 2 \quad \text{and} \\
&(a \otimes *)^{s\otimes *} = (a^s) \otimes * \quad \text{for } |s| = 1.
\end{aligned}\right\} \tag{2}$$

$$\left.\begin{aligned}
&(a + a') \otimes * = a \otimes * + a' \otimes *, \\
&(a + a') \otimes b = a \otimes b + a' \otimes b \quad \text{for } |a| \geq 2, \\
&(a + a') \otimes b = (a \otimes b)^{a'\otimes *} + a' \otimes b \quad \text{for } |a| = 1.
\end{aligned}\right\} \tag{3}$$

$$\left.\begin{aligned}
&* \otimes (b + b') = * \otimes b + * \otimes b', \\
&a \otimes (b + b') = a \otimes b + a \otimes b' \quad \text{for } |b| \geq 2, \\
&a \otimes (b + b') = a \otimes b' + (a \otimes b)^{*\otimes b'} \quad \text{for } |b| = 1.
\end{aligned}\right\} \tag{4}$$

$d(a \otimes *) = (da) \otimes *, \qquad d(* \otimes b) = * \otimes (db) \quad \text{and } d(a \otimes b) =$

$$\left.\begin{array}{ll} -a\otimes * - *\otimes b + a\otimes * + *\otimes b & \text{for } |a| = |b| = 1, \\ [-(*\otimes b)^{a\otimes *} + *\otimes b] - a\otimes db & \text{for } |a| = 1, |b| \geq 2, \\ (da)\otimes b + (-1)^{|a|}[-(a\otimes *)^{*\otimes b} + a\otimes *] & \text{for } |a| \geq 2, |b| = 1, \\ (da)\otimes b + (-1)^{|a|} a\otimes (db) & \text{for } |a| \geq 2, |b| \geq 2. \end{array}\right\} \tag{5}$$

For maps $F\colon A \to A'$, $G\colon B \to B'$ in **cross chain** the induced map

$$F \otimes G\colon A \otimes B \to A' \otimes B' \tag{6}$$

is defined by $(F \otimes G)(a \otimes b) = (Fa) \otimes (Gb)$. For the initial and final object $*$ in **cross chain** (which is the trivial group in each degree) one clearly has

$$A \otimes * = A = * \otimes A. \tag{7}$$

The inclusions and projections

$$A \underset{p_A}{\overset{i_A}{\rightleftarrows}} A \otimes B \underset{p_B}{\overset{i_B}{\leftrightarrows}} B \tag{8}$$

are defined by $i_A(a) = a \otimes *$, $i_B(b) = * \otimes b$, $p_A(a \otimes *) = a$, $p_B(* \otimes b) = b$ and $p_A(a \otimes b) = 0$, $p_B(a \otimes b) = 0$ otherwise. Moreover, one has natural isomorphisms

$$\left.\begin{array}{l} \alpha\colon A \otimes (B \otimes C) \cong (A \otimes B) \otimes C, \\ \alpha(a \otimes (b \otimes c)) = (a \otimes b) \otimes c, \end{array}\right\} \tag{9}$$

$$\left.\begin{array}{l} T\colon A \otimes B \cong B \otimes A, \\ T(a \otimes b) = (-1)^{|a|\,|b|} b \otimes a. \end{array}\right\} \tag{10}$$

It is convenient to identify $a \otimes * = a$ and $* \otimes b = b$ in $A \otimes B$ so that the inclusions in (8) carry a to a and b to b. In degree 1 the group

$$(A \otimes B)_1 = A_1 * B_1 \tag{11}$$

is the free product of the groups A_1 and B_1. By (5) we see that

$$\pi_1(A \otimes B) = \pi_1(A) \times \pi_1(B) \tag{12}$$

is the product of fundamental groups. The action of a pair $(\alpha, \beta) \in \pi_1(A \otimes B)$

is given by

$$(a \otimes b)^{(\alpha,\beta)} = a^s \otimes b^t \quad \text{for } |a \otimes b| \geq 3 \tag{13}$$

where $\alpha = \{s\}$ and $\beta = \{t\}$. This follows from (2) since the pair (α, β) is represented by $s \otimes * + * \otimes t \in (A \otimes B)_1$.

Remark. We point out that in (5) the following equation holds $|a| = 1, |b| = 2$ $d(a \otimes b) = u - v = -v + u$ with $u = -(* \otimes b)^{a \otimes *} + * \otimes b, v = a \otimes db$. Similarly one has for $|a| = 2, |b| = 1$ the equation $d(a \otimes b) = v' + u' = u' + v'$ with $v' = (da) \otimes b, u' = -(a \otimes *)^{* \otimes b} + a \otimes *$. These equations are needed for the proof that T in (10) is compatible with the differential d. One obtains the equations as follows:

$$\begin{aligned} -v + u + v &= u^{dv} \\ &= (-(* \otimes b)^{a \otimes *} + * \otimes b)^{-a \otimes * - * \otimes db + a \otimes * + * \otimes db} \\ &= -(* \otimes b)^{a \otimes *} + * \otimes b = u. \end{aligned}$$

Here we use the laws of a crossed module and (2)...(5) above.

Remark. The tensor product above is more generally defined for crossed complexes in Brown-Higgins (TP). The generators $a \tilde{\otimes} b$ of Brown Higgins, however, are different from the generators $a \otimes b$ above since we have $a \otimes * = a \tilde{\otimes} *, * \otimes b = * \tilde{\otimes} b$, but $a \otimes b = -a \tilde{\otimes} b$. Our choice of generators has the advantage that for $|a| = |b| = 1$ the boundary $d(a \otimes b) = -a - b + a + b$ is the commutator and not as in Brown-Higgins the anticommutator. Moreover in the first variable the function $a \mapsto a \otimes b$ for $|a| = 1$ is a crossed homorphism for any $b \in B$, see (3).

(9.2) **Lemma.** *If A and B are totally free crossed chain complexes then also $A \otimes B$ is totally free. Basis elements of $A \otimes B$ are all elements $e \otimes f$, $e \otimes *$, $* \otimes f$ where e and f are basis elements of A and B respectively.*

The Lemma shows that the tensor functor (9.1) yields as well a functor

$$\otimes : \mathbf{H} \times \mathbf{H} \to \mathbf{H}$$

where $\mathbf{H}$ is the full subcategory of totally free crossed chain complexes. For A, B in $\mathbf{H}$ we describe below the totally free crossed chain complex $A \otimes B$ by defining the boundary $d(e \otimes f)$ explicitly on basis elements, see (9.4). For computations this description is more suitable than the definition in (9.1).

Moreover we prove the following result on products of CW-complexes X, Y. Recall that the product $X \times Y$ has the CW-topology given by the product cells $e \times f = e \otimes f$ where e is a cell in X and f is a cell in Y. Clearly the dimension of product cells satisfies $|e \times f| = |e| + |f|$.

(9.3) **Theorem.** *For CW-complexes X, Y with $X^0 = *$ and $Y^0 = *$ there is an isomorphism $\rho(X \times Y) \cong \rho(X) \otimes \rho(Y)$ in $\mathbf{H}$ which is natural with respect to cellular maps on X and Y.*

Addendum. *For homotopy systems $\bar{X}$, $\bar{Y}$ of order 3 one has a natural isomorphism $\rho(\bar{X} \otimes \bar{Y}) = \rho\bar{X} \otimes \rho\bar{Y}$ where $\bar{X} \otimes \bar{Y}$ is the tensor product of homotopy systems in* (II.2.7) *and where $\rho\colon \mathbf{H}_3^c \xrightarrow{\sim} \mathbf{H}$ is the equivalence of categories in* (2.9).

Proof. We get by (2.11)(5) the chain complexes $C = C\rho X = C_*\hat{X}$ and $C' = C\rho Y = C_*\hat{Y}$. Moreover the universal covering of $X \times Y$ satisfies $(X \times Y)^\wedge = \hat{X} \times \hat{Y}$ so that $C\rho(X \times Y) = C_*(\hat{X} \times \hat{Y}) = C \otimes_{\mathbb{Z}} C'$, compare (II.2.7). Moreover it is well known that the 2-cell in the product $S^1 \times S^1$ has an attaching map given by a commutator. These facts imply the isomorphism in (9.3) since we can use the characterization of $\rho(X) \otimes \rho(Y)$ in (9.4) below. □

For objects A, B in $\mathbf{H}$ let $C^A = C(A)$ and $C^B = C(B)$ be the chain complexes as defined in (2.11) and let $*$ be the generator of $C_0^A = \mathbb{Z}[\pi_1 A]$ and of $C_0^B = \mathbb{Z}[\pi_1 B]$. Then the elements

$$e^\alpha \otimes f^\beta, \quad e^\alpha \otimes *^\beta, \quad *^\alpha \otimes f^\beta, \quad *^\alpha \otimes *^\beta \in C^A \otimes_{\mathbb{Z}} C^B \tag{1}$$

with $\alpha \in \pi_1(A)$, $\beta \in \pi_1(B)$ and with basis elements e, f as in (9.2) form a $\mathbb{Z}$-basis of the tensor product $C^A \otimes_{\mathbb{Z}} C^B$ of chain complexes. We define an action of $(\alpha, \beta) \in \pi_1 A \times \pi_1 B$ on $a \otimes b \in C^A \otimes_{\mathbb{Z}} C^B$ by

$$(a \otimes b)^{(\alpha,\beta)} = a^\alpha \otimes b^\beta. \tag{2}$$

Then the $\mathbb{Z}$-boundary d on $C^A \otimes_{\mathbb{Z}} C^B$ is actually equivariant with respect to this action and satisfies on generators the formula

$$d(e \otimes f) = (de) \otimes f + (-1)^{|e|} e \otimes df. \tag{3}$$

Here we have by (2.11)(3) the formula $d(e) = 1 - [i(e)] = * - *^e$ for $|e| = 1$. Whence for $|e| = 1$ we get

$$(de) \otimes f = (* - *^e) \otimes f = (* \otimes f) - (* \otimes f)^{(e,0)} \tag{4}$$

and similarly for $|f| = 1$

$$e \otimes df = e \otimes (* - *^f) = (e \otimes *) - (e \otimes *)^{(0,f)}. \tag{5}$$

The formulas (1)...(5) determine the structure of $C^A \otimes_{\mathbb{Z}} C^B$ as a chain complex of free $\pi_1(A) \times \pi_1(B)$-modules which is generated in degree 0 by $* \otimes *$. The next lemma characterizes the tensor product $A \otimes B$ in terms of this chain complex.

(9.4) **Lemma.** *There is a unique object $A \otimes B$ in* **H** *with basis elements $e \otimes f$ as in* (9.2) *such that the following properties are satisfied: One has inclusions $A \subset A \otimes B$ and $B \subset A \otimes B$ which carry e and f to $e \otimes *$ and $* \otimes f$ respectively. For $|e| = 1 = |f|$ the boundary of $A \otimes B$ satisfies the formula*

$$d_2(e \otimes f) = -e \otimes * - f \otimes * + e \otimes * + f \otimes *$$

in $(A \otimes B)_1 = A_1 * B_1$. Moreover there is an isomorphism of chain complexes

$$C(A \otimes B) \cong C^A \otimes_{\mathbb{Z}} C^B, \quad \text{see (1) above,}$$

which is the identity on basis elements and which is equivariant with respect to the isomorphism $\pi_1(A \otimes B) \cong \pi_1 A \times \pi_1 B$.

Proof. As in (2.11)(2) we consider the diagram

$$\begin{array}{ccccccc}
\longrightarrow & (A \otimes B)_3 & \xrightarrow{d_3} & (A \otimes B)_2 & \xrightarrow{d_2} & (A \otimes B)_1 & \\
 & \cong \downarrow h_3 & & \downarrow h_2 & & \downarrow h_1 & \\
\longrightarrow & (C^A \otimes C^B)_3 & \xrightarrow{d} & (C^A \otimes C^B)_2 & \xrightarrow{d} & (C^A \otimes C^B)_1 & \longrightarrow
\end{array}$$

where one readily checks that $h_1 d_2 = dh_2$. Since h_2 induces an isomorphism on kernels, see (2.4), there is a unique d_3 with $d_2 d_3 = 0$ and $h_2 d_3 = dh_3$. For an explicit definition of d_3 we define the function

$$\otimes\colon A_1 \times B_1 \to (A \otimes B)_2 \tag{1}$$

as follows. On basis elements this function carries (e, f) to the basis element $e \otimes f$ of the free crossed module d_2. Since A_1 and B_1 are free groups the function $\otimes$ in (1) is thus well defined by the "**commutator rules**"

$$\left.\begin{aligned}
(a + a') \otimes b &= (a \otimes b)^{a' \otimes *} + a' \otimes b \\
a \otimes (b + b') &= a \otimes b' + (a \otimes b)^{* \otimes b'}.
\end{aligned}\right\} \tag{2}$$

Since commutators satisfy similar rules we get

$$d_2(a \otimes b) = -a \otimes * - * \otimes b + a \otimes * + * \otimes b. \tag{3}$$

We now define d_3 by

$$d_3(e \times f) = (de) \otimes f - (e \otimes *)^{* \otimes f} + e \otimes * \tag{4}$$

for $|e| = 2, |f| = 1$ and by

$$d_3(e \times f) = -(* \otimes f)^{e \otimes *} + * \otimes f - e \otimes (df) \tag{5}$$

for $|e| = 1$ and $|f| = 2$. Then (3) immediately implies that $d_2 d_3 = 0$ and by (9.3)(3), (4), (5) we see that $h_2 d_3 = dh_3$. □

The proof above shows that the defining formulas in (9.1) mainly originate from corresponding properties of commutators.

§ 10 The action of the fundamental group and free homotopy classes

Up to now we only considered base point preserving maps and base point preserving homotopies. The homotopy classes of such maps $X \to U$ form the set $[X, U] = [X, U]^*$ of homotopy classes relative the basepoint $*$, see (4.6)(2). If the inclusion $* \to X$ is a cofibration one has an action of the fundamental group $\pi_1(U)$ on the set $[X, U]$ such that the set of orbits of this action is the set $[X, U]^\phi$ of "free homotopy classes" of maps $X \to U$ (which need not to preserve the basepoint); here we assume that X is path connected. In this section we show that the function

$$\rho\colon [X, U] \to [\rho X, \rho U]$$

given by the functor $\rho\colon \mathbf{CW}/\simeq \to \mathbf{H}/\simeq$ is equivariant with respect to the action of the fundamental group. This allows the computation of the set $[X, U]^\phi$ of free homotopy classes in case $\dim(X) \le 2$.

Let X be a path connected CW-complex with basepoint $* \in X^0$ and let U be a pointed space. We define the **action of the fundamental group**

(10.1) $$[X, U] \times \pi_1(U) \to [X, U], \quad (\xi, \alpha) \mapsto \xi^\alpha,$$

by the homotopy extension property of the inclusion $* \to X$. Let $x\colon X \to U$

be a map in **Top*** which represents ξ and let $a: I \to U$ be a path which represents α, $a(0) = a(1) = *$. Then we can choose a map $H: I \times X \to U$ with $Hi_0 = x$ and $H|I \times \{*\} = a$. Now the map Hi_1 represents $\xi^\alpha = \alpha^\# \xi$, compare (II.5.7), (II.5.16), (II.5.18) Baues (AH). If U is path connected one has the bijection

(10.2) $$[X, U]/\pi_1(U) = [X, U]^\phi$$

where the left hand side denotes the set of orbits of the action (10.1) and where the right hand side in the set of homotopy classes relative the empty space ϕ, see (4.6)(2).

We now consider the action of the fundamental group in the category **cross chain**. Let ρ be a totally free crossed chain complex and let ρ' be a crossed chain complex with fundamental group $\pi_1(\rho') = \rho'_1/d\rho'_2$. We observe that each element $a \in \rho'_1$ induces a map in **cross chain**,

(10.3) $$1^a: \rho' \to \rho' \quad \text{by } 1^a(x) = x^a, \quad x \in \rho_n, \quad n \geq 1.$$

These maps yield the **action of the fundamental group**

(10.4) $$[\rho, \rho'] \times \pi_1(\rho') \to [\rho, \rho'], \quad (\xi, \alpha) \mapsto \xi^\alpha,$$

as follows. Let $x: \rho \to \rho'$ be a map which represents ξ and assume $a \in \rho'_1$ represents α, then the map $x^a = (1^a) \circ x$ represents ξ^α. Recall that $[\rho, \rho']$ denotes the set of homotopy classes rel $*$ in the cofibration structure (4.10), these homotopy classes are as well defined by the cylinder (3.3). The action (10.4) corresponds exactly to the action (10.1) since one has the following fact which can be proved inductively using the skeletal filtration of X, see also Whitehead (CH).

(10.5) **Proposition.** *Let X and Y be CW-complexes in* **CW** *and let*

$$\rho: [X, Y] \to [\rho X, \rho Y]$$

be given by the functor ρ in (6.4). *Then ρ is equivariant with respect to the action of the fundamental group. That is $\rho(\xi^\alpha) = \rho(\xi)^\alpha$ for $\alpha \in \pi_1 Y = \pi_1 \rho Y$.*

We point out that a similar action as in (10.4) can be defined on the set of homotopy classes of chain maps, $[C\rho X, C\rho Y]$, in $\mathbf{H}_1/\simeq$. Whence (10.5) shows for $X^0 = Y^0 = *$ that

(10.6) $$\hat{C}_* = C\rho: [X, Y] \to [C\rho X, C\rho Y]$$

is equivariant with respect to the action of the fundamental group. Here $\alpha \in \pi_1 Y$ acts on $\{x\} \in [C\rho X, C\rho Y]$ by $\{x\}^\alpha = \{x^\alpha\} = \{(1^\alpha)x\}$. The next corollary is immediate by (7.2) and (10.2).

(10.7) **Corollary.** *Let X, Y be CW-complexes in* **CW** *with* $\dim(X) \leq 2$. *Then we have bijections*

$$[X, Y]^\phi \approx [\rho X, \rho Y]/\pi_1 Y \approx [C\rho X, C\rho Y]/\pi_1 Y.$$

The universal object for the action (10.1) is the quotient space

$$\hat{I}(X) = (I \times X)/(\partial I \times \{*\}) \tag{10.8}$$

where $\partial I = \{0, 1\} \subset I$. Since $a(0) = a(1) = *$ the map H in (10.1) yields a map $H\colon \hat{I}(X) \to U$ which as well can be used for the definition of $\xi^\alpha = \{Hi_1\}$. We have inclusions $i = (i_0, i_1) X \vee X \subset \hat{I}(X)$ and $s\colon S^1 = I \times \{*\}/\partial I \times \{*\} \subset \hat{I}(X)$. The reduced cylinder (I.1.1) is the quotient

$$\hat{I}(X)/S^1 = I_* X. \tag{1}$$

The **torus on** X is the push out

$$\begin{array}{ccc} \hat{I}(X) & \xrightarrow{q} & S^1 \times X \\ \uparrow & & \uparrow \\ X \vee X & \xrightarrow{(1,1)} & X \end{array} \tag{2}$$

where $(1, 1)$ is the folding map, this is the product of the 1-sphere S^1 and X. The space $\hat{I}X$ is a CW-complex with trivial 0-skeleton if X is a CW-complex with $X^0 = *$. We now compute $\rho(\hat{I}X)$.

(10.9) **Definition.** Let ρ be a totally free crossed chain complex. Then we obtain $\hat{I}(\rho)$ by the push out diagram in **H**

$$\begin{array}{ccc} I(\rho) & \xrightarrow{\hat{q}} & \hat{I}(\rho) \\ {\scriptstyle i_1}\uparrow\sim & \text{push} & \sim\uparrow{\scriptstyle (i_1, j)} \\ \rho & \xrightarrow[i^s]{} & \rho \vee \rho(S^1) \end{array} \tag{1}$$

where $I\rho$ is the cylinder. The object $\rho(S^1) \cong \mathbb{Z}$ is trivial in degree $\neq 1$ and is generated by s in degree 1; $\rho(S^1) = \langle s \rangle$. The map $i\colon \rho \to \rho \vee \rho(S^1)$ is the inclusion and i^s is defined by $i^s(x) = (ix)^s$, $x \in \rho$. We set $i_0 = \hat{q}i_0: \rho \to \hat{I}(\rho)$. Since i_1 is a weak equivalence also (i_1, j) is one.

Clearly we get the cylinder $I(\rho)$ by the quotient

$$I(\rho) = \hat{I}(\rho)/j\rho(S^1). \tag{2}$$

Moreover, the tensor product $\rho(S^1) \otimes \rho$, given by (9.2), is part of the push out diagram

$$\begin{array}{ccc} \hat{I}(\rho) & \xrightarrow{\quad q \quad} & \rho(S^1) \otimes \rho \\ {\scriptstyle (i_0, i_1)} \uparrow & & \uparrow {\scriptstyle i'} \\ \rho \vee \rho & \xrightarrow{\quad (1,1) \quad} & \rho \end{array} \tag{3}$$

where $(1, 1)$ is the folding map. The map q carries the generator $\hat{q}(sx)$ to

$$q(\hat{q}sx) = -(s \otimes x), \quad x \in Z_n, \quad n \geq 1. \tag{4}$$

The map i' in (3) is given by (9.1)(8). We leave it to the reader to check that (3) is actually a push out. For this one needs the equation $q\hat{q}S_1(x) = -s \otimes x$, $x \in \rho_1$, which follows from (4).

(10.10) **Proposition.** *For a* CW-*complex* X *with* $X^0 = *$ *there is a natural isomorphism* $\rho(\hat{I}X) = \hat{I}(\rho X)$ *in* **H** *which carries the product cell* $(0, 1) \times e$ *in* $\hat{I}X$ *to* $\hat{q}se$ *where* e *is a cell in* $X - *$; *for* $e = *$ *the isomorphism carries the cell* $(0, 1) \times *$ *to* s.

This result is a slight expansion of the corresponding result for cylinders in (3.4). Proposition (10.10) can also be proved by the tensor product of crossed complexes in Brown-Higgins (TP). Clearly $\hat{q}$ in (10.9) is a homotopy $i_0 \simeq i_1^s$. Therefore (10.5) follows as well from (10.10) by use of the functor ρ: **CW** $\to$ **H**.

Appendix A

Obstructions for the realizability of chain complexes

The cellular chain complex $\hat{C}_* X$ of the universal covering is a basic algebraic invariant of a CW-complex X. In this appendix we consider the realizability of chain complexes by CW-complexes, that is, we study the following question. Given a group π and a chain complex C of free $\mathbb{Z}[\pi]$-modules. Does there exist a CW-complex X, an isomorphism $\varphi\colon \pi_1 X \cong \pi$ and a φ-equivariant weak equivalence $F\colon \hat{C}_* X \xrightarrow{\sim} C$? We show that such a CW-complex X exists if certain "obstructions" vanish. This result is an immediate consequence of corresponding properties of the CW-tower which gives us the sequence of functor ($n \geq 3$)

(A.1) $$\mathbf{CW} \xrightarrow{r} \mathbf{H}^c_{n+1} \xrightarrow{\lambda} \mathbf{H}^c_n \xrightarrow{\lambda} \cdots \xrightarrow{\lambda} \mathbf{H}^c_3 \underset{\sim}{\xrightarrow{\rho}} \mathbf{H} \xrightarrow{C} \mathbf{H}_1 \subset \mathbf{Chain}^{\wedge}_{\mathbb{Z}}.$$

The composition of these functors is naturally isomorphic to the functor $\hat{C}_*$. A realization of the chain complex C can therefore be achieved by the following procedure. First realize C in $\mathbf{H}_1$, then in $\mathbf{H}$, and then inductively in $\mathbf{H}^c_n$, $n \geq 3$. Each step involves an obstruction of realizability as described in the following proposition.

(A.2) **Proposition.** *Let C be a chain complex of free $\mathbb{Z}[\pi]$-modules with $C_i = 0$ for $i < 0$, $H_0 C = \mathbb{Z}$, $H_1 C = 0$. Then there is a primary obstruction*

$$\mathcal{O}_5(C) \in H^5(C, \Gamma H_2 C)$$

which vanishes if and only if there is an object $X_{(4)} \in \mathbf{H}^c_4$ with $C\rho\lambda X_{(4)} \xrightarrow{\sim} C$. Assume now an object $X_{(n)} \in \mathbf{H}^c_n$, $n \geq 4$, exists with $C\rho\lambda \ldots \lambda X_{(n)} \xrightarrow{\sim} C$. Then there is an obstruction

$$\mathcal{O}_{n+2}(X_{(n)}) \in H^{n+2}(C, \Gamma_n X_{(n)})$$

which vanishes if and only if there is an object $X_{(n+1)} \in \mathbf{H}^c_{n+1}$ with $\lambda X_{(n+1)} \simeq X_{(n)}$ in $\mathbf{H}^c_n/\simeq$ and whence with $C\rho\lambda \ldots \lambda X_{(n+1)} \xrightarrow{\sim} C$. Morever if $\dim(C) \leq n + 2$ then $\mathcal{O}_{n+2}(X_{(n)}) = 0$ implies that there is a CW-complex X in $\mathbf{CW}$ with $\hat{C}_* X \xrightarrow{\sim} C$.

Proof. By (2.13) we know that there exists an object A in **H** together with a weak equivalence $C(A) \xrightarrow{\sim} C$ in $\mathbf{Chain}_{\mathbb{Z}}^{\wedge}$. The equivalence of categories $\rho\colon \mathbf{H}_3^c \to \mathbf{H}$ yields an object $X_{(3)}$ in $\mathbf{H}_3^c$ with

$$\Gamma_3(X_{(3)}) \cong \Gamma H_2 C(A) \cong \Gamma H_2 C.$$

Therefore we obtain the obstruction $\mathcal{O}_5(C) = \mathcal{O}(X_{(3)})$ by (II.3.13). More generally the obstruction $\mathcal{O}_{n+2}(X_{(n)}) = \mathcal{O}(X_{(n)})$ is given as well by (II.3.13), $n \geq 3$. □

A realization up to isomorphism of C can be obtained by the next result.

(A.3) **Proposition.** *Let C be a chain complex for which there exists an object A in* **H** *with $C(A) \cong C$. If there is a CW-complex X in* **CW** *which admits a weak equivalence $\hat{C}_* X \xrightarrow{\sim} C$ in $\mathbf{H}_1$ then there exists a CW-complex Y in* **CW** *which admits an isomorphism $\hat{C}_* Y \cong C$ in $\mathbf{H}_1$.*

Proof. By (2.12) we see that there is a homotopy equivalence $\rho(X) \simeq A$ in **H**. Therefore (A.3) follows from the strong sufficiency of the functor $\rho\colon \mathbf{CW}/\simeq \to \mathbf{H}/\simeq$, see (2.9)(a). □

Remark. The obstructions in (A.2) coincide with the obstructions defined in Smith in terms of the Postnikov decomposition of a space. This can be seen by the isomorphism $\Gamma_n X \cong \hat{H}_{n+1} P_{n-1} X$ in (II.4.8). The use of Postnikov decompositions, however, makes it much harder to define the obstructions. In particular the simple connection of the obstructions with the attaching maps was not obtained in Smith, see (II.3.13)(2)(3)(4).

In (V.§ 3) below we show the next result.

(A.4) **Theorem.** *Let π be a finite group and let $|\pi|$ be the number of elements in π. Moreover let C be a chain complex as in (A.2). Then the primary obstruction $\mathcal{O}_5(C)$ is a torsion element with $|\pi| \cdot \mathcal{O}_5(C) = 0$.*

We now describe a chain complex C which is not realizable. This example is due to Whitehead (CH)(15.4); we here give a short proof by use of the description of $\mathcal{O}_5(C)$ in (II.3.13).

(A.5) **Example.** Let C be the following chain complex, $\pi = \mathbb{Z}/2$. The 4-skeleton of C is

$$C^4 = \hat{C}_*(S^2 \times \mathbb{R}P_2) = \hat{C}_* S^2 \otimes \hat{C}_* \mathbb{R}P_2$$

where $\mathbb{R}P_2 = * \cup e^1 \cup e^2$ is the projective plane. The generators of C^4 are $s_2 = s \otimes *$, $s_3 = s \otimes e^1$, $s_4 = s \otimes e^2$, $e_1 = * \otimes e^1$, $e_2 = * \otimes e^2$ where s is the 2-cell of S^2. Let $e \in \pi$ be the generator. Then the boundary is given by $de_1 = 0$, $ds_2 = 0, de_2 = e_1 + e_1^e, ds_3 = s_2 - s_2^e, ds_4 = s_3 + s_3^e$. Now C_5 is the free $\mathbb{Z}[\pi]$-module generated by a single element a_5 with $da_5 = s_4 - s_4^e = x$, this element is the generator of $H_4 C^4 = \mathbb{Z}$. Since the universal covering of $S^2 \times \mathbb{R}P^2$ is $S^2 \times S^2$ the secondary boundary $b_4\colon H_4 C^4 \to \Gamma(\pi_2)$ satisfies $b_4(x) = [u, v]$ where $u, v \in \pi_2 = H_2 C^4$ are given by $u = s_2, v = e_2 - e_2^e$. We now show that

$$b_4 d\colon C_4 \to \Gamma(\pi_2)$$

where $b_4 d(a_5) = [u, v]$ is not a coboundary so that $\mathcal{O}_5(C) = \{b_4 d\} \neq 0$, see (II.3.13)(4). Assume there is $\alpha\colon C_3 \to \Gamma(\pi_2)$ with $b_4 d = \alpha d_4$, that is $[u, v] = \alpha s_4 - (\alpha s_4)^e$. We have $\alpha(s_4) = n_1 \gamma(u) + n_2 \gamma(v) + n_3 [u, v]$ with $u^e = u$, $v^e = -v$ so that $[u, v] = 2n_3 [u, v]$ with $n_i \in \mathbb{Z}$, $i \in \{1, 2, 3\}$. Whence α cannot exist but we clearly get $2b_4 d = \alpha d_4$ for an appropriate α. Thus $\mathcal{O}_5(C) = \{b_4 d\}$ is a non trivial element of order 2. We also see by (A.4) above that $2\mathcal{O}_5(C) = 0$.

In connection with the *Steenrod problem* one finds in the literature many examples of chain complexes which are not realizable.

Appendix B

The homotopy category of pseudo projective planes

In this section we give a simple description of the homotopy category of pseudo projective planes which seems to be new, compare Olum and Rutter. **Pseudo projective planes**, P_f, are the most elementary 2-dimensional CW-complexes. They are obtained by attaching a 2-cell e to a 1-sphere S^1 by an attaching map $f: S^1 \to S^1$ of degree $f \geq 1$, that is

(B.1) $$P_f = S^1 \cup_f e = D/\sim_f.$$

Here D is the unit disk of complex numbers with boundary $S^1 = \partial D$ and with basepoint $* = 1$. The equivalence relation $\sim_f$ is generated by the relations $x \sim_f y \Leftrightarrow x^f = y^f$ with $x, y \in S^1$. Clearly $P_2 = \mathbb{R}P_2$ is the **real projective plane**. Let $\mathbf{P}$ be the category consisting of pseudo projective planes P_f and of cellular maps; this is a full subcategory of $\mathbf{CW}$. We consider the quotient functors

(B.2) $$\mathbf{P} \to \mathbf{P}/\overset{0}{\simeq} \to \mathbf{P}/\simeq \to \mathbf{P}/\simeq \operatorname{rel} \phi$$

where we use 0-homotopies ($\overset{0}{\simeq}$) running through cellular maps, homotopies ($\simeq$) relative $*$ and homotopies ($\simeq \operatorname{rel} \phi$) relative the empty space ϕ which are also called free homotopies. Moreover, there is a canonical functor

(B.3) $$\tau: \mathbf{Pair}(\mathbb{N}) \to \mathbf{P}$$

where $\mathbf{Pair}(\mathbb{N})$ is the category of pairs in the monoid $\mathbb{N}$ of natural numbers. Objects are elements $f \in \mathbb{N}$ and morphisms $f \to g$ are pairs $(\xi, \eta) \in \mathbb{N} \times \mathbb{N}$ with $g\xi = \eta f$. Let $[f, g]$ be the set of such morphisms $(\xi, \eta): f \to g$. The functor τ carries f to P_f and (ξ, η) to the map $\tau_\xi: P_f \to P_g$ with $\tau_\xi\{x\} = \{x\xi\}$ for $x \in D$, see (B.1). The induced homomorphism

(B.4) $$\pi_1(\xi, \eta) = \pi_1(\tau_\xi): \pi_1(P_f) = \mathbb{Z}/f \to \pi_1(P_g) = \mathbb{Z}/g$$

on fundamental groups is given by the number $\eta = g\xi/f$ which carries the generator $\mathbf{1} \in \mathbb{Z}/f$ to $\eta \cdot \mathbf{1} \in \mathbb{Z}/g$. Clearly τ above is a faithful functor. We now introduce the natural equivalence relation $\simeq$ on $\mathbf{Pair}(\mathbb{N})$ which is generated by the relations

$$(\xi,\eta) \simeq (\xi',\eta') \quad \Leftrightarrow \eta, \eta' \equiv 0 \bmod g,$$
$$\Leftrightarrow \pi_1(\xi,\eta) = \pi_1(\xi',\eta') = 0.$$

(B.5) **Theorem.** *The functor τ induces faithful functors*

$$\tau\colon \mathbf{Pair}(\mathbb{N}) \rightarrowtail \mathbf{P}/\overset{0}{\simeq},$$
$$\tau\colon \mathbf{Pair}(\mathbb{N})/\simeq\ \rightarrowtail \mathbf{P}/\simeq,$$
$$\tau\colon \mathbf{Pair}(\mathbb{N})/\simeq\ \rightarrowtail \mathbf{P}/\simeq \operatorname{rel} \phi.$$

The image category of τ in $\mathbf{P}/\simeq$ is the subcategory of principal maps in the sense of (V.§ 3) in Baues (AH). We prove the theorem in (B.9) below. We now define a category $\mathbf{R}$ which is actually a simple algebraic model of the category $\mathbf{P}/\overset{0}{\simeq}$.

(B.6) **Definition.** The objects of the category $\mathbf{R}$ are the elements $f \in \mathbb{N}$. A morphism $\lambda \in \mathbf{R}(f,g)$ is an element $\lambda \in \mathbb{Z}[\mathbb{Z}/g]$ for which there is $\eta \in \mathbb{Z}$ with $g \cdot \varepsilon(\lambda) = f \cdot \eta$. Here $\varepsilon\colon \mathbb{Z}[\mathbb{Z}/g] \to \mathbb{Z}$ is the augmentation of the group ring. Composition $\lambda \circ \mu$ for $\mu \in \mathbf{R}(h,f)$ is defined by

$$\lambda \circ \mu = \lambda \cdot \lambda_{\#}(\mu) \tag{1}$$

where the right hand side is a product in the group ring $\mathbb{Z}[\mathbb{Z}/g]$. The homomorphism $\lambda_{\#}\colon \mathbb{Z}[\mathbb{Z}/f] \to \mathbb{Z}[\mathbb{Z}/g]$ with $\lambda_{\#}[x] = [\eta x]$ is induced by the homomorphism $\pi_1(\lambda) = \eta\colon \mathbb{Z}/f \to \mathbb{Z}/g$. Let

$$\partial_f = \sum_{x \in \mathbb{Z}/f} [x] \tag{2}$$

be the norm element in $\mathbb{Z}[\mathbb{Z}/f]$. We introduce natural equivalence relations $\simeq$ and $\simeq \operatorname{rel} \phi$ on the category $\mathbf{R}$ as follows $(\lambda, \mu \in \mathbf{R}(f,g))$:

$$\begin{aligned} \lambda \simeq \mu \Leftrightarrow \pi_1(\lambda) = \pi_1(\mu) \quad &\text{and} \quad \exists \beta \in \mathbb{Z}[\mathbb{Z}/g] \\ \text{with} \quad &\lambda - \mu = \lambda_{\#}(\partial_f) \cdot \beta, \end{aligned} \tag{3}$$

$$\begin{aligned} \lambda \simeq \mu \operatorname{rel} \phi \Leftrightarrow \pi_1(\lambda) = \pi_1(\mu) \quad &\text{and} \quad \exists x \in \mathbb{Z}/g,\ \beta \in \mathbb{Z}[\mathbb{Z}/g] \\ \text{with} \quad &\lambda^x - \mu = \lambda \cdot [x] - \mu = \lambda_{\#}(\partial_f) \cdot \beta. \end{aligned} \tag{4}$$

(B.7) **Theorem.** *There are isomorphisms of categories*

$$\rho\colon \mathbf{P}/\overset{0}{\simeq} \xrightarrow{\sim} \mathbf{R},$$

$$\rho\colon \mathbf{P}/\simeq \xrightarrow{\sim} \mathbf{R}/\simeq,$$

$$\rho\colon \mathbf{P}/\simeq \operatorname{rel}\phi \xrightarrow{\sim} \mathbf{R}/\simeq \operatorname{rel}\phi.$$

The proof of this result is based on the following commutative diagram in which the abelianization h_2 is an isomorphism.

$$\text{(B.8)}\qquad \begin{array}{ccc} \pi_2(P_f, S^1) & \xrightarrow{\partial} & \pi_1(S^1) = \mathbb{Z} \\ {\scriptstyle h_2}\downarrow\cong & \nearrow{\scriptstyle f\cdot\varepsilon} & \downarrow{\scriptstyle h_1} \\ \mathbb{Z}[\mathbb{Z}/f] & \xrightarrow[d_2]{} & \mathbb{Z}[\mathbb{Z}/f] \end{array}$$

This diagram is a special case of diagram (2.2). The boundary d_2 describes $\hat{C}_* P_f$. Since h_1 is a $(\mathbb{Z} \to \mathbb{Z}/f)$-crossed homomorphism we see that d_2 is given by the formula $d_2(x) = x \cdot \partial_f$ where ∂_f is the norm element in (B.6). The map ε in (B.8) is the augmentation. The map h_2 is actually an isomorphism since $\pi_2(P_f, S^1)$ is abelian. We see this since the generators e^n of this group, $n \in \mathbb{Z}$, satisfy the formula

$$-e^n - e^m + e^n + e^m = \langle e^n, e^m \rangle - \langle e^m, e^m \rangle = 0$$

in the crossed module ∂.

Proof of (B.7). We use the equivalences in (7.1) and (10.7). In fact $\mathbf{R}$ is a subcategory of $\mathbf{H}^2$ by identifying $\lambda \in \mathbf{R}(f, g)$ with the commutative diagram

$$\begin{array}{ccc} \mathbb{Z}[\mathbb{Z}/f] & \xrightarrow{f\cdot\varepsilon} & \mathbb{Z} \\ \downarrow{\scriptstyle\xi} & & \downarrow{\scriptstyle\eta} \\ \mathbb{Z}[\mathbb{Z}/g] & \xrightarrow[g\cdot\varepsilon]{} & \mathbb{Z} \end{array}$$

with $\xi[1] = \lambda$. Here ξ is η-equivariant so that (ξ, η) is a map in $\mathbf{H}^2$ by (B.8). It is clear that $\mathbf{R}$ is the full subcategory of $\mathbf{H}^2$ given by the crossed modules in (B.8). This proves that $\rho\colon \mathbf{P}/\overset{0}{\simeq} \xrightarrow{\sim} \mathbf{R}$ is the restriction of the equivalence in (7.1). A homotopy $\alpha\colon (\xi, \eta) \simeq (\xi', \eta')$ in $\mathbf{H}^2$ is an η-crossed homomorphism $\alpha\colon \mathbb{Z} \to \mathbb{Z}[\mathbb{Z}/g]$ which is determined by an element $\alpha(1) = \beta$ as in (B.6)(3). The equation $-\xi + \xi' = \alpha(f\varepsilon)$ is equivalent to the equation

$$-\lambda + \lambda' = -\xi[\mathbf{1}] + \xi'[\mathbf{1}] = \alpha(f) = \alpha(\mathbf{1})\cdot(\eta_{\#}\partial_f)$$

which is equivalent to the equation in (B.6)(3). □

(B.9) *Proof of* (B.5). The functor ρ in (B.7) carries τ_ξ to the element $\xi\cdot[0] \in \mathbf{R}(f,g)$ where $[0]$ is the unit of the ring $\mathbb{Z}[\mathbb{Z}/g]$. By (B.7) we know

$$\tau(\xi,\eta) \simeq \tau(\xi',\eta') \Leftrightarrow \varphi = \pi_1(\xi,\eta) = \pi_1(\xi',\eta') \quad \text{and}$$

$$\exists \beta \in \mathbb{Z}[\mathbb{Z}/g] \quad \text{with } (\xi - \xi')[0] = \beta\cdot\varphi_{\#}(\partial_f).$$

This implies $\eta - \eta' = \varepsilon(\beta)\cdot g$. We now observe

$$\varphi_{\#}\partial_f = \sum_{x\in\mathbb{Z}/f}[\varphi x] = t\cdot\sum_{y\in\varphi\mathbb{Z}/f}[y] \tag{1}$$

where t is the number of elements in the kernel of φ. For $\beta = \sum_{y\in\mathbb{Z}/g} a_y[y]$ in $\mathbb{Z}[\mathbb{Z}/g]$ we have

$$\begin{aligned}\beta\cdot\varphi_{\#}\partial_f &= t\cdot\sum_{y\in\mathbb{Z}/g,\, v\in\varphi\mathbb{Z}/f} a_y[y+v],\\ &= t\cdot\sum_{u\in\mathbb{Z}/g}\Bigg(\sum_{y\in u+\varphi\mathbb{Z}/f} a_y\Bigg)[u].\end{aligned} \tag{2}$$

Now $\beta\cdot\varphi_{\#}\partial_f = (\xi' - \xi)[0]$ with $\varphi = \pi_1\tau_\xi = \pi_1\tau_{\xi'}$ implies

$$0 = \sum_{y\in u+\varphi\mathbb{Z}/f} a_y \quad \text{for } u\in\mathbb{Z}/g, \quad u \neq 0. \tag{3}$$

The number of elements in $\varphi\mathbb{Z}/f$ is $g/gcd(\eta,g)$. If $gcd(\eta,g) < g$ we add up the equations in (3) for $u \in U = \{x\cdot 1; 1 \le y \le gcd(\eta,g)\} \subset \mathbb{Z}/g$. Since the union of all $u + \varphi\mathbb{Z}/f, u\in U$, is $\mathbb{Z}/g$ we get $\varepsilon(\beta) = \sum_y a_y = 0$ for $gcd(\eta,g) < g$. Whence in this case $\eta = \eta'$ and thus $(\xi,\eta) = (\xi',\eta')$. If $gcd(\eta,g) = g$, that is, if $\varphi = 0$ we see by (3) that $a_y = 0$ for $y \neq 0$ and $a_0 = \xi' - \xi$, so that in this case β exists. Similarly one gets the proposition for the relation $\simeq \operatorname{rel}\phi$. □

For $\varphi \in \operatorname{Hom}(\mathbb{Z}/f,\mathbb{Z}/g)$ let $[f,g]_\varphi$ and $[P_f,P_g]_\varphi$ be the set of all morphisms in $[f,g]$ and $[P_f,P_g]$ respectively which induce φ on fundamental groups, see (B.4). By (B.5) the function

(B.10) $$\tau\colon [f,g]_\varphi \to [P_f,P_g]_\varphi$$

is injective for $\varphi \neq 0$ and is identically 0 if $\varphi = 0$. The group of integers $\mathbb{Z}$ acts freely on $[f, g]$ by $(\xi, \eta) + k = (\xi + kf, \eta + kg)$ and $[f, g]_\varphi$ is the orbit of (ξ, η) with $\pi_1(\xi, \eta) = \varphi$. On the other hand the coaction $P_f \to P_f \vee S^2$ induces an action $+$ of the cohomology group

(B.11) $$E_\varphi = \hat{H}^2(P_f, \varphi^* \pi_2 P_g) = \pi_2 P_g / (\pi_2 P_g) \cdot \varphi_\# \partial_f, \text{ see (II.2.4)},$$

on the set $[P_f, P_g]_\varphi$ which is transitive and effective. The group $\pi_2 P_g$ can be described by each of the following equations (compare (B.8)).

(B.12) $$\begin{aligned} \pi_2 P_g &= H_2 \hat{P}_g = \{x \in \mathbb{Z}[\mathbb{Z}/g] \mid \partial_g \cdot x = 0\} \\ &= \text{kernel}(\varepsilon\colon \mathbb{Z}[\mathbb{Z}/g] \to \mathbb{Z}) = ([0] - [\mathbf{1}]) \cdot \mathbb{Z}[\mathbb{Z}/g]. \end{aligned}$$

Let $t\colon \mathbb{Z} \to E_\varphi$ be the homomorphism mapping **1** to the class of $t_\varphi = f \cdot [0] - \varphi_\# \partial_f \in \text{kernel}(\varepsilon) = \pi_2 P_g$.

(B.13) **Proposition.** *τ in* (B.10) *is t-equivariant or equivalently*

$$\tau(\xi, \eta) + k \cdot t_\varphi = \tau(\xi + kf, \eta + kg) \text{ in } [P_f, P_g].$$

This result follows easily from (7.4). We can define an inverse of the equivalence ρ in (B.7) by defining a functor

(B.14) $$r\colon \mathbf{R} \to \mathbf{P}/\overset{0}{\simeq}$$

as follows. The functor r carries the object f to P_f and carries the morphism λ to

$$r(\lambda) = \tau_{\varepsilon(\lambda)} + \alpha\colon P_f \to P_g.$$

Here ε is the augmentation and $\alpha = \lambda - \varepsilon(\lambda)[0]$ is an element in $\pi_2 P_g$ by (B.12). Addition is defined by the coaction $P_f \to P_f \vee S^2$ so that $r(\lambda)$ is well defined in $\mathbf{P}/\overset{0}{\simeq}$. We clearly have $\pi_1 r(\lambda) = \pi_1(\lambda)$, see (B.6). Moreover for $\lambda = \xi \cdot [0]$ we get $r(\lambda) = \tau_\xi$, see (B.3).

We next derive from (B.7) a result on the **group of homotopy equivalences** $\text{Aut}(P_f)^*$, resp. $\text{Aut}(P_f)^\phi$ in the category $\mathbf{P}/\simeq$, resp. $\mathbf{P}/\simeq$ rel ϕ. Let I be the ideal generated by the norm element ∂_f in $\mathbb{Z}[\mathbb{Z}/f]$ and let U_f be the group of units in the quotient ring $\mathbb{Z}[\mathbb{Z}/f]/I$. Moreover let $\tilde{U}_f$ be the group whose elements are those of U_f but with a multiplication

$$\{\lambda\} \circ \{\mu\} = \{\lambda \cdot \lambda_\#(\mu)\}$$

Here $\{\lambda\}$ denotes the class of $\lambda \in \mathbb{Z}[\mathbb{Z}/f]$ modulo I. There are obvious units in $\tilde{U}_f$ represented by the elements $[x]$, $x \in \mathbb{Z}/f$; they form an abelian and normal subgroup $T \cong \mathbb{Z}/f$ of $\tilde{U}_f$.

(B.15) **Proposition.** *There are isomorphisms of groups as in the commutative diagram*

$$\begin{array}{ccc} \mathrm{Aut}(P_f)^* & \xrightarrow{\cong} & \tilde{U}_f \\ \downarrow & & \downarrow \\ \mathrm{Aut}(P_f)^\phi & \xrightarrow{\cong} & \tilde{U}_f/T \end{array}$$

where the vertical arrows are the quotient maps.

Proof. Let $E(f)$ be the group of equivalences of the object f in $\mathbf{R}/\simeq$. Then we have $\{\lambda\} \in E(f)$ iff there is μ with $\lambda\mu \simeq [0]$, $\mu\lambda \simeq [0]$. This is equivalent to $\mu\cdot(\mu_\#\lambda) \simeq [0]$ and $\pi_1\mu = (\pi_1\lambda)^{-1}$. This is the case iff

$$\exists\beta \text{ with } \mu\cdot(\mu_\#\lambda) = [0] + \beta\cdot\mu_\#\partial_f$$

$$\Leftrightarrow \exists\beta \text{ with } (\lambda_\#\mu)\cdot\lambda = [0] + (\lambda_\#\beta)\cdot\partial_f \Leftrightarrow \{\lambda\} \in U_f.$$

Since the composition in $\mathrm{Aut}(P_f)^*$ corresponds to the composition in $\mathbf{R}$, see (B.6)(1), we get the isomorphism for $\mathrm{Aut}(P_f)^*$. We get the isomorphism for $\mathrm{Aut}(P_f)^\phi$ by (10.7). □

We do not know whether the functor

$$\pi_1\colon \mathbf{P}/\simeq \xrightarrow{\sim} \mathbf{R}/\simeq \twoheadrightarrow \mathbf{FCyc}$$

admits a splitting where **FCyc** is the category of finite cyclic groups $\mathbb{Z}/f$, $f \in \mathbb{N}$. However a splitting of the homomorphism

$$\pi_1\colon \mathrm{Aut}(P_f)^* \cong \tilde{U}_f \twoheadrightarrow \mathrm{Aut}(\mathbb{Z}/f)$$

can be constructed as follows. For this we consider the commutative diagram

$$\text{(B.16)} \qquad \begin{array}{ccc} [f,g] & \xrightarrow{\Gamma} & \mathbf{R}[f,g] \\ \downarrow{\scriptstyle \pi_1} & & \downarrow{\scriptstyle q} \\ \mathrm{Hom}(\mathbb{Z}/f,\mathbb{Z}/g) & \overset{\tilde{\Gamma}}{\dashrightarrow} & \mathbf{R}[f,g]/\simeq \end{array}$$

Here Γ is defined for $(\xi, \eta) \in [f, g]$ by

$$\Gamma(\xi, \eta) = \sum_{j=0}^{\xi-1} [j \cdot \varphi \mathbf{1}] \in \mathbb{Z}[\mathbb{Z}/g]$$

with $\varphi = \pi_1(\xi, \eta)$ and q is the quotient map.

(B.17) **Lemma.** *The function* Γ *induces a function* $\tilde{\Gamma}$ *such that* (B.16) *commutes.*

Proof. We have to check that $\pi_1(\xi,\eta) = \pi_1(\xi',\eta') = \varphi$ implies $\Gamma(\xi,\eta) \simeq \Gamma(\xi',\eta')$:

$$\begin{aligned}\Gamma(\xi + f, \eta + g) &= \sum_{j=0}^{\xi+f-1} [j \cdot \varphi \mathbf{1}] = \Gamma(\xi, \eta) + \sum_{j=\xi}^{\xi+f-1} [j \cdot \varphi \mathbf{1}] \\ &= \Gamma(\xi, \eta) + \varphi_{\#}\left([\xi \cdot \mathbf{1}] \cdot \sum_{j=0}^{f-1} [j \cdot \mathbf{1}]\right) \\ &= \Gamma(\xi, \eta) + \varphi_{\#}[\xi \cdot \mathbf{1}] \cdot \varphi_{\#}\partial_f \end{aligned}$$

□

(B.18) **Proposition.** *Let* U_f^1 *be the group of units x in the quotient ring* $\mathbb{Z}[\mathbb{Z}/f]/I$ *with* $\varepsilon(x) = 1$. *Then we have the split short exact sequence of groups*

$$0 \longrightarrow U_f^1 \longrightarrow \tilde{U}_f \xrightarrow{\pi_1} \mathrm{Aut}(\mathbb{Z}/f) \longrightarrow 0$$

where $\tilde{U}_f \cong \mathrm{Aut}(P_f)^*$ *by* (B.15). *The splitting carries* $\varphi \in \mathrm{Aut}(\mathbb{Z}/f)$ *to* $(\varphi^{-1})_{\#}\tilde{\Gamma}(\varphi)$. *Hence we also have a split short exact sequence*

$$0 \to U_f^1/T \to \tilde{U}_f/T \to \mathrm{Aut}(\mathbb{Z}/f) \to 0$$

where T is the subgroup in (B.15) *with* $\tilde{U}_f/T \cong \mathrm{Aut}(P_f)^\phi$.

As an example one gets for the real projective plane $\mathbb{R}P_2$ $\mathrm{Aut}(\mathbb{R}P_2)^* = \mathbb{Z}/2$ and $\mathrm{Aut}(\mathbb{R}P_2)^\phi = 0$.

(B.19) **Remark.** The first part of (B.18) is proved by different methods in (3.5) of Olum. It is known that there is an isomorphism of abelian groups

$$U_f^1 \cong \mathbb{Z}^\chi \oplus \mathbb{Z}/f, \quad T \cong \mathbb{Z}/F,$$

where χ is the number of all $i \in \mathbb{N}$, $1 \leq i \leq f/2$, for which i is not a divisor of f. It is, however, a deep number theoretic problem to determine the action of $\mathrm{Aut}(\mathbb{Z}/f)$ on $\mathbb{Z}^\chi \oplus \mathbb{Z}/f$ in terms of basis elements, compare Olum.

Next we consider the category of 'algebraic 2-types' of pseudo projective planes, see (B.23).

(B.20) **Definition.** Let F, G: $P_f \to P_g$ be in **P**; we write $F \approx G$ if the induced maps on π_1 and π_2 coincide, that is $\pi_1 F = \pi_1 G$ and $\pi_2 F = \pi_2 G$. Moreover for $\lambda, \mu \in \mathbf{R}[f, g]$ we write $\lambda \approx \mu$ if $\pi_1 \lambda = \pi_1 \mu$ and $\lambda - \mu = (\lambda - \mu) \cdot \lambda_\#[1]$ in $\mathbb{Z}[\mathbb{Z}/g]$.

(B.21) **Proposition.** *The functor* ρ *in* (B.7) *induces an isomorphism of quotient categories* $\rho: \mathbf{P}/\approx \xrightarrow{\sim} \mathbf{R}/\approx$.

For $\varphi \in \mathrm{Hom}(\mathbb{Z}/f, \mathbb{Z}/g)$ let $\bar{\bar{E}}_\varphi$ be the subgroup of E_φ in (B.11) given by

$$\bar{\bar{E}}_\varphi = \{\alpha \in \pi_2 P_g | \varphi_\#([0] - [1]) \cdot \alpha = 0\} / (\pi_2 P_g) \cdot \varphi_\# \partial_f.$$

This subgroup is a group cohomology of $\mathbb{Z}/f$, namely we have $i^*: \bar{\bar{E}}_\varphi = H^2(\mathbb{Z}/f, \varphi^* \pi_2 P_g) \subset E_\varphi = \hat{H}^2(P_f, \varphi^* \pi_2 P_g)$ with i^* induced by the canonical map $i: P_f \subset K(\mathbb{Z}/f, 1)$. The following lemma describes exactly all $(\approx)$-equivalence classes.

(B.22) **Proposition.** *The group* $\bar{\bar{E}}_\varphi$ *acts via* (B.11) *freely on the set* $[P_f, P_g]_\varphi$ *and the orbits are the* $(\approx)$*-equivalence classes. Moreover, we have* $\bar{\bar{E}}_\varphi = 0$ *if* φ *is surjective or injective. Otherwise we have*

$$\bar{\bar{E}}_\varphi \cong (\mathbb{Z}/t)^{(g \cdot t/f) - 1}$$

where t *is the order of* kernel(φ).

Proof. The first part is clear by the definition of $\approx$ and by (B.21). We now compute $\bar{\bar{E}}_\varphi$. We have

$$\bar{\bar{E}}_\varphi = A_\varphi / t \cdot A_\varphi \tag{1}$$

where A_φ is the following subgroup of $\mathbb{Z}[\mathbb{Z}/g]$.

$$A_\varphi = \left\{ \sum_{\Delta \in G} a_\Delta \left(\sum_{j \in \Delta} [j] \right) : \sum_\Delta a_\Delta = 0 \right\}. \tag{2}$$

Here G is the quotient group $(\mathbb{Z}/g)/\varphi(\mathbb{Z}/f)$ and $\Delta \in G$ is a coset which is a subset of $\mathbb{Z}/g$. We easily see that $\alpha \in \pi_2 P_g = \mathrm{kernel}(\varepsilon)$ and $\alpha = [\varphi 1]\alpha$ imply $\alpha \in A_\varphi$. Whence we get (1) by

$$A_\varphi = (\pi_2 P_g)\cdot(1/t)\varphi_\# \partial_f. \tag{3}$$

compare (B.9)(1). □

(B.23) **Remark.** The category $\mathbf{P}/\approx$ can be identified with the category of algebraic 2-types of pseudo projective planes; this is a full subcategory of $\mathbf{T}_2^2$ in (VI.8.11) of Baues (AH). The linear extension of categories in (VI.8.11) of Baues (AH) restricted to $\mathbf{P}/\approx$ yields the linear extension in example (3.3) of Baues Wirsching.

Appendix C

On the suspension and the James construction

The infinite reduced product $JX = X_\infty$ was originally used by James as a topological model of the loop space $\Omega\Sigma X$ of the suspension ΣX. We here introduce a James construction in general categories, in particular in the category of crossed chain complexes. This leads to interesting algebraic models of JX.

(C.1) **Definition.** Let $\mathbf{C}$ be a category with an initial object $*$. Moreover assume a bifunctor $\otimes: \mathbf{C} \times \mathbf{C} \to \mathbf{C}$ together with natural isomorphisms $X \otimes * = X = * \otimes X$, $X \in \mathbf{C}$, is given. Then we obtain the functor

$$J: \mathbf{C} \to \mathbf{C} \tag{1}$$

which we call the **James construction** in $\mathbf{C}$. The functor J carries an object X to the colimit of the diagram

$$* \to X \rightrightarrows X^{\otimes 2} \substack{\to\\\to\\\to} X^{\otimes 3} \dots \tag{2}$$

where we define $X^{\otimes n}$ inductively by $X^{\otimes 1} = X$ and $X^{\otimes n} = (X^{\otimes(n-1)}) \otimes X$, $n \geq 2$. The maps

$$i_t: X^{\otimes(n-1)} \to X^{\otimes n} \quad (t = 1, \dots, n) \tag{3}$$

in (2) are given by the composition

$$\begin{array}{ccc} X^{\otimes(n-1)} & = & (\dots((X^{\otimes(t-1)} \otimes *) \otimes X) \dots \otimes X) \\ {\scriptstyle i_t}\downarrow & & \downarrow{\scriptstyle (\dots(1 \otimes 0) \otimes 1) \dots \otimes 1)} \\ X^{\otimes n} & = & (\dots((X^{\otimes(t-1)} \otimes X) \otimes X) \dots \otimes X) \end{array}$$

Here $0: * \to X$ is given since $*$ is the initial object and 1 denotes the identity. We do not assume the functor $\otimes$ is associative. Since all maps i_t are natural with respect to maps $f: X \to Y$ in $\mathbf{C}$ we see that the functor J in (1) is well defined provided the colimit (2) exists in **C** for all X. We also get a natural filtered object

$$J_0X = * \to J_1X = X \to J_2X \to \cdots \to J_nX \to \tag{4}$$

where J_nX is the colimit of the subdiagram $* \to \ldots X^{\otimes n}$ of (2). Clearly JX is as well the colimit of (4).

The definition of the James construction in (C.1) is motivated by the classical example

(C.2) $$J\colon \mathbf{CW} \to \mathbf{CW}.$$

Here the bifunctor $\otimes$ on **CW** is given by the product $X \otimes Y = X \times Y$ of CW-complexes with the CW-topology. The space J_nX for $X \in \mathbf{CW}$ is also obtained by the quotient

$$J_nX = (X \times \cdots \times X)/\sim \tag{1}$$

where the equivalence relation $\sim$ on the n-fold product is generated by

$$(x_1, \ldots, x_{n-1}, *) \sim (x_1, \ldots, x_{t-1}, *, x_t, \ldots, x_{n-1}) \tag{2}$$

with $t = 1, \ldots, n$ and $x_t \in X$. As a set JX can be identified with the free monoid generated by the set $X - *$. The empty word in JX is denoted by $*$. Moreover JX is a CW-complex the cells of which are exactly all product cells

$$e_1 \times \cdots \times e_n, \quad e_i \quad \text{cell of } X - * \quad \text{for } i = 1, \ldots, n. \tag{4}$$

In this sense the free monoid generated by all cells of $X - *$ is the set of all cells in $JX - *$. The multiplication

$$\mu\colon JX \times JX \to JX \tag{5}$$

which carries the pair $(x_1 \ldots x_m, y_1 \ldots y_n)$ to $x_1 \ldots x_m y_1 \ldots y_n$ is a cellular map in **CW**. Here we use the CW-topology of the product $JX \times JX$. It is a classical result of James that there is a natural homotopy equivalence

$$g\colon JX \xrightarrow{\simeq} \Omega\Sigma X \tag{6}$$

for X with $X^0 = *$. The map g is obtained by the loop addition $+$ in the loop space $\Omega\Sigma X$ as follows. Let $i\colon X \to \Omega\Sigma X$ be adjoint to the identity of ΣX then one gets g by $g(x) = i(x)$ and $g(x_1 \ldots x_n) = g(x_1 \ldots x_{n-1}) + i(x_n)$.

Using the tensor product $\otimes$ in the category of crossed chain complexes one gets by (C.1) the functor

(C.3) $$J\colon \textbf{cross chain} \to \textbf{cross chain}.$$

We now describe this functor more explicitly.

(C.4) **Definition.** Let A be a crossed chain complex. The crossed chain complex JA is generated by all words $a_1 \ldots a_n$ ($a_i \in A$, $i = 1, \ldots, n$ and $n \geq 1$) with the following defining relations (plus, of course, the laws of crossed chain complexes). Let u, v be such words or empty words ϕ and let $a, a', t \in A$. Then (a) denotes the word given by $a \in A$.

$$|a_1 \ldots a_n| = |a_1| + \cdots + |a_n|. \tag{1}$$

$$(uav)^t = u(a^t)v \text{ for } |t| = 1, |a| \geq 2 \text{ and } (a)^t = (a^t) \text{ for } |t| = 1. \tag{2}$$

$$u(a + a')v = \begin{cases} uav + ua'v & \text{for } |a| \geq 2, \\ ua'v + (uav)^{a'} & \text{for } |a| = 1, |u| \geq 1, \\ (uav)^{a'} + ua'v & \text{for } |a| = 1, |v| \geq 1, \\ (a) + (a') & \text{for } u = \phi = v. \end{cases} \tag{3}$$

$d(a) = (da)$ and

$$d(uv) = \begin{cases} -u - v + u + v & \text{for } |u| = |v| = 1, \\ -v^u + v - u(dv) & \text{for } |u| = 1, |v| \geq 2, \\ (du)v + (-1)^{|u|}(-u^v + u) & \text{for } |u| \geq 2, |v| = 1, \\ (du)v + (-1)^{|u|}u(dv) & \text{for } |u| \geq 2, |v| \geq 2. \end{cases} \tag{4}$$

For a map $F\colon A \to B$ in **cross chain** the induced map $JF\colon JA \to JB$ is defined by $(JF)(a_1 \ldots a_n) = (Fa_1) \ldots (Fa_n)$. There is a well defined natural map

$$\mu\colon JA \otimes JA \to JA \tag{5}$$

in **cross chain** given by $\mu(u \otimes v) = uv$, $\mu(* \otimes v) = v$, $\mu(u \otimes *) = u$. Therefore JA is an example of a '*crossed chain algebra*'.

We leave it to the reader to check that JA, defined by (C.4), coincides with the definition of JA as a colimit in (C.1). Moreover one can check by (9.2) that JA is totally free if A is totally free. In fact, if Z is a basis of A then $\mathrm{Mon}(Z) - *$ is a basis of JA. Here $\mathrm{Mon}(Z)$ is the free monoid generated by Z. The restriction

(C.5) $$J: \mathbf{H} \to \mathbf{H}$$

of the functor J in (C.3) has by (9.3) and (6.12) the following important property.

(C.6) **Theorem.** *For the functor* ρ: $\mathbf{CW} \to \mathbf{H}$ *one has a natural isomorphism*

$$\rho(JX) \cong J(\rho X)$$

in $\mathbf{H}$ *where we use the functor* J *in* (C.2). *This isomorphism is compatible with multiplications* μ.

By (C.4) this result yields an explicit computation of all boundary maps in $\rho(JX)$. This as well gives us a formula for $\hat{C}_* JX = C\rho(JX)$ by use of the functor C in (2.11). As an application of (C.2)(6) and (7.2) we get

(C.7) **Corollary.** *Let* X, Y *be* CW-*complexes in* $\mathbf{CW}$ *with* $\dim(X) \leq 2$. *Then one has the binatural isomorphism of groups*

$$[\Sigma X, \Sigma Y] \cong [X, JY] \cong [\rho X, J\rho Y]$$

where the group structure in $[\rho X, J\rho Y]$ *is induced by* μ *in* (C.4)(5). *As a special case one gets* $\pi_3(\Sigma Y) = \pi_2(J\rho Y)$.

It follows from the cylinder in $\mathbf{H}$ that the functor J carries a homotopy equivalence in $\mathbf{H}$ to a homotopy equivalence. The functor J, however, does not take a weak equivalence in **cross chain** to a weak equivalence. For this we consider the following example which we studied in a joint work with D. Conduché.

(C.8) **Example.** We consider a group G as a crossed chain complex which is concentrated in degree 0. A cofibrant model

$$* \rightarrowtail \rho_G \xrightarrow{\sim} G \tag{1}$$

of G in **cross chain** can be obtained by

$$\rho_G = \rho K(G, 1) \tag{2}$$

where $K(G, 1)$ is an Eilenberg-Mac Lane complex in $\mathbf{CW}$ of the group G. The weak equivalence in (1) induces a canonical map

$$\alpha: \rho J K(G, 1) = J(\rho_G) \to J(G) \tag{3}$$

which in general is not a weak equivalence. This map, however, is always a 2-equivalence, that is, $\pi_i(\alpha)$ is an isomorphism for $i \leq 2$ and is surjective for $i = 3$. The map α in (3) has the following topological interpretation. Let $\underset{\sim}{\pi}(J_* K(G,1))$ be the crossed chain complex of the filtered object $J_* K(G,1)$, see (C.1)(4). Then one has the filtration preserving inclusion

$$\beta\colon (JK(G,1))^n \to J_n K(G,1) \tag{4}$$

which induces

$$\beta_*\colon \rho(JK(G,1)) \to \underset{\sim}{\pi}(J_* K(G,1)). \tag{5}$$

Now there is an isomorphism of crossed chain complexes such that the following diagram commutes.

$$\begin{array}{ccc} \rho(JK(G,1)) & \xrightarrow{\beta_*} & \underset{\sim}{\pi}(J_* K(G,1)) \\ \wr\| & & \wr\| \\ J(\rho_G) & \xrightarrow[\alpha]{} & J(G) \end{array} \tag{6}$$

In particular one has the isomorphism of groups

$$\pi_n(J_n K(G,1), J_{n-1} K(G,1)) \cong (JG)_n. \tag{7}$$

Here the left hand side is defined by a topological homotopy group while the right hand side is given purely algebraically by (C.4). For a *free group* G the space $K(G,1)$ can be chosen to be a one point union of 1-spheres, in this case α and β_* above are actually isomorphisms.

In chapter VI we shall use the following property of the 2-type of $J(G)$ where G is a free group. We consider the diagram

(C.9)
$$\begin{array}{ccc} J(G)_2 & \xrightarrow{q_2} & G^{ab} \otimes G^{ab} \\ \downarrow d & & \downarrow w \\ G & \xrightarrow[q_1]{} & G/\Gamma_3 G \end{array}$$

where q_1 is the quotient map and where w is the commutator map, see (I.4.16) and (IV.1.1). The map q_2 is the Hurewicz homomorphism

$$q_2\colon J(G)_2 = \pi_2(J_2 X, J_1 X) \to H_2(J_2 X, J_1 X) = G^{ab} \otimes G^{ab}$$

where X is a one point union of 1-spheres with $\pi_1 X = G$. The map q_2 carries the generator ab $(a, b \in G)$ to $\{a\} \otimes \{b\}$ with $\{a\} \in G^{ab}$. Clearly the diagram commutes since d as well is the commutator map with $d(ab) = -a - b + a + b$.

(C.10) **Theorem.** *Let G be a free group. Then the map w in (C.9) is a crossed module and (q_1, q_2) gives us a map $q\colon J(G) \to w$ in* **cross chain** *which induces an isomorphism $\pi_i(q)$ for $i \leq 2$.*

Proof. We clearly have $\pi_1 J(G) = G^{ab} = \operatorname{cok}(w)$ and we have

$$\pi_2 J(G) = \pi_3 \Sigma X = \Gamma(G^{ab}) = \operatorname{kernel}(w). \qquad \square$$

The theorem shows that the 2-type of the totally free crossed chain complex $J(G)$ for a free group G is actually given by the simple crossed module w which is not totally free. This fact is the reason why nilpotency degree 2 is sufficient for the algebraic models of 4-dimensional CW-complexes in chapter IV, compare the discussion in chapter VI.

If G is not a free group the 2-type of $J(G)$ is still given by the tensor product $G \mathbin{\overline{\otimes}} G$ of Brown-Loday; that is, one has an isomorphism of crossed modules

$$\text{(C.11)} \qquad \begin{array}{ccc} J(G)_2/dJ(G)_3 & = & G \mathbin{\overline{\otimes}} G \\ \downarrow{\scriptstyle d} & & \downarrow{\scriptstyle \overline{w}} \\ G & = & G \end{array}$$

where $\overline{w}$ is the commutator map. Whence we derive from the 2-equivalence α in (C.8)(3) the next result.

(C.12) **Theorem.** *Let G be a group. Then one has the isomorphism*

$$\pi_3 \Sigma K(G, 1) = \ker(\overline{w}\colon G \mathbin{\overline{\otimes}} G \to G)$$

More generally one has for any CW-complex X with $\dim X \leq 2$, $X^0 = *$ the bijection

$$[\Sigma X, \Sigma K(G, 1)] = [\rho X, \overline{w}]$$

where the right hand side is a set of homotopy classes in **cross chain**, see (4.6)(2) and (4.10). If G is free we can replace $\overline{w}$ in these formulas by $w\colon G^{ab} \otimes G^{ab} \to G/\Gamma_3 G$, in (C.9).

Proof. We have the 2-equivalence $J\rho K(G,1) \to J(G) \to \bar{w}$ in **cross chain**, see (C.8)(3). Such a 2-equivalence induces for $\dim(X) \leq 2$ the bijection

$$\begin{aligned}[\Sigma X, \Sigma K(G,1)] &= [X, JK(G,1)] \\ &= [\rho X, J\rho K(G,1)] \\ &= [\rho X, \bar{w}]. \end{aligned}$$ □

The formula for $\pi_3 \Sigma K(G,1)$ in (C.12) is due to Brown-Loday; their proof, however, is completely different since they do not use the James construction.

Appendix D

The homotopy category of suspended pseudo projective planes

In this appendix we apply the James construction for crossed chain complexes. We consider the suspensions

$$\Sigma^{n-1}P_f = M(\mathbb{Z}/f, n) = S^n \cup_f e^n \tag{D.1}$$

of pseudo projective planes, $n \geq 1$, which are Moore spaces of cyclic groups. Let $\mathbf{P}_n$ be the category consisting of the spaces $\Sigma^{n-1}P_f$, $f \geq 1$, and of cellular maps. This is a full subcategory of **CW**. In appendix B above we study the category $\mathbf{P} = \mathbf{P}_1$ of pseudo projective planes and its homotopy category $\mathbf{P}/\simeq$. We here compute the suspension functor $\Sigma\colon \mathbf{P}_n/\simeq \; \to \mathbf{P}_{n+1}/\simeq$, $n \geq 1$, which is an isomorphism of categories for $n \geq 3$. For this we consider the commutative diagram of functors

$$\begin{array}{ccccccc} \mathbf{Pair}(\mathbb{N}) & \xrightarrow{\tau} & \mathbf{P}/\simeq & \xrightarrow{\Sigma} & \mathbf{P}_2/\simeq & \xrightarrow{\Sigma} & \mathbf{P}_3/\simeq \\ & & \downarrow{\scriptstyle \pi_1} & \swarrow{\scriptstyle H_2} & & \swarrow{\scriptstyle H_3} & \\ & & \mathbf{FCyc} & & & & \end{array} \tag{D.2}$$

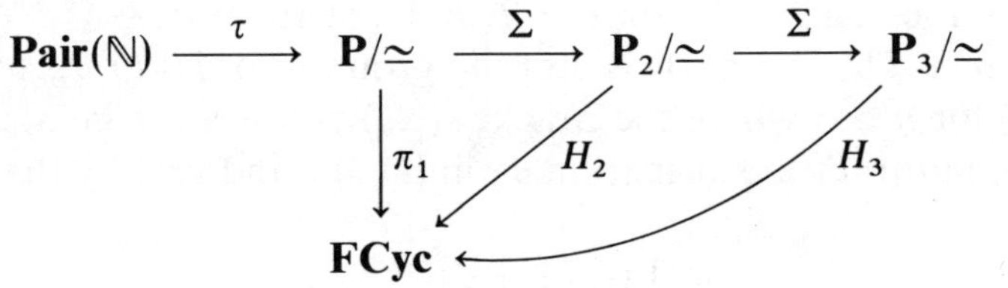

where H_2 and H_3 are the homology functors. The next results seems to be new; recall that $[f, g]$ is the set of morphisms $f \to g$ in $\mathbf{Pair}(\mathbb{N})$, $f, g \in \mathbb{N}$, see (B.3).

(D.3) **Theorem.** *Let $\varphi \in \operatorname{Hom}(\mathbb{Z}/f, \mathbb{Z}/g)$. Then there is a unique element $\bar{\varphi} = B_2(\varphi)$ in the image of*

$$\Sigma\tau\colon [f, g] \to [\Sigma P_f, \Sigma P_g]$$

with $H_2\bar{\varphi} = \varphi$. Moreover there is a unique element $\bar{\bar{\varphi}} = B_3(\varphi)$ in the image of

$$\Sigma\Sigma\colon [P_f, P_g] \to [\Sigma^2 P_f, \Sigma^2 P_g]$$

with $H_3\bar{\bar{\varphi}} = \varphi$.

(D.4) **Corollary.** *The functors H_n $(n = 2, 3)$ in* (D.2) *admit a splitting functor*

$$B_n\colon \mathbf{FCyc} \to \mathbf{P}_n/\simeq$$

with $H_n B_n = 1$

This follows immediately from (D.3) since the definition of $B_n(\varphi)$ is compatible with compositions. The splitting functor B_n, however, is not additive; below we describe the distributivity law for $B_n(\varphi + \varphi')$. The functors H_n in (D.2) are part of the following commutative diagram in which the rows are split linear extensions of categories (compare (II.1.2))

$$\text{(D.5)}\qquad \begin{array}{ccccc} E^2+ & \rightarrowtail & \mathbf{P}_2/\simeq & \overset{H_2}{\twoheadrightarrow} & \mathbf{FCyc} \\ \downarrow{\scriptstyle \sigma_*} & & \downarrow{\scriptstyle \Sigma} & & \| \\ E^3+ & \rightarrowtail & \mathbf{P}_3/\simeq & \underset{H_3}{\twoheadrightarrow} & \mathbf{FCyc} \end{array}$$

Here E^n is the bifunctor on **FCyc** given by

$$E^n(\mathbb{Z}/f, \mathbb{Z}/g) = \operatorname{Ext}(\mathbb{Z}/f, \Gamma_n^1 \mathbb{Z}/g) \tag{1}$$

where Γ_n^1 is the functor Γ for $n = 2$ and the functor $-\otimes \mathbb{Z}/2$ for $n \geq 3$, compare (I.6.9). The group (1) is a cyclic group of order $(f, 2g, g^2)$ for $n = 2$ and $(f, g, 2)$ for $n \geq 3$ where the bracket $(\ldots)$ denotes the greatest common divisor. The natural transformation σ_* in (D.5) is induced by the surjection

$$\sigma\colon \Gamma(\mathbb{Z}/g) \to \mathbb{Z}/g \otimes \mathbb{Z}/2 \tag{2}$$

compare (I.4.2). The action of E^n on $\mathbf{P}_n/\simeq$ is given by the well known central extension of groups

$$\operatorname{Ext}(\mathbb{Z}/f, \pi_{n+1} U) \overset{i}{\rightarrowtail} [\Sigma^{n-1} P_f, U] \xrightarrow{\pi_n} \operatorname{Hom}(\mathbb{Z}/f, \pi_n U) \tag{3}$$

which is known as the '*universal coefficient sequence*' compare Hilton or (V.3a) in Baues (AH). For $U = \Sigma^{n-1} P_g$ we have $\pi_{n+1} U = \Gamma_n^1(\mathbb{Z}/g)$. The splitting B_n gives us an identification $(n \geq 2)$

$$\operatorname{Hom}(\mathbb{Z}/f, \mathbb{Z}/g) \times E^n(\mathbb{Z}/f, \mathbb{Z}/g) = [\Sigma^{n-1} P_f, \Sigma^{n-1} P_g] \tag{4}$$

which carries (φ, α) to $B_n(\varphi) + i(\alpha)$. The composition in $\mathbf{P}_n/\simeq$ then satisfies the simple formula

$$(\varphi, \alpha) \circ (\Psi, \beta) = (\varphi\Psi, \varphi_* \alpha + \Psi^* \beta). \tag{5}$$

This indeed yields a very simple algebraic description of the category $\mathbf{P}_n/\simeq$. The suspension functor in (D.5) is given by $\Sigma(\varphi, \alpha) = (\varphi, \sigma_* \alpha)$. We now consider the image category of the functor $\Sigma\colon \mathbf{P}/\simeq \to \mathbf{P}_2/\simeq$, compare (I.0.2). Recall that $\mathbf{1} \in \mathbb{Z}/f$ denotes the canonical generator.

(D.6) **Definition.** For maps $u, v\colon P_f \to P_g$ in $\mathbf{P}$ we set $u \equiv v$ if $\Sigma f \simeq \Sigma g$. Whence the quotient category $\mathbf{P}/\equiv$ is the same as the image category $\Sigma(\mathbf{P}/\simeq)$. For morphisms $\lambda, \mu \in \mathbf{R}(f, g)$, see (B.6), we set $\lambda \equiv \mu$ if $\pi_1(\lambda) = \pi_1(\mu)$ and if for some $\beta \in \mathbb{Z}[\mathbb{Z}/g]$ with

$$\lambda - \mu - \varepsilon(\lambda - \mu)[0] = ([0] - [\mathbf{1}]) \cdot \beta$$

the greatest common divisor $(f, g^2, 2g)$ divides $g \cdot \varepsilon(\beta)$.

The following result shows that the image category $\Sigma(\mathbf{P}/\simeq)$ is surprisingly small. By (D.3) we know that the image category $\Sigma\Sigma(\mathbf{P}/\simeq)$ is isomorphic to **FCyc**.

(D.7) **Theorem.** *The isomorphism ρ in* (B.7) *induces an isomorphism of categories*

$$\rho\colon \Sigma(\mathbf{P}/\simeq) = \mathbf{P}/\equiv \xrightarrow{\cong} \mathbf{R}/\equiv.$$

Moreover one has a split linear extension of categories

$$\tilde{E} + \rightarrowtail \mathbf{P}/\equiv \underset{\pi_1}{\twoheadrightarrow} \mathbf{FCyc}$$

where $\tilde{E}$ is the quotient of E^2 above with $\tilde{E}(\mathbb{Z}/f, \mathbb{Z}/g) = g \cdot \operatorname{Ext}(\mathbb{Z}/f, \Gamma(\mathbb{Z}/g))$. This group is $\mathbb{Z}/2$ if $(f, g^2, 2g) = 2g$ and in 0 otherwise. The splitting is given by B_2 in (D.3).

We derive from (D.6) and (D.7) the following commutative diagram in which the rows are split extensions of groups.

$$\begin{array}{ccccc}
 & & \Sigma\operatorname{Aut}(P_f)^* & \cong & \operatorname{Aut}(\mathbb{Z}/f) \\
 & & \downarrow & & \| \\
\mathbb{Z}/f & \rightarrowtail & \operatorname{Aut}(\Sigma P_f)^* & \twoheadrightarrow & \operatorname{Aut}(\mathbb{Z}/f) \\
\downarrow & & \downarrow & & \| \\
\mathbb{Z}/(f,2) & \rightarrowtail & \operatorname{Aut}(\Sigma^2 P_f)^* & \twoheadrightarrow & \operatorname{Aut}(\mathbb{Z}/f)
\end{array} \tag{D.8}$$

Remark. Let S be the splitting in (B.18). Then a splitting of the exact sequences in (D.8) is given by ΣS. This splitting actually coincides with the restriction of B_2 since the top row of (D.8) is an isomorphism. Using different methods the split extension for $\mathrm{Aut}(\Sigma P_f)^*$ was obtained by Sieradski.

The crucial step for the proof of (D.3) and (D.7) is the computation of the suspension homomorphism Σ on $\pi_2 P_g$. For this we consider the diagram

(D.9)
$$\begin{array}{ccc} \pi_2(P_g) & \xrightarrow{\quad\Sigma\quad} & \pi_3\Sigma P_g \\ \wr\Vert & & \wr\Vert \\ ([0]-[\mathbf{1}])\mathbb{Z}[\mathbb{Z}/g] & \xrightarrow{\quad\Sigma'\quad} & \mathbb{Z}/(g^2, 2g) \end{array}$$

where Σ' is defined by the formula

$$\Sigma'(([0]-[\mathbf{1}])\beta) = g\cdot\varepsilon(\beta)\cdot\mathbf{1}$$

for $\beta \in \mathbb{Z}[\mathbb{Z}/g]$. Here ε is the augmentation. The right hand isomorphism carries the generator $\mathbf{1}$ to the Hopf element $S^3 \to S^2 \subset \Sigma P_g$. The left hand side isomorphism is described in (B.12).

(D.10) **Lemma.** *Diagram* (D.9) *commutes. This shows that $\Sigma\pi_2(P_g)$ is of order 2 if g is even, and is trivial if g is odd. Moreover the subgroup $\Sigma^2\pi_2(P_g) = 0$ is trivial in $\pi_4(\Sigma^2 P_g)$.*

Proof. We use (C.7) so that Σ in (D.9) corresponds to the map

$$i_*\colon \pi_2(\rho) \to \pi_2 J(\rho) = H_2 CJ(\rho) \tag{1}$$

where $\rho = \rho(P_g)$ and where $i\colon \rho \subset J(\rho)$ is the inclusion. Let $e = e_1$ and e_2 be the cells of P_g which are the generators of ρ with $d(e_2) = g\cdot e = e + \cdots + e$. The boundary $d\colon J(\rho)_3 \to J(\rho)_2$ is given on generators by the following formulas.

$$d(ee_2) = -e_2^e + e_2 - e(g\cdot e) = -e_2^e + e_2 - \sum_{i=0}^{g-1} (ee)^{ie} \tag{2}$$

$$d(e_2 e) = (g\cdot e)e - e_2^e + e_2 = \left(\sum_{i=1}^{g} (ee)^{(g-i)e}\right) - e_2^e + e_2 \tag{3}$$

$$d(eee) = -(ee)^e + ee \tag{4}$$

These equations are simple applications of (C.4) since $d(ee) = 0$. The same equations hold in $CJ(\rho)$. Let $B = d(CJ\rho)_3$ be the group of boundaries. Then we get modulo B the congruences (see (B.6)(2))

$$e_2([0] - [\mathbf{1}]) \equiv (ee)\cdot\partial_g \equiv (ee)(g\cdot[0]). \tag{5}$$

The Hopf element in (D.9) corresponds to the cycle ee in $CJ(\rho)$. Whence (5) shows that Σ' is defined correctly for $\beta = [0]$. For general β we can choose $\gamma \in \mathbb{Z}[\mathbb{Z}/g]$ such that $\beta = \varepsilon(\beta)[0] + \gamma([0] - [\mathbf{1}])$. Then we get

$$e_2([0] - [\mathbf{1}])\beta \equiv (ee)\partial_g\beta = (ee)\partial_g\varepsilon(\beta) \equiv (ee)(g\cdot\varepsilon(\beta)[0]) \tag{6}$$

since $([0] - [\mathbf{1}])\partial_g = 0$ for the norm element ∂_g. This proves (D.10). □

(D.11) *Proof of* (D.3) *and* (D.7). The second part of (D.3) and also (D.7) follow immediately from lemma (D.10) since Σ is compatible with the coaction on P_f. For the first part of (D.3) we have to check that $\Sigma' t_\varphi$ is trivial if considered as an element in the group

$$\operatorname{Ext}(\mathbb{Z}/f, \Gamma(\mathbb{Z}/g)) = \mathbb{Z}/(f, 2g, g^2). \tag{1}$$

Here we use (B.13). By (B.9)(1) we know

$$t_\varphi = f\cdot[0] - \varphi_\# \partial_f = t\cdot\left(v[0] - \sum_{x\in V}[x]\right) \tag{2}$$

where $t = |\ker\varphi|$, $t\cdot v = f$, $V = \operatorname{image}(\varphi)$ with $|V| = v$. Since $t_\varphi \in \ker(\varepsilon)$ there is β with

$$v[0] - \sum_{x\in V}[x] = ([0] - [\mathbf{1}])\beta. \tag{3}$$

This shows that there is an integer b with

$$\varepsilon(\beta) = g\cdot b - u(v(v-1))/2 \tag{4}$$

where $u\cdot v = g$. Whence $\Sigma'(t_\varphi)$ is given by the element

$$\Sigma'(t_\varphi) = t\cdot g\cdot\varepsilon(\beta)\cdot\mathbf{1} = -(tg^2(v-1)/2)\mathbf{1} \tag{5}$$

in $\mathbb{Z}/(g^2, 2g)$. Now it is clear that (5) represents the trivial element in the group (1). □

The morphism sets $[\Sigma P_f, \Sigma P_g]$ in $\mathbf{P}_2/\simeq$ are groups since the suspension ΣP_f is a co-H-group. As pointed out in (D.4) the splitting

(D.11) $$B_2\colon \operatorname{Hom}(\mathbb{Z}/f, \mathbb{Z}/g) \to [\Sigma P_f, \Sigma P_g]$$

is not additive. We now describe the distributivity law for $B_2(\varphi + \varphi')$. Let

$$\Delta\colon \operatorname{Hom}(\mathbb{Z}/f, \mathbb{Z}/g) \times \operatorname{Hom}(\mathbb{Z}/f, \mathbb{Z}/g) \to \operatorname{Ext}(\mathbb{Z}/f, \Gamma\mathbb{Z}/g) \cong \mathbb{Z}/(f, 2g, g^2)$$

be the linear map which carries the pair (φ, φ') to the element

$$\Delta(\varphi, \varphi') = (f(f-1)/2)\varphi_1 \cdot \varphi_1' \cdot \mathbf{1}$$

where $\varphi(\mathbf{1}) = \varphi_1 \cdot \mathbf{1}$, $\varphi'(\mathbf{1}) = \varphi_1' \cdot \mathbf{1}$. Then we get

(D.12) **Theorem.** $B_2(\varphi + \varphi') = B_2(\varphi) + B_2(\varphi') + \Delta(\varphi, \varphi')$

The splitting B_n, $n \geq 3$, satisfies the addition law $B_n(\varphi, \varphi') = B_n(\varphi) + B_n(\varphi') + \sigma_* \Delta(\varphi, \varphi')$. This follows from (D.5). The formula for Δ yields the following property.

(D.13) **Lemma.** *Let $f = 2^a f_0$, $g = 2^b g_0$ where f_0 and g_0 are odd. Then we have $\Delta \neq 0$ iff $a = b \geq 1$ or $a = b + 1 \geq 2$ and we have $\sigma_* \Delta \neq 0$ iff $a = b = 1$.*

Using the identification (D.5)(4) we can describe the group structure $+$ of the group $[\Sigma^{n-1} P_f, \Sigma^{n-1} P_g]$, $n \geq 2$, by the formula

(D.14) $$(\varphi, \alpha) + (\varphi', \alpha') = (\varphi + \varphi', \alpha + \alpha' + \Delta_n(\varphi, \varphi'))$$

where $\Delta_n = \Delta$ for $n = 2$ and $\Delta_n = \sigma_* \Delta$ for $n \geq 3$. This formula describes completely the additive structure of the category $\mathbf{P}_n/\simeq$. Since $\Delta(\varphi, \varphi') = \Delta(\varphi', \varphi)$ we see that also the group $[\Sigma P_f, \Sigma P_g]$ is abelian for all $f, g \in \mathbb{N}$. The cyclic summands and explicit generators of this group are described in the next result.

(D.15) **Corollary.** *Let $f = 2^a f_0$ and $g = 2^b g_0$ where f_0 and g_0 are odd. Then the homomorphism*

$$H_2\colon [\Sigma P_f, \Sigma P_g] \to \operatorname{Hom}(\mathbb{Z}/f, \mathbb{Z}/g) \cong \mathbb{Z}/d$$

has an additive splitting of abelian groups if and only if $(a, b) \neq (1, 1)$. Moreover for the greatest common divisors $d = (f, g)$ and $c = (f, g^2, 2g)$ one has

$$[\Sigma P_f, \Sigma P_g] = \begin{cases} \mathbb{Z}/d \oplus Z/c & \text{for } (a,b) \neq (1,1), \\ \mathbb{Z}/2d \oplus \mathbb{Z}/(c/2) & \text{for } (a,b) = (1,1). \end{cases}$$

The generator of the first summand is $(\varphi_0, (f/4)\mathbf{1})$ *if* $a > b = 1$ *and* $(\varphi_0, 0)$ *otherwise where* φ_0 *is a generator of* $\mathrm{Hom}(\mathbb{Z}/f, \mathbb{Z}/g)$. *The generator of the second summand is* $(0, \mathbf{1})$ *if* $(a,b) \neq (1,1)$ *and is* $(0, 2 \cdot \mathbf{1})$ *if* $(a,b) = (1,1)$. *Here we use again the identification in* (D.5)(4).

(D.16) **Addendum.** *The homomorphism*

$$H_3 \colon [\Sigma^2 P_f, \Sigma^2 P_g] \to \mathrm{Hom}(\mathbb{Z}/f, \mathbb{Z}/g) = \mathbb{Z}/d$$

has an additive splitting if and only if $(a,b) \neq (1,1)$. *Moreover for* $e = (f, g^2, 2)$ *one has*

$$[\Sigma^2 P_f, \Sigma^2 P_g] = \begin{cases} \mathbb{Z}/d \oplus \mathbb{Z}/e & \text{for } (a,b) \neq (1,1), \\ \mathbb{Z}/2d & \text{for } (a,b) = (1,1). \end{cases}$$

The generator of the first summand is $(\varphi_0, 0)$ *and the generator of the second summand* $\mathbb{Z}/e$ *is* $(0, \mathbf{1})$.

Remark. The result in (D.15), (D.16) is due to Barratt (T), (table 2 in 10.6). Barratt uses Whitney's tube system for proving this result; his arguments are highly geometrical and totally different from our method. Hilton (p. 125) presents a different approach for the stable groups $[\Sigma^2 P_f, \Sigma^2 P_g]$ and points out that a more simple minded proof of Barratt's result is needed. A further improvement in the results above is the fact that we describe explicitly generators of the cyclic summands. The algebraic description of the category $\mathbf{P}_n/\simeq$ by (D.5)(5) and (D.14) solves a problem of Barratt (HR) who used generators and relations for the description of $\mathbf{P}_n/\simeq$, $n \geq 3$. Our algebraic model of $\mathbf{P}_n/\simeq$ is simpler and also available for $n = 2$.

Proof of (D.15). H_2 has a splitting if and only if there is α such that (φ_0, α) has order d. By the group law in (D.14) we obtain the formula

$$(\varphi_0, \alpha) \cdot d = \left(0, \alpha \cdot d + \sum_{t=1}^{d-1} \Delta(\varphi_0, t\varphi_0)\right). \tag{1}$$

We choose the generator $\varphi_0 = \pi_1(\xi, \eta)$ with $\eta = g/d$, see (B.4). Then we have

$$\Delta(\varphi_0, t\varphi_0) = (f(f-1)/2)\eta \cdot t\eta \cdot \mathbf{1} \tag{2}$$

and therefore we have $(\varphi_0, \alpha) \cdot d = 0$ iff

$$\alpha d = (f(f-1)/2)\eta\eta(d(d-1)/2)\cdot \mathbf{1}. \tag{3}$$

If $a > b = 1$ we see that η and $d/2$ are odd. Thus $\alpha = (f/4)\mathbf{1}$ satisfies the equation (3). Otherwise $\alpha = 0$ satisfies (3) for $(a,b) \neq (1,1)$. For $(a,b) = (1,1)$ we have $\alpha d = 0$ for all α, however, the right hand side of (3) is a non trivial element of order 2 in this case. This proves the proposition. If we reduce equation (3) modulo 2 then both sides of (3) are zero for $a > b = 1$. This shows that $B_3(\varphi)$ yields an additive splitting for H_3 if $(a,b) \neq (1,1)$. □

The proof of (D.12) is based on the following lemma on commutators $(a,b) = -a - b + a + b$.

(D.17) **Lemma.** *Let $G = \langle a,b \rangle$ be the free group generated by elements a and b and let $f \in \mathbb{N}$. Then there exist elements $\xi_i \in G$ $(i = 1,2,\ldots,f(f-1)/2)$ such that*

$$(a+b)\cdot f = a\cdot f + b\cdot f - \sum_{i=1}^{f(f-1)/2} (a,b)^{\xi_i}.$$

Here we set $x\cdot f = x + \cdots + x$ (f times x) and we set $x^y = -y + x + y$. The sum is the ordered sum in the non commutative group G.

Proof. We show inductively that there are $\alpha_i \in G$ with

$$(a+b)\cdot f = a\cdot f + b\cdot f - \sum_{i=1}^{f-1} (a, b\cdot i)^{\alpha_i}. \tag{1}$$

This is true for $f = 1$. Now we get for $(a+b)(f+1)$:

$$\begin{aligned}(a+b)f + (a+b) &= a\cdot f + b\cdot f - \sum_{i=1}^{f-1} (a,bi)^{\alpha_i} + (a+b)\\ &= a\cdot f + b\cdot f + (a+b) - \sum_{i=1}^{f-1} (a,bi)^{\alpha_i+a+b}\\ &= a\cdot(f+1) + b(f+1) - (a,bf)^b - \sum_{i=1}^{f-1} (a,bi)^{\alpha_i+a+b}.\end{aligned}$$

This proves (1). Moreover there are $\beta_j \in G$ with

$$(a, b\cdot i) = \sum_{j=1}^{i} (a,b)^{\beta_j}. \tag{2}$$

This follows inductively from $(a, y + z) = (a, z) + (a, y)^z$ where we set $y = b \cdot i$ and $z = b$. From (1) and (2) we derive the proposition. $\square$

We derive from (D.17) the following algebraic description of the diagonal $\Delta: P_f \to P_f \times P_f$.

(D.18) **Corollary.** *Let e, e_2 be the generators of $\rho = \rho(P_f)$. Then $a = e \otimes *$, $b = * \otimes e$ are generators of $\rho \otimes \rho$ so that $(\rho \otimes \rho)_1 = \langle a, b\rangle$. Moreover we obtain a map $\Delta: \rho \to \rho \otimes \rho$ by $\Delta(e) = a + b$ and*

$$\Delta(e_2) = e_2 \otimes * + * \otimes e_2 - \sum_{i=1}^{f(f-1)/2} (e \otimes e)^{\xi_i}$$

where the ξ_i are the elements in (D.17). *The map Δ satisfies $p_1\Delta = 1$ and $p_2\Delta = 1$ where p_1, p_2 are the projections, see* (9.1)(8).

Since we have $[P_f, P_f \times P_f] = [\rho, \rho \otimes \rho]$ by (7.2) we see that Δ in (D.18) represents the homotopy class of the topological diagonal of P_f.

(D.19) *Proof of* (D.12). The group addition in the group

$$[\Sigma P_f, \Sigma P_g] = [\rho P_f, J\rho P_g] \tag{1}$$

can be described by the composition $F + F' = G$,

$$G: \rho P_f \xrightarrow{\Delta} \rho P_f \otimes \rho P_f \xrightarrow{F \otimes F'} J\rho P_g \otimes J\rho P_g \xrightarrow{\mu} J\rho P_g, \tag{2}$$

where Δ is the diagonal in (D.18). Now we choose (ξ, η), resp. $(\xi', \eta') \in [f, g]$ which induces φ, resp. φ', in (D.12). We may set $\eta = \varphi_1$, $\eta' = \varphi_1'$. Moreover let F, resp. F', be given by

$$\left.\begin{aligned} F(e) &= \eta e, \quad F(e_2) = \xi e_2, \\ F'(e) &= \eta' e, \quad F'(e_2) = \xi' e_2. \end{aligned}\right\} \tag{3}$$

Then the composition G in (2) represents $B_2(\varphi) + B_2(\varphi')$. Explicitly we get G by the formulas:

$$\left.\begin{aligned} G(e) &= (\eta + \eta')e, \\ G(e_2) &= (\xi + \xi')e_2 - \sum_{i=1}^{v} (\eta e)\cdot(\eta' e)^{\mu(\xi_i)} \end{aligned}\right\} \tag{4}$$

where $v = f(f-1)/2$ and $\eta e = e + \cdots e = \eta$-fold sum of e. On the other hand $B_2(\varphi + \varphi')$ is represented by the map $G'\colon \rho P_f \to J\rho P_g$ with $G'(e) = (\eta + \eta')e$ and $G'(e_2) = (\xi + \xi')e_2$. Whence (7.3) shows that $\Delta(\varphi, \varphi')$ is represented by the element

$$\alpha = \sum_{i=1}^{v} (\eta e)(\eta' e)^{\mu(\xi_i)} \in \text{kernel}(d_2) \tag{5}$$

where d_2 is the boundary of $J\rho(Pg)$. We know that $\pi_2 J\rho P_g = \text{kernel}(d_2)/\text{image}(d_3)$ is generated by the cycle ee. Whence we have to show

$$\alpha \equiv v \cdot \eta \cdot \eta'(ee) \quad \text{modulo image } (d_3). \tag{6}$$

This is easily checked by use of (D.10)(4) and (C.4)(3). Therefore the proof of (D.12) is complete. □

We also derive from (D.18) the following well known result on the reduced diagonal of P_f.

(D.20) **Corollary.** The following diagram homotopy commutes in **Top***.

$$\begin{array}{ccc} P_f & \xrightarrow{\Delta} & P_f \times P_f \\ \downarrow q & & \searrow q \\ S^2 & \xrightarrow{v} & S^1 \wedge S^1 \subset P_f \wedge P_f \end{array}$$

Here v is a map of degree $f(f-1)/2$ and q denotes the quotient maps. Recall that the smash product $A \wedge B$ is defined by the quotient $A \wedge B = A \times B/(A \vee B)$. We obtain (D.20) directly from (D.18) since $\rho(P_f \wedge P_f) = \rho(P_f) \otimes \rho(P_f)/\rho(P_f \vee P_f)$ by (6.12).

Finally we consider the group structure of the homotopy groups with coefficients in $\mathbb{Z}/f$. As in (D.5)(3) we have the central extension of groups

$$\text{(D.21)} \qquad \mathbb{Z}/f \otimes \pi_{n+1} U \rightarrowtail [\Sigma^{n-1} P_f, U] \xrightarrow{\pi_n} \text{Hom}(\mathbb{Z}/f, \pi_n U)$$

where we identify $\text{Ext}(\mathbb{Z}/f, \pi_{n+1} U) = \mathbb{Z}/f \otimes \pi_{n+1} U$. This extension is completely determined by the following result.

(D.22) **Proposition.** *For* x, $y \in [\Sigma P_f, U]$ *we have the commutator rule* ($v = f(f-1)/2$)

$$-x - y + x + y = v\mathbf{1} \otimes [i^* x, i^* y]$$

where $i: S^2 \subset \Sigma P_f$ is the inclusion and where $[i^*x, i^*y] \in \pi_3 U$ is the Whitehead product. Moreover let $\mathbb{Z}\varphi$ be the subgroup of $\mathrm{Hom}(\mathbb{Z}/f, \pi_n U)$ generated by an element φ. Then there is a function $T: \mathbb{Z}\varphi \to [\Sigma^{n-1} P_f, U]$ with $\pi_n T(x) = x$ for $x \in \mathbb{Z}\varphi$ and with $(r, s \in \mathbb{Z})$

$$-T((r+s)\varphi) + T(r\varphi) + T(s\varphi) = rtv\mathbf{1} \otimes (\eta^* \varphi(\mathbf{1}))$$

where $\eta: S^{n+1} \to S^n$ is the Hopf element.

Proof. The property of the commutator follows from the definition of the Whitehead product and (D.20), see for example II.1.12 in Baues (CC). Next let $\mathbb{Z}/g$ be the cyclic group generated by $\varphi(\mathbf{1})$ in $\pi_n U$. Then we can choose a map $F: \Sigma^{n-1} P_g \to U$ with $i^* F = \varphi(\mathbf{1})$. Moreover an element $t\varphi \in \mathbb{Z}\varphi$ corresponds to a homomorphism $t\varphi: \mathbb{Z}/f \to \mathbb{Z}/g$. Now we define T in (D.22) by $T(t\varphi) = F_* B_n(t\varphi)$ where B_n is the splitting in (D.4) and (D.12). □

Remark. Proposition (D.22) completes the partial results on the extension (D.21) in the book of Hilton (page 125–128).

Chapter IV

Quadratic modules and homotopy systems of order 4

This chapter contains some of the main new results of this book. We introduce algebraic objects which we call *quadratic modules*. They play a role in dimension 3 which is analogous to the role of crossed modules in dimension 2. Indeed most properties of crossed modules discussed in chapter III correspond to a similar feature of quadratic modules. Therefore this chapter is organized in the same manner as chapter III; we often only raise the dimension by one and replace crossed modules by quadratic modules and homotopy systems of order 3 by homotopy systems of order 4. In order to pin point the analogy we now discuss the contents of this chapter along the same lines as in the introduction of chapter III.

The category $\mathbf{H}_4^c$ of homotopy systems of order 4 is the second stage of the CW-tower in chapter II. Recall that a homotopy system of order 4 is a triple (C, f_4, X^3) consisting of a topological part X^3 and an algebraic part (C, f_4) with $f_4\colon C_4 \to \pi_3 X^3$. We associate with the 3-dimensional CW-complex X^3 a "totally free" quadratic module $\sigma = \sigma(X^3) = (\omega, d_3, d_2)$ which is a commutative diagram

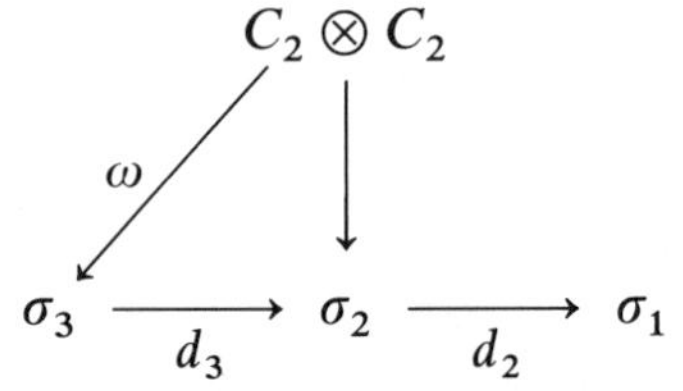

$$\begin{array}{ccccc} & & C_2 \otimes C_2 & & \\ & \overset{\omega}{\swarrow} & \downarrow & & \\ \sigma_3 & \xrightarrow[d_3]{} & \sigma_2 & \xrightarrow[d_2]{} & \sigma_1 \end{array} \tag{1}$$

of homomorphisms between σ_1-groups satisfying additional properties. In particular d_2 is a pre-crossed module of Peiffer nilpotency degree 2, that is, a "nil(2)-module". The associated crossed module of d_2 is the boundary map

$$\partial\colon \pi_2(X^2, X^1) \to \pi_1 X^1 = \sigma_1.$$

Moreover σ_3 is a central extension of C_3 and the "quadratic map" ω is a lift of the "Peiffer commutator map" $w = d_3\omega$. In addition one has an isomorphism

$$j\colon \pi_3(X^3) \cong \text{kernel}(d_3).$$

As an example we have the following quadratic module $\sigma(S^2)$ of the 2-sphere S^2 which is simply given by the diagram

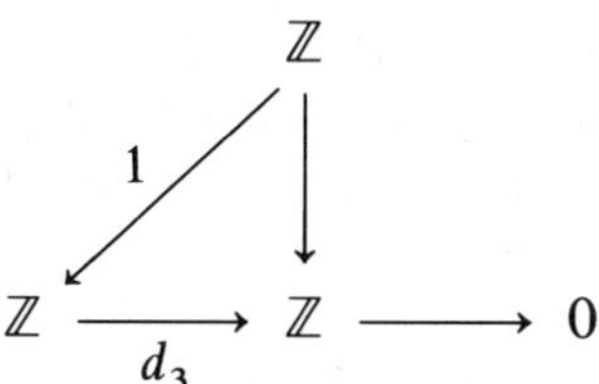

where $\omega = 1$ and $d_3 = 0$. Whence we get the result of Hopf that $\pi_3(S^2) \cong \text{kernel}(d_3) = \mathbb{Z}$. This example demonstrates the minimality of the algebraic model $\sigma(X^3)$ of the CW-complex X^3.

We now consider the function

$$(C, f_4, X^3) \mapsto (C, jf_4, \sigma) \tag{2}$$

which carries a homotopy system of order 4 to the triple (C, jf_4, σ) given by σ in (1). This triple encapsulates the algebraic data of a *quadratic chain complex* σ:

$$\begin{array}{ccccccccccc} & & & & & & & & C_2 \otimes C_2 & & \\ & & & & & & & \overset{\omega}{\swarrow} & \downarrow & & \\ \cdots \xrightarrow[d_6]{} & C_5 & \xrightarrow[d_5]{} & C_4 & \xrightarrow[d_4]{} & \sigma_3 & \xrightarrow[d_3]{} & & \sigma_2 & \xrightarrow[d_2]{} & \sigma_1 \end{array}$$

with $d_4 = jf_4$. Here σ is "totally free" in the sense that σ satisfies a freeness condition in each degree. We write $\sigma = \sigma X$ if the homotopy system $(C, f_4, X^3) = r_4 X$ is given by a CW-complex X. Then the "homotopy groups"

$$\pi_n(\sigma X) = \text{kernel}\, d_n / \text{image}\, d_{n+1}$$

of the quadratic chain complex σX satisfy the natural equation

$$\pi_n(\sigma X) = \begin{cases} \pi_n X & \text{for } n = 1, 2, 3 \\ \hat{h}\pi_4 X & \text{for } n = 4 \\ H_n \hat{X} & \text{for } n \geq 5 \end{cases}$$

where $\hat{h}\colon \pi_4 X \to H_4 \hat{X}$ is the Hurewicz homomorphism. In particular the quadratic chain complex σX can be used for the computation of the third homotopy group, $\pi_3 X$, of a space X.

The quadratic chain complexes may be thought of as chain complexes with operators from a group but with non abelian features in degree one, two and three and with a quadratic structure given by the quadratic map ω. The somewhat more intricate non abelian and quadratic part of a quadratic chain complex corresponds to the topological part X^3 of a homotopy system of order 4. The main result in this chapter states that the function (2) yields equivalences of categories (see (7.7))

$$\begin{aligned} \sigma\colon \mathbf{H}_4^c &\xrightarrow{\sim} \mathbf{Q}/\overset{0}{\simeq}, \\ \sigma\colon \mathbf{H}_4^c/\simeq &\xrightarrow{\sim} \mathbf{Q}/\simeq \end{aligned} \tag{3}$$

where $\mathbf{Q}$ is the category of totally free quadratic chain complexes and where $\overset{0}{\simeq}$ and $\simeq$ are explicitly described homotopy relations on the category $\mathbf{Q}$. This shows that the second stage $\mathbf{H}_4^c$ in the CW-tower can be replaced by the algebraic category $\mathbf{Q}/\overset{0}{\simeq}$. In fact, the functor

$$\sigma\colon \mathbf{CW} \xrightarrow{r_4} \mathbf{H}_4^c \xrightarrow{\sim} \mathbf{Q}/\overset{0}{\simeq} \tag{4}$$

which carries a CW-complex X to its quadratic chain complex σX is part of the CW-tower of categories so that the category $\mathbf{Q}/\simeq$ can be considered as the "second algebraic approximation" of the homotopy category $\mathbf{CW}/\simeq$ of connected CW-complexes. Moreover we obtain in § 9 a purely algebraic exact sequence

$$H^3\Gamma\pi_2 + \longrightarrow \mathbf{Q}/\simeq \xrightarrow{\lambda} \mathbf{H}/\simeq \xrightarrow{\mathcal{O}} H^4\Gamma\pi_2 \tag{5}$$

where the functor λ is obtained by dividing out the images of the quadratic maps ω and $d_3\omega$ in (1). The exact sequence (5) is equivalent to the corresponding exact sequence

$$H^3\Gamma_3 + \longrightarrow \mathbf{H}_4^c/\simeq \xrightarrow{\lambda} \mathbf{H}_3^c/\simeq \xrightarrow{\mathcal{O}} H^4\Gamma_3 \tag{6}$$

of the CW-tower via the functors σ in (3) and ρ in chapter III. This fact is valuable for explicit computations in homotopy classification problems.

We show that the functor σ in (4) has further nice properties. In particular σ carries homotopy push outs in **CW** to homotopy push outs in **Q**, see (6.7). This result generalizes the well known Van Kampen theorem for fundamental groups since it allows even the computation of the homotopy group π_3 of a union of spaces. For the definition of homotopy push outs in **Q** we observe that **Q** has in a canonical way the structure of a cofibration category. Therefore all basic notions of homotopy theory make sense in the category **Q**, see § 5. Moreover, the functor σ carries a cylinder in **CW** to an algebraic cylinder in **Q**. In addition σ carries a product $X \times Y$ of CW-complexes to the tensor product $\sigma X \otimes \sigma Y$ of quadratic chain complexes, see § 12. All these facts are very useful for computations, compare the examples in Appendix A of this chapter. Using the functor σ in (4) we obtain an equivalence of categories (see § 10)

$$\textbf{3-types} \xrightarrow{\sim} \mathrm{Ho}(\textbf{quad}) \tag{7}$$

where 3-**types** is the full homotopy category of 3-types and where Ho(**quad**) is the localization of the category of quadratic modules with respect to weak equivalences.

We now summarize the basic properties of quadratic modules which indeed show that they are an *optimal algebraic system* representing a 3-type. The following list of properties corresponds exactly to the list of properties (a) ... (h) of crossed modules in the introduction of chapter III.

(a) A quadratic module is an algebraic model of a 3-type by an equivalence of categories as in (7) above, see (10.1).
(b) A quadratic module is an algebraic model of a 3-type with minimal Peiffer nilpotency degree, see (III.1.10).
(c) A totally free quadratic module is an algebraic model of a 3-dimensional CW-complex by an equivalence of categories, see (8.1).
(d) Quadratic modules form the degree 3 part of quadratic chain complexes which are algebraic models of homotopy systems of order 4 by an equivalence of categories, see (7.7) and (5.7).
(e) The category of quadratic chain complexes satisfies the axioms of a cofibration structure and hence a homotopy theory is available in this category, see (5.5).
(f) The functor from spaces to quadraric chain complexes carries weak equivalences to weak equivalences and homotopy push outs to homotopy push outs, see (6.7). Moreover, this functor carries products to tensor products, see § 12.
(g) Homotopy types of 4-dimensional totally free quadratic chain complexes are in 1-1 correspondence with homotopy types of 4-dimensional connected CW-complexes. See (II.3.14)(2) with $N = 4$ and (7.7). Moreover the cells of the CW-complex are the generators of the corresponding totally free quadratic chain complex.
(h) Each 5-dimensional totally free quadratic chain complex σ can be realized, that is, there is a 5-dimensional CW-complex X^5 together with an isomorphism $\sigma \cong \sigma(X^5)$ of quadratic chain complexes.

There are various algebraic models of 3-types in the literature which, however, do not satisfy such a list of optimality conditions. We expect that there are also algebraic cubical models of 4-types which satisfy similar optimality conditions as above; the algebraic nature of such models, however, is completely unknown. The comparison of quadratic chain complexes with simplicial groups in Appendix B of this chapter might give hints how to find such models.

§ 1 Quadratic modules

In this section we introduce a new kind of algebraic object which we call a quadratic module. In section § 10 we shall see that a quadratic module is an algebraic equivalent of a 3-type. The algebraic outfit of a quadratic module is motivated on the one hand by certain properties (1.6) of a nil(2)-module, that is, of a pre-crossed module of Peiffer nilpotency class 2. On the other hand the geometric meaning of a quadratic module will become apparent in chapter VI below by use of a "track category" which is determined by nil(2)-modules; this way the author originally discovered quadratic modules. As pointed out in (III.§ 1) nil(2)-modules can be considered as generalizations of nil(2)-groups. In this section we first describe a well known exact sequence for nil(2)-groups and then we describe a corresponding (more intricate) exact sequence for nil(2)-modules which is a crucial fact used in the definition of a quadratic module. We do not yet discuss many examples of quadratic modules since we first have to develop the necessary algebraic theory describing the properties of quadratic modules and of quadratic chain complexes in § 3. In particular we show in the next section that free quadratic modules exist. Specific examples of quadratic modules derived from certain CW-complexes are described in Appendix A of this chapter.

Recall that a nil(2)-group G is a group which is nilpotent of class 2. Let $C = G^{ab}$ be the abelianization of G. Then one has the well known exact sequence of groups

(1.1) $$C \otimes C \xrightarrow{w} G \overset{p}{\twoheadrightarrow} C.$$

Here p is the projection with $p(x) = \{x\}$ and w is the **commutator map** defined by

$$w(\{x\} \otimes \{y\}) = (x, y) = -x - y + x + y$$

for x, $y \in G$. One readily checks that w is a well defined homomorphism since G is a nil(2)-group. Moreover, $w(C \otimes C)$ is central in G since

$$(w(\{x\} \otimes \{y\}), z) = ((x, y), z) \in \Gamma_3 G = 0.$$

Now let G be a free nil(2)-group, see (III.§ 1). Then we get the exact sequence

$$\Gamma(C) \overset{\tau}{\rightarrowtail} C \otimes C \xrightarrow{w} G \overset{p}{\twoheadrightarrow} C \tag{1.2}$$

where τ is the map in (I.4.2) which is injective since C is free abelian.

(1.3) **Remark.** The exact sequence (1.2) is obtained by the isomorphism of groups

$$w\colon F^{ab} \wedge F^{ab} \cong \Gamma_2(F)/\Gamma_3(F) \tag{1}$$

where F is a free group, see (III.1.8). The isomorphism w carries $\{x\} \wedge \{y\}$ to the class $\{(x, y)\}$ of the commutator (x, y), $(x, y \in F)$. This isomorphism is the 'degree 2 part' of the isomorphism of Witt

$$L(F^{ab}) \cong \bigoplus_{n \geq 1} \Gamma_n(F)/\Gamma_{n+1}(F). \tag{2}$$

Here $L(F^{ab})$ is the free **Lie algebra** generated by F^{ab}, compare Baues-Conduché or Magnus-Karrass-Solitar.

We now consider nil(2)-modules which by definition are pre-crossed modules of Peiffer nilpotency class 2, see (III.§ 1). The exact sequence (1.1) for nil(2)-groups has a natural generalization for nil(2)-modules. Let $\partial\colon M \to N$ be a nil(2)-module and let $C = (M^{cr})^{ab}$ be the abelianization of the associated crossed module $M^{cr} \to N$. Then there is an exact sequence of groups (actually of N-groups by (1.5) below)

$$C \otimes C \xrightarrow{w} M \overset{p}{\twoheadrightarrow} M^{cr}. \tag{1.4}$$

Here p is the projection and w is defined by the formula

$$w(\{x\} \otimes \{y\}) = \langle x, y \rangle = -x - y + x + y^{\partial x}$$

for $x, y \in M$. The element $\{x\} \in C$ denotes the class represented by $x \in M$ and $\langle x, y \rangle$ is the Peiffer commutator. We call w in (1.4) the **Peiffer commutator map** of the nil(2)-module $\partial\colon M \to N$.

By use of the following formulas on Peiffer commutators one readily checks that w is well defined. Let x, y, $z \in M$ and $\alpha \in N$.

$$\langle x, y \rangle^\alpha = \langle x^\alpha, y^\alpha \rangle \tag{P1}$$

(P2) (a) $$\langle x, y+z\rangle = \langle x,z\rangle - z^{\partial x} + \langle x,y\rangle + z^{\partial x}$$

(b) $$= \langle x,z\rangle + \langle x,y\rangle + \langle\!\langle \langle x,y\rangle, z^{\partial x}\rangle\!\rangle$$

(P3) (a) $$\langle x+y, z\rangle = -y + \langle x,z\rangle + y + \langle y, z^{\partial x}\rangle$$

(b) $$= \langle x,z\rangle + \langle\!\langle \langle x,z\rangle, y\rangle\!\rangle + \langle y, z^{\partial x}\rangle$$

(P4) Let $k \in M$ with $\partial k = 0$ then

(a) $$\langle k,x\rangle = (k,x)$$

(b) $$\langle k,x\rangle + \langle x,k\rangle = -k + k^{\partial x}$$

(P5) (a) $$-\langle x,y\rangle = -y^{\partial x} + \langle x,-y\rangle + y^{\partial x}$$

(b) $$= \langle x,-y\rangle + \langle\!\langle \langle x,-y\rangle, y^{\partial x}\rangle\!\rangle$$

(c) $$= -x + \langle -x, y^{\partial x}\rangle + x$$

(d) $$= \langle -x, y^{\partial x}\rangle + \langle\!\langle \langle -x, y^{\partial x}\rangle, x\rangle\!\rangle$$

Here (P5) follows from (P2) and (P3). We can compute $\langle y, z^{\partial x}\rangle$ in (P3)(b) by use of $z^{\partial x} = -x + z + x + \langle x,z\rangle$, compare the definition of $\langle x,z\rangle$. We leave the proof of the formulas above as an exercise.

The action of N on M induces an action of N on C by $\{x\}^\alpha = \{x^\alpha\}$. This yields an action of $\pi = N/\partial M$ on C since we have $\{x^{\partial y}\} = \{x\}$ in C. Moreover, we obtain an action of N (resp. of π) on $C \otimes C$ by the formula

(1.5) $$(\{x\} \otimes \{y\})^\alpha = \{x^\alpha\} \otimes \{y^\alpha\}.$$

Now (P1) above shows that the map w in (1.4) is equivariant with respect to the action of N.

The Peiffer commutator map w of any nil(2)-module satisfies the following formulas; they are similar to the formulas (1.10)(3), (4) in a quadratic module.

(1.6) **Lemma.** *Let $\partial: M \to N$ be a* nil(2)*-module. For $x, y \in M$ one gets*

(1) $$y^{\partial x} = y + w(\{y\} \otimes \{x\} + \{x\} \otimes \{y\}) \quad \textit{if} \quad \partial y = 0, \quad \textit{and}$$

(2) $$-x - y + x + y = w(\{x\} \otimes \{y\}) \quad \textit{if} \quad \partial x = 0.$$

Moreover the group $w(C \otimes C)$ is central in M.

Proof. The equations (1) and (2) follow from (P4)(b) and (a) respectively. Moreover, using $\partial w = 0$, the commutator $(w(\{x\} \otimes \{y\}), z) = \langle\langle x, y\rangle, z\rangle = 0$ is trivial since ∂ is a nil(2)-module. Thus $w(C \otimes C)$ is central. □

Now let $\partial\colon M \to N$ be a totally free nil(2)-module. Then $M^{cr} \to N$ is a free crossed module and we have by the chain functor in (III.2.10) the following commutative diagram with $C = C_2$.

(1.7)
$$\begin{array}{ccccc} & & M & & \\ & & \downarrow p & \searrow \partial & \\ \text{kernel}(\partial^{cr}) & \subset & M^{cr} & \xrightarrow{\partial^{cr}} & N \\ \downarrow \cong & & \downarrow h_2 & & \downarrow h_1 \\ K = \text{kernel}(d_2) & \subset & C & \xrightarrow[d_2]{} & C_1 \end{array}$$

The restriction of h_2 yields an isomorphism of the indicated kernels in the diagram. The kernel K of d_2 satisfies

$$K = \text{kernel}(d_2) = H_2\hat{X}^2 \cong \pi_2 X^2 = \text{kernel}(\partial^{cr})$$

where X^2 is a 2-dimensional CW-complex which realizes the totally free crossed module ∂, compare (III.7.1). The inclusion $K \subset C$ induces the injective map

$$\tau\colon \Gamma(K) \rightarrowtail K \otimes K \subset C \otimes C$$

by τ in (I.4.2). Using this injection we can formulate the following important theorem on the kernel of the Peiffer commutator map. This result is used in chapter VI for the definition of a certain track category.

(1.8) **Theorem.** *Let $\partial\colon M \to N$ be a totally free* nil(2)*-module. Then the sequence*

$$\Gamma(K) \overset{\tau}{\rightarrowtail} C \otimes C \xrightarrow{w} M \overset{p}{\twoheadrightarrow} M^{cr}$$

is exact.

The theorem generalizes (1.2); in fact, we get by (1.8) the exact sequence (1.2) when we set $M = G$, $N = 0$. Theorem (1.8) is proved in Baues-Conduché where we actually get a more general result as described in the following remark.

(1.9) **Remark.** In chapter VI we consider the geometric significance of the exact sequence in theorem (1.8) which as well was originally used for a proof of (1.8). The consideration of the algebraic background of the theorem led to the study of the Peiffer central series in Baues-Conduché. There we show that the result of Witt in (1.3) has a nice generalization for pre-crossed modules. In fact, we obtain for a totally free pre-crossed module $\partial\colon M \to N$ the isomorphism (see (III.1.9))

$$w\colon C \otimes C/\tau\Gamma K \cong P_2(\partial)/P_3(\partial) \tag{1}$$

which carries the class of $\{x\} \otimes \{y\}$ to the class of $\langle x, y\rangle$. The isomorphism is the 'degree 2 part' of the isomorphism

$$L(C, K) \cong C \oplus \bigoplus_{n\geq 2} P_{n+1}(\partial)/P_n(\partial) \tag{2}$$

Here the left hand side $L(C, K)$ is the '**free partial Lie algebra**' generated by the pair (C, K), see Baues-Conduché. Clearly (1) generalizes (1.3)(1), and (2) generalizes the isomorphism (1.3)(2) of Witt.

We use the Peiffer commutator map w in (1.4) for the following basic definition.

(1.10) **Definition. A quadratic module** $(\omega, \delta, \partial)$ is a diagram

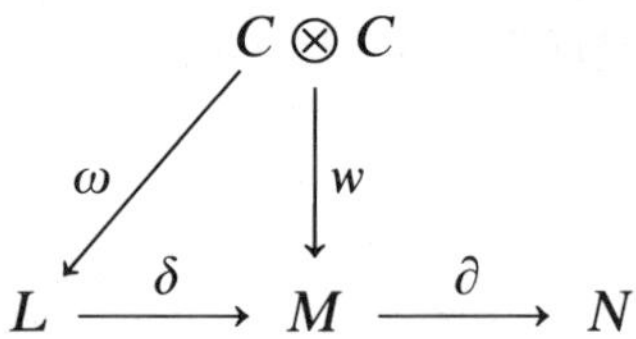

of homomorphisms between groups such that (1) ... (4) hold.

(1) The homomorphism $\partial\colon M \to N$ is a nil(2)-module with Peiffer commutator map w as in (1.4). The quotient map $M \twoheadrightarrow C = (M^{cr})^{ab}$ is denoted by $x \mapsto \{x\}$.
(2) The **boundary homomorphisms** ∂ and δ satisfy $\partial\delta = 0$ and the **quadratic map** ω is a lift of the Peiffer commutator map w, that is $\delta\omega = w$ or equivalently

$$\delta\omega(\{x\} \otimes \{y\}) = \langle x, y\rangle \qquad \text{for } x, y \in M.$$

(3) L is an N-group and all homorphisms of the diagram are equivariant with respect to the action of N. Moreover, the action of N on L satisfies the formula $(a \in L, x \in M)$

$$a^{\partial x} = a + \omega(\{\delta a\} \otimes \{x\} + \{x\} \otimes \{\delta a\}).$$

(4) Commutators in L satisfy the formula $(a, b \in L)$

$$(a, b) = -a - b + a + b = \omega(\{\delta a\} \otimes \{\delta b\}).$$

We point out that the equations in (3) and (4) are "lifts" of the corresponding equations in (1.6), that is, applying δ to the equation (3) and (4) yields exactly the equations (1) and (2) respectively in (1.6). A **map** $\varphi\colon (\omega, \delta, \partial) \to (\omega', \delta', \partial')$ between quadratic modules is given by a commutative diagram, $\varphi = (l, m, n)$,

$$\begin{array}{ccccccc} C \otimes C & \xrightarrow{\omega} & L & \xrightarrow{\delta} & M & \xrightarrow{\partial} & N \\ \downarrow{\scriptstyle \varphi_* \otimes \varphi_*} & & \downarrow{\scriptstyle l} & & \downarrow{\scriptstyle m} & & \downarrow{\scriptstyle n} \\ C' \otimes C' & \xrightarrow{\omega'} & L' & \xrightarrow{\delta'} & M' & \xrightarrow{\partial'} & N' \end{array} \tag{5}$$

where (m, n) is a map between pre-crossed modules which induces $\varphi_*\colon C \to C'$ and where l is an n-equivariant homomorphism. Let **quad** be the category of quadratic modules and of maps as in (5). We define the **homotopy groups** $\pi_n(\sigma)$ of a quadratic module $\sigma = (\omega, \delta, \partial)$ by

$$\left.\begin{aligned} \pi_1(\sigma) &= \text{cokernel}(\partial) \\ \pi_2(\sigma) &= \text{kernel}(\partial)/\text{image}(\delta) \\ \pi_3(\sigma) &= \text{kernel}(\delta) \end{aligned}\right\} \tag{6}$$

One readily checks that π_2 and π_3 are π_1-modules for which the action of π_1 is induced by the action of N. A map φ in **quad** is a **weak equivalence** if $\pi_n(\varphi)$ is an isomorphism for $n = 1, 2, 3$.

In the proof of (2.10) below we shall need the following construction of a nil(2)-module $\bar{\partial}\colon \bar{L} \to N$ defined in terms of the quadratic module $(\omega, \delta, \partial)$. We associate with $(\omega, \delta, \partial)$ in (1.10) the commutative diagram of unbroken arrows

(1.11)
$$\begin{array}{ccccc} L & \overset{i}{\rightarrowtail} & \bar{L} & \underset{\overset{s}{\dashleftarrow}}{\overset{p}{\twoheadrightarrow}} & M \\ & & {\scriptstyle \bar{\partial}}\searrow & & \swarrow{\scriptstyle \partial} \\ & & & N & \end{array}$$

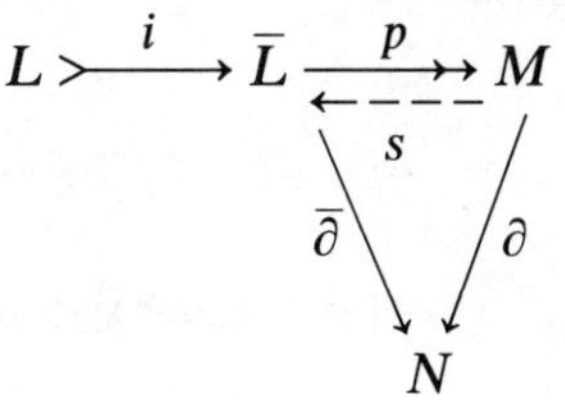

which is obtained as follows. There is a right action of M on L by the formula

$$a \cdot x = a + \omega(\{\delta a\} \otimes \{x\}) \tag{1}$$

for $a \in L$, $x \in M$. It is obvious that (1) is a well defined action, compare also (2.10) in Conduché. The top row in (1.11) is the semi direct product of L and M given by the action (1), that is $\bar{L}$ is the product of sets, $\bar{L} = M \times L$, with the group structure

$$(x, a) + (y, b) = (x + y, a \cdot y + b). \tag{2}$$

This implies $-(x, a) = (-x, -a \cdot (-x))$. The maps i and p in (1.11) are the obvious inclusion and projection respectively and s is the section of p with $s(x) = (x, 0)$. Let $\bar{\partial} = \partial p$, $\bar{\partial} i = 0$. We obtain an action of N on $\bar{L}$ by

$$(x, a)^\alpha = (x^\alpha, a^\alpha), \qquad \alpha \in N. \tag{3}$$

It is clear that $(a \cdot y)^\alpha = a^\alpha \cdot y^\alpha$ by (1) and whence $\bar{L}$ is a well defined N-group. Moreover, $\bar{\partial}$ is a pre-crossed module since ∂ is one. We now compute Peiffer commutators in $\bar{L}$. Let $L_\omega = L/\omega(C^{\otimes 2})$ be the quotient group given by ω. This group is abelian by (1.10)(4). We obtain an exact sequence

(1.12) $$(C \oplus L_\omega)^{\otimes 2} \xrightarrow{\bar{w}} \bar{L} \overset{p}{\twoheadrightarrow} \bar{L}^{cr}$$

where $\bar{w}$ satisfies the formula, $(x, a), (y, b) \in M \times L = \bar{L}$,

$$\bar{w}((\{x\} + \{a\}) \otimes (\{y\} + \{b\})) = \langle (x, a), (y, b) \rangle.$$

(1.13) **Lemma.** *The Peiffer commutator in $\bar{L}$ satisfies the formula*

$$\langle (x, a), (y, b) \rangle = (z, c) \in M \times L = \bar{L}$$

where z and c are given by $z = \langle x, y \rangle = w(\{x\} \otimes \{y\})$, $c = \omega(\{\delta a\} \otimes \{\delta b\} + \{\delta a\} \otimes \{y\} + \{x\} \otimes \{\delta b\})$.

The lemma shows that $\bar{L}$ is a nil(2)-module, that the map $\bar{w}$ in (1.12) is well defined and that (1.12) is exact.

Proof of (1.13). $\langle (x, a), (y, b) \rangle = -(x, a) - (y, b) + (x, a) + (y, b)^{\partial x}$

$$= -(y + x, b \cdot x + a) + (x + y^{\partial x}, a \cdot y^{\partial x} + b^{\partial x})$$

$$= (-x - y, -(b \cdot x + a) \cdot (-(y + x)))$$

$$+ (x + y^{\partial x}, a \cdot y^{\partial x} + b^{\partial x})$$

$$= (\langle x, y \rangle, u) \quad \text{where}$$

$$u = -(b \cdot x + a) \cdot \langle x, y \rangle + a \cdot y^{\partial x} + b^{\partial x}.$$

By (1.11)(1) we see that $a \cdot \langle x, y \rangle = a$ since $\{\langle x, y \rangle\} = 0$, and $a \cdot y^{\partial x} = a \cdot y$ since $\{y^{\partial x}\} = \{y\}$. Whence we get

$$u = -a - b \cdot x + a \cdot y + b^{\partial x}$$

$$= (-a - b + a + b) + (-b + b^{\partial x}) - \omega(\{\delta b\} \otimes \{x\}) + \omega(\{\delta a\} \otimes \{y\})$$

$$= \omega(\{\delta a\} \otimes \{\delta b\} + \{x\} \otimes \{\delta b\} + \{\delta a\} \otimes \{y\})$$

In the last equation we use (1.10)(3), (4). Moreover $\langle x, y \rangle = w(\{x\} \otimes \{y\})$ by (1.4). This completes the proof of (1.13). □

(1.14) **Example.** Any nil(2)-module $\partial: M \to N$ yields by (1.4) the quadratic module $\bar{\partial} = (1, w, \partial)$ given by

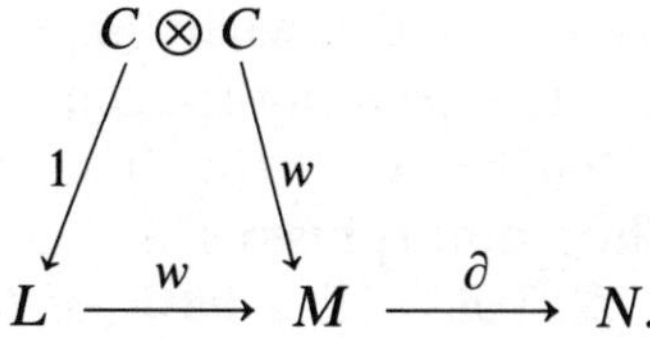

Here we set $L = C \otimes C$ and 1 is the identity. Clearly L is a $\pi_1(\partial)$-module and the equations (1.10)(3), (4) are satisfied since $\{w(a)\} = 0$ for $a \in L$. We call $\bar{\partial}$ the **quadratic module associated to the nil(2)-module** ∂. This yields the full embedding **cross**(2) ⊂ **quad**.

§ 2 Free quadratic modules

The definition of a free quadratic module by use of a universal property is similar to the corresponding definition of a free crossed module in (III.1.14). The explicit construction of a free quadratic module, however, is more complicated. In this section we obtain such a construction by the central push out in (2.9) below where we also use the sum $E \vee M$ of nil(2)-modules; the geometric significance of this central push out will become apparent by the use of certain tracks in chapter VI. The proof that our construction actually satisfies the universal property of a free quadratic module is fairly long and technical. We also consider a totally free quadratic module σ and we show that the third homotopy group $\pi_3(\sigma)$ is embedded in a short exact sequence, which later will turn out to be Whitehead's certain exact sequence of a 3-dimensional CW-complex, see (2.14) and (3.7) below.

(2.1) **Definition.** Let $\partial: M \to N$ be a nil(2)-module and let $f: F \to M$ be a homomorphism where F is a free group and for which $\partial f = 0$. We say that a quadratic module $(\omega, \delta, \partial)$ as in (1.10) is a **free quadratic module with basis** f if a homomorphism $i: F \to L$ is given such that $\delta i = f$ and such that the following universal property is satisfied. Consider any commutative diagram of unbroken arrows

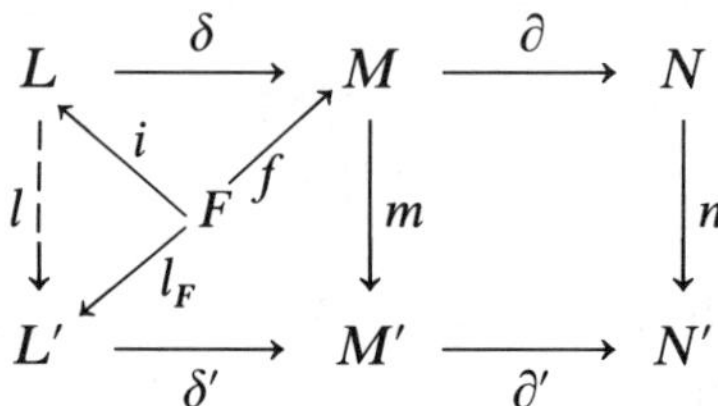

where $(\omega', \delta', \partial')$ is a quadratic module, where $(m, n): \partial \to \partial'$ is a map between nil(2)-modules, and where l_F is a homomorphism. Then there exists a unique map (l, m, n) in **quad** such that l extends the diagram commutatively, that is $li = l_F$. For $F = \langle Z \rangle$ the homomorphism f is determined by the restriction $f|Z$. In this case we call $f|Z$ as well a **basis** of $(\omega, \delta, \partial)$. We say that the quadratic module $(\omega, \delta, \partial)$ is **totally free** if it is free and if ∂ is a free nil(2)-module with a free group N of operators.

For the construction of a free quadratic module we need the following sums in the category **cross**(n) of nil(n)-modules.

(2.2) **Definition.** Let $\partial: M \to N$ and $\partial': M' \to N$ be nil(n)-modules. Then there exists the **sum under** N

$$\partial \vee \partial': M \vee M' \to N \tag{1}$$

in the category **cross**(n). We define $\partial \vee \partial'$ together with maps $\partial \to \partial \vee \partial' \leftarrow \partial'$ (which are the identity on N) by the usual universal property: for any diagram of unbroken arrows in **cross**(n)

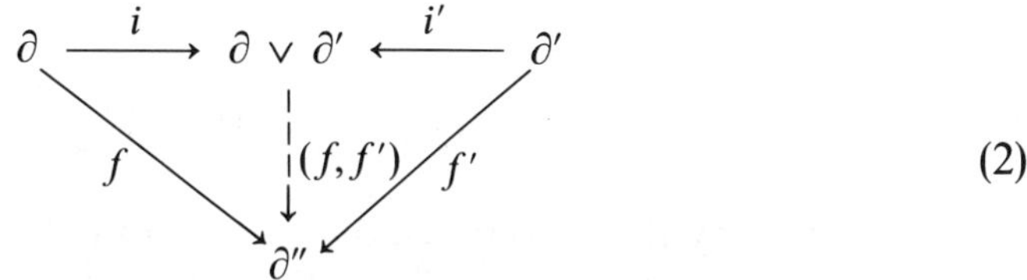

(2)

(where f and f' coincide on N) there exists a unique map (f, f') in **cross**(n) which extends the diagram commutatively. Let $M * M'$ be the free product of groups. Then

$$(\partial, \partial'): M * M' \to N \tag{3}$$

is a pre-crossed module with the obvious action of N on $M * M'$. Now $M \vee M'$ is given by

$$M \vee M' = M * M'/P_{n+1}(\partial, \partial') \tag{4}$$

or by $r_n(\partial, \partial')$ in (III.1.11).

Now let $\partial: M \to N$ and $\partial': M' \to N$ be nil(2)-modules. Then we have by the inclusion i, i' in (2) above the canonical surjection

(2.3) $$C \oplus C' \twoheadrightarrow (M \vee M')^{cr\,ab}$$

where $C = M^{cr\,ab}$, $C' = (M')^{cr\,ab}$. Whence one gets by (1.4) the exact sequence

(2.4) $$(C \oplus C')^{\otimes 2} \xrightarrow{w} M \vee M' \xrightarrow{p} (M \vee M')^{cr}.$$

We now consider a special case of a sum in **cross**(2). Let F be a free group generated by the set Z, $F = \langle Z \rangle$. Then we have the trivial nil(2)-module

$$0_E: E = \langle Z \times N \rangle / \Gamma_3 \langle Z \times N \rangle \to N$$

together with a homomorphism $\varepsilon: F \to E$ which is the identity on Z; here $0_E: E \to N$ is the free nil(2)-module with basis $0: F \to N, 0(z) = 0 \in N$ for $z \in Z$, see (III.1.12). Moreover, let

(2.5) $$0_E \vee \partial: E \vee M \to N$$

be the sum of 0_E and of a nil(2)-module $\partial: M \to N$. A group homomorphism $f: F \to M$ with $\partial f = 0$ yields a map $f: 0_E \to \partial$ in **cross**(2), also denoted by f, since 0_E is free. Hence we get by the universal property (2.2)(2) the group homomorphism

(2.6) $$(f,1)\colon E \vee M \to M$$

which is given by the map $(f,1)\colon 0_E \vee \partial \to \partial$ in **cross**(2). For $f = 0$ we get in the same way the projection $(0,1)\colon E \vee M \to M$ which is embedded in the following commutative diagram.

(2.7)
$$\begin{array}{ccccc}
\tau^{-1}J = \Gamma C_E \oplus K \otimes C_E & \rightarrowtail & \Gamma(C_E \otimes K) & \overset{\twoheadrightarrow}{\underset{\dashleftarrow}{}} & \Gamma K \\
\downarrow & & \downarrow{\scriptstyle \tau} & & \downarrow{\scriptstyle \tau} \\
J & \overset{j}{\rightarrowtail} & (C_E \oplus C)^{\otimes 2} & \overset{(0,1)^{\otimes 2}}{\underset{i^{\otimes 2}}{\rightleftarrows}} & C^{\otimes 2} \\
\downarrow{\scriptstyle w} & & \downarrow{\scriptstyle w} & & \downarrow{\scriptstyle w} \\
(E \vee M)_2 & \rightarrowtail & E \vee M & \overset{(0,1)}{\underset{i}{\rightleftarrows}} & M \\
\downarrow & & \downarrow{\scriptstyle p} & & \downarrow{\scriptstyle p} \\
C_E & \rightarrowtail & (E \vee M)^{cr} & \overset{(0,1)}{\underset{i}{\rightleftarrows}} & M^{cr}
\end{array}$$

Here we set (see (2.8) below for the definition of C_E)

$$\begin{cases} J = (C_E \otimes C_E) \oplus (C_E \otimes C) \oplus (C \otimes C_E), \\ C = (M^{cr})^{ab}, \\ (E \vee M)_2 = \text{kernel}(0,1). \end{cases}$$

The rows of the diagram are short exact sequences of groups which are split by the inclusions $i\colon \partial \to 0_E \vee \partial$. Moreover, the columns of the diagram correspond to the exact sequences given by (1.4) and (1.8) respectively. We use the top row of (2.7) only in case $\partial\colon M \to N$ is a totally free nil(2)-module. In this case also $0_E \vee \partial$ in (2.5) is totally free so that we can use (1.8) for ∂ and for $0_E \vee \partial$. In the next lemma we show that also the left hand column of (2.7) is an exact sequence.

(2.8) **Lemma.** *Let $\partial\colon M \to N$ be a* nil(2)*-module. Then there is a canonical isomorphism of N-groups*

$$(E \vee M)^{cr} = C_E \times M^{cr}$$

where C_E is the free $\pi_1(\partial)$-module generated by Z and where the right hand side is the product of groups. Moreover, the column at the left hand side of (2.7) *is exact.*

Proof. By the universal property we see that $(E \vee M)^{cr} = E^{cr} \vee M^{cr}$ is the sum in **cross**(1). On the other hand the map

$$\tilde{\partial}: C_E \times M^{cr} \to N \tag{1}$$

given by $\tilde{\partial}(x, y) = \partial^{cr}(y)$ is a crossed module. The diagram

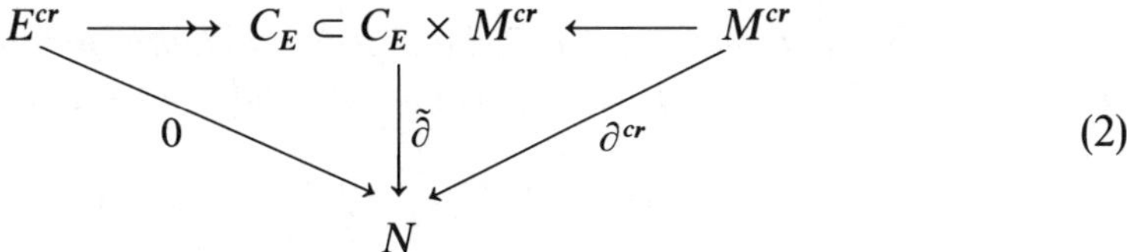

(2)

yields inclusion maps

$$0_E^{cr} \to \tilde{\partial} \leftarrow \partial^{cr} \tag{3}$$

for which the universal property (2.2)(2) is satisfied. This shows $\tilde{\partial} = (0_E \vee \partial)^{cr}$ and hence we get the equation in (2.8). This also shows that the bottom sequence of (2.7) is split short exact and that

$$(E \vee M)^{cr\,ab} = C_E \oplus C. \tag{4}$$

Whence the column in the middle of (2.7) is well defined by (1.4). Next we show that the column on the left hand side of (2.7) is exact. Let $x \in E \vee M$ with $px = 0$ and $(0, 1)x = 0$. Then there is $y \in (C_E \oplus C)^{\otimes 2}$ with $w(y) = x$. Let $z = y - i^{\otimes 2}(0, 1)^{\otimes 2}y$ where $i: C \subset C_E \oplus C$ is the inclusion. Then we have $z \in J$ and $w(z) = w(y) - i(0, 1)w(y) = x - 0 = x$. □

The map $(f, 1)$ in (2.6) yields the commutative diagram of group homomorphisms

(2.9)
$$\begin{array}{ccccc}
J \overset{j}{\subset} (C_E \oplus C)^{\otimes 2} & \xrightarrow{(f_*, 1)^{\otimes 2}} & C^{\otimes 2} & & \\
\downarrow w \quad c\text{-push} & & \omega \swarrow \quad \downarrow w & & \\
 & \overset{v}{\nearrow} \; L & \overset{\delta}{\searrow} & & \\
(E \vee M)_2 & \xrightarrow[(f,1)j]{} & M & \xrightarrow{\partial} & N
\end{array}$$

Here $(f_*, 1): C_E \oplus C \to C$ is the map $(f, 1)^{cr\,ab}$ induced by $(f, 1)$, in fact, f_* maps the generator $z \in Z$ to $\{f(z)\} = h_2 f(z)$ where $h_2\colon M \twoheadrightarrow C$ is the projection. Naturality of w shows that diagram (2.9) commutes. We define $\delta\colon L \to M$ by the central push out diagram of groups denoted by 'c-push', see (I.4.18).

(2.10) **Theorem.** *$(\omega, \delta, \partial)$ in (2.9) is the free quadratic module generated by $f\colon F \to M$. The map $i\colon F \to L$ is given by the composition $F \to E \to (E \vee M)_2 \to L$ which is the identity on generators in Z.*

We prove the theorem in (2.15) below. The next corollary is an easy consequence of (2.9) and (I.4.20).

(2.11) **Corollary.** *Let $\sigma = (\omega, \delta, \partial)$ be a free quadratic module as above. Then the* cokernel *C_E of ω is a free $\pi_1(\sigma)$-module and the* kernel *of ω is given by*

$$\mathrm{kernel}(\omega) = (f_*, 1)^{\otimes 2} j\, \mathrm{kernel}(w)$$

where f and $w\colon J \to (E \vee M)_2$ are given as in (2.9). Let $\delta_*\colon C_E = \mathrm{cok}(\omega) \to M^{cr} = \mathrm{cok}(w)$ be the map induced by δ. Then $f_* = h_2(\delta_*)$ where $h_2\colon M^{cr} \twoheadrightarrow C$ is the projection.

Now assume that $\sigma = (\omega, \delta, \partial)$ is a totally free quadratic module. Then we have the natural isomorphism

(2.12)
$$\pi_2\sigma \cong K/B \quad \text{with}$$
$$\begin{cases} K = h_2\,\mathrm{kernel}(\partial^{cr}\colon M^{cr} \to N) & \text{and} \\ B = \mathrm{image}(h_2\delta_*) & \text{where} \\ B \subset K \subset C. \end{cases}$$

The isomorphism is induced by $h_2 p$ in (1.7). Using (I.4.9) and (2.11) one gets the following commutative diagram of short exact sequences.

(2.13)
$$\begin{array}{ccccc} \Gamma B + [B, K] & \rightarrowtail & \Gamma K & \twoheadrightarrow & \Gamma(\pi_2\sigma) \\ \| & & \downarrow \tau & & \downarrow \tau_0 \\ \Delta_B & \overset{\tau}{\rightarrowtail} & C^{\otimes 2} & \twoheadrightarrow & C^{\otimes 2}/\tau\Delta_B \end{array}$$

Here Δ_B is the sum of subgroups in ΓK with $[B, K]$ defined by the Whitehead product (I.4.3). The injection τ_0 on the right hand side of (2.13) is induced by

(2.12) and by τ. The next corollary of (2.10) describes a crucial property of totally free quadratic modules.

(2.14) **Corollary.** *Let* $\sigma = (\omega, \delta, \partial)$ *be a totally free quadratic module. Then the* kernel *of* ω *is* $\tau\Delta_B$ *in* (2.13) *and* σ *yields the following natural commutative diagram in which all columns and rows are exact sequences.*

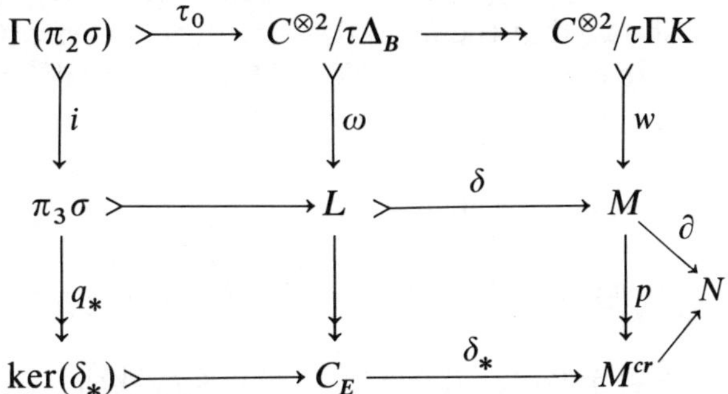

The group $C^{\otimes 2}/\tau\Gamma K$ is free as an abelian group so that the top row as an exact sequence of groups is split. We shall see that the left hand column of this diagram corresponds to Whitehead's certain exact sequence of a 3-dimensional CW-complex.

Proof of (2.14). We consider the kernel $J_0 = \tau^{-1}J$ of $w: J \to (E \vee M)_2$ in (2.7). This kernel is embedded in the commutative diagram

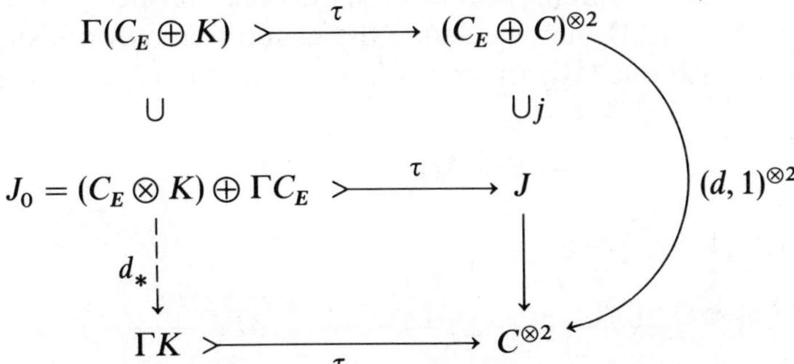

with $d = f_* = h_2\delta_*$, see (2.11). By definition of τ one can check that the induced map d_* satisfied $d_* = ([d, 1], \Gamma(d))$. Whence we get $\Delta_B = d_* J_0$ and thus by (2.11)

$$\text{kernel}(\omega) = (d, 1)^{\otimes 2} j \,\text{kernel}(w) = (d, 1)^{\otimes 2} j\tau J_0 = \tau d_* J_0 = \tau\Delta_B. \qquad \square$$

(2.15) *Proof of* (2.10). The group L is an N-group since all maps in (2.9) are equivariant with respect to the action of N. We now show that $(\omega, \delta, \partial)$ is actually a quadratic module. Clearly $\delta\omega = w$ and (1.9)(1) is satisfied. Moreover $\partial\delta = 0$ since $\partial w = 0$ and since

$$\partial(f, 1)g = (0_E \vee \partial)j = \partial(0, 1)j = 0. \tag{1}$$

Since $(0_E \vee \partial)(E \vee M)_2 = 0$ we can apply the formulas (1), (2) in (1.6) for x, $y \in (E \vee M)_2$. Moreover since $w(J)$ is central in $(E \vee M)_2$ also $\omega(C^{\otimes 2})$ is central in L. This shows that the formulas (1.10)(3), (4) are satisfied in L. We consider (1.10)(3). We can choose $y \in (E \vee M)_2$ and $z \in C^{\otimes 2}$ with $a = v(y) + \omega(z)$, see (2.9). Hence

$$\begin{aligned} a^{\partial x} &= v(y^{\partial x}) + \omega(z), \\ &= v(y + w(\{y\} \otimes \{x\} + \{x\} \otimes \{y\})) + \omega(z), \\ &= a + \omega(f_*\{y\} \otimes \{x\} + \{x\} \otimes f_*\{y\}), \\ &= a + \omega(\{\delta a\} \otimes \{x\} + \{x\} \otimes \{\delta a\}). \end{aligned} \tag{2}$$

Here we use

$$\{\delta a\} = \{\delta v(y) + \delta\omega(z)\} = \{(f, 1)j(y) + w(z)\} = f_*\{y\} \tag{3}$$

since $\{w(z)\} = 0$. In a similar way one obtains (1.10)(4).

It remains to check that $(\omega, \delta, \partial)$ has the universal property in (2.1). Let l_F, m, n be given as in (2.1) then we define the group homomorphism $l\colon L \to L'$ on the central push out L by the maps $lv = l_1$ and $l\omega = l_2$ as in the diagram

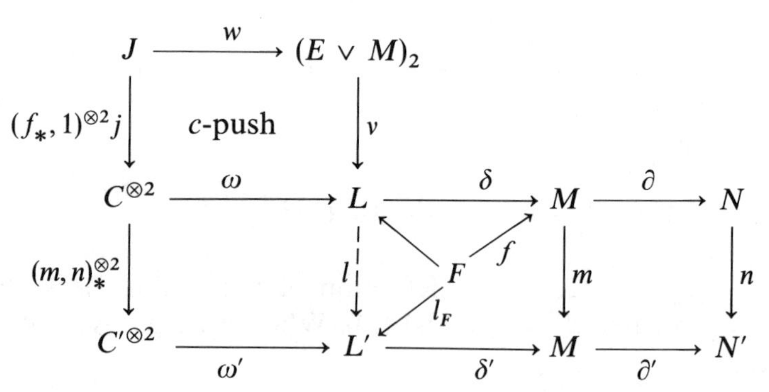

$$\begin{array}{ccccccc} J & \xrightarrow{w} & (E \vee M)_2 & & & & \\ \downarrow{\scriptstyle (f_*, 1)^{\otimes 2} j} & c\text{-push} & \downarrow{\scriptstyle v} & & & & \\ C^{\otimes 2} & \xrightarrow{\omega} & L & \xrightarrow{\delta} & M & \xrightarrow{\partial} & N \\ \downarrow{\scriptstyle (m,n)_*^{\otimes 2}} & & \downarrow{\scriptstyle l} & \nwarrow \; F \; \nearrow{\scriptstyle f} \; \swarrow{\scriptstyle l_F} & \downarrow{\scriptstyle m} & & \downarrow{\scriptstyle n} \\ C'^{\otimes 2} & \xrightarrow[\omega']{} & L' & \xrightarrow[\delta']{} & M & \xrightarrow[\partial']{} & N' \end{array} \tag{4}$$

Here we use the central map $l_2 = \omega'(m, n)_*^{\otimes 2}$ where $(m, n)_*\colon C \to C'$ is induced by the map $(m, n)\colon \partial \to \partial'$. For the definition of l_1 we need the nil(2)-module $\bar{\partial}'\colon \bar{L}' \to N'$ associated to $(\omega', \delta', \partial')$ as in (1.11). We obtain maps between nil(2)-modules

$$\begin{array}{ccccccc} F & \xrightarrow{l_F} & L' & & & & \\ \downarrow & & \downarrow & & & & \\ E & \xrightarrow{\bar{l}_F} & \bar{L}' & \xleftarrow{s} & M' & \xleftarrow{m} & M \\ {\scriptstyle 0_E}\downarrow & & {\scriptstyle \bar{\partial}'}\downarrow & \swarrow{\scriptstyle \partial'} & & & \downarrow{\scriptstyle \partial} \\ N & \xrightarrow[n]{} & N' & & \xleftarrow[n]{} & & N \end{array} \tag{5}$$

Here $\bar{l}_F$ is induced by l_F such that the diagram commutes. These maps induce a map $(t, n)\colon 0_E \vee \partial \to \bar{\partial}'$, see (2.2), such that the following diagram commutes

$$\begin{array}{ccc} (C_E \oplus C)^{\otimes 2} & \xrightarrow{t_*^{\otimes 2}} & (C' \oplus L'_\omega)^{\otimes 2} \\ {\scriptstyle w}\downarrow & & \downarrow{\scriptstyle \bar{w}} \\ E \vee M & \xrightarrow{t} & \bar{L}' \\ \downarrow & & \downarrow \\ N & \xrightarrow{n} & N' \end{array} \tag{6}$$

For $x \in Z \subset E$ we have $t(x) = (0, l_F x)$ and for $y \in M$ we have $t(y) = (my, 0)$. Hence we get for $x' \in Z$, $y' \in M$ the formulas

$$\begin{aligned} & tw((\{x\} \oplus \{y\}) \otimes (\{x'\} \oplus \{y'\}), \\ & = t\langle y + x, y' + x' \rangle \\ & = \langle ty + tx, ty' + tx' \rangle \\ & = \langle (my, l_F x), (my', l_F x') \rangle \\ & = \bar{w}((\{my\} \oplus \{l_F x\}) \otimes (\{my'\} \otimes \{l_F x'\})), \quad \text{see (1.12).} \end{aligned} \tag{7}$$

Therefore we define $t_*\colon C_E \oplus C \to C' \oplus L'_\omega$ by

$$\begin{aligned} t_*(\{x\} \oplus \{y\}) &= \{my\} \oplus \{l_F x\} \\ &= (m, n)_* \{y\} \oplus l_{F*}\{x\} \end{aligned} \tag{8}$$

so that (6) commutes. Moreover, the map t induces the commutative diagram

$$\begin{array}{ccc} J & \xrightarrow{t_*^{\otimes 2}} & J' \\ {\scriptstyle w}\downarrow & & \downarrow{\scriptstyle \bar{\bar{w}}} \\ (E \vee M)_2 & \xrightarrow{l_1} & L' \\ \downarrow & & \downarrow \\ E \vee M & \xrightarrow{t} & \bar{L}' \\ \downarrow & & \downarrow \\ M & \xrightarrow[m]{} & M' \end{array} \tag{9}$$

Here l_1 is the restriction of t and $\bar{\bar{w}}$ is the restriction of $\bar{w}$ in (6) for

$$J' = (L'_\omega \otimes L'_\omega) \oplus (C' \otimes L'_\omega) \oplus (L'_\omega \otimes C'), \tag{10}$$

compare the definition of J in (2.7). We deduce from (1.13) that $\bar{\bar{w}}$ satisfies the formulas ($a, b \in L', x, y \in M'$)

$$\left.\begin{aligned} \bar{\bar{w}}(\{a\} \otimes \{b\}) &= \omega'(\{\delta' a\} \otimes \{\delta' b\}) \\ \bar{\bar{w}}(\{a\} \otimes \{y\}) &= \omega'(\{\delta' a\} \otimes \{y\}) \\ \bar{\bar{w}}(\{x\} \otimes \{b\}) &= \omega'(\{x\} \otimes \{\delta' b\}) \end{aligned}\right\} \tag{11}$$

We now are ready to check the equation

$$l_1 w = \bar{\bar{w}} t_*^{\otimes 2} = \omega'(m, n)_*^{\otimes 2} (f_*, 1)^{\otimes 2} j = l_2 (f_*, 1)^{\otimes 2} j \tag{12}$$

which shows that the central push map l in (4) is well defined by the pair of maps (l_1, l_2). It is clear that l is uniquely defined by (l_F, m, n) in (4); therefore the proof of (2.10) is complete by (12). We check (12) on $\{x\} \otimes \{x'\} \in C_E \otimes C_E$, $x, x' \in Z$; by the formulas

$$\begin{aligned} \bar{\bar{w}}(t_*\{x\} \otimes t_*\{x'\}) &= \bar{\bar{w}}(\{l_F x\} \otimes \{l_F x'\}), \quad \text{see (8)}, \\ &= \omega'(\{\delta' l_F x\} \otimes \{\delta' l_F x'\}), \quad \text{see (11)}, \\ &= \omega'(\{mfx\} \otimes \{mfx'\}) \end{aligned} \tag{13}$$

where we use the commutativity of (4). One readily checks (12) on $C_E \otimes C$ and $C \otimes C_E$ in a similar way. □

§ 3 Quadratic chain complexes

We introduce the category of quadratic chain complexes and we define a functor λ which carries quadratic chain complexes to crossed chain complexes. This functor satisfies a kind of Whitehead theorem, see (3.8). Moreover we show that the homotopy groups of a totally free quadratic chain complex are embedded in an exact sequence which corresponds to Whitehead's certain exact sequence, see (3.7). In this section and in the following two sections § 4, § 5 we study the basic algebraic properties of quadratic chain complexes. The connection of quadratic chain complexes with spaces is described in § 6.

(3.1) **Definition.** A **quadratic chain complex** $\sigma = \{\sigma_n, d_n, \omega\}$ is a diagram of homorphisms between groups

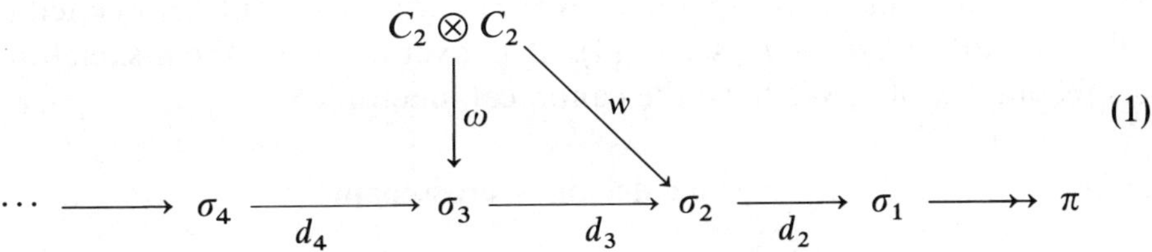

(1)

such that $d_{n-1}d_n = 0$ for $n \geq 3$ and such that the following properties are satisfied. The triple (ω, d_3, d_2) is a quadratic module with $\pi = \text{cokernel}(d_2)$; hence kernel$(d_3)$ is a π-module. Moreover, σ_n is a right π-module, $n \geq 4$, and d_n with $n \geq 5$ and $d_4\colon \sigma_4 \to \text{kernel}(d_3)$ are homomorphisms between right π-modules. A **quadratic chain map** $f\colon \sigma \to \sigma'$ between quadratic chain complexes is a family of homomorphisms between groups $(n \geq 1)$

$$f_n\colon \sigma_n \to \sigma'_n \quad \text{with} \quad f_{n-1}d_n = d_n f_n \tag{2}$$

such that (f_3, f_2, f_1) is a map between quadratic modules and such that f_n is $\bar{f}_1$-equivariant, where $\bar{f}_1\colon \pi \to \pi'$ is induced by f_1, $n \geq 4$. Let **quadchain** be the category of quadratic chain complexes and of quadratic chain maps.

We define the **homotopy groups**

$$\left.\begin{aligned} \pi_1(\sigma) &= \pi = \text{cokernel}(d_2) \\ \pi_n(\sigma) &= \text{kernel}(d_n)/\text{image}(d_{n+1}), \quad n \geq 2. \end{aligned}\right\} \tag{3}$$

Here $\pi_n(\sigma)$ is a $\pi_1(\sigma)$-module for $n \geq 2$. A quadratic chain map $f: \sigma \to \sigma'$ is a **weak equivalence** if f induces isomorphisms

$$\pi_n(f): \pi_n(\sigma) \cong \pi_n(\sigma') \quad \text{for } n \geq 1. \tag{4}$$

Clearly the category **quad** of quadratic modules is the full subcategory of **quadchain** consisting of objects σ with $\sigma_i = 0$ for $i > 3$. Moreover, the category **cross** of crossed modules is the full subcategory of **quadchain** consisting of objects σ with $\sigma_i = 0$ for $i > 2$.

The n-**skeleton** σ^n of a quadratic chain complex σ is a quadratic chain complex given by

$$(\sigma^n)_i = \begin{cases} \sigma_i & i \leq n \\ C_2 \otimes C_2 & i = 3, n = 2 \\ 0 & \text{otherwise.} \end{cases} \tag{5}$$

Here $\sigma^1 = \sigma_1$ is just a group and σ^2 is the quadratic module associated to the nil(2)-module $d_2: \sigma_2 \to \sigma_1$, see (1.14). Moreover for $n \geq 3$ the n-skeleton σ^n is a subcomplex of σ. We have the canonical functor

(3.2) λ: **quadchain** $\to$ **crosschain**

which carries $\sigma = \{\sigma_n, d_n, \omega\}$ to the crossed chain complex $\lambda(\sigma) = \rho = \{\rho_n, d_n\}$ given by the commutative diagram

(3.3)
$$\begin{array}{ccccccccccc}
 & & & & C_2 \otimes C_2 & \overset{w}{\searrow} & & & & & \\
 & & & & \downarrow \omega & & & & & & \\
\cdots \longrightarrow & \sigma_4 & \longrightarrow & & \sigma_3 & \longrightarrow & \sigma_2 & \longrightarrow & \sigma_1 & \longrightarrow & \pi \\
 & \| q_4 & & & \downarrow q_3 & & \downarrow q_2 & & \| q_1 & & \| \\
\cdots \longrightarrow & \rho_4 & \longrightarrow & & \rho_3 & \longrightarrow & \rho_2 & \longrightarrow & \rho_1 & \longrightarrow & \pi
\end{array}$$

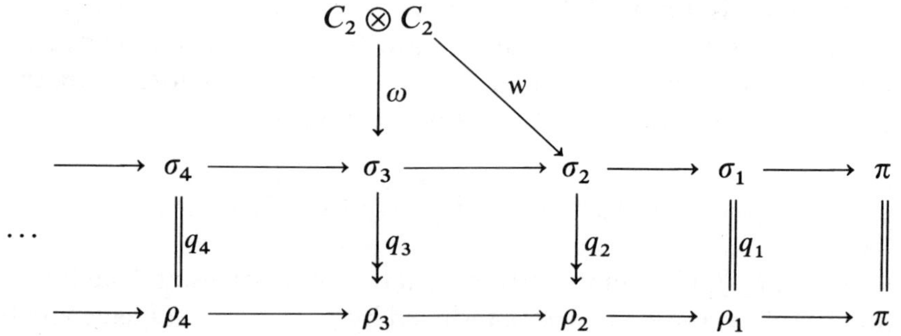

where $\rho_3 = \sigma_3/\omega(C_2 \otimes C_2)$ and where $\rho_2 = \sigma_2/w(C_2 \otimes C_2)$ are the quotient groups. The map q_n is the projection for $n = 2, 3$ and q_n is the identity otherwise. One readily checks that the bottom row of the diagram is a well defined crossed chain complex, compare (III.2.6), with $C_2 = \rho_2^{ab}$.

Similarly as in (III.2.7) we define totally free objects.

(3.4) **Definition.** We call a quadratic chain complex σ **totally free** if σ_1 is a free group, if $d_2: \sigma_2 \to \sigma_1$ is a free nil(2)-module, if $d_3: \sigma_3 \to \sigma_2$ is given by a free quadratic module (ω, d_3, d_2), and if σ_n is a free π-module for $n \geq 4$. Let $\mathbf{Q} \subset$ **quadchain** be the full subcategory consisting of totally free quadratic chain complexes.

A 3-skeleton of a totally free quadratic chain complex is a totally free quadratic module for which we have all the properties described in (2.9)...(2.14). In particular, we have by (2.9) the central pushout diagram

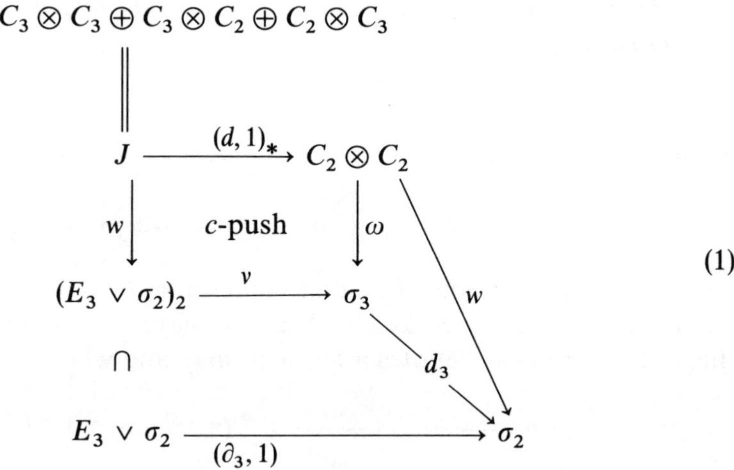

(1)

Here we use the chain complex $(C_*, d) = C\lambda(\sigma)$ which yields $(d, 1)_*$, see (2.11) and (III.2.10). Moreover,

$$0: E_3 \to \sigma_1 = \rho_1 = \langle Z_1 \rangle \tag{2}$$

is the free nil(2)-module with trivial boundary generated by $Z_3 \subset E_3$ and $d_2: \sigma_2 \to \sigma_1$ is a free nil(2)-module generated by Z_2 determined by a function $d_2|Z_2: Z_2 \to \sigma_1$. The map

$$\partial_3: E_3 \to \sigma_2 \tag{3}$$

is the map between nil(2)-modules defined on generators Z_3 by $d_3|Z_3$.

(3.5) **Lemma.** *The functor λ in* (3.2) *carries totally free objects to totally free objects.*

Proof. Let $Z_n \subset \sigma_n$, $n \geq 1$, be given by a basis of $\sigma \in \mathbf{Q}$. Then $q_n Z_n \subset \rho_n$ yields a basis of $\rho = \lambda(\sigma)$, see (3.3). In fact, ρ_3 is a free π-module generated by $q_3 Z_3$ by (2.11). □

The lemma shows that the functor λ in (3.2) yields the functor

(3.6) $$\lambda\colon \mathbf{Q} \to \mathbf{H}.$$

Next we derive from a quadratic chain complex an exact sequence which corresponds to Whitehead's certain exact sequence, see (7.3) below. Let σ be a quadratic chain complex and let $\rho = \lambda(\sigma)$ be the crossed chain complex defined by the functor λ in (3.2). For the homotopy groups $\pi_n(\rho)$ and $\pi_n(\sigma)$ in (III.2.6)(5) and (3.1)(3) we get the following result.

(3.7) **Proposition.** *Assume σ is totally free, then the map $q\colon \sigma \to \lambda\sigma = \rho$ in (3.3) induces isomorphisms*

$$q_*\colon \pi_n\sigma \xrightarrow{\cong} \pi_n\rho$$

for $n = 1$, $n = 2$, and $n \geq 5$. Moreover, there is a natural exact sequence

$$0 \longrightarrow \pi_4\sigma \xrightarrow{q_*} \pi_4\rho \xrightarrow{b} \Gamma(\pi_2\rho) \xrightarrow{i} \pi_3\sigma \xrightarrow{q_*} \pi_3\rho \longrightarrow 0$$

Proof. We only consider the exact sequence in (3.7). For this we use (2.14). Let σ^3 and ρ^3 be the 3-skeleta. Then we have the following commutative diagram where $\twoheadrightarrow$ denotes a quotient map and where $\rightarrowtail$ is injective.

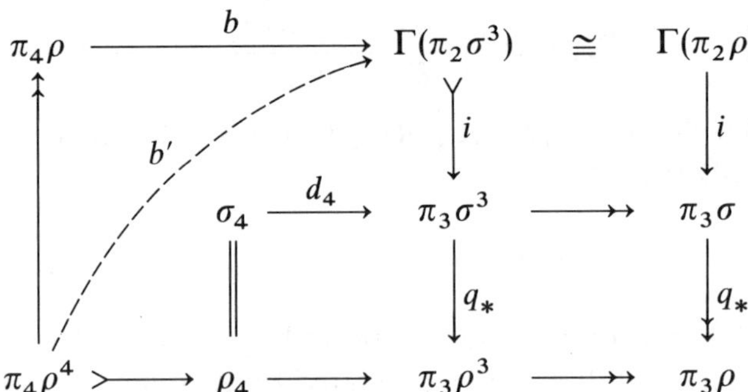

Here the isomorphism is given by (I.4.9). The exact column (i, q_*) is induced by the one in (2.14). The bottom row of (3) is exact so that b' is well defined. Moreover b' induces b. This completes the definition of the exact sequence (3.7). □

We derive from (3.7) the following result which is a kind of a 'Whitehead theorem' for the functor $\lambda\colon \mathbf{Q} \to \mathbf{H}$.

(3.8) **Proposition.** *Let $f\colon \sigma \to \sigma'$ be a map in* $\mathbf{Q}$: *Then* (A) *and* (B) *holds.*
(A) *f is a weak equivalence in* $\mathbf{Q}$ *if and only if λf is one in* $\mathbf{H}$.
(B) *f is an isomorphism in* $\mathbf{Q}$ *if and only if λf is an isomorphism in* $\mathbf{H}$.

Proof. (A) is clear by (3.7). We now prove (B). Assume λf is an isomorphism. We have to show that $f_3\colon \sigma_3 \to \sigma'_3$ and $f_2\colon \sigma_2 \to \sigma'_2$ are isomorphisms. Now f_2 is an isomorphism by the naturality of the exact sequence (1.8) and by the five-lemma. Moreover, f_3 is an isomorphism, since we have the natural exact sequence (see (2.14))

$$\Delta_B \overset{\tau}{\rightarrowtail} C_2 \otimes C_2 \xrightarrow{\omega} \sigma_3 \longrightarrow \rho_3 \longrightarrow 0. \qquad \square$$

§ 4 Homotopies for quadratic chain maps

Recall that we described already in (III.2.6) homotopies for maps between crossed chain complexes in **crosschain**. In this section we introduce homotopies $\alpha\colon f \simeq g$ for maps between quadratic chain complexes in **quadchain**. The functor

$$\lambda\colon \textbf{quadchain} \to \textbf{crosschain}$$

in § 3 above carries the homotopy α to a homotopy $\lambda\alpha\colon \lambda f \simeq \lambda g$ in **crosschain**. For the subcategories of totally free objects λ induces the functor

$$\lambda\colon \mathbf{Q}/\!\simeq\; \to \mathbf{H}/\!\simeq$$

between homotopy categories. We also consider 0-homotopies $\alpha\colon f \overset{0}{\simeq} g$ which induce the trivial homotopy $\lambda\alpha\colon \lambda f = \lambda g$. These homotopies yield the functor

$$\lambda\colon \mathbf{Q}/\overset{0}{\simeq}\; \to \mathbf{H}.$$

The definition of a 0-homotopy in (4.8) is fairly simple compared with the complicated formulas (4.1) which define a homotopy $\alpha\colon f \simeq g$ in **quadchain**. These formulas describe explicitly the distributivity laws of the function α with respect to addition and actions in the quadratic chain complex σ. The distributivity laws lead to the notion of a "quadratic operator" in (4.3) which is the quadratic analogue of a crossed homomorphism. The computations needed in this section are fairly extensive and long winded but they form a good exercise in getting familiar with quadratic modules. The computations in (4.13) below show how we originally found the distributivity laws of a quadratic operator. We first define the notion of homotopy in the category **quadchain** of quadratic chain complexes.

(4.1) **Definition.** Let $f, g\colon \sigma \to \sigma'$ be quadratic chain maps. A **homotopy** $\alpha\colon f \simeq g$ is given by a sequence of functions

$$\alpha_n\colon \sigma_n \to \sigma'_{n+1} \quad (n \geq 1) \tag{1}$$

satisfying

$$\left.\begin{aligned} -f_1 + g_1 &= d'_2\alpha_1 \\ -f_n + g_n &= d'_{n+1}\alpha_n + \alpha_{n-1}d_n \quad (n \geq 2). \end{aligned}\right\} \tag{2}$$

Here α_n is a homomorphism between groups for $n \geq 3$, which is f_1-equivariant for $n = 3$, and which is $\bar{f}_1$-equivariant for $n > 3$, $\bar{f}_1 = \pi_1(f)$. Moreover, α_1 is an f_1-crossed homomorphism and $\alpha_2\colon \sigma_2 \to \sigma'_3$ is a function which satisfies the following formulas (3) and (5) where ω' is the quadratic map of σ', $(x, y \in \sigma_2, \beta \in \sigma_1)$.

$$\alpha_2(x + y) = \alpha_2(x) + \alpha_2(y) + \Delta(x, y) \tag{3}$$

where $\Delta(x, y) = \Delta_1 + \Delta_2 \in \omega'(C'_2 \otimes C'_2)$ is given by $(d = d_2)$

$$\left.\begin{aligned} \Delta_1 &= \omega'(\{-f_2x + g_2x\} \otimes \{f_2y\}) \\ \Delta_2 &= \omega'(\{\alpha_1 dy\} \otimes \{\alpha_1 dx\} - \{\alpha_1 dx\} \otimes \{f_2y\} - \{g_2y\} \otimes \{\alpha_1 dx\}). \end{aligned}\right\} \tag{4}$$

We point out that Δ_2 vanishes if x is a boundary in $d_3\sigma_3$.

$$\alpha_2(x^\beta) = (\alpha_2 x)^{f_1\beta} + \nabla(x, \beta). \tag{5}$$

Here $\nabla(x, \beta) = \nabla_1 + \nabla_2 \in \omega'(C'_2 \otimes C'_2)$ is defined by the formulas

$$\left.\begin{aligned} \nabla_1 &= \omega'(\{g_2x\}^{\bar{f}_1\beta} \otimes \{\alpha_1\beta\} + \{\alpha_1\beta\} \otimes \{g_2x\}^{\bar{f}_1\beta}) \\ \nabla_2 &= \omega'(-\{\alpha_1 d_2 x\}^{\bar{f}_1\beta} \otimes \{\alpha_1\beta\}). \end{aligned}\right\} \tag{6}$$

The term ∇_2 vanishes for $x \in d_3\sigma_3$.

(4.2) **Remark.** The notion of homotopy for quadratic chain maps is compatible with the notion of homotopy for crossed chain maps in (III.2.6). In fact, $\alpha\colon f \simeq g$ in (4.1) induces $\lambda\alpha\colon \lambda f \simeq \lambda g$ where λ is the functor in (3.2). Here $\lambda\alpha$ is given by the unique function $\lambda\alpha_n$, $n \geq 1$, for which the diagram

$$\begin{array}{ccc} \sigma_n & \xrightarrow{\alpha_n} & \sigma'_{n+1} \\ {\scriptstyle q_n}\downarrow & & \downarrow{\scriptstyle q_{n+1}} \\ (\lambda\sigma)_n = \rho_n & \xrightarrow{\lambda\alpha_n} & \rho'_{n+1} = (\lambda\sigma')_{n+1} \end{array}$$

commutes. It is clear by (4.1)(3), (5) that $\lambda\alpha_2$ is an $\bar{f}_1$-equivariant homorphism. Moreover, one readily checks that $\lambda\alpha$ is a well defined homotopy in the sense of (III.2.6).

(4.3) **Definition.** Let $\sigma = (d_2: \sigma_2 \to \sigma_1)$ be a nil(2)-module and let $\sigma' = (\omega', d_3', d_2')$ be a quadratic module and let f, g: $d_2 \to d_2'$ be maps between nil(2)-modules. Moreover let $\alpha_1: \sigma_1 \to \sigma_2'$ be an f_1-crossed homomorphism with

$$-f_1 + g_1 = d_2'\alpha_1. \tag{1}$$

Then we call a function $\alpha_2: \sigma_2 \to \sigma_3'$ an (α_1, f, g)-**quadratic operator** if α_2 satisfies the quadratic distributivity formulas (4.1)(3) and (4.1)(5) and if

$$-f_2 + g_2 = d_3'\alpha_2 + \alpha_1 d_2. \tag{2}$$

We point out that f, g cannot be interpreted as being maps in **quadchain**; therefore (α_2, α_1) is not a homotopy as in (4.1). The following result describes a crucial property of quadratic operators.

(4.4) **Proposition.** *Let α_2 be an (α_1, f, g)-quadratic operator as in (4.2) and let x, $y \in \sigma_2$. Then the Peiffer commutator $\langle x, y\rangle \in \sigma_2$ satisfies the equation*

$$\alpha_2(\langle x, y\rangle) = \omega'(-\{f_2 x\} \otimes \{f_2 y\} + \{g_2 x\} \otimes \{g_2 y\})$$

Here ω' is the quadratic map of σ'.

The proof is a good exercise in handling the quadratic distributivity laws (4.1)(3), (5), and (1.10)(3), (4). Using these laws for $\alpha_2\langle x, y\rangle$ one gets an equation with a fair amount of summands. Most of these summands cancel so that only equation (4.4) remains.

Proof of (4.4). We have for $d = d_2$

$$\alpha_2(\langle x, y\rangle) = \alpha_2((-x-y) + (x + y^{dx})) \tag{1}$$

$$= \alpha_2(-x-y) + \alpha_2(x + y^{dx}) \tag{2}$$

$$+ \Delta(-x-y, x + y^{dx}) \tag{3}$$

$$= \alpha_2(-x) + \alpha_2(-y) + \alpha_2(x) + \alpha_2(y^{dx}) \tag{4}$$

$$+ \Delta(-x, -y) + \Delta(x, y^{dx}) + (3) \tag{5}$$

$$= -\alpha_2(x) - \alpha_2(y) + \alpha_2(x) + \alpha_2(y)^{f_1 dx} \tag{6}$$

$$- \Delta(x, -x) - \Delta(y, -y) + \nabla(y, dx) + (5) \tag{7}$$

$$= -\alpha_2(x) - \alpha_2(y) + \alpha_2(x) + \alpha_2(y) + (7) \tag{8}$$

$$+ \omega'(\{d'_3\alpha_2 y\} \otimes \{f_2 x\} + \{f_2 x\} \otimes \{d'_3\alpha_2 y\}) \tag{9}$$

$$= \omega'(\{d'_3\alpha_2 x\} \otimes \{d'_3\alpha_2 y\}) + (7) + (9) \tag{10}$$

Now we use $d'_3\alpha_2 = -\alpha_1 d + (-f_2 + g_2)$. Moreover, we observe that the functions $x \mapsto \{\alpha_1 dx\}$, $x \mapsto \{-f_2 x + g_2 x\}$ are f_1-equivariant homomorphisms and that $\Delta(x, y)$ is bilinear. Whence we get

$$-\Delta(y, x) + \Delta(x, y) + \nabla(y, dx) = (7). \tag{11}$$

In writing equation (10) explicitly we see that most terms cancel. This yields (4.4). □

We derive from (4.4) the next additional freeness property of a free nil(2)-module.

(4.5) **Proposition.** *Let α_2 be an (α_1, f, g)-quadratic operator as in* (4.2) *and assume σ is a free* nil(2)-*module with basis $Z_2 \subset \sigma_2$. Then $\alpha_2\colon \sigma_2 \to \sigma'_3$ is completely determined by the restriction $\alpha_2|Z_2\colon Z_2 \to \sigma'_3$. Moreover, let a function $a\colon Z_2 \to \sigma'_3$ be given. Then there is a unique (α_1, f, g)-quadratic operator α_2 with $\alpha_2|Z_2 = a$ provided the equation*

$$-f_2 x + g_2 x = d'_3 a(x) + \alpha_1 d_2(x)$$

holds for $x \in Z_2$.

Proof of (4.5). σ_2 is a quotient of the free σ_1-group $\langle Z_2 \times \sigma_1 \rangle$, namely

$$\sigma_2 = \langle Z_2 \times \sigma_1 \rangle / P_3(\partial_f), \tag{1}$$

see (III.1.16). Here f is a basis of σ. The quadratic distributivity laws (4.1)(3), (5) show that the function α_2 is determined by $\alpha_2|Z_2$. On the other hand let $a\colon Z_2 \to \sigma'_3$ be given. This gives us a function

$$a_1\colon (Z_2 + \bar{Z}_2) \times \sigma_1 \to \sigma'_3 \tag{2}$$

where $\bar{Z}_2 = -Z_2$ and where $+$ is the disjoint union. Here we set ($x \in Z_2$,

$\beta \in \sigma_1$)

$$\left.\begin{aligned} a_1(x, \beta) &= a(x) + \nabla(x, \beta) \\ a_1(-x, \beta) &= -a(x) - \Delta(x, -x) + \nabla(-x, \beta). \end{aligned}\right\} \tag{3}$$

Now the formula (4.1)(3) yields a well defined function

$$a_2\colon \mathrm{Mon}((Z_2 + \bar{Z}_2) \times \sigma_1) \to \sigma_3' \tag{4}$$

where Mon(Z) denotes the free monoid generated by Z. Clearly a_2 is given on generators by a_1. One can check that a_2 factors over the free group $\langle Z_2 \times \sigma_1 \rangle$ which is a quotient of $\mathrm{Mon}((Z_2 + \bar{Z}_2) \times \sigma_1)$. Moreover (4.3) shows that the factorization

$$a_3\colon \langle Z_2 \times \sigma_1 \rangle \to \sigma_3' \tag{5}$$

induces a unique map

$$\alpha_2\colon \langle Z_2 \times \sigma_1 \rangle / P_3(\partial_f) \to \sigma_3' \tag{6}$$

and that this map is an (α_1, f, g)-quadratic operator. For this we have to check equation (4.3)(2), namely

$$\alpha_1 d_2 = -d_3' \alpha_2 - f_2 + g_2. \tag{7}$$

In fact, both sides of (7) coincide on a basis $Z_3 \subset \sigma_3$ by the assumption in (4.5). Moreover, both sides of (7) satisfy the same distributivity laws, see (4.13)(14) below. Whence both sides of (7) coincide as functions on σ_2. □

The next lemma is an extension of (4.5).

(4.6) **Lemma.** *Let σ be a totally free quadratic chain complex and let $\alpha\colon f \simeq g$ be a homotopy for $f, g\colon \sigma \to \sigma'$. Then α is completely determined by its values on a basis $Z_n \subset \sigma_n$.*

The lemma follows from the construction of the cylinder below since this cylinder is a totally free quadratic chain complex. A similar lemma holds for homotopies on crossed chain complexes. We do not claim that the relation of homotopy is a natural equivalence relation on the categories **crosschain** or **quadchain**. However, on the subcategories **H**, resp. **Q**, of totally free objects we get this property:

(4.7) **Lemma.** *The relation of homotopy is a natural equivalence relation on the categories* **H** *and* **Q** *respectively.*

Proof. We check the result only for **Q**; the proof for **H** is similar. Let $\alpha\colon f \simeq g$, $\alpha'\colon g \simeq h$ be homotopies where $f, g, h\colon \sigma \to \sigma'$ are quadratic chain maps. Assume that σ is a totally free quadratic chain complex and let $Z_n \subset \sigma_n$ be given by a basis of σ. Then we define the homotopy $\alpha'' = \alpha \oplus \alpha'\colon f \simeq h$ by the unique homotopy $\alpha''\colon f \simeq h$ which satisfies $\alpha''_n(x) = \alpha(x) + \alpha'(x)$ for $x \in Z_n$. Here we use (4.6). Similarly we obtain a homotopy $\bar{\alpha} = \Theta\alpha\colon g \simeq f$ by the unique homotopy which satisfies $\bar{\alpha}_n(x) = -\alpha_n(x)$ for $x \in Z_n$. We leave it to the reader to check naturality of the homotopy relation. □

Lemma (4.7) shows that the homotopy categories $\mathbf{H}/\simeq$ and $\mathbf{Q}/\simeq$ are well defined. Moreover, we need the category $\mathbf{Q}/\overset{0}{\simeq}$ defined by 0-homotopies as follows, compare also (VI.6.2)(7) below.

(4.8) **Definition.** Let $f, g\colon \sigma \to \sigma'$ be quadratic chain maps. A 0-**homotopy** $\alpha_2\colon f \overset{0}{\simeq} g$ is a $\pi_1(f)$-equivariant homomorphism

$$\alpha_2\colon C_2 \to C'_2 \otimes C'_2$$

with the following properties:

$$-f_2 + g_2 = d'_3(\omega'\alpha_2 q),$$

$$-f_3 + g_3 = (\omega'\alpha_2 q)d_3,$$

$$-f_n + g_n = 0 \quad \text{for} \quad n = 1 \quad \text{and} \quad n \geq 4.$$

Here $\omega'\colon C'_2 \otimes C'_2 \to \sigma'_3$ is the quadratic map of σ' and $q\colon \sigma_2 \twoheadrightarrow C_2$ is the quotient map for $C_2 = (\sigma_2^{cr})^{ab}$.

The relation $\overset{0}{\simeq}$ is a natural equivalence relation on **Q**. A 0-homotopy corresponds to a homotopy $\alpha\colon f \simeq g$ which is filtration preserving with respect to skeleta, that is, α restricts to homotopies $\alpha^n\colon f^n \simeq g^n$ for $n \geq 1$, compare (III.1.5) in Baues (AH). In fact, let α in (4.1) be filtration preserving. Then we see that $\alpha_i = 0$ for $i \neq 2$ and $\alpha_2\colon \sigma_2 \to C'_2 \otimes C'_2$ by the definition of skeleta in (3.1)(5). This shows that the functor λ in (3.2) satisfies $\lambda f = \lambda g$. Therefore we also have $\{f_2 x\} = \{g_2 x\}$. Whence the elements $\Delta_1, \Delta_2, \nabla_1, \nabla_2$ in (4.1) are all trivial. Moreover, also the element (4.4) is trivial. Therefore α_2 factors over C_2 and yields a map as in (4.8).

The functor λ in (3.6) induces the following commutative diagram of functors

(4.9)
$$\begin{array}{ccc} Q/\overset{0}{\simeq} & \xrightarrow{\lambda} & \mathbf{H} \\ \downarrow & & \downarrow \\ Q/\simeq & \xrightarrow{\lambda} & \mathbf{H}/\simeq \end{array}$$

where the vertical arrows denote the quotient functors. Here (4.2) shows that a homotopy in **Q** induces a homotopy in **H**, see also (4.12) below.

Next we construct a **cylinder $I\sigma$ for a totally free quadratic chain complex** σ. We proceed in a similar way as in the definition of a cylinder of a totally free crossed chain complex, see (III.3.3). The cylinder $I\sigma$ is obtained together with structure maps

(4.10)
$$\sigma \vee \sigma \overset{i}{\rightarrowtail} I\sigma \xrightarrow{p} \sigma$$

as follows. Here $\sigma \vee \sigma$ is the sum in the category **quadchain.** Let $Z_n = Z_n(\sigma) \subset \sigma_n$ be a basis of σ and let $Z_0 = \phi$. Then we set

$$Z_n(I\sigma) = Z_n' \cup sZ_{n-1} \cup Z_n'' \tag{1}$$

where we use the same notation as in (III.3.1). The set (1) is a basis of the totally free quadratic chain complex $I\sigma$ and the quadratic chain maps $i = (i_0, i_1)$ and p in (4.10) are defined in terms of this basis by

$$\left.\begin{aligned} &i_0 x = x', \quad i_1 x = x'', \\ &px' = px'' = x, \quad psx = 0 \end{aligned}\right\} \tag{2}$$

where $x \in Z_n$, compare (III.3.2)(2). The boundary d of $I\sigma$ is defined on the basis (1) by

$$\left.\begin{aligned} &dx' = i_0(dx), \quad dx'' = i_1(dx), \\ &dsx = x' + x'' - S_{n-1}\,dx \end{aligned}\right\} \tag{3}$$

for $x \in Z_n$. Here we set $S_0 dx = 0$ and

$$S_1\colon \sigma_1 \to (I\sigma)_2 \tag{4}$$

is the i_0-crossed homorphism with $S_1 x = sx$ for $x \in Z_1$. The function

$$S_2\colon \sigma_2 \to (I_\sigma)_3 \tag{5}$$

is the (S_1, i_0, i_1)-quadratic operator with $S_2x = sx$ for $x \in Z_2$, see (4.3). Moreover for $n \geq 3$

$$S_n\colon \sigma_n \to (I\sigma)_{n+1} \tag{6}$$

is the i_0-equivariant homomorphism with $S_n(x) = sx$ for $x \in Z_n$. It is clear that i and p in (4.10) are quadratic chain maps. We derive from (3.12) that p is a weak equivalence, see (4.12) below. Moreover $S\colon i_0 \simeq i_1$ is a homotopy in the sense of (4.1). A quadratic chain map $H\colon I\sigma \to \sigma'$ with $Hi = (f, g)$ yields a homotopy $\alpha\colon f \simeq g$ by $\alpha = HS$. Vice versa a homotopy $\alpha\colon f \simeq g$ as in (4.1) yields a quadratic chain map H by defining H on generators as in (III.3.2)(6). Using (4.3) one readily checks

(4.11) **Lemma.** *The boundary d in $I(\sigma)$ above is well defined and satisfies $dd = 0$.*

(4.12) **Lemma.** *The functor λ: $\mathbf{Q} \to \mathbf{H}$ yields the natural isomorphism $\lambda(I\sigma) = I(\lambda\sigma)$ where the right hand side is the cylinder in* (III.3.3).

A similar kind of compatibility of cylinders was obtained in (III.3.4).

We now do some computations which illustrate the properties of a homotopy α as in (4.1). In fact, these computations describe the method which we originally used to find the quadratic distributivity laws in (4.1).

(4.13) **Distributivity properties of a homotopy.** Let f, $g\colon \sigma \to \sigma'$ be quadratic chain maps and assume functions $\alpha_n\colon \sigma_n \to \sigma'_{n+1}$, $n \geq 1$, are given which satisfy pointwise the equations

$$\left.\begin{aligned} -f_1 + g_1 &= d'_2\alpha_1 \\ -f_n + g_n &= d'_{n+1}\alpha_n + \alpha_{n-1}d_n. \end{aligned}\right\} \tag{1}$$

Then certain distributivity laws of α_n with respect to addition in σ_n and with respect to the action of σ_1 on σ_n are implied by (1). Let

$$B_n = -f_n + g_n\colon \sigma_n \to \sigma'_n, \quad n \geq 1, \tag{2}$$

be the function with $B_n(x) = -f_n(x) + g_n(x)$ for $x \in \sigma_n$. For $n = 1$ we have the formula

$$\begin{aligned} B_1(x + y) &= -f_1y - f_1x + g_1x + g_1y \\ &= -f_1y + B_1(x) + f_1y + B_1(y) \end{aligned} \tag{3}$$

Whence B_1 is an f_1-crossed homomorphism with respect to the action by inner automorphisms. Now let $\alpha_1\colon \sigma_1 \to \sigma_2'$ be an f_1-crossed homomorphism. Then $A_1 = d_2'\alpha_1$ again is an f_1-crossed homomorphism. Whence the equation $B_1 = A_1$ or equivalently $-f_1 + g_1 = d_2'\alpha_1$ is an equation of f_1-crossed homomorphisms. This equation implies that the induced homomorphisms

$$\bar{f}_1 = \bar{g}_1\colon \pi = \operatorname{coker}(d_2) \to \pi' = \operatorname{coker}(d_2') \tag{4}$$

coincide. We also write $\pi = \pi_1(\sigma)$ and $\bar{f}_1 = \pi_1(f)$. Since σ_n is a π_1-module for $n \geq 4$ we see that B_n is an $\bar{f}_1$-equivariant homomorphism for $n \geq 4$. Therefore we may assume that also α_n is an $\bar{f}_1$-equivariant homomorphism for $n \geq 3$, more precisely α_3 is equivariant with respect to $\bar{f}_1 q\colon \sigma_1 \to \pi \to \pi'$. We now consider B_3. The group structure $+$ in σ_3 yields the equations $(x, y \in \sigma_3)$:

$$\begin{aligned}
B_3(x + y) &= -f_3 y - f_3 x + g_3 x + g_3 y \\
&= B_3(x) - g_3 x + f_3 x - f_3 y - f_3 x + g_3 x + f_3 y + B_3(y) \\
&= B_3(x) + (B_3 x, f_3 y) + B_3(y) \qquad\qquad (5)\\
&= B_3(x) + \omega'(\{d_3 B_3 x\} \otimes \{d_3 f_3 y\}) + B_3(y) \\
&= B_3(x) + B_3(y) + \omega'(\{B_2 d_3 x\} \otimes \{f_2 d_3 y\})
\end{aligned}$$

Here we use (1.10)(4) and the fact that the image of $\omega'\colon C_2' \otimes C_2' \to \sigma_3'$ lies in the center of σ_3'. The action of $\beta \in \sigma_1$ on $x \in \sigma_3$ gives us the following equations.

$$\begin{aligned}
B_3(x^\beta) &= -f_3(x^\beta) + g_3(x^\beta) \\
&= -f_3(x)^{f_1(\beta)} + g_3(x)^{g_1(\beta)} \qquad\qquad (6)\\
&= (-f_3(x) + g_3(x)^{g_1(\beta) - f_1(\beta)})^{f_1(\beta)}
\end{aligned}$$

Here we use:

$$\begin{aligned}
g_1(\beta) - f_1(\beta) &= f_1\beta + (-f_1\beta + g_1\beta) - f_1\beta \\
&= f_1\beta + d_2'\alpha_1(\beta) - f_1\beta \qquad\qquad (7)\\
&= d_2'(\alpha_1(\beta)^{-f_1\beta})
\end{aligned}$$

Whence we get by (6) and (1.10)(3)

$$B_3(x^\beta) = B_3(x)^{f_1\beta} + \omega'(\Delta)^{f_1\beta} \quad \text{with}$$

$$\Delta = \{d_3'g_3x\} \otimes \{\alpha_1(\beta)^{-f_1\beta}\} + \{\alpha_1(\beta)^{-f_1\beta}\} \otimes \{d_3'g_3x\} \quad \text{so that}$$

$$B_3(x^\beta) = (B_3x)^{f_1\beta} + \omega'(\{g_2d_3x\}^{\bar{f}_1\beta} \otimes \{\alpha_1\beta\} + \{\alpha_1\beta\} \otimes \{g_2d_3x\}^{\bar{f}_1\beta}). \quad (8)$$

We now consider the equation $B_3 = d_4'\alpha_3 + \alpha_2d_3$ in (1). Here $d_4'\alpha_3$ is an $\bar{f}_1$ equivariant homomorphism which maps σ_3 to the center of σ_3'. Whence the function

$$A_3 = -d_4'\alpha_3 + B_3 \quad (9)$$

satisfies similar formulas as in (5) and (8). Since $A_3 = \alpha_2d_3$ we get by (5)

$$\alpha_2d_3(x+y) = \alpha_2d_3x + \alpha_2d_3y + \omega'(\{(-f_2+g_2)d_3x\} \otimes \{f_2d_3y\}). \quad (10)$$

Whence we assume that α_2 satisfies for $x, y \in \sigma_2$ the equation

$$\left.\begin{aligned} \alpha_2(x+y) &= \alpha_2x + \alpha_2y + \Delta_1(x,y) + \Delta_2(x,y) \\ \Delta_1(x,y) &= \omega'(\{(-f_2+g_2)x\} \otimes \{f_2y\}) \end{aligned}\right\} \quad (11)$$

with $\Delta_2(x,y) = 0$ for $x, y \in d_3\sigma_3$. Then (11) implies (10). Similarly we get by (8)

$$\alpha_2d_3(x^\beta) = (\alpha_2d_3x)^{f_1\beta} + \omega'(\{g_2d_3x\}^{\bar{f}_1\beta} \otimes \{\alpha_1\beta\} + \{\alpha_1\beta\} \otimes \{g_2d_3x\}^{\bar{f}_1\beta}). \quad (12)$$

Therefore we assume that α_2 satisfies for $x \in \sigma_2$, $\beta \in \sigma_1$ the equation

$$\begin{aligned} \alpha_2(x^\beta) &= (\alpha_2x)^{f_1\beta} + \nabla_1(x,\beta) + \nabla_2(x,\beta), \\ \nabla_1(x,\beta) &= \omega'(\{g_2x\}^{\bar{f}_1\beta} \otimes \{\alpha_1\beta\} + \{\alpha_1\beta\} \otimes \{g_2x\}^{\bar{f}_1\beta}) \end{aligned} \quad (13)$$

with $\nabla_2(x,\beta) = 0$ for $x \in d_3\sigma_3$. Then (13) implies (12). We point out that (11) and (13) are already the formulas for a quadratic operator in (4.1), but we still have to determine Δ_2 and ∇_2. For this we consider the distributivity properties of the function

$$A_2 = -d_3'\alpha_2 + B_2 = -d_3'\alpha_2 - f_2 + g_2. \quad (14)$$

We get for $x, y \in \sigma_2$ the formula

$$\begin{aligned} A_2(x+y) = &-d_3'\Delta_2(x,y) - \langle A_2x, f_2y\rangle \\ &+ \langle A_2y, A_2x\rangle - \langle g_2y, A_2x\rangle + (A_2x)^{f_1dy} + A_2y \end{aligned} \quad (15)$$

with $d = d_2$. We prove (15) in (4.14) below. Here $\langle\ ,\ \rangle$ denotes the Peiffer commutator in σ_2' which is central in σ_2' and which satisfies the formula $(u, v \in \sigma_2')$

$$\langle u,v\rangle = w'(\{u\} \otimes \{v\}) = d_3'\omega'(\{u\} \otimes \{v\}) \tag{16}$$

see (1.4) and (1.10). On the other hand the function $\alpha_1 d$ satisfies

$$\alpha_1 d(x + y) = (\alpha_1\, dx)^{f_1\, dy} + \alpha_1 dy \tag{17}$$

since α_1 is an f_1-crossed homomorphism, see (3). Hence (17) and (15) yield an equation for Δ_2 since $A_2 = \alpha_1 d$, namely

$$d_3'\Delta_2(x,y) = \langle \alpha_1\, dy, \alpha_1\, dx\rangle - \langle \alpha_1\, dx, f_2 y\rangle - \langle g_2 y, \alpha_1\, dx\rangle. \tag{18}$$

By use of (16) we may assume

$$\begin{aligned}\Delta_2(x,y) = {}& \omega'(\{\alpha_1\, dy\} \otimes \{\alpha_1\, dx\}) - \omega'(\{\alpha_1\, dx\} \otimes \{f_2 y\}) \\ & - \omega'(\{g_2 y\} \otimes \{\alpha_1\, dx\}).\end{aligned} \tag{19}$$

Then (18) is satisfied and, in fact, $\Delta_2(x,y) = 0$ for x and y in $d_3\sigma_3$, see (11). This shows that Δ_2 in (19) is suitable for (18) and for (11). In a similar way we obtain ∇_2. For this we have on the one hand the formula, $(x \in \sigma_2, \beta \in \sigma_1)$

$$\begin{aligned}A_2(x^\beta) &= -d_3'\alpha_2(x^\beta) + B_2(x^\beta) \\ &= -d_3'((\alpha_2 x)^{f_1\beta} + \nabla_1(x,\beta) + \nabla_2(x,\beta)) + B_2(x^\beta) \\ &= -d_3'\nabla_2(x,\beta) - \langle \alpha_1\beta, g_2 x^{f_1\beta}\rangle - \langle g_2 x^{f_1\beta}, \alpha_1\beta\rangle \\ &\quad - d_3'(\alpha_2 x)^{f_1\beta} + B_2(x^\beta)\end{aligned} \tag{20}$$

Here we use the definition of ∇_1 in (13) and we use (16). On the other hand we prove in (4.15) below the formula

$$\begin{aligned}\alpha_1 d_2(x^\beta) = {}& \langle(\alpha_1\, dx)^{f_1\beta}, \alpha_1\beta\rangle - \langle \alpha_1\beta, (g_2 x)^{f_1\beta}\rangle \\ & - \langle (g_2 x)^{f_1\beta}, \alpha_1\beta\rangle - (d_3'\alpha_2 x)^{f_1\beta} + B_2(x^\beta).\end{aligned} \tag{21}$$

Now the equation $A_2 = \alpha_1 d_2$ yields by (20) and (21) the following condition on ∇_2.

$$-d_3'\nabla_2(x,\beta) = \langle(\alpha_1 d_2 x)^{f_1\beta}, \alpha_1\beta\rangle \tag{22}$$

Therefore we set by use of (16)

$$-\nabla_2(x,\beta) = \omega'(\{\alpha_1 d_2 x\}^{\bar{f}_1\beta} \otimes \{\alpha_1\beta\}). \tag{23}$$

Then (22) is satisfied. Clearly $\nabla_2(x,\beta) = 0$ for $x \in d_3\sigma_3$. Whence formula (23) is compatible with (22) and (11). It remains to check the equations (21) and (15).

(4.14) *Proof of* (4.13)(15). We here denote d_3' also by d_3.

$$A_2(x+y) = -d_3\alpha_2(x+y) - f_2(x+y) + g_2(x+y) \tag{1}$$

$$\begin{aligned} &= -d_3(\alpha_2 x + \alpha_2 y + \Delta_1(x,y) + \Delta_2(x,y)) \\ &\quad - f_2 y - f_2 x + g_2 x + g_2 y \end{aligned} \tag{2}$$

$$= -d_3\Delta_2(x,y) - \langle B_2 x, f_2 y\rangle - d_3\alpha_2 y \tag{3}$$

$$- d_3\alpha_2 x - f_2 y \tag{4}$$

$$- f_2 x + g_2 x + g_2 y \tag{5}$$

$$= (3) + \langle d_3\alpha_2 x, f_2 y\rangle - f_2 y - d_3\alpha_2 x + (5) \tag{6}$$

$$= (3) + \langle d_3\alpha_2 x, f_2 y\rangle - f_2 y + A_2 x + g_2 y \tag{7}$$

$$= -d_3\Delta_2(x,y) - \langle A_2 x, f_2 y\rangle \tag{8}$$

$$- d_3\alpha_2 y - f_2 y + A_2 x + g_2 y \tag{9}$$

$$= (8) - d_3\alpha_2 y - f_2 y + A_2 x + f_2 y \tag{10}$$

$$+ d_3\alpha_2 y + A_2 y \tag{11}$$

$$= (8) + \langle d_3\alpha_2 y, -f_2 y - A_2 x + f_2 y\rangle \tag{12}$$

$$- f_2 y + A_2 x + f_2 y - d_3\alpha_2 y + (11) \tag{13}$$

$$= (8) + \langle d_3\alpha_2 y, -A_2 x\rangle \tag{14}$$

$$- f_2 y + A_2 x + f_2 y + A_2 y \tag{15}$$

$$= (14) + \langle f_2 y, -A_2 x\rangle + (A_2 x)^{d' f_2 y} + A_2 y \tag{16}$$

Hence we get

$$A_2(x,y) = -d_3\Delta_2(x,y) - \langle A_2 x, f_2 y\rangle \tag{17}$$

$$- \langle d_3\alpha_2 y, A_2 x\rangle - \langle f_2 y, A_2 x\rangle \tag{18}$$

$$+ (A_2 x)^{f_1 dy} + A_2 y \tag{19}$$

$$= (17) + \langle A_2 y, A_2 x\rangle - \langle g_2 y, A_2 x\rangle + (19) \tag{20}$$

This proves (4.13)(15). □

(4.15) *Proof of* (4.13)(21). We set $d = d_2$.

$$\alpha_1 d(x^\beta) = \alpha_1(-\beta + dx + \beta) \tag{1}$$

$$= \alpha_1(-\beta + dx)^{f_1\beta} + \alpha_1\beta \tag{2}$$

$$= (\alpha_1(-\beta)^{f_1\,dx} + \alpha_1\,dx)^{f_1\beta} + \alpha_1\beta \tag{3}$$

$$= -(\alpha_1\beta)^{-f_1\beta + f_1\,dx + f_1\beta} + (\alpha_1\,dx)^{f_1\beta} + \alpha_1\beta \tag{4}$$

$$= -(\alpha_1\beta)^{f_1 d(x^\beta)} + \langle -(\alpha_1\,dx)^{f_1\beta}, -\alpha_1\beta\rangle \tag{5}$$

$$+ (\alpha_1\beta)^{-d(\alpha_1\,dx)^{f_1\beta}} + (\alpha_1\,dx)^{f_1\beta} \tag{6}$$

$$= (5) + (\alpha_1\beta)^{-d(B_2 x)^{f_1\beta}} + (\alpha_1\,dx)^{f_1\beta} \tag{7}$$

Here we use

$$\begin{aligned} d(\alpha_1\,dx)^{f_1\beta} &= -f_1\beta + d\alpha_1\,dx + f_1\beta \\ &= -f_1\beta + (-f_1 + g_1)\,dx + f_1\beta \\ &= -f_1\beta + d(-f_2 + g_2)x + f_1\beta \\ &= d(B_2 x)^{f_1\beta} \end{aligned} \tag{8}$$

We now continue the equation in (7):

$$\alpha_1 d(x^\beta) = -(\alpha_1\beta)^{d(f_2 x^\beta)} + (\alpha_1\beta)^{-d(B_2 x)^{f_1\beta}} \tag{9}$$

$$+ (\alpha_1\,dx)^{f_1\beta} + \langle(\alpha_1\,dx)^{f_1\beta}, \alpha_1\beta\rangle \tag{10}$$

$$= -f_2x^\beta - \alpha_1\beta + f_2x^\beta + \langle f_2x^\beta, -\alpha_1\beta\rangle \quad (11)$$

$$+ (B_2x)^{f_1\beta} + \alpha_1\beta - (B_2x)^{f_1\beta} + \langle -(B_2x)^{f_1\beta}, \alpha_1\beta\rangle + (10) \quad (12)$$

$$= -f_2x^\beta \quad (13)$$

$$- \alpha_1\beta + (g_2x)^{f_1\beta} + \alpha_1\beta \quad (14)$$

$$- (g_2x)^{f_1\beta} + (f_2x)^{f_1\beta} \quad (15)$$

$$+ (10) - \langle f_2x^\beta, \alpha_1\beta\rangle - \langle (B_2x)^{f_1\beta}, \alpha_1\beta\rangle \quad (16)$$

$$= (13) + \langle \alpha_1\beta, -(g_2x)^{f_1\beta}\rangle + (g_2x)^{f_1\beta + d\alpha_1\beta} + (15) + (16) \quad (17)$$

$$= -f_2x^\beta + (g_2x)^{g_1\beta} + (15) \quad (18)$$

$$+ (16) - \langle \alpha_1\beta, (g_2x)^{f_1\beta}\rangle \quad (19)$$

$$= B_2(x^\beta) - (B_2x)^{f_1\beta} + (19) \quad (20)$$

$$= B_2(x^\beta) - (B_2x)^{f_1\beta} + (\alpha_1\,dx)^{f_1\beta} \quad (21)$$

$$+ \langle (\alpha_1\,dx)^{f_1\beta}, \alpha_1\beta\rangle - \langle \alpha_1\beta, (g_2x)^{f_1\beta}\rangle \quad (22)$$

$$- \langle (B_2x)^{f_1\beta}, \alpha_1\beta\rangle - \langle f_2x^\beta, \alpha_1\beta\rangle \quad (23)$$

$$= B_2(x^\beta) + (-\alpha_1\,dx - d_3'\alpha_2x + \alpha_1\,dx)^{f_1\beta} + (22) + (23) \quad (24)$$

$$= B_2(x^\beta) + (\langle d_3'\alpha_2x, -\alpha_1\,dx\rangle - d_3'\alpha_2x)^{f_1\beta} + (22) + (23) \quad (25)$$

$$= -\langle d_3'\alpha_2x, \alpha_1\,dx\rangle^{f_1\beta} + (22) + (23) \quad (26)$$

$$+ B_2(x^\beta) - d_3'(\alpha_2x)^{f_1\beta} \quad (27)$$

$$= (26) - \langle d_3'(\alpha_2x)^{f_1\beta}, -B_2x^\beta\rangle \quad (28)$$

$$- d_3'(\alpha_2x)^{f_1\beta} + B_2x^\beta \quad (29)$$

$$= \langle d_3'(\alpha_2x)^{f_1\beta}, -(\alpha_1\,dx)^{f_1\beta} + B_2x^\beta\rangle \quad (30)$$

$$+ \langle (\alpha_1\,dx)^{f_1\beta}, \alpha_1\beta\rangle \quad (31)$$

$$- \langle (g_2x)^{f_1\beta}, \alpha_1\beta\rangle - \langle \alpha_1\beta, (g_2x)^{f_1\beta}\rangle + (29) \quad (32)$$

Here (30) is trivial since $\bar{f}_1 = \bar{g}_1$ and

$$\{-(\alpha_1\, dx)^{f_1\beta} - f_2 x^\beta + g_2 x^\beta\} = \{-(\alpha_1\, dx)^{f_1\beta} - (f_2 x)^{f_1\beta} + (g_2 x)^{g_1\beta}\}$$

$$= \{-\alpha_1\, dx - f_2 x + g_2 x\}^{\bar{f}_1\beta} = \{d_3'(\alpha_2 x)^{f_1\beta}\}.$$

Whence (30) is the trivial commutator $(u, u) = 0$ with $u = d_3'(\alpha_2 x)^{f_1\beta}$. This completes the proof of (4.13)(21). □

§ 5 Cofibrations in the category of quadratic chain complexes

In (III.§ 4) we introduced cofibrations in the category **crosschain** of crossed chain complexes by use of the universal properties of free extensions. In the same way we now define cofibrations in the category **quadchain** of quadratic chain complexes and we show that these cofibrations give as well rise to a cofibration structure in the category **quadchain**, see (III.§ 4). The subcategory of cofibrant quadratic chain complexes is exactly the category **Q** of totally free quadratic chain complexes. Therefore one gets the equivalence of categories

$$\mathbf{Q}/\simeq \ \xrightarrow{\sim} \mathrm{Ho}(\mathbf{quadchain})$$

where Ho denotes the localization with respect to weak equivalences, see (5.7).

(5.1) **Definition.** A map $f: \lambda \to \rho$ in **quadchain** is a **cofibration** if f is a free extension in each degree n, $n \geq 1$. Here we define a **free extension** in degree n with basis ∂_n literally in the same way as in (III.4.7), where we simply replace the word "crossed" by the word "quadratic".

The cofibrant objects in **quadchain** are exactly the totally free quadratic chain complexes; hence we get with the notation in (III.4.3) and (3.4)

(5.2) $$\mathbf{Q} = (\mathbf{quadchain})_c.$$

The next lemma shows that free extensions exist.

(5.3) **Lemma:** *Let $\lambda = \lambda^n$ be an n-skeleton and assume $f^{n-1}: \lambda^{n-1} \to \rho^{n-1} \in$ **quadchain** and a function $\partial_n: Z_n \to \rho_{n-1}$ are given. Then a free extension $f: \lambda \to \rho = \rho^n$ as in (5.1) with basis ∂_n exists provided $d_{n-1}\partial_n = 0$. In this case we write $\rho_n = \lambda_n(Z_n)$.*

Proof. As in (III.4.9) we set for $n = 1$,

$$\rho_1 = \lambda_1 * \langle Z_1 \rangle. \tag{1}$$

Moreover, for $n = 2$ let $d_2: \rho_2 \to \rho_1$ be the nil(2)-module given by

$$d_2 = r_3(\partial) \tag{2}$$

where ∂ is defined in the same way as ∂ in (III.4.9)(4). Next we construct for $n = 3$ the quadratic module ρ^3 as follows. Let

$$C_2 \otimes C_2 \xrightarrow{\bar{\omega}} \bar{\rho}_2 \xrightarrow{\bar{\delta}} \rho_2 \xrightarrow{d_2} \rho_1 \tag{3}$$

be the free quadratic module with basis $(f_2 d_3, \partial_3): \lambda_3 \cup Z_3 \to \rho_2$. The inclusion $i: \lambda_3 \to \bar{\rho}_2$ is not a map between quadratic modules. Let $\mathscr{U}$ be the normal ρ_1-subgroup of $\bar{\rho}_2$ generated by the relations

$$\left.\begin{aligned} &i(x) + i(y) - i(x+y) \sim 0, \\ &i(x^\alpha) - (ix)^{f_1\alpha} \sim 0, \\ &i\omega(v) - \bar{\omega} f_*^2 v \sim 0 \end{aligned}\right\} \tag{4}$$

for $x, y \in \lambda_3$, $\alpha \in \lambda_1$ and $v \in C' \otimes C'$ where $C' = (\lambda_2^{cr})^{ab}$. Then (3) induces the commutative diagram

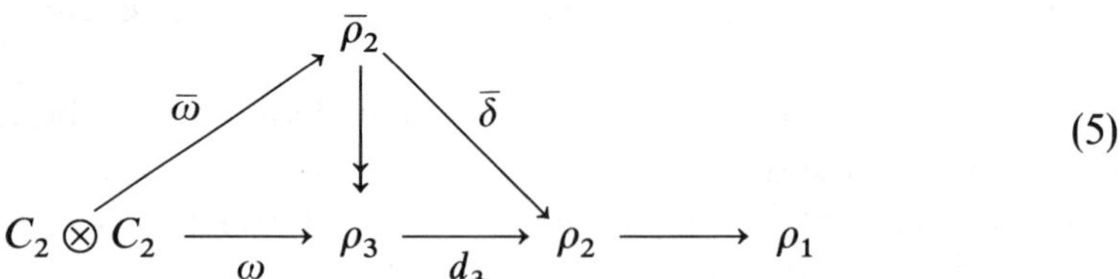

where $\rho_3 = \bar{\rho}_2/\mathscr{U}$ is the quotient group with the induced action of ρ_1. The bottom row of (5) is a well defined quadratic module and one readily checks that the universal property in (5.1), $n = 3$, is satisfied. In particular the map $f^3 = (f_3, f_2, f_1)$ with $f_3: \lambda_3 \subset \bar{\rho}_3 \twoheadrightarrow \rho_3$, given by i above, is a map between quadratic modules. Finally, for $n \geq 4$ let ρ_n be defined as in (III.4.9)(5). □

(5.4) **Lemma.** *The functor λ in (3.2) carries free extensions in degree n to free extensions in degree n.*

Proof. We only consider $n = 3$. We know for (5.3)(3) that $\bar{\rho}_2/\bar{\omega}(C_2 \otimes C_2)$ is the free $\pi_1(\rho^2)$-module generated by the set $\lambda_3 \cup Z_3$. Now the relations (5.3)(4) show that $\rho_2/\omega(C_2 \otimes C_2)$ is the direct sum of the tensor product

$$(\lambda_2/\omega(C_2' \otimes C_2')) \otimes_{\mathbb{Z}[\pi_1\lambda]} \mathbb{Z}[\pi_1\rho^2]$$

and of the free $\mathbb{Z}[\pi_1\rho^2]$-module M generated by Z_3. □

(5.5) **Theorem.** *The category* **quadchain** *with cofibration as in* (5.1) *and weak equivalences as in* (3.1)(4) *is a cofibration structure for which all objects are fibrant.*

Proof. We use almost literally the same arguments as in the proof of (III.4.10), see (III.4.17). Again we proof (C3) by choosing 'enough' generators. Moreover, we prove that all objects are fibrant by showing that $i\colon \lambda \rightarrowtail^{\sim} \rho$ is a **strong deformation retract** morphism. A retraction r and a homotopy $\alpha\colon ir \simeq 1$ is obtained by the same formulas as in (III.4.16)(2)...(5). Moreover we construct **push outs** in **quadchain**

$$\begin{array}{ccc} \rho & \xrightarrow{\bar{f}} & \rho' \\ {\scriptstyle i}\uparrow & & \uparrow{\scriptstyle \bar{i}} \\ \lambda & \xrightarrow[f]{} & \lambda' \end{array} \tag{1}$$

as follows. Let $\rho_n = \lambda_n(Z_n)$ as in (5.3). Then we set $\rho_n' = \lambda_n'(Z_n)$. The basis of ρ' is given in degree n by the composition

$$f_{n-1}d|Z_n\colon Z_n \to \lambda_{n-1} \to \rho_{n-1}, \quad n \geq 2. \tag{2}$$

The map $\bar{f}$ is the identity on Z_n. Now we prove C2(b) in the same way as in (III.4.17)(6). □

The lemmas (III.4.4), (III.4.5) and (5.2) imply the following corollaries of (5.5).

(5.6) **Corollary.** *The category* **Q** *is a cofibration category. Moreover, the cylinder* (4.10) *is a cylinder in the cofibration category* **Q**, *see* (I.1.5) *Baues* (AH).

Proof. For the proposition on cylinders we use (3.8)(A) and (4.12); these lemmas show that p in (4.10) is a weak equivalence. Moreover, one readily checks that i in (4.10) is a cofibration. □

(5.7) **Corollary.** *There is an equivalence of categories*

$$M\colon \mathrm{Ho}(\textbf{quadchain}) \xrightarrow{\sim} \mathrm{Ho}(\mathbf{Q}) = \mathbf{Q}/\simeq$$

where the homotopy relation on **Q** *is defined as in* (4.1).

(5.8) **Corollary.** *A weak equivalence in* **Q** *is a homotopy equivalence in* **Q**.

(5.9) **Remark.** By (3.8) we know that a map f in **Q** is a weak equivalence iff $\lambda(f)$ is one. We do not know whether the same holds for f in **quadchain**. Therefore we do not know whether λ in (3.2) induces a functor $\mathrm{Ho}(\lambda)$. We have, however, the composition

$$\mathrm{Ho}(\mathbf{quadchain}) \xrightarrow{\sim} \mathbf{Q}/\simeq \xrightarrow{\lambda} \mathbf{H}/\simeq \xleftarrow{\sim} \mathrm{Ho}(\mathbf{crosschain})$$

by (5.7), (4.9) and (III.4.12) which is certainly the 'correct' functor between the localized categories.

(5.10) **Remark.** The cylinder $I\sigma$ in (4.10) is natural. In fact, a map $f\colon \sigma \to \sigma'$ in **Q** induces a unique map If: $I\sigma \to I\sigma'$ in **Q** with the properties

$$i_t f = (If)i_t \quad \text{for } t = 0, 1 \tag{1}$$

and

$$Sf = (If)S, \quad \text{see (4.10)(5), (7), (9).} \tag{2}$$

Now (5.6) can be used to show that **Q** with the cylinder (4.10) and with cofibrations (5.1) is an I-category in the sense of (I.3.1) Baues (AH).

Next we describe the ***n*-ball** $D(n)$ and $(n-1)$-**sphere** $S(n-1)$ in **quadchain**. Let $D(n)$, $n \geq 2$, be the totally free quadratic chain complex generated by elements c_n and $d_n c_n$ in degree n and $n-1$ respectively. Hence $D(n)$ is in degree n and $n-1$ an infinite cyclic group and is otherwise trivial, moreover the quadratic map of $D(n)$ is trivial for $n \geq 2$. (For $n = 3$ one can show that the quadratic map is trivial by use of the central push out (3.4)(1)). Let $S(n-1)$ be the $(n-1)$-skeleton of $D(n)$, with inclusion $S(n-1) \rightarrowtail D(n)$. Hence $S(n)$, $n \geq 1$, is in degree n an infinite cyclic group and is otherwise trivial provided $n \neq 2$. The '2-sphere' $S(2)$, however, is infinite cyclic in degree 2 and 3 and is otherwise trivial; the differential is trivial and the quadratic map of $S(2)$ is the identity, compare (3.1)(5). We say that a cofibration $\lambda \rightarrowtail \rho$ in **quadchain** is given by **attaching an *n*-cell** to λ if for $n \geq 2$ there exists a push out diagram

$$\begin{array}{ccc} D(n) & \longrightarrow & \rho \\ \uparrow & \text{push} & \uparrow \\ S(n-1) & \longrightarrow & \lambda \end{array} \tag{5.11}$$

in the category **quadchain** and if for $n = 1$ there is an isomorphism $\rho \cong \lambda \vee$

$S(1)$ where $\lambda \vee S(1)$ is the sum in **quadchain**. Now one can check that a cofibration as in (5.1) is always given by inductively attaching cells to λ.

The functor λ: **quadchain** $\to$ **crosschain** carries cells and spheres to cells and spheres respectively, that is

(5.12) $$\lambda D(n) = D(n), \quad \lambda S(n-1) = S(n-1),$$

compare (III.4.15). Moreover, λ carries a push out as in (5.11) to a push out as in (III.4.15). This shows

(5.13) **Lemma.** *The functor λ carries cofibrations to cofibrations and push outs as in* (5.5)(1) *to push outs.*

This shows by (3.8), see (III.6.2):

(5.14) **Proposition.** *The functor λ: **Q** $\to$ **H** is a model functor.*

§ 6 The secondary homotopy addition lemma and a model functor from spaces to quadratic chain complexes

Recall that a model functor is a functor between homotopy theories which carries weak equivalences to weak equivalences and homotopy push outs to homotopy push outs. In this section we introduce a model functor σ_S from spaces to quadratic chain complexes. For the construction of this functor we consider the quadratic chain complex $\sigma(V)$ of a (0-reduced) simplicial set V; the secondary homotopy addition lemma describes just the boundary formulas in $\sigma(V)$. To this end recall that the classical homotopy addition lemma (III.5.2) gave us just the boundary formulas in the crossed chain complex $\rho(V)$ of the simplicial set V, see (III.6.8). Now the quadratic chain complex $\sigma(V)$ is determined (up to natural isomorphism) by the condition that $V \mapsto \sigma(V)$ is a functor on simplicial sets and that the equation

$$\lambda\sigma(V) = \rho(V)$$

holds. Here λ is the functor in (3.2) or (3.6) which divides out the quadratic part of $\sigma(V)$. For a space X we obtain $\sigma_S(X)$ by the quadratic chain complex $\sigma(V)$ where V is the singular set of X. This commonly used 'singular trick' yields indeed a nice functor $\sigma_S X$ from spaces to quadratic chain complexes; for computations, however, the object $\sigma_S X$ is too big. Therefore we show that for a CW-complex X there is actually a 'small' quadratic chain complex $\sigma(X)$ together with a weak equivalence

$$\sigma(X) \xrightarrow{\sim} \sigma_S(X)$$

such that $\lambda\sigma(X) = \rho(X)$. Here $\rho(X)$ is the crossed chain complex of the CW-complex X given by relative homotopy groups $\rho_n(X) = \pi_n(X^n, X^{n-1})$. Moreover $\sigma(X)$ is a totally free quadratic chain complex in **Q** generated by the cells of X. The existence of the object $\sigma(X)$, which is well defined up to isomorphism by the cell structure of X, is very important for explicit computations. Examples are described in section § 12 below. There is, however, a slight disadvantage of the object $\sigma(X)$; namely we have no direct geometrical interpretation of the groups $\sigma_n(X)$ as in the case of $\rho(X)$ where we know that $\rho_n(X)$ is a relative homotopy group. Such an interpretation might not exist since the construction of $\sigma(X)$ does not give us a functor from the category **CW** of cellular maps to the category **Q** but gives only a functor σ such that the diagram

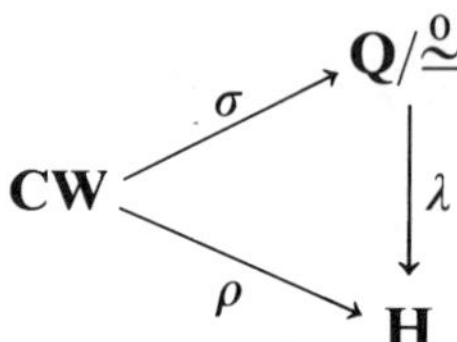

commutes. Here we use the 0-homotopies, $\overset{0}{\simeq}$, in (4.8) and the functor λ in (4.9). On the other hand we shall see in chapter VI how the boundary operator in $\sigma(X)$ can be interpreted geometrically by use of tracks.

We use the notation on simplicial sets V as in (III.6.5), (III.6.6). Let Z_n be the set of non degenerate elements in V_n. For $\xi \in Z_n$ we set

(6.1) $$\xi_{b_1 \ldots b_r} = |\partial_{a_1} \ldots \partial_{a_k} \xi| = {}_{a_1 \ldots a_k}\xi$$

if $(b_1 < \cdots < b_r, a_1 < \cdots < a_k)$, $r + k = n + 1$, is a partition of the set $\{0, 1, \ldots, n\}$.

(6.2) **Definition.** For a 0-reduced simplicial set V let $\sigma(V) = \sigma$ be the following totally free quadratic chain complex in **Q**. The bases of σ_n is the set, Z_n, of non degenerate elements in V_n and the boundary $d_n\colon \sigma_n \to \sigma_{n-1}$ is defined for $\xi \in Z_n$ by

$$d_n(\xi) = \begin{cases} |\partial_2\xi| + |\partial_0\xi| - |\partial_1\xi|, & n = 2 \\ -|\partial_1\xi| - |\partial_3\xi| + |\partial_0\xi|^{u(\xi)} + |\partial_2\xi|, & n = 3 \\ |\partial_0\xi|^{u(\xi)} + \sum_{i=1}^{n} (-1)^i |\partial_i\xi|, & n \geq 5 \end{cases} \qquad (1)$$

Here $u(\xi) = -\xi_{01} = -|\partial_2\partial_3\ldots\partial_n\xi|$ is an element in the group $\langle Z_1\rangle = \sigma_1$. Moreover, for $\xi \in Z_4$ we have the following boundary formula (2) in σ_3 where we use the quadratic map ω: $C_2 \otimes C_2 \to \sigma_3$. We call this formula the *secondary homotopy addition lemma.*

$$d_4(\xi) = d_4'\xi + \omega(d_4''\xi) \tag{2}$$

with

$$d_4'\xi = -|\partial_1\xi| + |\partial_4\xi| + |\partial_2\xi| + |\partial_0\xi|^{u(\xi)} - |\partial_3\xi| \tag{3}$$

$$d_4''\xi = \xi_{012} \otimes (\xi_{234}{}^{-\xi_{02}}) + (d\xi_{0123}) \otimes \xi_{034} + (d\xi_{1234}^{u(\xi)}) \otimes \xi_{014} \tag{4}$$

Here d: $C_3 \to C_2$ is the boundary of the chain complex $C\lambda\sigma^3$, that is, C_n is the free π_1-module generated by Z_n and $d\xi$, $\xi \in Z_3$, is defined in the same way as $d_3\xi$ in (1). A simplicial map f: $V \to W$ yields a quadratic chain map $\sigma(f)$: $\sigma(V) \to \sigma(W)$ in $\mathbf{Q}$ by

$$(\sigma f)(\xi) = |f\xi| \tag{5}$$

for $\xi \in Z_n$, $n \geq 1$.

One readily checks that the formulas in (6.2) are compatible with the formulas in (III.6.8). This shows

(6.3) $$\lambda(\sigma V) = \rho V$$

where λ is the functor in (3.2) and where ρV is the crossed chain complex of the simplicial set V.

In order to understand the **secondary homotopy addition lemma**, (that is, formula (6.2)(2) above) one has to do the following computations. Consider the simplex Δ^4 in which we divide out the 0-skeleton to obtain a CW-complex $X = \Delta^4/\Delta_0^4$ with trivial 0-skeleton $X^0 = *$. From (III.5.2) we obtain explicit formulas for the boundaries in the crossed chain complex $\rho(X)$. Now there is up to isomorphism a unique totally free quadratic chain complex $\sigma(X)$ with $\lambda\rho(X) = \sigma(X)$; this follows from (9.7) below since $\pi_2 X = 0$. Moreover the cells of X are just the generators of $\sigma(X)$. For the 4-cell ξ in X we know that $d_4'\xi$ in (6.2)(3) corresponds to $d_4\xi$ in $\rho(X)$. In the quadratic chain complex $\sigma(X)$ the element $d_3 d_4'\xi$, however, needs not to be trivial. This element has to be of the form $d_3 d_4'\xi = \omega(\eta)$ since $d_3 d_4\xi = 0$ in $\rho(X)$. Now we computed such an element η and we found that $\eta = -d_4''\xi$ in (6.2)(4) is a good choice. Hence we get $d_3 d_4\xi = 0$ in $\sigma(X)$. This method is similar to the argument in the proof of (III.5.2). The proof that $d_3 d_4\xi = 0$ actually is satisfied can be found in the proof of (6.5) below. For this we use the following lemma which simplifies the formula for $d_4\xi$ in (6.2) by use of the maps v and ω in (3.4)(1).

(6.4) **Lemma.** $d_4\xi = \nu(a) + \omega(b)$ *with*

$$a = -|\partial_1\xi| - \xi_{034} + |\partial_4\xi| + \xi_{034} + |\partial_2\xi| - \xi_{014} + |\partial_0\xi|^{u(\xi)} + \xi_{014} - |\partial_3\xi|$$

$$b = \xi_{012} \otimes \xi_{234}{}^{-\xi_{02}}$$

We point out that b is part of the Alexander Whitney diagonal.

Proof. The Peiffer commutator in the free nil(2)-module $E_3 \vee \sigma_2$ satisfies the formula

$$-x + y + x = \langle x, -y\rangle + y^{\partial x} \tag{1}$$

where we set $x = \xi_{034}$ and $y = |\partial_4\xi|$ (or $x = \xi_{014}$, $y = |\partial_0\xi|^{u(\xi)}$). We have by (3.4)(1)

$$\begin{aligned} \nu\langle x, -y\rangle &= -\nu w(\{x\} \otimes \{y\}) \\ &= -\omega(\{x\} \otimes \{d_3 y\}). \end{aligned} \tag{2}$$

On the other hand we get by (1.10)(3)

$$\nu(y^{\partial x}) = y^{\partial x} = y + \omega(\{d_3 y\} \otimes \{x\} + \{x\} \otimes \{d_3 y\}). \tag{3}$$

Hence by (1) we have

$$\nu(-x + y + x) = y + \omega(\{d_3 y\} \otimes \{x\}). \tag{4}$$

This yields the proposition in (6.3) since $|\partial_0\xi| = \xi_{1234}$, $|\partial_4\xi| = \xi_{0123}$. □

(6.5) **Lemma.** *The construction* (6.2) *yields a well defined functor from the category of simplicial sets to the category* **Q** *of totally free quadratic modules such that* (6.3) *holds.*

Proof. We first check that the quadratic chain complex $\sigma(V)$ is well defined, for this we have to show that

$$d_{n-1}d_n = 0, \quad n \geq 3. \tag{1}$$

For $n = 3$ we obtain (1) by the same equations as in (III.5.2)(2). For $n \geq 6$ equation (1) holds by (6.3). It remains to check (1) for $n = 4, 5$. Let $\xi \in Z_4$. Then we get by (6.4)

$$\begin{aligned} d_3 d_4(\xi) &= d_3 v(a) + d_3 \omega(b) \\ &= (\partial_3, 1)(a) + w(b) \end{aligned} \tag{2}$$

where we use (3.4)(1). Using the definition of $(\partial_3, 1)$ by d_3 in (6.2)(1) we get

$$\begin{aligned} (\partial_3, 1)(a) = &-(-\xi_{034} - \xi_{023} + \xi_{234}{}^{-\xi_{02}} + \xi_{024}) - \xi_{034} \\ &+ (-\xi_{023} - \xi_{012} + \xi_{123}{}^{-\xi_{01}} + \xi_{013}) + \xi_{034} \\ &+ (-\xi_{034} - \xi_{013} + \xi_{134}{}^{-\xi_{01}} + \xi_{014}) - \xi_{014} \\ &+ (-\xi_{134} - \xi_{123} + \xi_{234}{}^{-\xi_{12}} + \xi_{124})^{-\xi_{01}} + \xi_{014} \\ &- (-\xi_{024} - \xi_{012} + \xi_{124}{}^{-\xi_{01}} + \xi_{014}) \end{aligned} \tag{3}$$

Here many terms cancel so that we get

$$(\partial_3, 1)(a) = -\xi_{024} - \xi_{234}{}^{-\xi_{02}} \tag{4}$$

$$- \xi_{012} + \xi_{234}{}^{-\xi_{12} - \xi_{01}} + \xi_{012} \tag{5}$$

$$+ \xi_{024} \tag{6}$$

The term (5) of this sum can be replaced by the sum (7) + (8) with $\partial = d_2$,

$$\text{term}(5) = \langle \xi_{012}, -\xi_{234}{}^{-\xi_{12} - \xi_{01}} \rangle \tag{7}$$

$$+ \xi_{234}{}^{-\xi_{12} - \xi_{01} + \partial \xi_{012}}. \tag{8}$$

Here we have

$$\text{term}(8) = \xi_{234}{}^{-\xi_{02}}, \quad \text{and} \tag{9}$$

$$\begin{aligned} \text{term}(7) &= \langle \xi_{012}, -\xi_{234}{}^{-\xi_{02} + \partial \xi_{012}} \rangle \\ &= -w(\xi_{012} \otimes \xi_{234}{}^{-\xi_{02}}) \\ &= -w(b), \quad \text{see (4).} \end{aligned} \tag{10}$$

Since w maps to the center of σ_2 we see by (4)...(10) that $(\partial_3, 1)(a) = -w(b)$. This shows by (2) that (1) is satisfied for $n = 4$.

Next let $\xi \in Z_5$. The element $d'_4 d_5 \xi$ is a sum of commutators. Each commutator yields an element in the image of the quadratic map by (1.10)(4).

This shows that $d'_4 d_5 \xi + \omega d''_4 d_5 \xi = 0$ though the calculation is tedious, see Weick p. 105, here we also use the exact sequence in (2.14) which determines the kernel of ω: $C_2 \otimes C_2 \to \sigma_3$. The equation $d_4 d_5 \xi = 0$ corresponds to the 'cocycle condition' in (6.9). Finally we have to show that $\sigma(f)$ in (6.2)(5) is a well defined quadratic map. For this it is enough to check that $(\sigma f)(d_n \xi) = 0$ if $|f\xi| = 0$, $\xi \in Z_n$. We leave this as an exercise. □

We now are ready to define the functor

(6.6) $$\sigma_S\colon \textbf{CW-spaces}_0^* \to \mathbf{Q}$$

which is an analogue of the functor ρ_S in (III.6.3). For a path connected CW-space X let

$$\sigma_S(X) = \sigma(SX) \tag{1}$$

be the quadratic chain complex of the 0-reduced singular set SX of X, see (III.6.7). By (6.3) we have the natural isomorphism

$$\lambda\sigma_S(X) = \rho_S(X). \tag{2}$$

The next result is a generalization of the Van Kampen theorem, see (III.6.12), (III.6.13).

(6.7) **Theorem.** *The functor σ_S is a model functor.*

Proof. Clearly σ_S carries weak equivalences to weak equivalences since we can apply (6.6)(2) and (3.8)(A). Moreover, consider a homotopy push out in $\mathbf{C} = \textbf{CW-spaces}_0^*$ as in (III.6.2). Then we have to show that

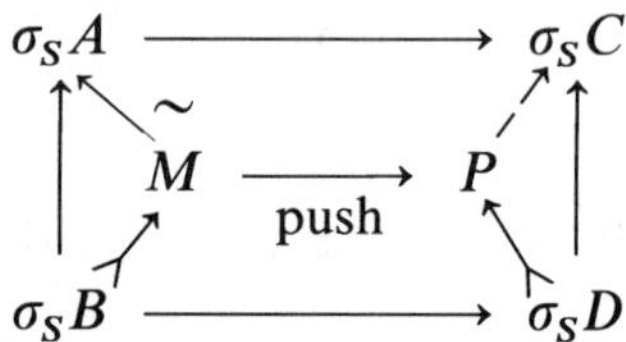

is a homotopy push out. Here α: $P \to \sigma_3 C$ is a weak equivalence if and only if the induced map $\lambda\alpha$: $\lambda P \to \lambda\sigma_S C$ is a weak equivalence. By (5.13) and (6.6)(2) the map $\lambda\alpha$ is part of the diagram

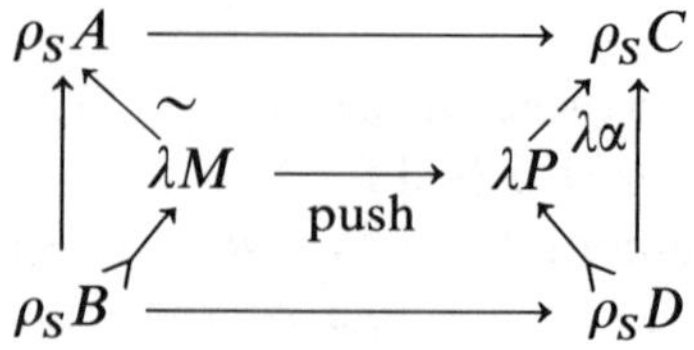

where $\lambda\alpha$ is a weak equivalence since ρ_S is a model functor, see (III.6.12). This proves that α is a weak equivalence by (3.8)(A). □

For a cofibration category **C** let Fil(**C**) be the category of **filtered objects**

$$A = (A_0 \to A_1 \cdots \to A_n \to \cdots)$$

in **C**. This is a cofibration category by (III.1.2) Baues (AH). We now construct a functor σ such that the following diagram of functors commutes.

(6.8)

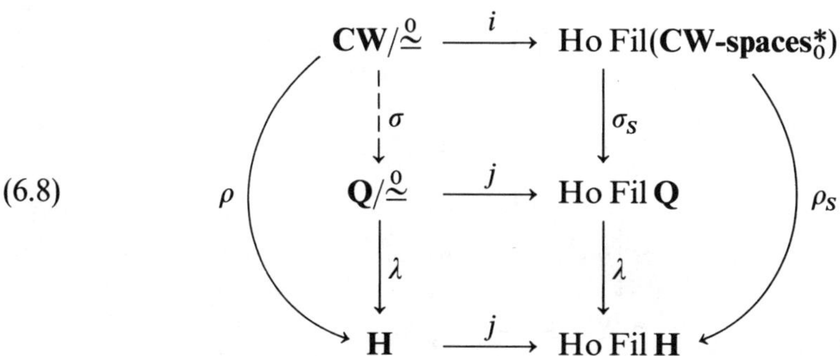

Here ρ is the functor in (III.2.8) and σ_S is induced by σ_S in (6.6). We consider a CW-complex X as a filtered object filtered by skeleta $* \rightarrowtail X^1 \rightarrowtail X^2 \rightarrowtail \cdots$. This yields the full inclusion i of categories in (6.8). Similarly an object σ in **Q** (or **H**) is filtered by skeleta $* \rightarrowtail \sigma^1 \rightarrowtail \sigma^2 \rightarrowtail \cdots$ and this yields the full inclusion j in (6.8). We construct the functor σ together with natural isomorphisms

$$a\colon \lambda\sigma(X) \cong \rho(X) \quad \text{in } \mathbf{H}, \tag{1}$$

$$b\colon j\sigma(X) \simeq \sigma_S i(X) \quad \text{in Ho Fil } \mathbf{Q}. \tag{2}$$

Here the homotopy equivalence b induces the natural isomorphism

$$j\rho(X) \simeq \rho_S i(X) \quad \text{in Ho Fil } \mathbf{H} \tag{3}$$

which is given by $(p_X)_*$ in (III.6.11). These properties of σ are summarized in the following definition.

(4) **Definition.** Let X be a CW-complex in **CW**. We call an object $\sigma = \sigma(X)$ in **Q** a **quadratic chain complex of** X if isomorphisms a and b with the properties above are given. Here $\sigma(X)$ is 'minimal' with respect to the choice of generators since the generators are exactly the cells of X.

(5) **Lemma.** *A quadratic chain complex* $(\sigma(X), a, b)$ *of* X *exists and is well defined up to isomorphism in* **Q**.

Proof. We construct $\sigma(X)$ inductively by attaching cells. Here we use (6.7). The triple $(\sigma(X), a, b)$ is well defined up to isomorphism since we can use (3.8)(B). □

(6) **Definition of the functor** σ. We choose for each X in **CW** a quadratic chain complex $(\sigma(X), a, b)$. For a map $f: X \to Y$ in $\mathbf{CW}/\overset{0}{\simeq}$ we obtain the commutative diagram

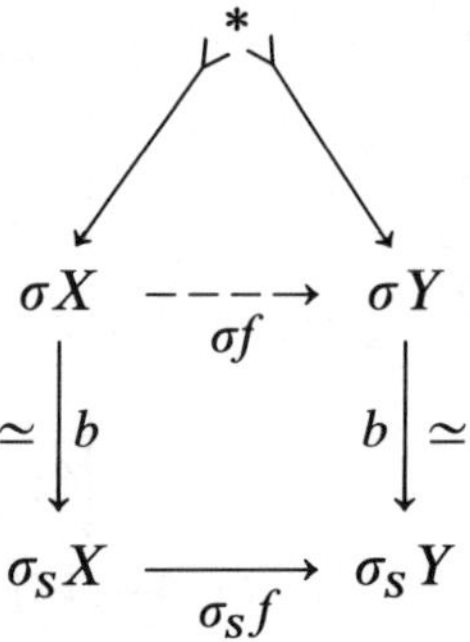

in Ho Fil **Q**. Hence σf is a well defined map in $\mathbf{Q}/\overset{0}{\simeq}$. This completes the definition of the functor σ. Clearly σ is a functor since σ_S is one, compare (III.4.6)(3).

Finally we observe that the functor σ in (6.8) induces a functor

$$\sigma: \mathbf{CW}/\simeq \;\to \mathbf{Q}/\simeq \tag{7}$$

between homotopy categories. This follows since $\sigma(I_* X)$ is a cylinder in **Q** of σX. In fact, we have by the uniqueness property in (5) an isomorphism of cylinders

$$I\sigma X = \sigma I_* X. \tag{8}$$

Compare (III.3.4) and (4.12).

§7 Homotopy systems of order 4

Recall that we obtained in (6.8) a functor

$$\sigma: \mathbf{CW} \to \mathbf{Q}/\overset{0}{\simeq}$$

which carries a CW-complex X to its quadratic chain complex $\sigma(X)$. In this section we first show that one has a natural homomorphism of $\pi_1 X$-modules

$$\sigma\colon \pi_n X \to \pi_n \sigma X$$

which is an isomorphism for $n \leq 3$ and which is surjective for $n = 4$. This implies in particular that the third homotopy group $\pi_3 X$ can be computed in terms of the quadratic chain complex σX. Moreover we show that the 4-dimensional part of Whiteheads exact sequence for X coincides with the corresponding exact sequence for σX, see (3.7). Using the isomorphism σ: $\pi_3 X \cong \pi_3 \sigma X$ we can define a functor

$$\sigma\colon \mathbf{H}_4^c \xrightarrow{\sim} \mathbf{Q}/\overset{0}{\simeq}$$

which is actually an equivalence of categories. Here $\mathbf{H}_4^c$ is the category of homotopy systems of order 4 which is part of the CW-tower of categories. Hence we may consider totally free quadratic chain complexes to be the appropriate algebraic models of homotopy systems of order 4. This is one of the main new results in this book.

Recall that $S(n) = \sigma(S^n)$ is the quadratic chain complex of the sphere $S^n = * \cup e^n$, see (5.11). It is easy to see that there is a natural identification, $\sigma \in$ **quadchain**,

(7.1) $$[S(n), \sigma] = \pi_n(\sigma), \quad n \geq 1.$$

Here the right hand side is defined in (3.1)(3) and the left hand side is the set of homotopy classes relative $*$ in the cofibration structure **quadchain**, see (II.§ 2) Baues (AH). We derive from (7.1) the natural transformation

(7.2) $$\pi_n X = [S^n, X] \xrightarrow{\sigma} [S(n), \sigma X] = \pi_n \sigma X$$

where X is a CW-complex and where σ is given by the functor in (6.8)(7). The map σ for $n \leq 4$ is part of the following natural diagram of $\pi_1(X)$-modules:

(7.3) $$\begin{array}{ccccccccc} \pi_4 X & \longrightarrow & H_4 \hat{X} & \longrightarrow & \Gamma(\pi_2 X) & \longrightarrow & \pi_3 X & \twoheadrightarrow & H_3 \hat{X} \\ \sigma\downarrow & & \rho\downarrow\cong & & \Gamma(\rho)\downarrow\cong & & \sigma\downarrow\cong & & \rho\downarrow\cong \\ \pi_4 \sigma & \rightarrowtail & \pi_4 \rho & \longrightarrow & \Gamma(\pi_2 \rho) & \longrightarrow & \pi_3 \sigma & \twoheadrightarrow & \pi_3 \rho \end{array}$$

Here we set $\sigma = \sigma(X)$ and $\rho = \lambda\sigma(X) = \rho(X)$. The bottom row is the exact sequence in (3.7) and the top row is **Whitehead's certain exact sequence** (I.3.7) where we set $n = 3$. The morphisms ρ are induced by the Hurewicz maps in (III.2.2).

(7.4) **Proposition.** *Diagram* (7.3) *commutes and the vertical maps are homomorphisms of* $\pi_1(X)$*-modules.*

Proof of (7.4). In fact diagram (7.3) is an example of (III.10.13)(3) in Baues (AH) where we set $A = S^1$, $n = 2$. For this we observe that the relative homotopy groups in **Q** satisfy

$$\pi_{n-1}^{S(1)}(\sigma^n, \sigma^{n-1}) = \rho_n, \quad n \geq 2. \tag{1}$$

Here the left hand side is defined by (II.7.7) Baues (AH). Moreover, we have

$$\Gamma(\pi_2\rho) = \text{image}(\pi_3(\sigma^2) \to \pi_3(\sigma^3)) \tag{2}$$

similarly as in (I.4.11). Finally we consider the action of $\pi_1 X$ which is induced by an action map μ: $S^n \to S^n \vee S^1$, $n \geq 2$. Now the composition

$$\pi_n(S^n \vee S^1) \xrightarrow{\sigma} \pi_n(\sigma(S^n \vee S^1)) \overset{q*}{=} \pi_n\rho(S^n \vee S^1) \tag{3}$$

carries μ to the corresponding action map in **H** and hence $\sigma\mu$ is the action map in **Q**. Clearly $\sigma\mu$ induces the action of $\pi_1 X = \pi_1 \sigma X$ on $\pi_n \sigma X$ given by the action of σ_1 on σ_n. Compare also § 13 below. □

(7.5) **Corollary.** *There is a natural isomorphism,*

$$\pi_n \sigma X \cong \begin{cases} \pi_n X & \textit{for } n \leq 3 \\ \text{image}(\hat{h}\colon \pi_4 X \to H_4 \hat{X}) & \textit{for } n = 4 \\ H_n \hat{X} & \textit{for } n \geq 5 \end{cases}$$

Here $\hat{h}$ is the Hurewicz homomorphism in (I.3.3). The isomorphism is given by σ in (7.2). This is a consequence of (7.3) since ρ is an isomorphism in (7.3), for $n = 3$ we use the five lemma for exact sequences.

We now are ready to describe the quadratic analogue of diagram (III.2.8). For this we consider the following diagram of functors.

(7.6)

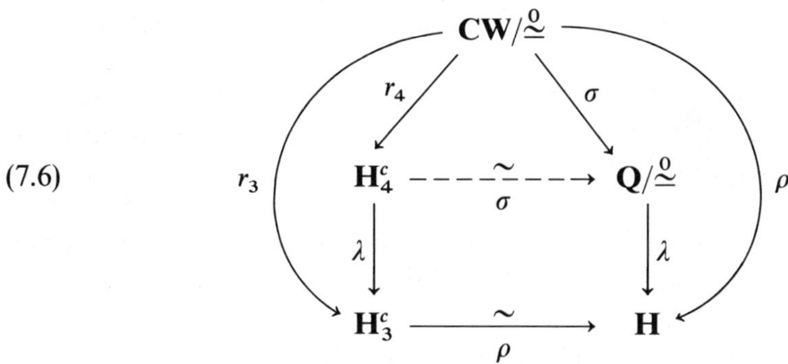

Here the right hand side is given by (6.8) and the left hand side is given by the CW-tower of categories (II.3.3) and the horizontal arrow ρ denotes the corresponding functor in (III.2.8).

(7.7) **Theorem.** *There is an equivalence of categories* $\sigma\colon \mathbf{H}_4^c \xrightarrow{\sim} \mathbf{Q}/\overset{0}{\simeq}$ *which induces an equivalence* $\sigma\colon \mathbf{H}_4^c/\simeq \xrightarrow{\sim} \mathbf{Q}/\simeq$ *of homotopy categories. Moreover* σ *extends diagram* (7.6) *commutatively.*

(7.8) **Remark.** All functors in (7.6) induce functors between homotopy categories. Moreover, the functors ρ, λ, σ on the right hand side of (7.6) satisfy the strong sufficiency condition with respect to $\simeq$. This follows from (II.3.15).

We prove theorem (7.7) in (8.5) below. The functor σ in theorem (7.7) is defined as follows.

(7.9) **Definition.** Let $A = (C, f_4, X^3)$ be a homotopy system of order 4. We define the quadratic chain complex $\sigma(A)$ by

$$\cdots \longrightarrow C_5 \xrightarrow{\alpha} C_4 \xrightarrow{if_4} \sigma_3(X^3) \xrightarrow{d} \sigma_2(X^3) \xrightarrow{d} \sigma_1(X^3)$$

where $\sigma(X^3)$ is defined by σ in (6.8) and where if_4 is given by the injection

$$i\colon \pi_3 X^3 \cong \pi_3 \sigma X^3 \subset \sigma_3 X^3.$$

Here we use the isomorphism (7.5). The object $\sigma(A)$ in $\mathbf{Q}$ is well defined since the cocycle condition $f_4 d = 0$ is satisfied, see (II.2.1)(3). Clearly the quadratic map of $\sigma(A)$ is the one of $\sigma(X^3)$. A map $f = (\xi, \eta)\colon A \to B$ in $\mathbf{H}_4^c$ induces a map $\sigma(f)\colon \sigma(A) \to \sigma(B)$ in $\mathbf{Q}/\overset{0}{\simeq}$. Here $\sigma(f)$ is given by $\sigma(\eta)$ in degree ≤ 3 and by ξ in degree ≥ 4. One readily checks that $\sigma\colon \mathbf{H}_4^c \to \mathbf{Q}/\overset{0}{\simeq}$ is well defined and extends diagram (7.6) commutatively.

§8 The homotopy category of 3-dimensional CW-complexes

In this section we show that the homotopy category, $\mathbf{CW}^3/\simeq$, of 3-dimensional CW-complexes is equivalent to the homotopy category $\mathbf{Q}^3/\simeq$ of 3-dimensional totally free quadratic chain complexes. In addition we show that for CW-complexes X, Y with $\dim(X) \leq 3$ the set of homotopy classes $[X, Y]$ can be computed by the corresponding set of homotopy classes in the algebraic category $\mathbf{Q}/\simeq$. That is, one has the natural bijection

$$\sigma\colon [X, Y] \approx [\sigma X, \sigma Y] \quad \text{if } \dim(X) \leq 3.$$

There is an action of the cohomology $\hat{H}^3(X, \varphi^*\pi_3 Y)$ on the subset $[X, Y]_\varphi$ of $[X, Y]$. Using the bijection σ we obtain explicit formulas for the computation of the isotropy groups of this action. This corresponds to the computation of a differential d_2 in a spectral sequence, see (8.10). As an important application we consider the group of homotopy equivalences $\mathrm{Aut}(X)^*$ of a 3-dimensional CW-complex X, see (8.12). Using this result we shall compute as an example the group $\mathrm{Aut}(\mathbb{R}P_3 \# \mathbb{R}P_3)^* = \mathbb{Z}/2$ where $\mathbb{R}P_3 \# \mathbb{R}P_3$ is the connected sum of two 3-dimensional real projective spaces $\mathbb{R}P_3$, see Appendix A. Recall that $\mathbf{CW}^n$, $\mathbf{H}^n$, and $\mathbf{Q}^n$ denote the full subcategories of $\mathbf{CW}$, $\mathbf{H}$, and $\mathbf{Q}$ respectively consisting of n-dimensional objects. In this section we proceed similarly as in (III.§7), for example the next result corresponds to (III.7.1).

(8.1) **Theorem.** *The functor σ in* (6.8) *induces equivalences of categories*

$$\sigma\colon \mathbf{CW}^3/\overset{0}{\simeq} \xrightarrow{\sim} \mathbf{Q}^3/\overset{0}{\simeq},$$

$$\sigma\colon \mathbf{CW}^3/\simeq \xrightarrow{\sim} \mathbf{Q}^3/\simeq.$$

We prove this result in (8.6)...(8.9) below. In addition to (8.1) the functors σ_S and $\sigma\colon \mathbf{CW}/\simeq \to \mathbf{Q}/\simeq$ yield the following result where $[A, B]$ denotes the set of homotopy classes.

(8.2) **Theorem.** *Let X, Y be* CW*-complexes in* $\mathbf{CW}$. *Then we have the bijection of homotopy sets $\sigma\colon [X, Y] \approx [\sigma X, \sigma Y]$ for* $\dim X \leq 3$. *More generally let A be a subcomplex of X with $\dim(X/A) \leq 3$ and $* \in A$ and let $a\colon A \to Y$ be a cellular map which corresponds to a map $\alpha = \sigma(a)\colon \sigma A \to \sigma Y$ in* $\mathbf{Q}$. *Then we have the bijection of relative homotopy sets*

$$[X, Y]^A \approx [\sigma X, \sigma Y]^{\sigma A}$$

under a and α respectively, see (III.4.6)(2).

Proof. The first part is a consequence of (II.3.3) and (7.7). The second part follows from the exact sequence (II.13.10) in Baues (AH) which is compatible with the model functor σ_S in (6.7). The bijection depends on a track for diagram (6.8)(6) where we set $f = a$. This track induces a bijection via (II.5.7) in Baues (AH). $\square$

Next we consider the action in (II.2.4)(3) for $n = 3$ which yields the top row of the following commutative diagram ($\dim X \leq 3$)

$$\begin{array}{ccc} [X, Y]_\varphi \times \hat{H}^3(X, \varphi^*\pi_3 Y) & \xrightarrow{+} & [X, Y]_\varphi \\ \approx \Big\downarrow \sigma \times \sigma_* & & \approx \Big\downarrow \sigma \\ [\sigma, \sigma']_\varphi \times \hat{H}^3(C, \varphi^*\pi_3 \sigma') & \xrightarrow{+} & [\sigma, \sigma']_\varphi \end{array} \tag{8.3}$$

Here we set $\sigma = \sigma(X)$, $\sigma' = \sigma(Y)$ and $C = C\rho(\sigma), C' = C\rho(\sigma')$. We now describe a formula for the action in the bottom row of (8.3). Let $f\colon \sigma \to \sigma'$ be a map in **Q** which induces φ on fundamental groups. We have the following diagram which is an analogue of diagram (III.7.3)(1).

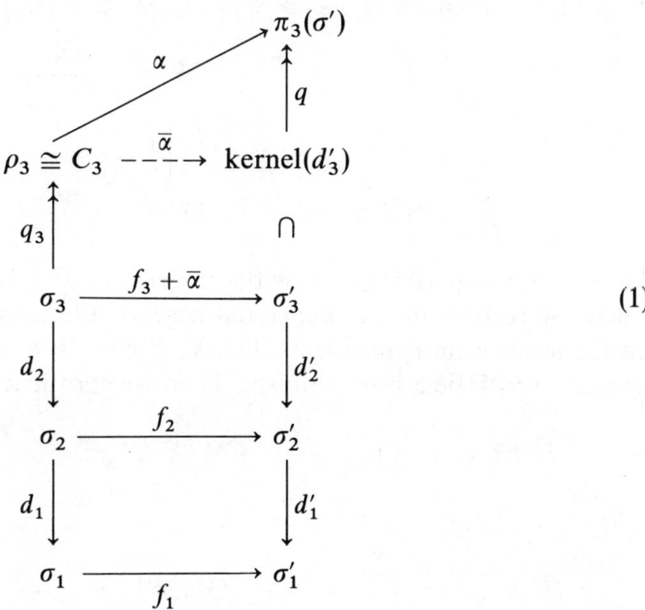

The maps q and q_3 are the quotient maps, see (3.3). Let α in (1) be a φ-equivariant homomorphism which represents the cohomology class $\{\alpha\} \in H^3(C, \varphi^*\pi_3\sigma')$. Then we can choose a φ-equivariant homomorphism $\bar{\alpha}$ with $q\bar{\alpha} = \alpha$. The f_1-equivariant homomorphism $f_3 + \bar{\alpha}$ in (1) is defined by

$$(f_3 + \bar{\alpha})(x) = f_3(x) + \bar{\alpha} q_3(x), \quad x \in \sigma_3. \tag{2}$$

One readily checks that

$$f + \bar{\alpha} = (f_3 + \bar{\alpha}, f_2, f_1)\colon \sigma \to \sigma' \tag{3}$$

is a well defined map in **Q** which induces φ on fundamental groups. Now we get the action in the bottom row of (8.3) by:

$$\{f\} + \{\alpha\} = \{f + \bar{\alpha}\}. \tag{4}$$

(8.4) **Proposition.** *Diagram* (8.3) *commutes.*

Proof. This is a consequence of (II.8.27)(3) in Baues (AH). □

(8.5) *Proof of theorem* (7.7). Using the definitions of the categories $\mathbf{H}_4^c$ and $\mathbf{H}_4^c/\simeq$ we see that theorem (7.7) is a consequence of theorem (8.1), of (8.4), and of (7.5). Compare also the remark in (III.7.5). □

Thus it remains to prove theorem (8.1). For this we consider the following diagram in which the rows are weak linear extensions of categories

$$(8.6)\qquad \begin{array}{ccccc} H^3\Gamma_3 & \xrightarrow{+} & \mathbf{CW}^3/\overset{0}{\simeq} & \xrightarrow{\lambda} & (\mathbf{H}_3^c)^3 \\ \cong\big\downarrow j_* & & \big\downarrow \sigma & & \sim\big\downarrow \rho \\ H^3\Gamma\pi_2 & \xrightarrow{+} & \mathbf{Q}^3/\overset{0}{\simeq} & \xrightarrow{\lambda} & \mathbf{H}^3 \end{array}$$

The top row of (8.6) is given by the top row of (II.3.3) where we set $n = 3$ and where we restrict to 3-dimensional objects. We now consider diagram (8.6) on the level of morphism sets. Let X, Y be CW-complexes in $\mathbf{CW}^3$ and let $\varphi\colon \pi_1 X \to \pi_1 Y$ be a homorphism. Then we get the diagram

$$(8.7)\qquad \begin{array}{ccccc} \hat{H}^3(X, \varphi^*\Gamma_3 Y) & \xrightarrow{+} & \mathbf{CW}(X, Y)_\varphi/\overset{0}{\simeq} & \xrightarrow{\lambda} & \mathbf{H}_3^c(\lambda X, \lambda Y)_\varphi \\ \cong\big\downarrow j_* & & \big\downarrow \sigma & & \approx\big\downarrow \rho \\ \hat{H}^3(C, \varphi^*\Gamma\pi_2\rho') & \xrightarrow{+} & \mathbf{Q}(\sigma, \sigma')_\varphi/\overset{0}{\simeq} & \xrightarrow{\lambda} & \mathbf{H}(\rho, \rho')_\varphi \end{array}$$

The action in the top row is defined by (II.3.7)(3). One can check that this action is actually effective. This shows that the top row of (8.6) is a weak linear extension of categories. We now define the action in the bottom row of (8.7) similarly as in (8.3)(4) by

$$\{f\}^0 + \{\alpha\} = \{f + \overline{i\alpha}\}^0. \tag{1}$$

Here $\{f\}^0$ denotes the equivalence class of f in $\mathbf{Q}(\sigma,\sigma')/\overset{0}{\simeq}$ and $\{\alpha\} \in \hat{H}^3(C, \varphi^*\Gamma\pi_2\rho')$ is represented by a φ-equivariant homorphism $\alpha\colon C_3 \to \Gamma\pi_2\rho'$, $C = C\rho(X)$, which yields $i\alpha\colon C_3 \to \pi_3\sigma'$ by i in (3.7). Hence we get $f + \overline{i\alpha}$ in the same way as in (8.3)(3). One can check that (1) yields a well defined action with trivial isotropy groups. Moreover, the bottom row of (8.7) is exact. This shows, that the bottom row of (8.6) is a weak linear extension of categories. The isomorphism

$$j\colon \Gamma_3 Y \cong \Gamma(\pi_2 Y) \cong \Gamma(\pi_2 \rho Y) \tag{2}$$

induces the isomorphism j_* in (8.7). As in (8.4) we see that σ in (8.7) is equivariant with respect to the action. Now it is clear that σ in (8.7) is a bijection since diagram (8.7) commutes and since j_* and ρ are bijections. This also proves that σ in (8.6) *is an equivalence of categories.* Next we consider the following diagram which is obtained from (8.6) by dividing out the homotopy relation $\simeq$.

(8.8)
$$\begin{array}{ccccc}
H^3\Gamma_3/I & \xrightarrow{+} & \mathbf{CW}^3/\simeq & \xrightarrow{\lambda} & (\mathbf{H}_3^c)^3/\simeq \\
\downarrow j_* & & \downarrow \sigma & & \sim \downarrow \rho \\
H^3\Gamma\pi_2/J & \xrightarrow{+} & \mathbf{Q}^3/\simeq & \xrightarrow{\lambda} & \mathbf{H}^3/\simeq
\end{array}$$

Here the top row is a weak linear extension deduced from the bottom row in (II.3.3), compare (II.1.6). The bottom row of (8.8) is a weak linear extension derived from the bottom row of (8.6). Now we describe diagram (8.8) on the level of morphism sets similarly as in (8.7). This yields the commutative diagram

(8.9)
$$\begin{array}{ccccc}
\hat{H}^3(X, \varphi^*\Gamma_3 Y) & \xrightarrow{+} & [X, Y]_\varphi & \xrightarrow{\lambda} \!\!\!\!\rightarrow & [\lambda X, \lambda Y]_\varphi \\
\cong \downarrow j_* & & \downarrow \sigma & & \approx \downarrow \rho \\
\hat{H}^3(C, \varphi^*\Gamma\pi_2\rho') & \xrightarrow{+} & [\sigma X, \sigma Y]_\varphi & \xrightarrow{\lambda} \!\!\!\!\rightarrow & [\rho X, \rho Y]_\varphi
\end{array}$$

Here the top row is a special case of (II.3.9). Now the actions in (8.9) have non trivial isotropy groups. Let $F\colon X \to Y$ be a cellular map which induces φ on fundamental groups and let $f\colon \sigma X \to \sigma Y$ be a quadratic chain map which represents $\sigma F \in \mathbf{Q}/\overset{0}{\simeq}$, see (8.6). For $\{\alpha\} \in \hat{H}^3(X, \varphi^*\Gamma_3 Y)$ we choose $F + i\alpha\colon X \to Y$ as in (II.3.7)(2) and we choose $f + \overline{i\alpha}\colon \sigma X \to \sigma Y$ as in (8.3)(3). Now the isotropy groups $I(F)$ and $I(f)$ respectively of the actions in (8.9) are given by

$$I(F) = \{\{\alpha\}: \exists F + \alpha \simeq F\} \tag{1}$$

$$I(f) = \{\{\alpha\}: \exists f + \overline{i\alpha} \simeq f\} \tag{2}$$

The map in (II.5a.4) Baues (AH) and (6.8)(8) show that the isomorphism j_* in (8.9) satisfies

$$j_* I(F) = I(f). \tag{3}$$

This proves that σ in (8.9) is a bijection since ρ is one. Whence σ in (8.8) is an equivalence of categories and the proof of (8.1) is complete.

(8.10) **Computation of the isotropy group.** By (II.3.8) there is a differential $d_2 = d_2(F)$ of a spectral sequence,

$$\hat{H}^1(X, *; \varphi^*\pi_2 Y) \xrightarrow{d_2} \hat{H}^3(X, \varphi^*\pi_3 Y) \xleftarrow{i_*} \hat{H}^3(X, \varphi^*\Gamma_3 Y), \tag{1}$$

such that

$$I(F) = i_*^{-1}(\text{image}\, d_2(F)). \tag{2}$$

Here i_* is induced by $i: \Gamma_3 Y \to \pi_3 Y$. In fact, equation (2) is a special case of (VI.5.16) in Baues (AH). The homorphism $d_2(F)$ and the group $I(F)$ actually depend only on $\rho\lambda\{F\} = \lambda\{f\} = \{\xi\}$ in $[\rho X, \rho Y]_\varphi$. We now describe a formula for d_2 in terms of the quadratic chain complexes $\sigma = \sigma X$, $\sigma' = \sigma Y$ and in terms of a map $\xi: \rho = \lambda\sigma \to \rho' = \lambda\sigma'$. For $\{\alpha\} \in \hat{H}^1(C, C^0; \varphi^*\Gamma\pi_2\sigma'\}$ we choose a diagram

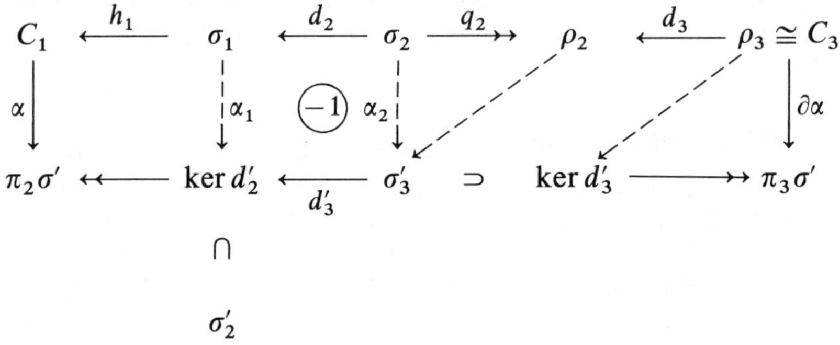

Here α_1 is a ξ_1-crossed homomorphism and α_2 is an $(\alpha_1, \bar{\xi}^2, \bar{\xi}^2)$-quadratic operator where $\bar{\xi}^2: d_2 \to d_2'$ is any map which induces ξ^2, see (4.3). The subdiagram (-1) satisfies $-\alpha_1 d_2 = d_3'\alpha_2$, the other subdiagrams commute. One readily checks that for α there exist such maps α_1 and α_2. Moreover (4.4) shows that α_2 yields a well defined map $\partial\alpha$ as indicated in the diagram. Now

d_2 in (1) is given by the formula

$$d_2\{\alpha\} = \{-\partial\alpha\}. \tag{3}$$

Here we use the isomorphisms σ: $\pi_3 Y = \pi_3\sigma'$ in (7.5).

Proof of (3). Recall that the torus on X, $\Sigma_* X$, is the push out $I_* X \cup_\nabla X$ where ∇: $X \vee X \to X$ is the folding map. The topological definition of d_2 in (1) is given by the following diagram

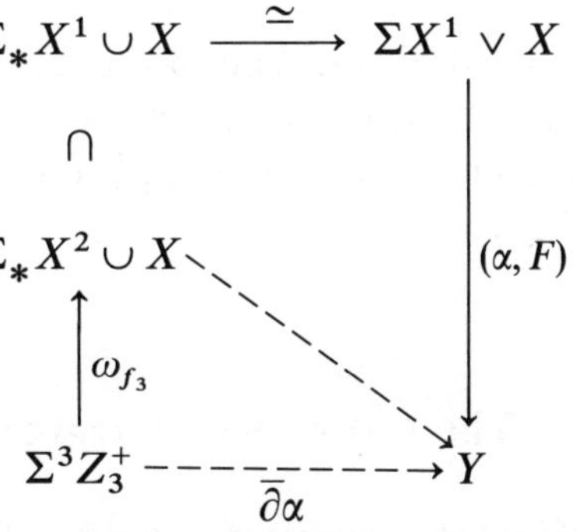

$$\begin{array}{ccc} \Sigma_* X^1 \cup X & \xrightarrow{\simeq} & \Sigma X^1 \vee X \\ \cap & & \Big\downarrow (\alpha, F) \\ \Sigma_* X^2 \cup X & \searrow & \\ \Big\uparrow \omega_{f_3} & & \\ \Sigma^3 Z_3^+ & \overset{\bar\partial\alpha}{\dashrightarrow} & Y \end{array} \tag{4}$$

Here ω_{f_3} is the attaching map of 4-cells in the CW-complex $\Sigma_* X$. The map (α, F) given by α and F above has an extension as indicated in (4). This yields a map $\bar\partial\alpha$ which represents the element $d_2\{\alpha\}$ in (1). We now can apply (6.8)(8) and we see that $\bar\partial\alpha$ as well is given by $-\partial\alpha$ in (3). □

Since $X^0 = *$ we can replace the set $[\rho X, \rho Y]_\varphi$ in (8.9) by the homotopy set of chain maps in $\mathbf{H}_1$

(8.11) $$C\colon [\rho X, \rho Y]_\varphi \approx [\hat C_* X, \hat C_* Y]_\varphi$$

This follows from (III.2.12). In addition to (8.9) we have the following result on the group of homotopy equivalences which is a special case of the tower of groups in (II.3.11).

(8.12) **Proposition.** *Let X be a 3-dimensional CW-complex with $X^0 = *$. Then we have isomorphic short exact sequences of groups*

$$\begin{array}{ccccc} \hat H^3(X, \Gamma_3 X)/I(1) & \overset{1^+}{\rightarrowtail} & \operatorname{Aut}(X)^* & \overset{\hat C_*}{\twoheadrightarrow} & \operatorname{Aut}(\hat C_* X) \\ \cong \Big\downarrow j & & \cong \Big\downarrow \sigma & & \Big\| \\ \hat H^3(C, \Gamma H_2 C)/I(1) & \rightarrowtail & \operatorname{Aut}(\sigma X) & \overset{C\lambda}{\twoheadrightarrow} & \operatorname{Aut}(C) \end{array}$$

where $C = C\rho(X) = \hat{C}_* X$. *Here* Aut *denotes the group of homotopy equivalences in* $\mathbf{Q}/\simeq$ *and* $\mathbf{H}_1/\simeq$ *respectively. The group* $I(1)$ *can be computed by* (8.10)(2) *with* $F = 1$.

Recall that a weak linear extension is a detecting functor, see (II.1.9). Thus we get by (8.8) the following result.

(8.13) **Proposition.** *The homotopy types of 3-dimensional objects in the categories* $\mathbf{CW}/\simeq$, $\mathbf{Q}/\simeq$, *and* $\mathbf{H}/\simeq$ *respectively are in* $1-1$ *correspondence to each other.*

Moreover (III.2.1) shows that the cellular chain complex $\hat{C}_* X$ of a 3-dimensional CW-complex $X = X^3$ determines the homotopy type of X. The homotopy type of a 4-dimensional CW-complex, however, is not determined by the chain complex $\hat{C}_* X$.

§9 The CW-tower in degree ≤ 4

Recall that the CW-tower of categories consists of exact sequences for the functors

$$\lambda\colon \mathbf{H}^c_{n+1}/\simeq \;\to \mathbf{H}^c_n/\simeq, \quad n \geq 3.$$

For $n = 3$ this functor can be replaced by the functor $\lambda\colon \mathbf{Q}/\simeq \;\to \mathbf{H}/\simeq$ between algebraic categories which is therefore as well embedded in an exact sequence of the form

$$H^3\Gamma\pi_2 + \longrightarrow \mathbf{Q}/\simeq \xrightarrow{\lambda} \mathbf{H}/\simeq \xrightarrow{\mathcal{O}} H^4\Gamma\pi_2.$$

In this section we give a purely algebraic description of this sequence. In particular we obtain a formula for the obstruction operator $\mathcal{O}$, see (9.4) and (9.6). Moreover we obtain in (9.7) a criterion showing that certain crossed chain complexes ρ determine the corresponding quadratic chain complex σ with $\lambda\sigma = \rho$ up to isomorphism. This fact is very useful for computations.

We consider the following two diagrams in which the rows are exact sequences for the functor λ.

$$(9.1)\qquad \begin{array}{ccccccc} H^3\Gamma_3 + & \longrightarrow & \mathbf{H}^c_4 & \xrightarrow{\lambda} & \mathbf{H}^c_3 & \xrightarrow{\mathcal{O}} & H^4\Gamma_3 \\ \cong\downarrow j_* & & \sim\downarrow\sigma & & \sim\downarrow\rho & & \cong\downarrow j_* \\ H^3\Gamma\pi_2 + & \longrightarrow & \mathbf{Q}/\overset{0}{\simeq} & \xrightarrow{\lambda} & \mathbf{H} & \xrightarrow{\mathcal{O}} & H^4\Gamma\pi_2 \end{array}$$

(9.2)
$$\begin{array}{ccccccc} H^3\Gamma_3 + & \longrightarrow & \mathbf{H}_4^c/\simeq & \xrightarrow{\lambda} & \mathbf{H}_3^c/\simeq & \xrightarrow{\mathcal{O}} & H^4\Gamma_3 \\ \cong \downarrow j_* & & \sim \downarrow \sigma & & \sim \downarrow \rho & & \cong \downarrow j_* \\ H^3\Gamma\pi_2 + & \longrightarrow & \mathbf{Q}/\simeq & \xrightarrow{\lambda} & \mathbf{H}/\simeq & \xrightarrow{\mathcal{O}} & H^4\Gamma\pi_2 \end{array}$$

Diagram (9.2) is obtained by dividing out the homotopy relation $\simeq$ in the categories of diagram (9.1). The functors ρ and σ are equivalences of categories by (III.2.9) and (7.7). The top rows of (9.1) and (9.2) respectively are the exact sequences in (II.3.3), $n = 3$. We now describe the obstruction $\mathcal{O}$ and the action $+$ in the bottom row by algebraic formulas. The natural system

(9.3) $$H^m\Gamma\pi_2 \colon F(\mathbf{H}/\simeq) \to \mathbf{Ab}$$

carries $(f\colon \rho \to \rho') \in \mathbf{H}$ to the abelian group

$$(H^m\Gamma\pi_2)(f) = H^m(C\rho, \varphi^*\Gamma\pi_2\rho')$$

where $\varphi = \pi_1(f)$. The isomorphism j in (8.7)(2) induces the natural isomorphisms j_* in diagram (9.1) and (9.2). The definition of the **obstruction** $\mathcal{O}$ in the bottom row corresponds exactly to the definition (II.3.4). Let σ, σ' be objects in $\mathbf{Q}$ and let $f\colon \rho = \lambda\sigma \to \lambda\sigma' = \rho'$ be a map in $\mathbf{H}$ which induces $\varphi = \pi_1(f)$. Then an element

(9.4) $$\mathcal{O}_{\sigma,\sigma'}(f) \in \hat{H}^4(C\rho, \Gamma\pi_2\rho')$$

is defined such that $\mathcal{O}_{\sigma,\sigma'}(f) = 0$ if and only if there exists a map $F\colon \sigma \to \sigma'$ in $\mathbf{Q}$ with $\lambda F = f$. Let $f^3\colon \rho^3 \to \rho'^3$ be the 3-skeleton of f. One readily verifies that there is a map $F^3\colon \sigma^3 \to \sigma'^3$ in $\mathbf{Q}$ with $\lambda F^3 = f^3$. Whence we obtain the diagram

$$\begin{array}{ccc} \sigma_4 = \rho_4 & \xrightarrow{f_4} & \rho'_4 = \sigma'_4 \\ d_4 \downarrow & & \downarrow d'_4 \\ \pi_3\sigma^3 & \xrightarrow[(F^3)_*]{} & \pi_3\sigma'^3 \end{array} \tag{1}$$

which needs not to be commutative. The difference

$$\mathcal{O}(F^3) = -d'_4 f_4 + F^3_* d_4 \tag{2}$$

maps $\rho_4 = C_4$ to $\Gamma\pi_2\rho' \subset \pi_3\sigma'^3$, see (3.7). Moreover, this difference is a cocycle in $\mathrm{Hom}_\varphi(C\rho, \Gamma\pi_2\rho')$. The obstruction

$$\mathcal{O}_{\sigma,\sigma'}(f) = \{\mathcal{O}(F^3)\} \tag{3}$$

is the cohomology class represented by the cocycle $\mathcal{O}(F^3)$. The class (3) depends only on the homotopy class of f in $\mathbf{H}/\simeq$ so that (3) as well yields the obstruction in (9.2). A further description of the obstruction (3) can be obtained by the cellular boundary invariants β_σ, $\beta_{\sigma'}$ in (V.1.8) below, compare (II.4.21).

We now define the action in the bottom row of (9.1). Let $f: \sigma \to \sigma'$ be a map in $\mathbf{Q}$ and let $\{\alpha\} \in \hat{H}^3(C\rho, \varphi^*\Gamma\pi_2\rho')$ with $\rho = \lambda(\sigma)$, $\rho' = \lambda(\sigma')$ and $\varphi = \pi_1(f)$. Then we define

(9.5) $$\{f\}^0 + \{\alpha\} = \{f + \overline{i\alpha}\}^0$$

similarly as in (8.7)(1). Here $g = f + \overline{i\alpha}$ is given by $g_n = f_n$ for $n \neq 3$ and by $g_3 = f_3 + \overline{i\alpha}$, see (8.3)(2). The action in the bottom row of (9.2) is given in the same way by

$$\{f\} + \{\alpha\} = \{f + \overline{i\alpha}\}. \tag{1}$$

Clearly $\{f\}$ denotes the homotopy class of f in $\mathbf{Q}/\simeq$ and $\{f\}^0$ denotes the 0-homotopy class of f in $\mathbf{Q}/\overset{0}{\simeq}$. One can check directly that the functor $\lambda: \mathbf{Q}/\overset{0}{\simeq} \to \mathbf{H}$ satisfies

$$\lambda\{f\}^0 = \lambda\{g\}^0 \Leftrightarrow \exists\{\alpha\} \quad \text{with } \{f\}^0 + \{\alpha\} = \{g\}^0. \tag{2}$$

Similarly the functor $\lambda: \mathbf{Q}/\simeq \to \mathbf{H}/\simeq$ satisfies

$$\lambda\{f\} = \lambda\{g\} \Leftrightarrow \exists\{\alpha\} \quad \text{with } \{f\} + \{\alpha\} = \{g\}. \tag{3}$$

These facts are part of the following result.

(9.6) **Theorem.** *The bottom sequences of* (9.1) *and* (9.2) *respectively are exact sequences in the sense of* (II.1.5). *Moreover, the vertical arrows in* (9.1), (9.2) *describe equivalences of exact sequences, in particular,* σ *is* j_**-equivariant and* j_* *induces an isomorphism of isotropy groups. Moreover, the obstruction satisfies*

$$j_*\mathcal{O}_{X,Y}(F) = \mathcal{O}_{\sigma X,\sigma Y}(\rho F).$$

The next result allows the computation of $\sigma(X)$ in certain cases.

(9.7) **Corollary.** *Let $\rho \in \mathbf{H}$ be realizable with respect to $\lambda: \mathbf{Q} \to \mathbf{H}$ and assume*

$$\hat{H}^4(C\rho, \Gamma\pi_2\rho) = 0. \tag{1}$$

Then there is up to isomorphism only one realization σ in $\mathbf{Q}$ with $\lambda\sigma \cong \rho$.

Proof. Let σ, σ' be realizations of ρ with $f: \lambda\sigma \cong \rho \cong \lambda\sigma'$. Since the group (1) is trivial there exists by (9.4) a map $F: \sigma \to \sigma'$ with $\lambda F = f$. Since f is an isomorphism also f is one by (3.8)(B). □

In particular, if a CW-complex X satisfies

$$\hat{H}^4(X, \Gamma\pi_2 X) = 0 \tag{2}$$

then we can choose any realization σ of $\rho(X)$ with $\lambda\sigma \cong \rho(X)$. Proposition (9.7) implies $\sigma \cong \sigma(X)$. This result and the compatibility of σ with push outs, see (6.7) are important tools for the computations of $\sigma(X)$.

§ 10 The homotopy category of 3-types

Quadratic modules are algebraic models of 3-types; in fact, in this section we show that the localized category Ho(**quad**) of quadratic modules is equivalent to the homotopy category of 3-types. This result is an application of theorem (8.1). In (III.§ 8) we have already seen that crossed modules are algebraic models of 2-types. The result in this section on quadratic modules is obtained completely analogously to the corresponding result on crossed modules in (III.§ 8).

Recall that Ho(**quad**) denotes the localization of the category **quad** of quadratic modules with respect to weak equivalences as defined in (1.10)(6). Moreover, 3-**types** denotes the full homotopy category of 3-types.

(10.1) **Theorem.** *There is an equivalence of categories*

$$\bar{\sigma}: 3\text{-}\mathbf{types} \xrightarrow{\sim} \mathrm{Ho}(\mathbf{quad}).$$

This generalizes the result in (III.8.2) on the category 2-**types**. The proof of (10.1) proceeds almost literally in the same way as the proof in (III.8.9). We leave this as an exercise to the reader.

The Postnikov functor P_2 in (II.4.1) yields the following diagram of functors

(10.2)
$$\begin{array}{ccc} \textbf{3-types} & \xrightarrow{P_2} & \textbf{2-types} \\ \sim\downarrow\bar{\sigma} & & \sim\downarrow\bar{\rho} \\ \mathrm{Ho}(\mathbf{quad}) & \xrightarrow{\mu} & \mathrm{Ho}(\mathbf{cross}) \end{array}$$

Here the functor μ carries a quadratic module σ to the associated crossed module $\mu(\sigma)$: $\sigma_2^{cr} \to \sigma_1$ given by d_2: $\sigma_2 \to \sigma_1$.

(10.3) **Proposition.** *Diagram* (10.2) *commutes, that is, there is a natural isomorphism* $\bar{\rho}P_2 \cong \mu\bar{\sigma}$ *in* Ho(**cross**).

The proposition follows readily from the definitions of $\bar{\rho}$ in (III.8.3) and from the

(10.4) **Definition of** $\bar{\sigma}$. Let X be a 3-type which is a CW-complex in **CW**. Then we get the commutative diagram

$$\begin{array}{ccccccc} \pi_3(X^3) & \overset{j}{\rightarrowtail} & \sigma_3(X) & \xrightarrow{d_3} & \sigma_2(X) & \xrightarrow{d_2} & \sigma_1(X) \\ i\downarrow & \text{push} & \downarrow\bar{i} & & \| & & \| \\ \pi_3(X) & \rightarrowtail & \bar{\sigma}_3 & \xrightarrow{\delta} & \bar{\sigma}_2 & \xrightarrow{\partial} & \bar{\sigma}_1 \end{array} \tag{1}$$

where $\sigma(X)$ is the quadratic chain complex of X and where 'push' is a push out diagram of groups. The bottom row of (1) yields a quadratic module $\tilde{\sigma} = (\bar{i}\omega, \delta, \partial)$ where ω is the quadratic map of $\sigma(X)$. The homotopy groups of $\tilde{\sigma}$ are

$$\pi_n(\tilde{\sigma}) = \pi_n(X) \quad \text{for } n = 1, 2, 3. \tag{2}$$

The functor $\bar{\sigma}$ carries X to $\bar{\sigma}(X) = \tilde{\sigma}$. Since diagram (1) is natural with respect to the functor σ: $\mathbf{CW}/\overset{0}{\simeq} \to \mathbf{Q}/\overset{0}{\simeq}$ we obtain induced maps for $\bar{\sigma}(X)$ similarly as in (III.8.3). This completes the definition of $\bar{\sigma}$.

§ 11 The action of the fundamental group for quadratic chain complexes

Recall that the fundamental group $\pi_1 U$ acts on the set of homotopy classes $[X, U]$ such that the orbit set

$$[X, U]/\pi_1 U = [X, U]^\phi$$

is the set of free homotopy classes, see (III.§ 10); here we assume that X and U are path connected CW-complexes. We now define the action of the fundamental group on homotopy sets in the category of quadratic chain complexes and we show that the natural map

$$\sigma\colon [X, U] \to [\sigma X, \sigma U]$$

is equivariant with respect to the action of $\pi_1 U$. This allows the computation of $[X, U]^\phi$ in case $\dim(X) \leq 3$. We proceed similarly as in (III.§ 10) where we studied the action of the fundamental group on homotopy sets in the category of crossed chain complexes.

Let σ be a totally free quadratic chain complex, that is $\sigma \in \mathbf{Q}$, and let σ' be a quadratic chain complex with fundamental group $\pi_1\sigma' = \sigma_1'/d\sigma_2'$. Each element $a \in \sigma_1'$ induces a map in **quadchain**

(11.1) $$1^a\colon \sigma' \to \sigma' \quad \text{by} \quad 1^a(x) = x^a, \quad x \in \sigma_n, \quad n \geq 1.$$

These maps yield the **action of the fundamental group**

(11.2) $$[\sigma, \sigma'] \times \pi_1(\sigma') \to [\sigma, \sigma'], \quad (\xi, \alpha) \mapsto \xi^\alpha$$

as follows. Let $x\colon \sigma \to \sigma'$ be a quadratic chain map which represents ξ and assume $a \in \sigma_1'$ represents α. Then the map $x^a = (1^a)x$ represents ξ^α. Recall that $[\sigma, \sigma']$ denotes the set of homotopy classes rel$*$ in the cofibration structure (5.5), these homotopy classes are as well defined by the cylinder (4.10). Clearly the functor λ in (3.2) gives us a function

(11.3) $$\lambda\colon [\sigma, \sigma'] \to [\lambda\sigma, \lambda\sigma']$$

which is equivariant with respect to the action of the fundamental group $\pi_1\sigma' = \pi_1\lambda\sigma'$, see (III.10.5). Moreover, we have the following result which corresponds to (III.10.6).

(11.4) **Proposition.** *Let X, Y be* CW-*complexes in* **CW**. *Then the function*

$$\sigma\colon [X, Y] \to [\sigma X, \sigma Y]$$

(*given by the functor σ in* (6.8)) *is equivariant with respect to the action of the fundamental group $\pi_1 Y = \pi_1 \sigma Y$.*

The next corollary is immediately obtained by (III.10.2) and (8.2).

(11.5) **Corollary.** *Let X, Y be* CW*-complexes in* **CW**. *Then we have the bijection*

$$[X, Y]^{\phi} = [\sigma X, \sigma Y]/\pi_1 Y \quad \textit{for } \dim X \leq 3.$$

For the proof of (11.4) we compute the quadratic chain complex $\sigma(\hat{I}X)$ where $\hat{I}X$ is the CW-complex in (III.10.3).

(11.6) **Definition.** Let σ be a totally free quadratic chain complex. We define $\hat{I}(\sigma)$ by the push out diagram

$$\begin{array}{ccc} I(\sigma) & \xrightarrow{\hat{q}} & \hat{I}(\sigma) \\ i_1 \uparrow \sim & \text{push} & \sim \uparrow (i_1, s) \\ \sigma & \xrightarrow[i^s]{} & \sigma \vee \sigma(S^1) \end{array} \tag{1}$$

where $I(\sigma)$ is the cylinder. The object $\sigma(S^1) = \rho(S^1) = \langle s \rangle$ is defined as in (III.10.9). The map $i: \sigma \to \sigma \vee \sigma(S^1)$ is the inclusion and i^s is defined by $i^s(x) = (ix)^s$. We set $i_0 = \hat{q}i_0: \sigma \to \hat{I}(\sigma)$. Since i_1 is a weak equivalence also (i_1, s) is one.

Clearly we get the cylinder $I(\sigma)$ by the quotient

$$I(\sigma) = \hat{I}(\sigma)/s\sigma(S^1). \tag{2}$$

Moreover, we now define the 'special' tensor product $\sigma(S^1) \hat{\otimes} \sigma$ by the push out diagram

$$\begin{array}{ccc} \hat{I}(\sigma) & \xrightarrow{q} & \sigma(S^1) \hat{\otimes} \sigma \\ (i_0, i_1) \uparrow & \text{push} & \uparrow i' \\ \sigma \vee \sigma & \xrightarrow{(1,1)} & \sigma \end{array} \tag{3}$$

where $(1, 1)$ is the folding map. The map q carries the generator $sx = \hat{q}(sx)$ to

$$q(sx) = -(s \times x), \quad x \in Z_n, \quad n \geq 1. \tag{4}$$

In the next section we will introduce a more general tensor product $\otimes$ of totally free quadratic chain complexes. The remarks (2), (3), and (4) correspond exactly to those in (III.10.9) (2), (3), (4). Clearly $\hat{q}$ in (11.6) is a homotopy $\hat{q}: i_0 \simeq i_1^s$. Therefore (11.4) follows from the next result which corresponds to (III.10.10).

(11.7) **Proposition.** *For a CW-complex X with $X^0 = *$ there is an isomorphism*

$$\sigma(\hat{I}X) \cong \hat{I}(\sigma X) \quad \textit{in } \mathbf{Q}$$

which induces the isomorphism (III.10.10) *and which is compatible with the inclusions i_0, i_1, and s.*

Proof. The inclusion $X \vee S^1 \subset \hat{I}X$ and the inclusion $\sigma(X) \vee \sigma(S^1) \subset \hat{I}\sigma(X)$ are homotopy equivalences. This implies (11.7) by using $\lambda\hat{I}\sigma = \hat{I}\lambda\sigma$ and by use of (III.10.10). □

Proposition (11.7) yields as well the isomorphism

(11.8) $$\sigma(S^1 \times X) \cong \sigma(S^1)\,\hat{\otimes}\,\sigma(X) \quad \text{in } \mathbf{Q}$$

where the right hand side is the tensor product in (11.6)(3) above. We obtain this isomorphism by the push out (III.10.8)(2) since the functor σ is compatible with push outs. In the next section we obtain an isomorphism

(11.9) $$\sigma(S^1) \otimes \sigma \cong \sigma(S^1)\,\hat{\otimes}\,\sigma \quad \text{in } \mathbf{Q}$$

where the left hand side is given by a tensor product of objects in $\mathbf{Q}$. The isomorphism (11.9) is defined on generators by $s \times * \mapsto s$, $* \times e \mapsto e$, $s \times e \mapsto s \times e$ where e is a generator of σ, compare (4.10) and (11.6)(4) above.

§ 12 The quadratic chain complex of a product

We have seen in (III.§ 9) that the crossed chain complex of a product $X \times Y$ in **CW** is given by the isomorphism $\rho(X \times Y) \cong \rho(X) \otimes \rho(Y)$ where the right hand side is the tensor product of Brown-Higgins. In this section we describe the quadratic chain complex $\sigma(X \times Y)$ in terms of σX and σY by a similar isomorphism. For this we introduce the tensor product of totally free quadratic chain complexes in the category $\mathbf{Q}$. This tensor product gives us the functor (see (12.12))

$$\otimes: \mathbf{Q}/\overset{0}{\simeq} \times \mathbf{Q}/\overset{0}{\simeq} \to \mathbf{Q}/\overset{0}{\simeq}$$

for which the following result holds.

(12.1) **Theorem.** *For CW-complexes X, Y in* **CW** *there is an isomorphism*

$$\sigma(X \times Y) \cong \sigma(X) \otimes \sigma(Y)$$

in **Q** *which is natural in* $\mathbf{Q}/\overset{0}{\simeq}$ *with respect to cellular maps on X and Y.*

Addendum. *For homotopy systems* $\bar{X}$, $\bar{Y}$ *of order* 4 *one has an isomorphism*

$$\sigma(\bar{X} \otimes \bar{Y}) \cong \sigma(\bar{X}) \otimes \sigma(\bar{Y})$$

in **Q** *where* $\bar{X} \otimes \bar{Y}$ *is the tensor product in* (II.2.7) *and where* $\sigma\colon \mathbf{H}_4^c \xrightarrow{\sim} \mathbf{Q}/\overset{0}{\simeq}$ *is the equivalence of categories in* (7.7). *The isomorphism is natural in* $\mathbf{Q}/\overset{0}{\simeq}$ *with respect to maps on* $\bar{X}$ *and* $\bar{Y}$ *in* $\mathbf{H}_4^c$.

The following definition of the tensor product of totally free quadratic chain complexes in **Q** is more complicated than the corresponding definition of tensor products in **H** in (III.§ 9). We use a notation which is compatible with the notation in (III.§ 9).

(12.2) **Definition.** For A, $B \in \mathbf{Q}$ we define the **tensor product** $A \otimes B \in \mathbf{Q}$ as follows. We first choose a basis

$$Z_n^A \subset A_n, \quad Z_n^B \subset B_n, \quad n \geq 1 \tag{1}$$

and we set $Z_0^A = \{*\}$, $Z_0^B = \{*\}$. Then $A \otimes B$ is the totally free quadratic chain complex with basis elements

$$e \times f \in (A \otimes B)_n \quad \text{for} \quad e \in Z_i^A, \quad f \in Z_j^B, \quad i, j \geq 0 \text{ and } i + j = n. \tag{2}$$

We also write $|a| = i$ if $a \in A$ is an element of degree i, that is $a \in A_i$. Hence we have $|e \times f| = |e| + |f|$ where we set $|*| = 0$. We define below the boundary d of $A \otimes B$ in such a way that one has inclusions

$$A \xrightarrow[i_A]{} A \otimes B \xleftarrow[i_B]{} B \tag{3}$$

in **Q** which are given on basis elements by $i_A(e) = e \times *$ and $i_B(f) = * \times f$.

We also define the boundary d of $A \otimes B$ in such a way that the functor $\lambda\colon \mathbf{Q} \to \mathbf{H}$ satisfies

$$\lambda(A \otimes B) = \lambda A \otimes \lambda B \tag{4}$$

where the right hand side is the tensor product of crossed chain complexes in (III.§ 9). Hence we have as in (3.3) the commutative diagram

$$\begin{array}{ccccccccc}
\longrightarrow & (A\otimes B)_4 & \xrightarrow{d} & (A\otimes B)_3 & \xrightarrow{d} & (A\otimes B)_2 & \xrightarrow{d} & (A\otimes B)_1 \\
& =\Big\downarrow q & & \Big\downarrow q & & \Big\downarrow q & & =\Big\downarrow q \\
\longrightarrow & (\lambda A\otimes \lambda B)_4 & \longrightarrow & (\lambda A\otimes \lambda B)_3 & \longrightarrow & (\lambda A\otimes \lambda B)_2 & \longrightarrow & (\lambda A\otimes \lambda B)_1
\end{array} \tag{5}$$

where q is the identity in degree n for $n = 1$ und $n \geq 4$. Moreover q carries the basis element $e \times f$ in (2) to the basis element $e \otimes f$ in $\lambda A \otimes \lambda B$, see (III.9.2); clearly the basis (1) is as well a basis in λA and λB respectively. The bottom row of diagram (4) is already known by the crossed chain complex $\lambda A \otimes \lambda B$. By (4) we see that the quadratic map ω *of* $A \otimes B$ is a homomorphism

$$\omega\colon (C^A \otimes C^B)_2^{\otimes 2} \to (A \otimes B)_3 \tag{6}$$

where $C^A = C\lambda A$, $C^B = C\lambda B$ and where

$$C^A \otimes C^B = C^A \otimes_{\mathbb{Z}} C^B = C(\lambda A \otimes \lambda B) \tag{7}$$

is given by the functor $C\colon \mathbf{H} \to \mathbf{H}_1$ in (III.2.11). Recall that we have the quotient maps

$$A_2 \underset{q}{\twoheadrightarrow} (\lambda A)_2 = A_2^{cr} \underset{q_2}{\twoheadrightarrow} C_2^A = (A_2^{cr})^{ab} \tag{8}$$

which carry an element $a \in A_2$ to the class $\{a\} \in C_2^A$. Here we set $e = \{e\}$ if $e \in A_2$ is a basis element, moreover $* \in C_0^A = \mathbb{Z}[\pi_1 A]$ is the generator. Using the operators $\bar{\otimes}$, $\bar{\bar{\otimes}}$ and Θ below we now can define the boundary

$$d = d_n\colon (A \otimes B)_n \to (A \otimes B)_{n-1} \tag{9}$$

in $A \otimes B$ by the following formulas where $e \in A$ and $f \in B$ are basis elements as in (2).

(i) $$d_n(e \times *) = i_A(de) \quad \text{for } |e| = n \geq 2,$$
$$d_n(* \times f) = i_B(df) \quad \text{for } |f| = n \geq 2.$$

(ii) $d_2(e \times f) = -e \times * - * \times f + e \times * + * \times f \quad \text{for } |e| = |f| = 1.$

(iii) $$d_3(e \times f) = \begin{cases} (de) \bar{\otimes} f - (e \times *)^{* \times f} + e \times * & \text{for } |e| = 2, |f| = 1, \\ -e \bar{\bar{\otimes}} (df) - (* \times f)^{e \times *} + * \times f & \text{for } |e| = 1, |f| = 2. \end{cases}$$

(iv)

$$d_4(e \times f) = \begin{cases} e \mathbin{\bar{\otimes}} (df) + (de) \mathbin{\bar{\bar{\otimes}}} f + \omega(e \otimes * \otimes * \otimes f + * \otimes f \otimes e \otimes * \\ \quad - \Theta(de, df)) & \text{for } |e| = |f| = 2, \\ -e \underset{=}{\times} * + (e \times *)^{* \times f} + (de) \mathbin{\bar{\otimes}} f & \text{for } |e| = 3, |f| = 1, \\ -e \mathbin{\bar{\bar{\otimes}}} (df) - (* \times f)^{e \times *} + * \times f & \text{for } |e| = 1, |f| = 3. \end{cases}$$

(v) d_n, $n \geq 5$ is defined as in $\lambda A \otimes \lambda B$, see (5).

We now define the operators $\bar{\otimes}$, $\bar{\bar{\otimes}}$ and Θ needed in the definition of the boundary d of $A \otimes B$ above.

(12.3) **Definition.** Let $d_2\colon (A \otimes B)_2 \to (A \otimes B)_1$ be the free nil(2)-module given by d_2 above. We define operators

$$\bar{\otimes}, \bar{\bar{\otimes}}\colon A_1 \times B_1 \to (A \otimes B)_2, \tag{1}$$

which satisfy

$$q(a \mathbin{\bar{\otimes}} b) = a \otimes b = q(a \mathbin{\bar{\bar{\otimes}}} b) \tag{2}$$

in $(\lambda A \otimes \lambda B)_2$ as follows. Let $e \in Z_1^A, f \in Z_1^B, a, a' \in A_1, b, b' \in B_1$. Then we set

$$e \mathbin{\bar{\otimes}} f = e \times f = e \mathbin{\bar{\bar{\otimes}}} f, \tag{3}$$

$$\left.\begin{aligned} (a + a') \mathbin{\bar{\bar{\otimes}}} f &= (a \mathbin{\bar{\bar{\otimes}}} f)^{a' \otimes *} + a' \mathbin{\bar{\bar{\otimes}}} f \\ a \mathbin{\bar{\bar{\otimes}}} (b + b') &= a \mathbin{\bar{\bar{\otimes}}} b' + (a \mathbin{\bar{\bar{\otimes}}} b)^{* \otimes b'} \end{aligned}\right\} \tag{4}$$

$$\left.\begin{aligned} e \mathbin{\bar{\otimes}} (b + b') &= e \mathbin{\bar{\otimes}} b' + (e \mathbin{\bar{\otimes}} b)^{* \otimes b'} \\ (a + a') \mathbin{\bar{\otimes}} b &= (a \mathbin{\bar{\otimes}} b)^{a' \otimes *} + a' \mathbin{\bar{\otimes}} b \end{aligned}\right\} \tag{5}$$

By (4) and (5) we see that

$$e \mathbin{\bar{\otimes}} b = e \mathbin{\bar{\bar{\otimes}}} b \quad \text{and} \quad a \mathbin{\bar{\otimes}} f = a \mathbin{\bar{\bar{\otimes}}} f. \tag{6}$$

We also derive from (2) that

$$\{a \mathbin{\bar{\otimes}} b\} = \{a \mathbin{\bar{\bar{\otimes}}} b\} = h_1 a \otimes h_1 b \in (C^A \otimes C^B)_2 \tag{7}$$

where $h_1\colon A_1 \to C_1^A$, $h_1\colon B_1 \to C_1^B$ is defined as in (III.2.11), compare also the notation in (12.2)(8) above.

(12.4) **Definition.** We define the operator Θ by use of **square maps** ϑ^A, ϑ^B, see Baues (*M*). Let $a, a' \in A_1$ and $b, b' \in B_1$ and let $e \in Z_1^A$, $f \in Z_1^B$. We obtain

$$\vartheta^A\colon A_1 \to C_1^A \otimes C_1^A \tag{1}$$

by the formulas

$$\begin{cases} \vartheta^A(e) = 0, \\ \vartheta^A(a + a') = \vartheta^A(a)^{a'} + \vartheta^A(a') + (h_1 a)^{a'} \otimes h_1 a'. \end{cases}$$

Here the action of $a' \in A_1$ is given by the action of $\pi_1 A$ on C_1^A, that is we set $(x \otimes y)^{a'} = x^{\{a'\}} \otimes y^{\{a'\}}$ where $\{a'\} \in \pi_1 A$ is represented by a', $x, y \in C_1^A$. In the same way we define the square map

$$\vartheta^B\colon B_1 \to C_1^B \otimes C_1^B. \tag{2}$$

Now let

$$T\colon C_1^A \otimes C_1^A \otimes C_1^B \otimes C_1^B \cong C_1^A \otimes C_1^B \otimes C_1^A \otimes C_1^B \tag{3}$$

be the interchange map given by

$$T(x_1 \otimes x_2 \otimes y_1 \otimes y_2) = x_2 \otimes y_2 \otimes x_1 \otimes y_1.$$

Then one can check that the Peiffer commutator map w for $A \otimes B$ satisfies

$$w\bar{\bar{\Theta}}(a, a')(b) = (a \bar{\bar{\otimes}} b)^{a' \otimes *} + a' \bar{\bar{\otimes}} b - (a + a') \bar{\bar{\otimes}} b \tag{4}$$

with $\bar{\bar{\Theta}}(a, a')(b) = -T((h_1 a)^{a'} \otimes h_1 a' \otimes \vartheta^B(b))$.

$$w\bar{\Theta}(b, b')(a) = -a \bar{\otimes} (b + b') + a \bar{\otimes} b' + (a \bar{\otimes} b)^{* \otimes b'} \tag{5}$$

with $\bar{\Theta}(b, b')(a) = T(\vartheta^A(a) \otimes (h_1 b)^{b'} \otimes h_1 b')$. Here we use the inclusion $(C_1^A \otimes C_1^B)^{\otimes 2} \subset (C^A \otimes C^B)_2^{\otimes 2}$. Moreover we get the operator

$$\left.\begin{aligned} &\Theta\colon A_1 \times B_1 \to (C^A \otimes C^B)_2^{\otimes 2}, \\ &\Theta(a, b) = -T(\vartheta^A(a) \otimes \vartheta^B(b)) \end{aligned}\right\} \tag{6}$$

which satisfies

$$w\Theta(a, b) = -a \bar{\bar{\otimes}} b + a \bar{\otimes} b. \tag{7}$$

One can check (7) by use of (4) and (5). Further properties of the square map ϑ^A are studied in Baues (*M*).

(12.5) **Definition.** Let $d_3: (A \otimes B)_3 \to (A \otimes B)_2$ be given by the free quadratic module defined by the boundary (12.2)(9)(iii) where we use (12.3). We now define the operators

$$\bar{\otimes}: A_2 \times B_1 \to (A \otimes B)_3, \tag{1}$$

$$\bar{\bar{\otimes}}: A_1 \times B_2 \to (A \otimes B)_3, \tag{2}$$

which are used in the definition of d_4 in (12.9)(9)(iv). Let $e \in Z_2^A$, $f \in Z_2^B$, $a, a \in A_1$, $b, b' \in B_1$. We define the function

$$S(e): B_1 \to (A \otimes B)_3 \tag{3}$$

by setting

$$\begin{cases} S(e)(f') = e \times f' \quad \text{for } f' \in Z_1^B, \\ S(e)(b + b') = S(e)(b') + S(e)(b)^{*\otimes b'} \\ \qquad\qquad + \omega(\{d_3 S(e)(b)^{*\otimes b'}\} \otimes \{-(e \times *)^{*\otimes b'} + e \times *\}) \end{cases}$$

One can check that

$$d_3 S(e)(b) = (d_2 e) \bar{\bar{\otimes}}\, b - (e \times *)^{*\otimes b} + e \times *. \tag{4}$$

We have the i_A-crossed homomorphism

$$\left.\begin{aligned} &S_1^b: A_1 \to (A \otimes B)_2, \\ &S_1^b(a) = a \bar{\otimes} b \end{aligned}\right\} \tag{5}$$

which gives us the $(S_1^b, i_A, i_A^{*\otimes b})$-quadratic operator (see (4.3))

$$\left.\begin{aligned} &S_2^b: A_2 \to (A \otimes B)_3 \quad \text{with} \\ &S_2^b(e) = S(e)(b) - \omega\Theta(de, b). \end{aligned}\right\} \tag{6}$$

In terms of this operator we define $\bar{\otimes}$ in (1) by

$$x \bar{\otimes} b = -S_2^b(x) \quad \text{for } x \in A_2. \tag{7}$$

By (4.3), (4) and (12.4)(6) we get

$$d_3(x \mathbin{\bar{\otimes}} b) = (d_2 x) \mathbin{\bar{\otimes}} b - (i_A x)^{*\otimes b} + i_A x. \tag{8}$$

In a similar way we get the operator $\bar{\bar{\otimes}}$ in (2). For this we define the function

$$S(f)\colon A_1 \to (A \otimes B)_3 \tag{9}$$

by setting

$$\begin{cases} S(f)(e') = e' \times f \quad \text{for } e' \in Z_1^A, \\ S(f)(a + a') = S(f)(a') + S(f)(a)^{a'\otimes *} \\ \qquad\qquad + \omega(\{d_3 S(f)(a)^{a'\otimes *}\} \otimes \{-(* \times f)^{a'\otimes *} + * \times f\}) \end{cases}$$

One can check that

$$d_3 S(f)(a) = -a \mathbin{\bar{\otimes}} (d_2 f) - (* \times f)^{a\otimes *} + * \times f. \tag{10}$$

We have the i_B-crossed homomorphism

$$\left.\begin{aligned} &S_1^a\colon B_1 \to (A \otimes B)_2 \\ &S_1^a(b) = -a \mathbin{\bar{\bar{\otimes}}} b \end{aligned}\right\} \tag{11}$$

which gives us the $(S_1^a, i_B, i_B^{a\otimes *})$-quadratic operator

$$\left.\begin{aligned} &S_2^a\colon B_2 \to (A \otimes B)_3 \quad \text{with} \\ &S_2^a(f) = -S(f)(a) - \omega\Theta(a, df). \end{aligned}\right\} \tag{12}$$

In terms of this operator we define $\bar{\bar{\otimes}}$ in (2) by

$$a \mathbin{\bar{\bar{\otimes}}} y = -S_2^a(y) \quad \text{for } y \in B_2. \tag{13}$$

As in (8) one gets

$$d_3(a \mathbin{\bar{\bar{\otimes}}} y) = -a \mathbin{\bar{\bar{\otimes}}} (d_2 y) - (i_B y)^{a\otimes *} + i_B y. \tag{14}$$

We leave it to the reader to check that $A \otimes B$ is a well defined quadratic chain complex, in particular, that $dd = 0$. Moreover, one can check that the **interchange map**

(12.6) $$T: A \otimes B \to B \otimes A,$$

defined on generators by $T(e \times f) = (-1)^{|e||f|} f \times e$, is a well defined map between quadratic chain complexes. Clearly $TT = 1$. The operators $\bar{\otimes}$, $\bar{\bar{\otimes}}$ above satisfy

$$T(a \bar{\otimes} b) = -b \bar{\bar{\otimes}} a \quad \text{for } |a| = |b| = 1, \tag{1}$$

$$T(a \bar{\otimes} b) = b \bar{\bar{\otimes}} a \quad \text{for } |a| = 1, |b| = 2 \text{ or } |a| = 2, |b| = 1. \tag{2}$$

Moreover T induces the interchange maps T on $\lambda A \otimes \lambda B$ and $C^A \otimes C^B$ respectively. The functions in (12.4) satisfy

$$T^{\otimes 2} \bar{\bar{\Theta}}(a, a')(b) = -\bar{\Theta}(a, a')(b), \tag{3}$$

$$T^{\otimes 2} \Theta(a, b) = \Theta(b, a). \tag{4}$$

Remark. Our definition of $A \otimes B$ above was found by the following conditions. First of all (12.2)(3) and (12.2)(4) should hold, moreover T in (12.6) should be a well defined isomorphism and the isomorphism

$$\sigma(S^1) \otimes \sigma \cong \sigma(S^1) \hat{\otimes} \sigma$$

in § 11 should hold. These conditions already determine all boundary formulas in (12.2)(9) except the formula for $d_4(e \times f)$ with $|e| = |f| = 2$ which was motivated by the condition $d_3 d_4 = 0$.

There are canonical **projections**

(12.7) $$A \xleftarrow{p_A} A \otimes B \xrightarrow{p_B} B$$

which are maps in **Q** defined on generators by

$$p_A(e \times f) = \begin{cases} e & \text{for } f = *, \\ 0 & \text{otherwise,} \end{cases}$$

$$p_B(e \times f) = \begin{cases} f & \text{for } e = *, \\ 0 & \text{otherwise.} \end{cases}$$

Clearly we have $p_A i_A = 1$, $p_B i_B = 1$, $p_A i_B = 0$ and $p_B i_A = 0$. The inclusions and projections in (12.2)(1) and (12.7) correspond via the isomorphism in (12.1) to the inclusions and projections of the product $X \times Y$.

The tensor product $A \otimes B$ above of quadratic chain complexes was studied in the Diplomarbeit of my student A. Hohmann where one can find the partly laborious proofs of the following properties of $A \otimes B$. Recall that we denote the map $A \xrightarrow{q} \lambda A \xrightarrow{h} C\lambda A = C^A$ by $a \mapsto \{a\}$. The boundary d of $A \otimes B$ satisfies the following formulas.

$$(12.8)\quad \begin{aligned} d_3(a \bar{\otimes} b) &= (da) \bar{\otimes} b - (i_A a)^{*\otimes b} + i_A a \quad \text{for } |a| = 2, |b| = 1, \\ d_3(a \bar{\bar{\otimes}} b) &= -a \bar{\bar{\otimes}} (db) - (i_B b)^{a\otimes *} + i_B b \quad \text{for } |a| = 1, |b| = 2. \end{aligned}$$

$$d_4(\{a\} \otimes f) = \begin{cases} i_A a + (i_A a)^{*\otimes f} + (da) \bar{\otimes} f & \text{for } a \in A_3, f \in Z_1^B, \\ a \bar{\otimes} (df) + (da) \bar{\bar{\otimes}} f + \omega(\{a\} \otimes * \otimes * \otimes f \\ \quad + * \otimes f \otimes \{a\} \otimes * - \Theta(da, df)) & \text{for } a \in A_2, f \in Z_2^B, \\ -a \bar{\bar{\otimes}} (df) - (* \times f)^{a\otimes *} + * \times f & \text{for } a \in A_1, f \in Z_3^B. \end{cases}$$

$$d_4(e \otimes \{b\}) = \begin{cases} -e \bar{\bar{\otimes}} (db) - (i_B b)^{e\otimes *} + i_B b & \text{for } b \in B_3, e \in Z_1^A, \\ e \bar{\otimes} (db) + (de) \bar{\bar{\otimes}} b + \omega(e \otimes * \otimes * \otimes \{b\} \\ \quad + * \otimes \{b\} \otimes e \otimes * - \Theta(de, db)) & \text{for } b \in B_2, e \in Z_2^A, \\ -e \times * + (e \times *)^{*\otimes b} + (de) \bar{\otimes} b & \text{for } b \in B_1, e \in Z_3^A. \end{cases}$$

We also remark that the operators $\bar{\bar{\otimes}}$, $\bar{\otimes}$ in (12.5) satisfy

$$(12.9)\quad \begin{cases} (a + a') \bar{\bar{\otimes}} b = (a \bar{\bar{\otimes}} b)^{a'\otimes *} + a' \bar{\bar{\otimes}} b & \text{for } a, a' \in A_1, b \in B_2, db = 0, \\ a \bar{\otimes} (b + b') = (a \bar{\otimes} b)^{*\otimes b'} + a \bar{\otimes} b' & \text{for } b, b' \in B_1, a \in A_2, da = 0. \end{cases}$$

Next we define the induced maps on the tensor product.

(12.10) **Definition.** Let $F\colon A \to A'$ and $G\colon B \to B'$ be maps in $\mathbf{Q}$ and let $e \in A$, $f \in B$ be basis elements. Then we obtain the **induced maps**

$$F \otimes 1\colon A \otimes B \to A' \otimes B \quad \text{and}$$

$$1 \otimes G\colon A \otimes B \to A \otimes B'$$

in $\mathbf{Q}$ by setting

$$(F \otimes 1)(e \times f) = \begin{cases} (Fe) \bar{\otimes} f & \text{for } |e| \le 2, |f| = 1, \\ (Fe) \bar{\bar{\otimes}} f & \text{for } |e| = 1, |f| = 2, \\ (Fe) \otimes f & \text{otherwise,} \end{cases}$$

$$(1 \otimes G)(e \times f) = \begin{cases} e \mathbin{\bar{\otimes}} (Gf) & \text{for } |e| \le 2, |f| = 1, \\ e \mathbin{\bar{\bar{\otimes}}} (Gf) & \text{for } |e| = 1, |f| = 2, \\ e \otimes (Gf) & \text{otherwise.} \end{cases}$$

One can check that the definitions yield functors $- \otimes B\colon \mathbf{Q} \to \mathbf{Q}$ and $A \otimes -\colon \mathbf{Q} \to \mathbf{Q}$.

The composition $(F \otimes 1)(1 \otimes G)$ and $(1 \otimes G)(1 \otimes F)$, however, do not coincide but there is a canonical 0-homotopy between these maps, see (4.8). Let

$$\alpha_2\colon (C^A \otimes C^B)_2 \to (C^{A'} \otimes C^{B'})_2^{\otimes 2}$$

be the $(\pi_1 F \times \pi_1 G)$-equivariant homomorphism defined on generators by

$$\alpha_2(e \otimes f) = \begin{cases} 0 & \text{for } e = *, |f| = 2 \text{ or } |e| = 2, f = *, \\ \Theta(Fe, Gf) & \text{for } |e| = |f| = 1. \end{cases}$$

Then one can check that α_2 is a 0-homotopy

(12.11) $$\alpha_2\colon (F \otimes 1)(1 \otimes G) \overset{0}{\simeq} (1 \otimes G)(F \otimes 1).$$

This fact allows the definition of the functor

(12.12) $$\otimes\colon \mathbf{Q}/\overset{0}{\simeq} \times \mathbf{Q}/\overset{0}{\simeq} \to \mathbf{Q}/\overset{0}{\simeq}$$

which is defined on objects by the tensor product in (12.2) and which is defined on morphisms by the 0-homotopy class $\{F\} \otimes \{G\} = \{(F \otimes 1)(1 \otimes G)\} = \{(1 \otimes G)(F \otimes 1)\}$. Finally we consider the **associativity** of the tensor product in $\mathbf{Q}$. Let A, B, C be objects in $\mathbf{Q}$ with basis Z^A, Z^B and Z^C respectively. We define the isomorphism

(12.13) $$\phi\colon (A \otimes B) \otimes C \cong A \otimes (B \otimes C)$$

together with its inverse ϕ^{-1} by the following formulas ($e \in Z^A, f \in Z^B, g \in Z^C$)

$$\phi((e \times f) \times g) = \begin{cases} e \times (f \times g) + \omega\xi(e, f, g) & \text{for } |e| = |f| = |g| = 1, \\ e \times (f \times g) & \text{otherwise,} \end{cases}$$

$$\phi^{-1}(e \times (f \times g)) = \begin{cases} (e \times f) \times g - \omega\xi(e, f, g) & \text{for } |e| = |f| = |g| = 1, \\ (e \times f) \times g & \text{otherwise.} \end{cases}$$

Here the function

$$\xi\colon Z_1^A \times Z_1^B \times Z_1^C \to (C^A \otimes C^B \otimes C^C)_2^{\otimes 2}$$

is given as follows. For $(e, f, g) \in Z_1^A \times Z_1^B \times Z_1^C$ let $a = e \otimes f \otimes *$, $b = e \otimes * \otimes g$, $c = * \otimes f \otimes g$. Then we set

$$\xi(e, f, g) = a \otimes (b + c^{e\otimes * \otimes *}) - c \otimes (b^{*\otimes f \otimes *} + a) - q \otimes p \quad \text{with}$$

$$q = b + a^{*\otimes * \otimes g} - b^{*\otimes f \otimes *} - a - c + c^{e\otimes * \otimes *},$$

$$p = a + b - c + c^{e\otimes * \otimes *}.$$

The function ξ was found by Hohmann. The isomorphisms ϕ and ϕ^{-1} fit into the following commutative diagram where we use the interchange map T in (12.6).

(12.14)
$$\begin{array}{ccc}
(A \otimes B) \otimes C & \xrightarrow{\ \phi\ } & A \otimes (B \otimes C) \\
\uparrow T & & \downarrow T \\
C \otimes (A \otimes B) & & (B \otimes C) \otimes A \\
\uparrow 1 \otimes T & & \downarrow T \otimes 1 \\
C \otimes (B \otimes A) & \xrightarrow{\ \phi^{-1}\ } & (C \otimes B) \otimes A
\end{array}$$

(12.15) *Proof of* (12.1). Let $A = \sigma(X)$, $B = \sigma(Y)$. The remark following (12.6) above shows that we have chosen in $A \otimes B$ the correct boundary $d(e \times f)$ except possibly for $|e| = |f| = 2$. This shows that the 3-skeleton of $A \otimes B$ is the 3-skeleton of $\sigma(X \times Y)$. A 4-cell $e \times f$ in $X \times Y$ is attached by a map

$$F\colon S^1 \times D^2 \cup D^2 \times S^1 \to (X \times Y)^3$$

Now we know that $\sigma(D^2 \times D^2) = \sigma(D^2) \otimes \sigma(D^2)$ by (12.2)(4) and (9.7). The map $\sigma(F)$ can be computed by gluing together the restrictions $\sigma(F|S^1 \times S^1)$, $\sigma(F|S^1 \times D^2)$ and $\sigma(F|D^2 \times S^1)$ respectively. These restrictions are essentially determined by the corresponding maps between crossed chain complexes. This implies that also $d(e \times f)$ is chosen correctly. The naturality of the isomorphism (12.6) in $\mathbf{Q}/\overset{0}{\simeq}$ follows from the compatibility of the maps in (12.10) with the projections (12.7). □

A more careful discussion of the naturality of the isomorphism (12.2) and of (12.11) involves the concept of **pseudo functor** of Fantham-Moore, see Baues (M).

Appendix A

Some diverse examples and applications of quadratic chain complexes

The theory of quadratic chain complexes offers an effective concreteness for calculations for many topological and geometric problems in dimensions ≤ 4. In this appendix we describe some diverse examples and applications which indicate the utility of our methods though we have chosen examples of a fairly simple nature.

One of the main advantages of the quadratic chain complex $\sigma(X)$ is its cellular construction. In fact, the cells of $X - *$ are the generators of the totally free object $\sigma(X)$ and for any subcomplex Y of X we can assume that $\sigma(Y)$ is a subchain complex of $\sigma(X)$. That is, the formulas for boundaries $d(e)$ in $\sigma(Y)$ are the same as the formulas for the boundaries $d(e)$ in $\sigma(X)$ where e is a cell in Y. This also gives us immediately the formulas for boundaries in the quotient

$$\sigma(X/Y) = \sigma(X)/\sigma(Y). \tag{A.1}$$

Here the right hand side is a quotient in the category **Q**, that is, a push out of $* \leftarrow \sigma(Y) \rightarrow \sigma(X)$. The formula for the boundary $d(e)$, $e \subset X - Y$, in the quotient $\sigma(X)/\sigma(Y)$ is obtained from the formula for $d(e)$ in $\sigma(X)$ simply by forgetting all generators in this formula which are cells in Y, this means that such generators are set to be 0. Below we describe some examples of such quotients. More generally we know that constructions for CW-complexes like products and push outs correspond to similar constructions for quadratic chain complexes. This, in fact, leads to the effective computability of $\sigma(X)$ for various examples X. We also use the result in (9.7) for the computation of $\sigma(X)$ which shows that $\sigma(X)$ for special X is completely determined by the crossed chain complex $\rho(X)$.

We first consider some simple examples. Since a **sphere** S^n has only one cell e with $e \neq *$ we clearly get $\sigma(S^n)$ by the unique totally free quadratic chain complex which is generated by a single element e. Hence we have $\sigma(S^1) = \langle e \rangle = \mathbb{Z}$ and $\sigma(S^2)$ is given by

(A.2)

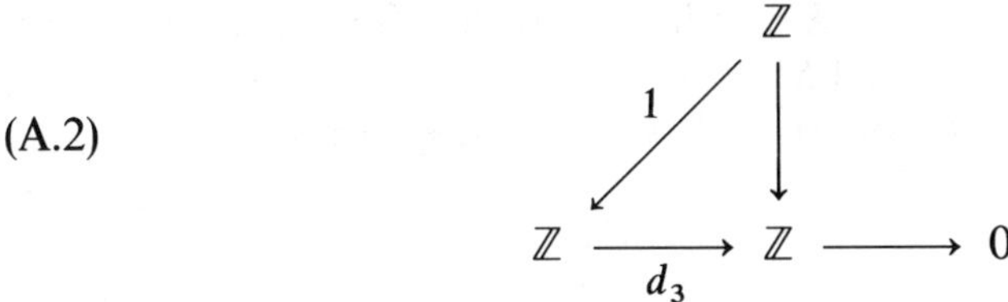

with $d_3 = 0$, $\omega = 1$. More generally we get for a **one point union** $\bigvee_Z S^2$ **of 2-spheres** the quadratic chain complex generated by the set Z in degree 2, that is

(A.3)

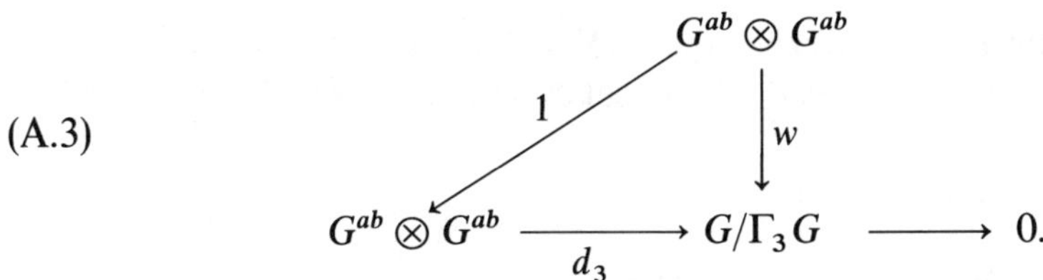

Here $G = \langle Z \rangle$ is the free group generated by the set Z and $d_3 = w$ is the commutator map in (1.1).

In general it is not necessary to compute explicitely the terms σ_i in $\sigma = \sigma(X)$ as this is done in the example (A.3). The totally free quadratic chain complex $\sigma(X)$ is completely determined by writing down formulas for the boundaries $d(e)$ where e is a cell in $X - *$. For example we consider the **complex projective plane** $X = \mathbb{C}P_2$ which has a cell structure $X = S^2 \cup_\eta e^4$ where η is the Hopf map, $S^2 = e^2 \cup *$, then $\sigma(\mathbb{C}P_2)$ is determined by the formula

(A.4) $$de^4 = \omega(e^2 \otimes e^2).$$

This is an example where $\rho(X)$ is not sufficient since $de^4 = 0$ in $\rho(X)$. Similarly we get for $X = S^2 \times S^2$ the quadratic chain complex $\sigma(S^2 \times S^2)$ by the formula

(A.5) $$d(e^2 \times e^2) = \omega(e^2 \times * \otimes * \times e^2 + * \times e^2 \otimes e^2 \times *).$$

Now let M be a **1-connected 4-dimensional manifold**. The M is homotopy equivalent to a mapping cone C_f where $f\colon S^3 \to \bigvee_Z S^2$ is determined by the intersection form of M. This form can be considered to be the element $b[M] \in \Gamma(H_2 M)$ with $H_2 M = \bigoplus_Z \mathbb{Z}$. Now $\sigma(M) = \sigma(C_f)$ is given by the formula

(A.6) $$d(e^4) = \omega(\tau b[M])$$

where e^4 is the 4-cell of M and where $\tau\colon \Gamma(H_2 M) \to H_2 M \otimes H_2 M$ is the homomorphism in (I.4.2). Clearly (A.4) and (A.5) are examples of this formula.

Next we consider examples of $\sigma(X)$ with non trivial fundamental group $\pi_1 X$. First we consider the product $S^1 \times S^3$ of spheres with cells $e = e^1 \times *$, $e^3 = * \times e^3$, $e^4 = e^1 \times e^3$. We obtain $\sigma(S^1 \times S^3)$ by the formulas

(A.7) $$de^3 = 0, \qquad de^4 = (-e^3)^e + e^3.$$

Since there are no 2-cells in $X = S^1 \times S^3$ the quotient map $q\colon \sigma(X) \to \rho(X)$ is actually an isomorphism. The next example is more complicated but typical for the computation of $\sigma(X)$.

(A.8) **Example.** The **real projective space** $X = \mathbb{R}P_\infty$ is a CW-complex with exactly one cell e_n in each degree $n \geq 0$, we set $e_1 = e$. Then $\rho(\mathbb{R}P_\infty)$ is given by

$$\left.\begin{aligned} de_2 &= e + e \\ de_n &= e_{n-1} + (-1)^n e_{n-1}^e, \quad n \geq 3. \end{aligned}\right\} \tag{1}$$

We define the boundary in $\sigma(\mathbb{R}P_\infty)$ by (1) for $n \neq 4$. For $n = 4$ we set $de_4 = e_3 + e_3^e + \omega(\xi)$ where we can compute ξ by the condition $dde_4 = 0$. In fact, we get

$$\begin{aligned} 0 = dde_4 &= de_3 + (de_3)^e + w(\xi) \\ &= e_2 - e_2^e + (e_2 - e_2^e)^e + w(\xi) \\ &= e_2 - e_2^{e+e} + w(\xi) \\ &= \langle e_2, -e_2 \rangle + w(\xi) \\ &= w(-e_2 \otimes e_2 + \xi). \end{aligned} \tag{2}$$

Hence the choice $\xi = e_2 \otimes e_2$ gives us the boundary

$$de_4 = e_3 + e_3^e + \omega(e_2 \otimes e_2) \tag{3}$$

which satisfies $dd = 0$. Thus $\sigma(\mathbb{R}P_\infty)$ with d in (3) is a well defined quadratic chain complex with $\lambda\sigma(\mathbb{R}P_\infty) = \rho(\mathbb{R}P_\infty)$. Since $\pi_2(\mathbb{R}P_\infty) = 0$ we see by (9.7) that (3) is a correct boundary for $\sigma(\mathbb{R}P_\infty)$.

(A.9) **Example.** We consider the quotient $\mathbb{R}P_4/S^1$ where $S^1 = \mathbb{R}P_1$ is the 1-skeleton. By (A.1) we obtain $\sigma(\mathbb{R}P_4/S^1)$ simply by forgetting the generator e in $\sigma(\mathbb{R}P_4)$ in (A.8). Whence $\sigma(\mathbb{R}P_4/S^1)$ is generated by e_2, e_3, e_4 with

$$\left.\begin{aligned} &de_2 = 0, \quad de_3 = e_2 + (-1)^3 e_2 = 0 \quad \text{and} \\ &de_4 = e_3 + e_3 + \omega(e_2 \otimes e_2), \quad \text{see (A.8)(3).} \end{aligned}\right\} \tag{1}$$

This shows that $\mathbb{R}P_4/S^1$ admits a homotopy equivalence

$$\mathbb{R}P_4/S^1 \simeq (S^2 \vee S^3) \cup_f e^4 \tag{2}$$

where the attaching map of the 4-cell is the sum $f = i_2\eta + 2i_3$. Here η is the Hopf map and $i_2(i_3)$ denote the inclusions of $S^2(S^3)$ into $S^2 \vee S^3$. In fact the right hand side of (2) has the same quadratic chain complex as $\mathbb{R}P^4/S^1$ above, see (A.4).

(A.10) **Example.** The **product** $X = S^1 \times \mathbb{R}P_\infty$ has exactly two cells e_n and $s_n = s \times e_{n-1}$ in each degree $n \geq 1$, we set $e = e_1$, $s = s_1$ with $S^1 = * \cup s$. We compute

$$\sigma(S^1 \times \mathbb{R}P_\infty) = \sigma(S^1) \otimes \sigma(\mathbb{R}P_\infty)$$

by use of the tensor product in § 12. Thus we get

$$de_2 = e + e$$

$$ds_2 = -s - e + s + e$$

$$de_3 = e_2 - e_2^e$$

$$ds_3 = -s_2^e - s_2 - e_2^s + e_2$$

$$de_4 = e_3 + e_3^e + \omega(e_2 \otimes e_2), \quad \text{see (A.8),}$$

$$ds_4 = -s_3 + s_3^e - e_3^s + e_3 + \omega(\xi) \quad \text{where}$$

$$\xi = -e_2 \otimes s_2 - s_2 \otimes e_2 - (e_2 + e_2^e) \otimes (s_2 + s_2^e) + (s_2 + s_2^e) \otimes s_2$$

We also can try to find this boundary d in $\sigma(X)$ by use of $\rho(X)$ and by (9.4). This however, turns out to be fairly complicated. In degree $n \geq 5$ we have by (A.8) and (III.9.3)

$$de_n = e_{n-1} + (-1)^n e_{n-1}^e$$

$$ds_n = (-1)^n s_{n-1}^e - s_{n-1} - e_{n-1}^s + e_{n-1}.$$

(A.11) **Example.** We claim in example (I.5.4) that the Pontrjagin square is non trivial on the suspension $\Sigma\mathbb{R}P_3$. We now prove this result. For this we show that there is a homotopy equivalence $\Sigma\mathbb{R}P_3 \simeq C_f$ where C_f is the mapping cone of

$$f: S^3 \vee S^2 \to S^2 \tag{1}$$

with $f|S^3 = 2\eta$, $f|S^2 = -2$. Equivalently one has a homotopy commutative diagram

$$\begin{array}{ccc} \Sigma S^2 & \xrightarrow{\Sigma h} & \Sigma\mathbb{R}P_2 \\ \| & & \cup \\ S^3 & \xrightarrow[2\eta]{} & S^2 \end{array} \tag{2}$$

where $h: S^2 \to \mathbb{R}P_2$ is the quotient map and where η is the Hopf map. A fairly complicated proof of this result is given in the Appendix to chapter 8 in Hilton. We now prove this result by the methods of this book. In fact, the suspension $\Sigma\mathbb{R}P_3$ is the quotient space

$$\Sigma\mathbb{R}P_3 = (S^1 \times \mathbb{R}P_3)/S^1 \vee \mathbb{R}P_3 \tag{3}$$

so that we can use (A.10) and (A.1). This gives us for the cells s_2, s_3, s_4 of $\Sigma\mathbb{R}P_3$ the following boundary formulas in $\sigma(\Sigma\mathbb{R}P_3)$. For this we set all elements s, e, e_2, e_3, e_4 in (A.10) to be 0.

$$\left.\begin{aligned} ds_2 &= 0 \\ ds_3 &= -2s_2 \\ ds_4 &= \omega(2s_2 \otimes s_2) \end{aligned}\right\} \tag{4}$$

Here $s_2 \otimes s_2 = \gamma(s_2) \in \Gamma(H_2)$ corresponds to the Hopf map η. This proves (2).

(A.12) **Example.** The product $X = \mathbb{R}P_2 \times \mathbb{R}P_2$ has the cells $e = e \times *$, $f = * \times e$, $e_2 = e_2 \times *$, $f_2 = * \times e_2$ and $e \times e$, $e \times e_2$, $e_2 \times e$, $e_2 \times e_2$. We compute the quadratic chain complex

$$\sigma(\mathbb{R}P_2 \times \mathbb{R}P_2) = \sigma(\mathbb{R}P_2) \otimes \sigma(\mathbb{R}P_2)$$

by the tensor product in § 12. This yields the formulas:

$$de_2 = e + e$$

$$df_2 = f + f$$

$$d(e \times e) = -e - f + e + f$$

$$d(e \times e_2) = -(e \times e)^f - e \times e - f_2^e + f_2$$

$$d(e_2 \times e) = (e \times e)^e + e \times e - e_2^f + e_2$$

$$d(e_2 \times e_2) = e_2 \times e + (e_2 \times e)^f + e \times e_2 + (e \times e_2)^e + \omega(\xi)$$

$$\xi = e_2 \otimes f_2 + f_2 \otimes e_2 - (e \times e) \otimes (e \times e)^{e+f}$$

$$+ d(e \times e_2)^f \otimes (-e_2^f + e_2) + d(e_2 \times e)^e \otimes (-f_2 + f_2^e)$$

In the formula for ξ we use the formulas for $d(e \times e_2)$ and $d(e_2 \times e)$ above.

(A.13) **Example.** We consider the smash product

$$\mathbb{R}P_2 \wedge \mathbb{R}P_2 = (\mathbb{R}P_2 \times \mathbb{R}P_2)/\mathbb{R}P_2 \vee \mathbb{R}P_2$$

of real projective planes. The space has the cells $e \wedge e$, $e \wedge e_2$, $e_2 \wedge e$ and $e_2 \wedge e_2$ which are the images of the corresponding product cells. Now we can use (A.1) and (A.12) for the computation of $\sigma(\mathbb{R}P_2 \wedge \mathbb{R}P_2)$. For this we set e, f, e_2, f_2 in (A.12) to be 0. This yields the boundary formulas

$$d(e \wedge e) = 0$$

$$d(e_2 \wedge e) = 2(e \wedge e)$$

$$d(e \wedge e_2) = -2(e \wedge e)$$

$$d(e_2 \wedge e) = 2(e_2 \wedge e) + 2(e \wedge e_2) - \omega(e \wedge e \otimes e \wedge e)$$

We can choose a new basis g, g' of σ_3 by $g = e_2 \wedge e + e \wedge e_2$, $g' = e_2 \wedge e$. Then we get $dg = 0$, $dg' = 2e \wedge e$ and $d(e_2 \wedge e_2) = 2g - \omega(\gamma(e \wedge e))$. Here $\gamma(e \wedge e)$ corresponds to the Hopf map η. Thus one obtains the well known homotopy equivalence

$$\mathbb{R}P_2 \wedge \mathbb{R}P_2 \simeq (S^3 \vee S^2 \cup_2 e^3) \cup e^4$$

where e^4 is attached by $2i_3 + i_2\eta$. Here i_n: $S^n \subset S^3 \vee S^2 \cup_2 e^3$ is the inclusion for $n = 2, 3$.

(A.14) **Example.** We have the quotient map $q: \mathbb{R}P_2 \to \mathbb{R}P_2/S^1 = S^2$ which carries the cell e_2 to the cell s_2 of dimension 2. This yields as well the map

$$q \times 1: \mathbb{R}P_2 \times \mathbb{R}P_2 \to S^2 \times \mathbb{R}P_2$$

which carries some cells homeomorphically to cells. The quadratic chain complex $\sigma(S^2 \times \mathbb{R}P_2)$ thus can be obtained by reducing the formulas in (A.12). This yields the boundaries

$$ds_2 = 0$$

$$de_2 = e + e$$

$$ds_3 = -s_2^e + s_2$$

$$ds_4 = s_3 + s_3^e + \omega(s_2 \otimes e_2 + e_2 \otimes s_2)$$

where $e = * \times e_1, e_2 = * \times e_2, s_2 = s_2 \times *, s_3 = s_2 \times e$ and $s_4 = s_2 \times e_2$ are the cells of $S^2 \times \mathbb{R}P_2$. The map $\sigma(q \times 1)$ carries a cell in $\sigma(\mathbb{R}P_2 \times \mathbb{R}P_2)$ to 0 or to the corresponding cell in $\sigma(S^2 \times \mathbb{R}P_2)$. We know that the universal covering of $Y = S^2 \times \mathbb{R}P_2$ is $\hat{Y} = S^2 \times S^2$ so that $b_4: H_4\hat{Y} = \mathbb{Z} \to \Gamma(\pi_2 Y)$ is given by the Whitehead product $b_4(1) = [u, v]$ where u, v are generators of $\pi_2 Y = \mathbb{Z} \oplus \mathbb{Z}$. We used this fact in example (III.A.5). Here we want to compute b_4 by the method in (7.4) and in the proof of (3.7). The group $H_4\hat{Y} = \mathbb{Z}$ is generated by the homology class of $s_4 - s_4^e = x$. Moreover $\pi_2 Y = H_2\hat{Y} = \mathbb{Z} \oplus \mathbb{Z}$ is generated by the homology classes $\{s_2\} = u$ and $\{e_2 - e_2^e\} = v$. We now have to find an element $\beta \in \Gamma(K)$, $K = \ker(C_2 \to C_1)$, with $dx = \omega\tau\beta$. Then

$$b_4(x) = \Gamma(p)(\beta)$$

where $p: K \twoheadrightarrow \pi_2$ is the quotient map. Using the formulas above we get for $\xi = s_2 \otimes e_2 + e_2 \otimes s_2 = \tau[s_2, e_2]$

$$dx = ds_4 - (ds_4)^e = s_3 + s_3^e - s_3^{e+e} - s_3^e + \omega(\xi - \xi^e)$$

Here we use (1.10)(3) so that

$$s^{e+e} = s_3^{de_2} = s_3 + \omega((ds_3) \otimes e_2 + e_2 \otimes ds_3) = s_3 + \omega\tau(y) \text{ with } y = [ds_3, e_2]$$

These formulas yield by (1.10)(4) the equations

$$dx = \omega\tau\beta = (-s_3, -s_3^e) + \omega(\xi - \xi^e + \tau y) = \omega(ds_3 \otimes ds_3^e) + \omega(\xi - \xi^e + \tau y)$$

where $ds_3^e = -ds_3$ in C_2 so that β can be chosen to be

$$\begin{aligned}\beta &= -\gamma(ds_3) + [s_2, e_2] - [s_2, e_2] + [ds_3, e_2] \\ &= -\gamma(ds_3) + [ds_3, e_2] + [s_2, e_2] + [ds_3, e_2^e] - [s_2, e_2^e] \\ &= [s_2, e_2 - e_2^e] + z\end{aligned}$$

Since $p\,ds_3 = 0$ we get $\Gamma(p)(z) = 0$ for $p\colon K \to \pi_2$. This shows

$$b_4(x) = \Gamma(p)(\beta) = [s_2, e_2 - e_2^e] = [u, v].$$

Hence the method in (3.7) yields indeed the element $b_4(x)$ which we already know by the geometric argument $\hat{Y} = S^2 \times S^2$.

(A.15) **Example.** We consider the **connected sum** $X \# Y$ of two n-dimensional Poincaré complexes. We assume that X, resp. Y, are CW-complexes with a single n-cell e_n^X, resp. e_n^Y, and with an attaching map $f_X\colon S^{n-1} \to X^{n-1}$, resp. f_Y. Then $X \# Y$ is given by

$$X \# Y = (X^{n-1} \vee Y^{n-1}) \cup e_n \tag{1}$$

where $X^{n-1} \vee Y^{n-1}$ is the one point union. The attaching map of the n-cell e_n is the sum $i_1 f_X - i_2 f_Y$ where i_1, resp. i_2 is the inclusion of X^{n-1}, resp. Y^{n-1}, into the one point union $X^{n-1} \vee Y^{n-1}$. Let $j\colon X^{n-1} \vee Y^{n-1} \subset X \# Y$ be the inclusion. Then we get the canonical element

$$\varphi = ji_1 f_X = ji_2 f_Y \in \pi_{n-1}(X \# Y). \tag{2}$$

The quadratic chain complex $\sigma(X \# Y)$ is simply given by

$$\left.\begin{aligned}&\sigma(X \# Y)^{n-1} = (\sigma X)^{n-1} \vee (\sigma Y)^{n-1} \\ &d(e_n) = i_1(de_n^X) - i_2 d(e_n^Y)\end{aligned}\right\} \tag{3}$$

This follows from (6.7). Here the boundaries $d(e_n^X)$ and $d(e_n^Y)$ are given by $\sigma(X)$ and $\sigma(Y)$ respectively. The canonical element (2) corresponds to

$$\{ji_1 de_n^X\} = \{ji_1 de_n^Y\} \in \pi_{n-1}\sigma(X \# Y). \tag{4}$$

(A.16) **Example.** Recall that $\mathrm{Aut}(X)^*$ is the **group of homotopy equivalences** of X in $\mathbf{Top}^*/\simeq$. For 3-dimensional closed manifolds X this group was studied by Hendricks. We here use quadratic chain complexes for the computation of such a group. We consider the connected sum $X = \mathbb{R}P_3 \# \mathbb{R}P_3$ of two real projective 3-spaces, see (A.15). This is one of the four Poincaré 3-complexes X for which $\pi_1(X)$ has 2 ends, see theorem 4.4 in Wall(P).

Theorem. $\mathrm{Aut}(\mathbb{R}P_3 \# \mathbb{R}P_3)^* = \mathbb{Z}/2$.

The generator is clearly the interchange map for the connected sum $\mathbb{R}P_3 \# \mathbb{R}P_3$. We have chosen this example since, a priori, the group $\mathrm{Aut}(X)^*$ for $X = \mathbb{R}P_3 \# \mathbb{R}P_3$ looks more complicated. In fact, the universal cover $\hat{X}$ is homotopy equivalent to the 2-sphere S^2 and the generator of

$$\pi_2 = \pi_2 X = H_2 \hat{X} \cong \mathbb{Z} \tag{1}$$

is the canonical element $\varphi\colon S^2 \to X$ in (A.15)(2). We now have the following geometric problem. Let 1 be the identity of X and let

$$1 + \varphi\eta\colon X \xrightarrow{\mu} X \vee S^3 \xrightarrow{(1,\varphi\eta)} X \tag{2}$$

be given by the coaction for the mapping cone (A.15)(1) and by the Hopf map η. The map $1 + \varphi\eta$ is a homotopy self equivalence of the space X and there is the problem to decide whether $1 + \varphi\eta$ is homotopic to the identity 1 or not. In fact, we have by (8.12) the exact sequence of groups

$$\begin{array}{ccccccc} \hat{H}^1(X,*;\pi_2) & \xrightarrow{d_2} & \hat{H}^3(X,\Gamma\pi_2) & \xrightarrow{1^+} & \mathrm{Aut}(X)^* & \twoheadrightarrow & \mathrm{Aut}(\hat{C}_* X) \\ & & \wr\| & & & & \\ & & \mathbb{Z} & & & & \end{array} \tag{3}$$

where 1^+ carries the generator to $1 + \varphi\eta$. The fundamental group of X is the free product of groups

$$\pi_1 X = \pi_1 \mathbb{R}P_2 * \pi_1 \mathbb{R}P_2 = \mathbb{Z}/2 * \mathbb{Z}/2 \tag{4}$$

This group acts on π_2 in (1) non trivially by

$$\varphi^{\{e\}} = \varphi^{\{e'\}} = -\varphi \tag{5}$$

where $\{e\} = i_1\xi$, $\{e'\} = i_2\xi$ for $\xi \in \pi_1 \mathbb{R}P_2 = \mathbb{Z}/2$, $\xi \neq 0$. However the action on $\Gamma(\pi_2) = \mathbb{Z}$ is trivial since $\Gamma(-1) = 1$. Therefore we have in (3)

$$\hat{H}^3(X,\Gamma\pi_2) = H^3(X,\mathbb{Z}) = \mathbb{Z} \tag{6}$$

since X is an oriented manifold. We now prove that 1^+ in (3) is trivial or equivalently that the element (2) satisfies

$$1 + \varphi\eta \simeq 1. \tag{7}$$

This shows by (3) that $\mathrm{Aut}(X)^* = \mathrm{Aut}(\hat{C}_* X)$. Here $\mathrm{Aut}(\hat{C}_* X)$ is the group of homotopy equivalences of the chain complex $\hat{C}_* X$ in the category $\mathbf{H}_1/\simeq$. One can check that the canonical homomorphism

$$\mathrm{Aut}(\hat{C}_* X) \xrightarrow{\cong} \mathrm{Aut}(\pi_1 X) = \mathbb{Z}/2$$

is an isomorphism, see Weick. This proves the theorem above. For the proof of (7) we use the isomorphism σ of (8.12) and we use (8.3) for the computation of $\sigma(1 + \varphi\eta)$. The cells of $X = \mathbb{R}P_3 \# \mathbb{R}P_3$ are e, e', e_2, e_2' and e_3, see (A.8), and $\sigma(X)$ is given via (A.15)(3) by

$$\left.\begin{aligned} de_2 &= e + e \\ de_2' &= e' + e' \\ de_3 &= (e_2 - e_2^e) - (e_2' - e_2'^{e'}). \end{aligned}\right\} \tag{8}$$

Here the element

$$\varphi = \{e_2 - e_2^e\} = \{e_2' - e_2'^{e'}\} \in H_2 C, \quad C = \hat{C}_* X, \tag{9}$$

represents the generator φ in (1), (this immediately implies (5) since $2\xi = 0$). see also (A.15)(4). The map $1 + \varphi\eta$ corresponds to the map

$$g = 1 + \varphi\eta \colon \sigma X \to \sigma X \tag{10}$$

which is the identity in degree ≤ 2 and which satisfies

$$\left.\begin{aligned} g(e_3) &= e_3 + \omega(\bar{\eta}) \quad \text{with} \\ \bar{\eta} &= \{e_2 - e_2^e\} \otimes \{e_2 - e_2^e\} \end{aligned}\right\} \tag{11}$$

in degree 3, see (8.3)(1). We define a homotopy $\alpha\colon 1 \simeq g$ in the sense of (4.1) by the formulas

$$\left.\begin{aligned} \alpha_1(e) &= -e_2^e + e_2 \\ \alpha_1(e') &= 0 \\ \alpha_2(e_2) &= \omega(e_2 \otimes e_2) \\ \alpha_2(e_2') &= 0. \end{aligned}\right\} \tag{12}$$

We have to check that $-1 + g = d\alpha + \alpha d$ is satisfied, or equivalently that

$$0 = d\alpha_1(e)$$

$$0 = d\alpha_1(e')$$

$$0 = d\alpha_2(e_2) + \alpha_1 d(e_2) \qquad (13)$$

$$0 = d\alpha_2(e_2') + \alpha_1 d(e_2')$$

$$\omega(\bar{\eta}) = \alpha_2 d(e_3), \quad \text{see (11).} \qquad (14)$$

We only have to show (13) and (14), the other equations are obvious. For (13) we have

$$\begin{aligned} d\alpha_2(e_2) &= w(e_2 \otimes e_2) \\ &= \langle e_2, e_2 \rangle \\ &= -e_2 + e_2^{e+e} \\ &= -(-e_2^{e+e} + e_2^e - e_2^e + e_2) \\ &= -(\alpha_1(e)^e + \alpha_1(e)) \\ &= -\alpha_1(e+e) = -\alpha_1 de_2 \end{aligned} \qquad (15)$$

Moreover we get (14) by using the rules in (4.1):

$$\alpha_2 d(e_3) = \alpha_2(-(e_2^e - e_2) + (e_2'^{e'} - e_2')) \qquad (16)$$

$$= \alpha_2(e_2) - \alpha_2(e_2)^e + \alpha_2(e_2')^{e'} - \alpha_2(e_2') \qquad (17)$$

$$+ \begin{cases} -\Delta_2(-e_2, e_2)^e + \Delta_2(e_2, -e_2^e) \\ +\nabla(-e_2, e) + \nabla(e_2', e') \\ -\Delta_2(-e_2', e_2') + \Delta_2(e_2'^{e'}, -e_2') \end{cases} \qquad (18)$$

In (18) only $\nabla(-e_2, e)$ is non trivial since $\{\alpha_1 de_2\} = 0$. Hence we get $\alpha_2 d(e_3) = \omega(\xi)$ with

$$\xi = e_2 \otimes e_2 - (e_2 \otimes e_2)^e + (-e_2)^e \otimes (\alpha_1 e) + (\alpha_1 e) \otimes (-e_2)^e$$

$$= e_2 \otimes e_2 - e_2^e \otimes e_2^e - e_2^e \otimes e_2 + e_2^e \otimes e_2^e + e_2^e \otimes e_2^e - e_2 \otimes e_2^e \quad (19)$$

$$= \bar{\eta}, \quad \text{see (11).}$$

This completes the proof that α in (12) is a well defined homotopy $1 \simeq g$. Hence also (7) is proved.

(A.17) **Example.** We consider **homotopy groups of function spaces**. Let U^X be the space of all maps $X \to U$ with the compact open topology. For a map $u: X \to U$ and for a subspace $Y \subset X$ let

$$(U^{X|Y}, u) \subset U^X \quad (1)$$

be the subspace consisting of all functions $f: X \to U$ with $f|Y = u|Y$; the basepoint of this subspace is u. In II.§ 10 of Baues (AH) we describe an exact sequence which is valuable for the computation of the homotopy groups

$$\pi_n(U^{X|Y}, u) = [\Sigma_Y^n X, U]^u \quad (2)$$

of the function space (1). Here we assume that the inclusion $Y \subset X$ is a cofibration in **Top**. The space $\Sigma_Y^n X$ is given by the push out diagram

$$\begin{array}{ccc} S^n \times X & \longrightarrow & \Sigma_Y^n X \\ \uparrow & & \uparrow \\ S^n \times Y & \xrightarrow{p} & Y \end{array} \quad (3)$$

where p is the projection. For $Y = \phi$ this is the product $\Sigma_\phi^n X = S^n \times X$. If X is in **CW** and if $Y \subset X$ is a subcomplex then $\Sigma_Y^n X$ is again in **CW** for $n \geq 1$ or $n = 0$ and $* \in Y$. Moreover assume that $u: X \to U$ is a cellular map in **CW** which corresponds to a map $\sigma(u): \sigma(X) \to \sigma(U)$ between quadratic chain complexes. Then we get a homomorphism

$$\sigma: \pi_n(U^{X|Y}, u) \to [\Sigma_{\sigma Y}^n \sigma X, \sigma U]^{\sigma(u)} \quad (4)$$

which is an isomorphism if $n + \dim X \leq 3$, see (8.2). The quadratic chain complex

$$\sigma(\Sigma_Y^n X) = \Sigma_{\sigma Y}^n \sigma X \quad (5)$$

is obtained as a push out

$$\begin{array}{ccc}
\sigma(S^n)\otimes\sigma X & \longrightarrow & \Sigma^n_{\sigma Y}\sigma X \\
\uparrow & & \uparrow \\
\sigma(S^n)\otimes\sigma Y & \xrightarrow[p]{} & \sigma Y
\end{array} \tag{6}$$

in the category **Q**, $n \geq 1$. The homomorphism (4) can be used for explicit computations in case $\dim(X) \leq 2$ and $n = 1$. For example let $X = P_2 = \mathbb{R}P_2$ be the real projective plane and $Y = \phi$. Then we get

$$\pi_1(\mathbb{R}P_2^{\mathbb{R}P_2}, 1) = [\sigma(S^1)\otimes\sigma, \sigma]^{\sigma} = \mathbb{Z}/4 \tag{7}$$

where $\sigma = \sigma\mathbb{R}P_2$. Using the methods in (II.§10) of Baues (AH) one can get the short and exact sequence

$$\begin{array}{ccccc}
\pi_2(X,*) & \overset{\partial}{\rightarrowtail} & \pi_1(X^{X|*},1) & \twoheadrightarrow & \pi_1(X^X,1) \\
\| & & \wr\| & & \\
\mathbb{Z} & & \mathbb{Z}\oplus\mathbb{Z}/2 & &
\end{array} \tag{8}$$

But for the computation of the boundary map ∂ the formula in (7) is needed. The calculations which show that the group (7) is actually $\mathbb{Z}/4$ are worked out by Hohmann. We point out that the methods in Baues (AH) readily yield the isomorphisms

$$\pi_2(X^X,1) = \pi_2(X^{X|*},1) = \pi_2(X^{X|S^1},1) = \mathbb{Z}/2 \tag{9}$$

for $X = \mathbb{R}P_2$.

(A.18) **Example.** We define the **James construction** $J(\sigma)$ of a totally free quadratic chain complex σ similarly as in (III.C.1). For this we define first

$$J_n(\sigma) = \sigma^{\otimes n}/\sim \tag{1}$$

by the following equivalence relation on the n-fold tensor product $\sigma^{\otimes n}$ with $\sigma^{\otimes 1} = \sigma$ and $\sigma^{\otimes n} = \sigma^{\otimes(n-1)}\otimes\sigma$ for $n \geq 2$. The generators of $\sigma^{\otimes n}$ are the elements $e_1 \times \cdots \times e_n$ where e_i $(i = 1, \ldots, n)$ is a generator in σ or where $e_i = *$. The equivalence relation in (1) is given by identifying

$$e_1 \times \cdots \times e_{n-1} \times * \sim e_1 \times \cdots \times e_{i-1} \times * \times e_i \times \cdots \times e_{n-1}. \tag{2}$$

We have the canonical maps

$$\sigma = J_1(\sigma) \to J_2\sigma \to \cdots \to J_n\sigma \tag{3}$$

with $J(\sigma) = \lim J_n\sigma$. The James construction $J(\sigma)$ is again a totally free quadratic chain complex. The basis of $J(\sigma)$ is just the free monoid generated by the basis elements in σ. We clearly have for the functor λ: $\mathbf{Q} \to \mathbf{H}$ the equation

$$\lambda J(\sigma) = J(\lambda(\sigma)) \tag{4}$$

where the crossed chain complex $J(\lambda(\sigma))$ is defined in (III.C.4). Moreover J and J_n yield functors

$$J, J_n: \mathbf{Q}/\overset{0}{\simeq} \to \mathbf{Q}/\overset{0}{\simeq}, \tag{5}$$

compare (12.12). The James construction J, however, is not a functor on $\mathbf{Q}$ but a pseudo functor in the sense of Fantham Moore; it would be of interest to describe explicitely the 0-homotopies for this pseudo functor. With the same arguments as in (III.C.6) we get

Theorem. *For a* CW-*complex X in* **CW** *one has an isomorphism*

$$\sigma(JX) \cong J\sigma(X) \tag{6}$$

in **Q** *which is natural in* $\mathbf{Q}/\overset{0}{\simeq}$ *with respect to maps in* **CW**.

Corollary. *Let X, Y be* CW-*complexes in* **CW** *with* $\dim(X) \leq 3$. *Then one has the binatural isomorphism of groups*

$$[\Sigma X, \Sigma Y] \cong [X, JY] = [\sigma X, J\sigma Y]. \tag{7}$$

Compare (III.C.7) and (8.2). The group structure on the right hand group is given by the map

$$J\sigma Y \otimes J\sigma Y \cong \sigma(JY \times JY) \xrightarrow{\sigma(\mu)} \sigma(JY) = J\sigma Y \tag{8}$$

where μ is the multiplication in (III.C.2)(5). This map can be computed by the explicit associativity isomorphism in (12.13). For example M. Hennes used (7) for the computation of

$$\pi_4 \Sigma K(\mathbb{Z}/n, 1) = \begin{cases} \mathbb{Z}/n & \text{for } n \text{ odd}, \\ \mathbb{Z}/n \oplus \mathbb{Z}/2 & \text{for } n \equiv 0(4), \\ \mathbb{Z}/2n & \text{for } n \equiv 2(4). \end{cases} \tag{9}$$

Compare (III.C.12).

Appendix B

Quadratic chain complexes and simplicial groups

It is a fundamental result of Kan that a simplicial group is an algebraic model representing a connected homotopy type. Moreover a CW-complex X with $X^0 = *$ corresponds to a free simplicial group G_X generated by the cells of X. This free simplicial group in general is too big to do calculations since G_X in degree n is given by the free group generated by all $(n+1)$-cells of X and all degeneracies of k-cells of X with $k \leq n$. Already the computation of the homotopy group $\pi_3(S^2) \cong \mathbb{Z}$ of the 2-sphere S^2 in terms of the simplicial group of S^2, as achieved in Kan, is fairly complicated. Indeed, it seems to be true that whenever one has an algebraic system (like the category of simplicial groups) which yields algebraic models of all homotopy types then explicit calculations in such a system are almost impossible. For this reason we restrict in this book our attention to the category of homotopy systems of order n, $n \geq 3$, and to the problem of finding suitable algebraic models of homotopy systems which satisfy optimality conditions as described in the introductions of this chapter and of chapter III.

We have seen that crossed chain complexes and quadratic chain complexes are optimal algebraic models of homotopy systems of order 3 and order 4 respectively. The computation of the homotopy group $\pi_3(S^2)$ in terms of the quadratic chain complex of S^2 is a triviality, see Appendix A. It is a challenging problem of combinatorial homotopy to find similar optimal algebraic models of homotopy systems of order 5. In this appendix we describe connections between quadratic chain complexes and simplicial groups which might give hints for a solution of this problem.

In fact, one can deduce a quadratic chain complex from the Moore complex of a free simplicial group. This yields a functor from pointed connected spaces to quadratic chain complexes which is naturally homotopy equivalent to the singular functor σ_S in § 6. Moreover we describe a connection of quadratic chain complexes with the connectivity result in Curtis (SR) concerning the lower central series of simplicial groups. The results in this appendix are not needed in the proofs of this book. We use the same notation on simplicial sets as in the survey article 'Simplicial homotopy theory' of Curtis (SH).

Recall that a simplicial set G is called a **simplicial group** if each G_n is a group and all $d_i\colon G_n \to G_{n-1}, s_i\colon G_n \to G_{n+1}$ for $0 \leq i \leq n$ are homomorphisms. A **chain**

complex of groups (C, ∂) is a sequence of groups and homomorphisms

(B.1) $$\cdots \longrightarrow C_n \xrightarrow{\partial_n} C_{n-1} \longrightarrow \cdots$$

with image (∂_{n+1}) a normal subgroup of $\ker(\partial_n)$. For each integer n, the homology group $H_n(C, \partial)$ is defined to be the quotient group $\ker(\partial_n)/\operatorname{image}(\partial_{n+1})$. For the simplicial group G the **Moore chain complex** $(N(G), \partial)$ is defined by the intersection

(B.2) $$\begin{cases} N(G)_n = \bigcap_{i<n} \ker(d_i), \\ \partial_n = d_n | \text{restricted to } N(G)_n. \end{cases}$$

One has the natural isomorphism

$$\pi_n(G) = H_n(NG, \partial),$$

here the left hand side denotes the n-th **homotopy group of** G, compare for example 3.7 Curtis (SH). Now let X be a reduced simplicial set (that is $X_0 = *$). The **loop group** GX is the simplicial group defined by

(B.3) (1) $(GX)_n$ is the group (written additively) which has one generator $\bar{x}$ for every $x \in X_{n+1}$ and one relation $\overline{s_0 x}$ = neutral element 0 for every $x \in X_n$;

(2) the face and degeneracy operators are given by

$$d_0\bar{x} = \overline{d_1 x} - \overline{d_0 x},$$

$$d_i\bar{x} = \overline{d_{i+1} x} \quad \text{for } i > 0,$$

$$s_i\bar{x} = \overline{s_{i+1} x}.$$

Thus $G(\_)$ is a functor from reduced simplicial sets to simplicial groups. One has a (natural) homotopy equivalence

$$|GX| \simeq \Omega|X| \tag{3}$$

where $\Omega|X|$ is the loop space of the realization $|X|$. Hence the homotopy groups (B.3) satisfy

$$\pi_n(GX) = \pi_{n+1}(X) = \pi_{n+1}|X|. \tag{4}$$

The loop group GX is actually a **free** simplicial group. In addition to (B.5) Kan proved that the number of generators in a loop group G can be chosen to be small in the following sense. Let X be a CW-complex with $X^0 = *$. Then there is a free simplicial group G_X generated by the cells of $X - *$ together with a homotopy equivalence

(B.4) $$|G_X| \simeq \Omega X.$$

Here $(G_X)_n$ contains the set Z_{n+1} of $(n+1)$-cells of X, $n \geq 0$. Moreover the boundary maps of G_X satisfy $d_i(e) = 0$ for $e \in Z_{n+1}$ and $0 \leq i < n$ so that actually

(B.5) $$Z_{n+1} \subset N(G_X)_n.$$

Let G_{X^n} be the simplicial subgroup of G_X generated by all i-cells, $i \leq n$; this is a loop group for the n-skeleton X^n of X. For an $(n+1)$-cell $e \in Z_{n+1}$ the boundary $d_n e = \partial_n e$ yields the class $\{\partial_n e\} \in \pi_n(X^n)$ by the composition

(B.6) $$\begin{array}{ccccc} \partial_n e & \in & (NG_X)_{n-1} & = & (NG_{X^n})_{n-1} \\ \big\downarrow & & \big\downarrow & & \big\downarrow q \\ \{\partial_n e\} & \in & \pi_n(X^n) & \subset & \operatorname{cokernel}(\partial_n^n) \end{array}$$

Here $\partial_n^n = \partial_n$ is the boundary of NG_{X^n} and q is the quotient map, moreover, the inclusion is given via (B.2) and (B.3). Clearly, the element $\{\partial_n e\}$ is exactly the homotopy class $f_{n+1}(e)$ of the attaching map of e, see (I.3.13). We now are ready to associate with G_X the following chain complex of groups.

(B.7) **Definition.** Let $n \geq 2$. **The Moore chain complex of order** $(n+1)$ of G_X is the chain complex $M = M^{(n+1)}$ with

$$M_i = \begin{cases} C_i = \hat{C}_i X, & i \geq n+1 \\ \operatorname{cokernel} \partial_n^n, & i = n \\ N_{i-1} = N_{i-1} G_X, & i < n \end{cases}$$

The boundary maps of M are given by

$$\cdots \xrightarrow{d} C_{n+2} \xrightarrow{d} C_{n+1} \xrightarrow{f_{n+1}} \operatorname{cok} \partial_n^n \xrightarrow{\partial} N_{n-2} \xrightarrow{\partial} N_{n-3} \longrightarrow \cdots$$

where d is the boundary of the cellular chain complex of the universal cover and where ∂ is induced by the boundary in the Moore chain complex NG_X. Moreover, f_{n+1} is the attaching map of $(n+1)$-cells with $f_{n+1}(e) = \{\partial_n e\}$ as in

(B.6). We have the fundamental group

$$\pi_1 X = \text{cokernel}(\partial_1 \colon N_1 \to N_0) \tag{2}$$

where $N_0 = (G_X)_0 = \langle Z_1 \rangle$ is the free group generated by the one cells of X. The group N_0 acts via $\pi_1 X$ on C_i and acts via inner automorphisms on N_i, that is

$$x^\alpha = -(s_0^i \alpha) + x + (s_0^i \alpha) \tag{3}$$

for $\alpha \in N_0$, $x \in N_i$, (the map s_0^i denotes the i-fold composition $s_0 \dots s_0$). Clearly all boundary maps of M are equivariant with respect to the action of $(G_X)_0$.

We point out that the definition of the Moore chain complex of order $(n+1)$ is similar to the definition of a homotopy system of order $n+1$ in (II.§ 2). Similarly as in (7.5) we have the following homology of the Moore chain complex of order $n+1$.

(B.8) $$H_i M^{(n+1)} \cong \begin{cases} \pi_i X & \text{for } i \le n \\ \text{image}(\hat{h} \colon \pi_{n+1} X \to H_{n+1} \hat{X}) & \text{for } i = n+1 \\ H_i \hat{X} & \text{for } i > n+1 \end{cases}$$

This suggests that actually $M^{(3)}$ is related to $\rho(X)$ and that $M^{(4)}$ is related to the quadratic chain complex $\sigma(X)$. In fact, we deduce immediately from the definitions the

(B.9) **Lemma.** *The Moore chain complex of order* 3 *of* G_X *is naturally isomorphic to the crossed chain complex* $\rho(X)$, *see* (III.§ 3).

Next we derive the quadratic chain complex $\sigma(X)$ from the Moore chain complex of order 4 of G_X. For this, however, we need the following extra structure of the Moore chain complex of order 4. Let $N = N(G_X)$ be the Moore chain complex of G_X. We observe that the action of N_0 in (B.7)(3) gives us the totally free pre-crossed module $\partial = \partial_1 \colon N_1 \to N_0$. We consider the following diagram of functions in which the bottom row is the Moore chain complex of order 4 of G_X.

(B.10)
$$\begin{array}{ccccccccc} & & & & N_1 \times N_1 & & & & \\ & & & \omega \swarrow & & \searrow w & & & \\ \cdots \xrightarrow{d} & C_4 & \xrightarrow{f_4} & \text{cok}\,\partial_3^3 & & \xrightarrow{\partial_2} & N_1 & \xrightarrow{\partial_1} & N_0 \end{array}$$

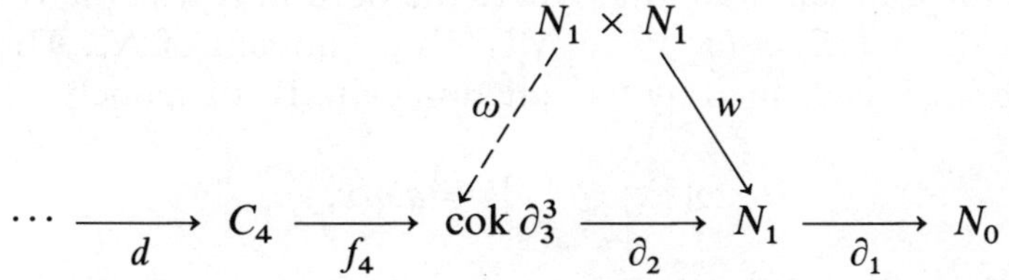

The function w carries a pair (x, y) to the **Peiffer commutator**

$$w(x, y) = \langle x, y \rangle = -x - y + x + y^{\partial x} \in N_1. \tag{1}$$

It was observed by D. Conduché that the function w has actually a canonical lift ω such that diagram (B.10) commutes. The function ω is defined by

$$\omega(x, y) = \{s_1(-x - y + x) - s_0 x + s_1 y + s_0 x\} \tag{2}$$

for $x, y \in N_1$; here the right hand side denotes a coset in $\operatorname{cok}(\partial_3^3)$ represented by an element in N_2. We call the element $\omega(x, y) = \langle x, y \rangle \in \operatorname{cok}(\partial_3^3)$ the **formal Peiffer commutator** of x, y. One readily checks (by the simplicial identities) that $\omega(x, y)$ is well defined and that diagram (B.10) commutes. We derive from diagram (B.10) a quadratic chain complex σ as follows. Recall that $P_3(\partial)$ is the subgroup of N_1 generated by triple brackets

$$\langle\langle x, y\rangle, z\rangle \quad \text{and} \quad \langle x, \langle y, z\rangle\rangle \tag{3}$$

with x, y, $z \in N_1$. Moreover, let $P_3'(\partial)$ be the subgroup of $\operatorname{cok}(\partial_3^3)$ generated by formal triple brackets as in (3). Then we have the quotient groups

$$\left.\begin{aligned} \sigma_2 &= N_1/P_3(\partial), \\ \sigma_3 &= \operatorname{cok}(\partial_3^3)/P_3'(\partial) \end{aligned}\right\} \tag{4}$$

The boundary maps of σ are given by the commutative diagram

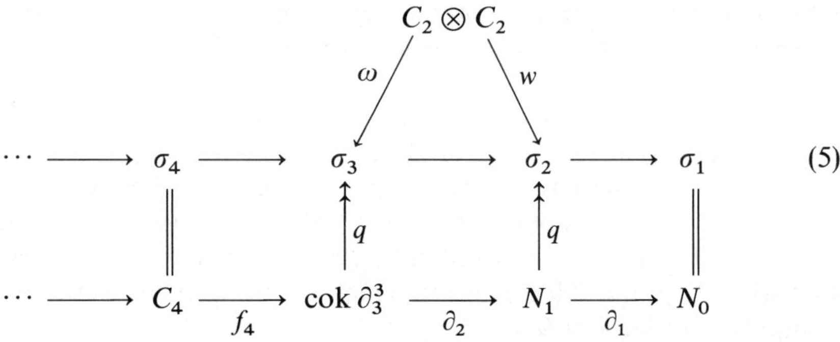

$$\begin{array}{ccccccccc} & & & & C_2 \otimes C_2 & & & & \\ & & & \omega \swarrow & & \searrow w & & & \\ \cdots \longrightarrow & \sigma_4 & \longrightarrow & \sigma_3 & \longrightarrow & \sigma_2 & \longrightarrow & \sigma_1 & \\ & \| & & \uparrow q & & \uparrow q & & \| & \\ \cdots \longrightarrow & C_4 & \xrightarrow{f_4} & \operatorname{cok} \partial_3^3 & \xrightarrow{\partial_2} & N_1 & \xrightarrow{\partial_1} & N_0 & \end{array} \tag{5}$$

where q is the quotient map and where the bottom row is the row in (B.10) above. Recall that $C_2 = (\sigma_2^{cr})^{ab} = (N_1^{cr})^{ab}$ is a quotient of N_1. The **quadratic map** ω of σ in (5) is defined by the function ω in (B.10), namely

$$\omega(\{x\} \otimes \{y\}) = q\omega(x, y). \tag{6}$$

Using the formulas of Conduché one can check that (σ, ω) is a well defined quadratic chain complex. In fact, (σ, ω) is a totally free quadratic chain complex with basis $Z_n \subset \sigma_n$, $n \geq 1$. We call (σ, ω) the **quadratic chain complex associated** to G_X. Using the uniqueness (3.8)(B) one gets the following result:

(B.11) **Lemma.** *The quadratic chain complex σ associated to G_X above is isomorphic to the quadratic chain complex $\sigma(X)$ in* (6.8)(6). *The isomorphism is natural in* $\mathbf{Q}/\simeq$.

The lemma implies that the quotient map q in (B.10)(5) induces an isomorphism of homology groups

(B.12) $$q_*: H_i(M^{(4)}) \cong \pi_i(\sigma(x))$$

where $\pi_i(\sigma(X))$ is given by (7.5) and where $H_i(M^{(4)})$ is given by (B.8) above. In a similar way as in (B.10) we obtain as well a quadratic chain complex σGX associated to the functorial loop group GX in (B.3). This yields a functor from reduced simplicial sets to totally free quadratic chain complexes. Using the reduced singular set SX of a pointed space X, see (III.6.7), we thus get a functor

(B.13) $$\sigma GS\colon \mathbf{CW\text{-}spaces}_0^* \to \mathbf{Q}$$

which carries a space X to the quadratic chain complex σGSX. By (B.11) one gets a natural homotopy equivalence $\sigma GS(X) \simeq \sigma_S(X)$ in $\mathbf{Q}/\simeq$ where σ_S is the functor in (6.6).

The lemmas (B.9) and (B.11) suggest that there might be as well algebraic models of homotopy systems of order n, $n \geq 5$, based on extra structure of the Moore chain complex of order n. Already for $n = 5$, however, it is not so easy to see, what extra structure actually could be used and what quotients of the Moore chain complex are good? Clearly, the Moore chain complex itself is not an appropriate model since the groups appearing in $N(G_X)$ are too big.

At this point we want to mention that lemma (B.11) has a connection with the **connectivity result of Curtis** who showed that the quotient map

(B.14) $$q\colon G_X \to G_X/\Gamma_r G_X$$

induces an isomorphism of homotopy groups in all dimensions $\leq \{N + \log_2 r\}$. Here X is a CW-complex with trivial skeleton $X^{N+1} = *$, $N \geq 0$, and $\Gamma_r G_X$ is the r-th term of the lower central series, $r \geq 2$. Moreover, $\{a\}$ denotes the least integer $\geq a$. We can define the Moore chain complex of order $(n + 1)$ of the simplicial group $G_X/\Gamma_r G_X$ similarly as in (B.7). Then we get:

(B.15) **Lemma.** *Let X be a* CW-*complex with $X^1 = *$. Then the Moore chain complex order order* 4 *of $G_X/\Gamma_3 G_X$ is isomorphic to the quadratic chain complex $\sigma(X)$.*

Hence for $X^1 = *$ the quotient map q in (B.10)(5) corresponds to the quotient map q in (B.14), $r = 3$. Since q in (B.10)(5) is as well a weak equivalence for $X^1 \neq *$, see (B.12), we may conjecture that there is an appropriate extension of the connectivity result of Curtis for $X^1 \neq *$ (though the lower central series won't be the right filtration of G_X for $X^1 \neq *$).

Appendix C

Reduced and stable quadratic modules

In this appendix we consider certain quadratic modules which lead to algebraic models of $(n+2)$-dimensional CW-complexes X with $X^{n-1} = *$, $n \geq 2$. In (I.§ 8) we have already seen that the homotopy types of such CW-complexes are determined by the invariants

$$\mathscr{P}(X) = (C_*(X,*), \wp_n(X)) \quad \text{or} \quad \mathscr{B}(X) = (H_* X, b_{n+2}, \{\pi_{n+1}\})$$

given by the Pontrjagin-Steenrod square and by the Γ-sequence of Whitehead respectively. The algebraic model $\sigma(X)$ introduced below is more subtle since it gives more information on maps between CW-complexes. Moreover we shall see in appendix D that the model $\sigma(X)$ admits a generalization for CW-complexes with semi free actions while this is not the case for the models $\mathscr{P}(X)$ and $\mathscr{B}(X)$.

(C.1) **Definition.** A **reduced quadratic module** (ω, δ) is a diagram

$$M^{ab} \otimes M^{ab} \xrightarrow{\omega} L \xrightarrow{\delta} M$$

of homomorphism between groups such that (1)...(4) hold.

(1) The group M is a nil(2)-group and the quotient map $M \twoheadrightarrow M^{ab}$ to the abelianization M^{ab} of M is denoted by $x \mapsto \{x\}$.
(2) The composition $\delta\omega = w$ is the commutator map, that is

$$\delta\omega(\{x\} \otimes \{y\}) = -x - y + x + y = (x, y)$$

for $x, y \in M$.
(3) For $a \in L$, $x \in M$ we have

$$\omega(\{\delta a\} \otimes \{x\} + \{x\} \otimes \{\delta a\}) = 0$$

(4) Commutators in L satisfy the formula $(a, b \in L)$

$$(a, b) = -a - b + a + b = \omega(\{\delta a\} \otimes \{\delta b\}).$$

We say that (ω, δ) is a **stable quadratic module** if in addition

(5) $$\omega(\{x\} \otimes \{y\} + \{y\} \otimes \{x\}) = 0$$

is satisfied for $x, y \in M$. This is stronger than (3). A **map** (l, m): $(\omega, \delta: L \to M) \to (\omega', \delta': L' \to M')$ is a pair of homomorphism $l: L \to L'$, $m: M \to M'$ with $m\delta = \delta' l$ and $l\omega = \omega'(m^{ab} \otimes m^{ab})$. Let **rquad** (resp. **squad**) be the corresponding category of reduced (resp. stable) quadratic modules.

(C.2) **Remark.** A reduced quadratic module (ω, δ) is the same as a quadratic module $(L \to M \to N, \omega)$ as in (1.10) with the property that $N = 0$ is the trivial group. This is easily seen by comparing the properties in (C.1) with the corresponding properties in (1.10). Therefore we have inclusions of categories **squad** $\subset$ **rquad** $\subset$ **quad**.

We also have the **stabilization functor**

(C.3) $$\Sigma: \mathbf{rquad} \to \mathbf{squad}$$

which carries the object $(\omega, \delta: L \to M)$ to the object $(\omega_\Sigma \bar{\sigma}, \delta_\Sigma: L_\Sigma \to M)$ obtained by the central push out diagram

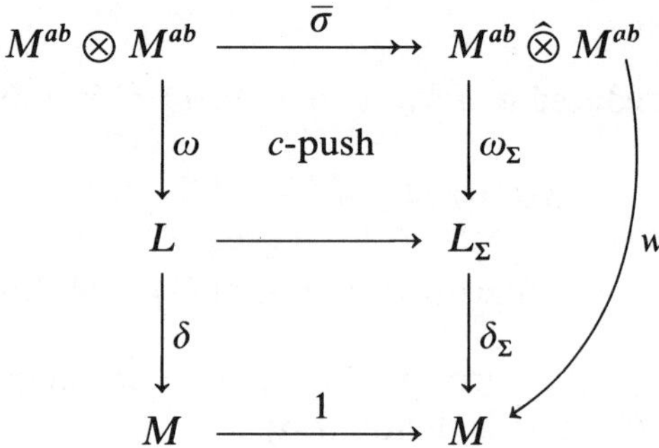

Here we use $\bar{\sigma}$ in (I.4.2), the map ω is the commutator map. One readily checks that Σ is a well defined functor.

The theory of quadratic modules and quadratic chain complexes in chapter IV above leads to a similar theory in the reduced (resp. stable) case. We now study some special features of such a reduced (resp. stable) theory. For example we define totally free objects similarly as in (2.1) as follows.

(C.4) **Definition.** Let M be a nil(2)-group and let $f: F \to M$ be a homomorphism where F is a free group. We say that the object $(\omega, \delta: L \to M)$ in **rquad** (resp. **squad**) is a free object with basis f if a homomorphism $i: F \to L$ is given such that $\delta i = f$ and such that the following universal property is satisfied. Consider any commutative diagram in the category of groups

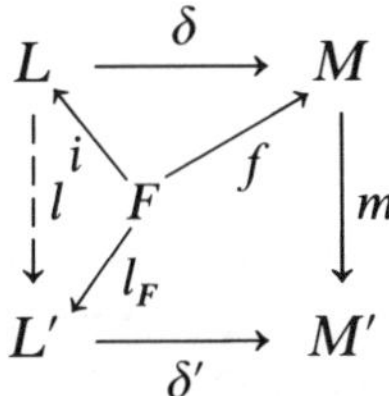

where $(\omega', \delta') \in$ **rquad** (resp. $\in$ **squad**). Then there exists a unique map (l, m): $(\omega, d) \to (\omega', \delta')$ as in (C.2) such that l extends the diagram commutatively, that is $li = l_F$. The object (ω, δ) is **totally free** if (ω, δ) is free and if M is a free nil(2)-group.

The construction of free objects as in (C.4) is simpler than the construction in (2.10). For this we consider the central push out diagram

(C.5)

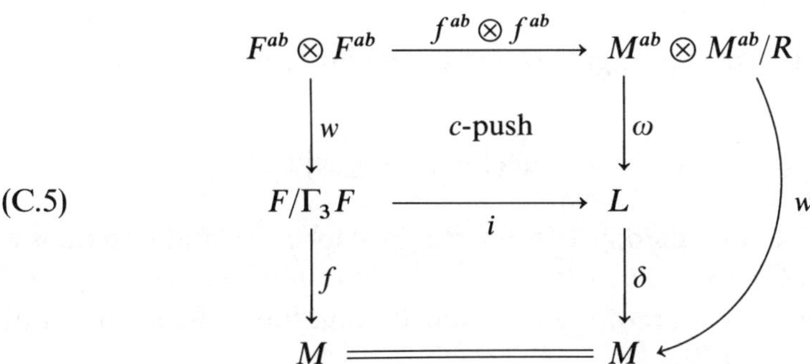

where f is induced by the basis f in (C.4) and where $R = \tau[f^{ab}F^{ab}, M^{ab}]$ is the subgroup generated by all elements $\{fa\} \otimes \{x\} + \{x\} \otimes \{fa\}$ with $x \in M$, $a \in F$. The map w denotes the commutator map.

(C.6) **Lemma.** *(ω, δ) in (C.5) is the free object in* **rquad** *with basis f and the stabilization of this object is the free object in* **squad** *with basis f.*

We point out that the free object (ω, δ) in **rquad** actually coincides with the free object $(\omega, \delta, 0)$ in **quad** where 0: $M \to 0$ is the trivial map. This follows from the universal properties. The construction of the free object $(\omega, \delta, 0)$ in (2.9), however, looks more complicated than the construction in (C.5). A totally free reduced quadratic module has a similar property as described in (2.14) (where we set $N = 0$), in particular, $M^{ab} = C$ and $C_E = \operatorname{cok}(\omega)$ are free abelian groups and δ induces d: $C_E \to C$ with image$(\omega) = C^{\otimes 2}/\tau\Delta_B, B = dC_E, K = C$.

(C.7) **Definition.** For $n \geq 2$ let $\mathbf{RQ}_n$ be the following category. Objects are **reduced quadratic chain complexes $\sigma = (\sigma, \omega)$ with bottom degree** n which are given by diagrams of groups

$$\begin{array}{ccccccc} & & & & C_n \otimes C_n & & \\ & & & & \downarrow \omega & & \\ \cdots \longrightarrow \sigma_{n+3} & \xrightarrow{d_{n+3}} & \sigma_{n+2} & \xrightarrow{d_{n+2}} & \sigma_{n+1} & \xrightarrow{d_{n+1}} & \sigma_n \end{array} \tag{1}$$

with the following properties. The pair (ω, d_{n+1}) is a reduced quadratic module for $n = 2$ and is a stable quadratic module for $n \geq 3$, in particular $C_n = \sigma_n^{ab}$. Moreover (ω, d_{n+1}) is totally free and all groups σ_i, $i \geq n+2$, are free abelian groups. The boundary maps $d = d_i$ satisfy $dd = 0$. Maps $f: \sigma \to \sigma'$ are chain maps compatible with the quadratic map ω. A **homotopy** $\alpha: f \simeq g$ is a sequence of functions $\alpha_i: \sigma_i \to \sigma'_{i+1}$ satisfying $\alpha_i = 0$ for $i < n$ and

$$-f_i + g_i = d_{i+1}\alpha_i + \alpha_{i-1}d_i, \quad i \in \mathbb{Z}. \tag{2}$$

Here α_i is a homomorphism of groups for $i > n$ and α_n is a function which satisfies

$$\alpha_n(x + y) = \alpha_n x + \alpha_n y + \omega'(\{-f_n x + g_n x\} \otimes \{f_n y\}) \tag{3}$$

for $x, y \in \sigma_n$. This is a **0-homotopy** $\alpha: f \overset{0}{\simeq} g$ if $\alpha_i = 0$ for $i > n$ and if there is a homomorphism $\beta: C_n \to C'_n \otimes C'_n$ with $\alpha_n = \omega'\beta q$; this implies $\{f_n x\} = \{g_n x\}$. One readily checks that homotopy, $\simeq$, and 0-homotopy, $\overset{0}{\simeq}$, are natural equivalence relations on $\mathbf{RQ}_n$.

Let $\mathbf{RH}_n$ be the category of chain complexes C of free abelian groups with $C_i = 0$ for $i < n$. Then we have the functor

(C.8) $$\lambda: \mathbf{RQ}_n \to \mathbf{RH}_n$$

which carries $\sigma = (\sigma, \omega)$ to the chain complex C for which the following diagram commutes

$$\begin{array}{ccccc} & & C_n \otimes C_n & & \\ & & \downarrow \omega & & \\ \cdots \longrightarrow \sigma_{n+3} & \longrightarrow & \sigma_{n+1} & \longrightarrow & \sigma_n \\ \| & & \downarrow q & & \downarrow q \\ \cdots \longrightarrow C_{n+3} & \longrightarrow & C_{n+1} & \longrightarrow & C_n \end{array}$$

Here we set $C_n = \sigma_n^{ab}$ and $C_{n+1} = \text{cokernel}(\omega)$; the maps q denote the quotient maps, compare also (3.3). The functor λ induces functors $\lambda\colon \mathbf{R}Q_n/\overset{0}{\simeq} \to \mathbf{RH}_n$ and $\lambda\colon \mathbf{RQ}_n/\simeq \to \mathbf{RH}_n/\simeq$.

The homology H_iC with $C = \lambda(\sigma)$ is the **homology** of (σ, ω) in $\mathbf{RQ}_n$. The **homotopy groups** of (σ, ω) are the abelian groups

$$\pi_i(\sigma) = \ker(d_i)/\text{im}(d_{i+1}) \tag{C.9}$$

which are the homology groups of the chain complex of groups given by σ. As in (3.7) one gets the natural exact Γ-sequence

$$0 \to \pi_{n+2}\sigma \to H_{n+2}C \to \Gamma_n^1 H_n C \to \pi_{n+1}\sigma \to H_{n+1}C \to 0 \tag{C.10}$$

where Γ_n^1 is the functor in (I.6.9) with $\Gamma_2^1 = \Gamma$ and $\Gamma_n^1 = -\otimes\mathbb{Z}/2$ for $n \geq 3$. Moreover we observe that the stabilization Σ in (C.3) yields a functor

$$\Sigma\colon \mathbf{RQ}_2 \to \mathbf{RQ}_3 \tag{C.11}$$

as follows. We set $(\Sigma\sigma)_i = \sigma_{i-1}$ for $i = 3$ and for $i \geq 5$. Moreover we obtain $((\Sigma\sigma)_4 \to (\Sigma\sigma)_3, \omega_\Sigma)$ by stabilization of $(d\colon \sigma_3 \to \sigma_2, \omega)$. Similarly one gets for $n \geq 3$ stabilization functors $\Sigma\colon \mathbf{RQ}_n \xrightarrow{\sim} \mathbf{RQ}_{n+1}$ which are actually isomorphisms of categories since they are simply given by shifting all degrees by $+1$. The Γ-sequence for $\Sigma\sigma$ in (C.10) is (up to a shift of degrees) given as a quotient of the Γ-sequence for σ via the natural transformation $\Gamma A \to A \otimes \mathbb{Z}/2$ in (I.4.2).

We now are ready to consider the topological interpretation of the algebraic categories $\mathbf{RQ}_n$. For this we introduce the following notation.

(C.12) **Definition.** For $n \geq 1$ let $R_n\mathbf{CW}$ be the full subcategory of $\mathbf{CW}$ consisting of CW-complexes X with trivial $(n-1)$-skeleton $X^{n-1} = *$. Thus n refers to the dimension of bottom spheres of **$(n-1)$-reduced** CW-complexes in $R_n\mathbf{CW}$. More generally let $\mathbf{C}$ be a category with an initial object $*$ for which each object X of $\mathbf{C}$ has a "skeletal filtration" X^i, $i \geq 1$. Then $R_n\mathbf{C}$ is the full subcategory of $\mathbf{C}$ consisting of objects X with $X^{n-1} = *$. For example we get from the categories $\mathbf{H}_m^c$ of homotopy systems of order m the subcategory $R_n\mathbf{H}_m^c$, compare (II.2.1). Or we get from the categories $\mathbf{Q}$ in (3.4) and $\mathbf{H}$ in (III.2.7) the subcategories $R_n\mathbf{Q}$ and $R_n\mathbf{H}$ respectively. One has canonical isomorphisms of categories $R_2\mathbf{Q} = \mathbf{RQ}_2$ and $R_n\mathbf{H} = \mathbf{RH}_n$ for $n \geq 2$. For $n \geq 3$, however, $R_n\mathbf{Q}$ does not coincide with $\mathbf{RQ}_n$, see (C.7).

(C.13) **Theorem.** *Let $n \geq 2$. Then one has the commutative diagram of functors*

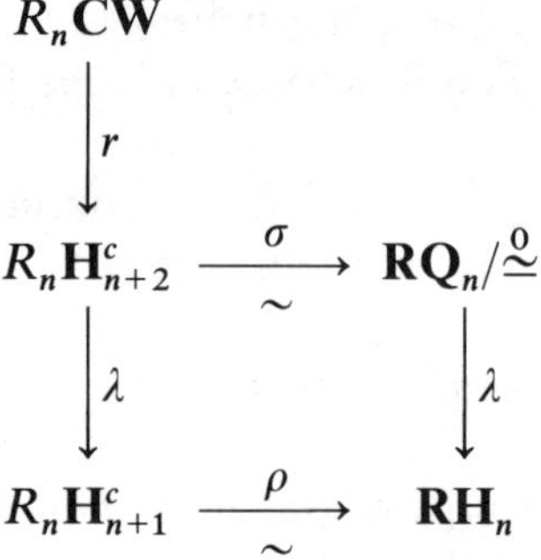

where σ and ρ are equivalences of categories. The left hand side is given by restriction of the CW*-tower of categories in* (II.3.3). *Moreover this diagram is compatible with the suspension functor Σ on each category. All functors are also compatible with homotopy relations, $\simeq$, so that one gets the diagram*

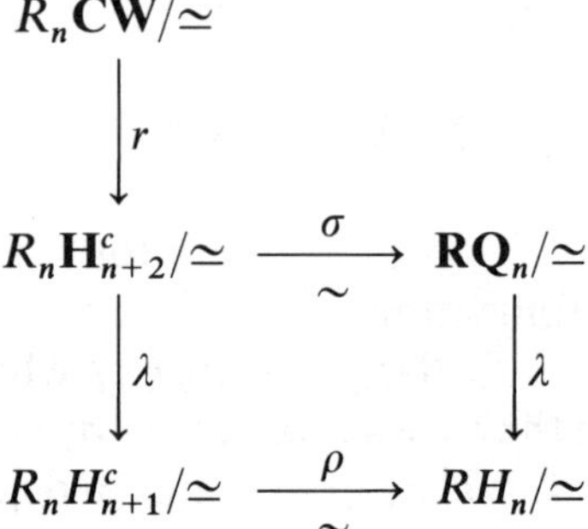

where σ and ρ are again equivalences of categories. We also write $\sigma(X) = \sigma r(X)$ and $\tilde{C}_* X = \rho\lambda r(X)$ for $X \in R_n\mathbf{CW}$.

Proof. For $n = 2$ the result is just the restriction of the corresponding result (7.7) to 1-reduced objects. For $n \geq 3$ we obtain the result by use of properties of the suspension functor Σ. □

Using (II.3.14) we get from (C.13) for example the following result on the homotopy types of A_n^2-polyhedra, (recall that A_n^2-polyhedra are $(n-1)$-connected CW-complexes of dimension $\leq n+2$, see (I.7.1)).

(C.14) **Corollary.** *Homotopy types of A_n^2-polyhedra* $(n \geq 2)$ *are in* 1-1 *correspondence with homotopy types of* $(n+2)$*-dimensional objects in* $\mathbf{RQ}_n$. *Moreover the homotopy category* $\mathbf{A}_n^1$ *is equivalent to the full subcategory of* $\mathbf{RQ}_n/\simeq$ *consisting of* $(n+1)$*-dimensional objects.*

The result on $\mathbf{A}_n^1$ gives us a further possibility to compute the homotopy category of suspended projective planes in appendix (III.D). We have seen in

(I.§ 8) that the homotopy types of A_n^2-polyhedra are as well determined by the invariants

$$(C.15)\qquad \begin{cases} \mathscr{B}(X) = (H_*X, b_{n+2}, \{\pi_{n+1}\}) \quad \text{or} \\ \mathscr{P}(X) = (\tilde{C}_*X, \wp_n(X)). \end{cases}$$

These invariants as well can be deduced from $\sigma(X)$, in fact, $\mathscr{B}(X)$ is given by the Γ-sequence of σX in (C.10) and $\mathscr{P}(X)$ is given by $\tilde{C}_*X = \lambda\sigma(X)$ and by a construction as in (V.1.8)(1) below applied to $\sigma(X)$, compare (V.4.3) below. In the next appendix we shall see that there is a major disadvantage of the invariants $\mathscr{B}(X)$ and $\mathscr{P}(X)$ since they are not suitable for complexes X with semi free actions of groups G while the model $\sigma(X)$ is available for such complexes and corollary (C.14) has a nice extension to the equivariant case.

Appendix D

On the homotopy classification of semi free group actions

Semi free group actions on CW-complexes arise naturally by the quotient spaces ($n \geq 0$)

$$Q_n(X) = (\pi_1(X), \hat{X}/\hat{X}^{n-1})$$

where $\hat{X}$ is the universal covering of X and where $\hat{X}^{n-1}$ is the $(n-1)$-skeleton of $\hat{X}$. We consider the 0-homotopy type of $Q_n(X)$ as an invariant of the 0-homotopy type of X. In particular the Pontrjagin square and the Steenrod square with local coefficients in degree n actually depend only on this invariant $Q_n(X)$. Therefore the equivariant homotopy type of $Q_n(X)$ is relevant for the problem considered in (I.5.3)(7). In this appendix we study more generally the combinatorial homotopy theory of semi free group actions on CW-complexes. It turns out that actually most results of this book have a canonical equivariant generalization for such semi free group actions.

Let **C** be a category. An **action** (from the right) of a (discrete) group G on an object X of **C** is given by a homomorphism

(D.1) $$h\colon G^{\text{op}} \to \text{Aut}_{\mathbf{C}}(X)$$

where G^{op} is the opposite group. If there is a forgetful functor from **C** to the category of sets we can describe the action also by a function

$$\bar{h}\colon X \times G \to X$$

with $\bar{h}(x,g) = h(g)(x) = x \cdot g$. Here we write the group $G = (G, \cdot, {}^{-1}, 1)$ multiplicatively so that $x \cdot 1 = x$, $(x \cdot g) \cdot g' = x \cdot (g \cdot g')$. We call $X = (G, X)$ a G-object in **C** if an action of G on X is given. Let **GrC** be the following **category of action objects** in **C**. Objects are pairs (G, X) where G is a group and where X is a G-object. Morphisms $(\varphi, f)\colon (G, X) \to (H, Y)$ are given by a homomorphism $\varphi\colon G \to H$ between groups and by a **φ-equivariant** map $f\colon X \to Y$, that is $f(x \cdot g) = f(x) \cdot \varphi(g)$, $x \in X$, $g \in G$.

A ***G*-set** is a group object (G, X) in the category of sets. A **free *G*-set** with basis Z is a G-set X for which one has an equivariant bijection $Z \times G \approx X$ where G acts on the product set $Z \times G$ by $(x, g) \cdot g' = (x, gg')$ for $x \in Z, g \in G$.

We are mainly interested in the following example.

(D.2) **Definition.** Let $\mathbf{Top}_0^*$ be the category of path connected topological spaces with basepoint $*$. An action object (G, X) in $\mathbf{Top}_0^*$ is a path connected ***G*-space** X with a basepoint $*$ which is also a fixpoint of the action of G. Let $\mathbf{Gr}\,\mathbf{Top}_0^*$ be the category of such G-spaces defined as above for $\mathbf{C} = \mathbf{Top}_0^*$. We say that a map $(\varphi, f)\colon (G, X) \to (H, Y)$ in $\mathbf{Gr}\,\mathbf{Top}_0^*$ is a **weak equivalence** if φ is an isomorphism and if f induces isomorphisms $\pi_n(f)\colon \pi_n(X) \cong \pi_n(Y)$, $n \geq 1$, between the homotopy groups of the spaces X and Y. A **spherical object** in $\mathbf{Gr}\,\mathbf{Top}_0^*$ is the one point union

$$S_G^n = \bigvee_G S^n \tag{1}$$

of n-spheres S^n where G acts by permutation of the spheres. We call $(1, i)\colon (G, X) \to (G, Y)$ an **elementary cofibration** if there exists a push out diagram

$$\begin{array}{ccc} CS_G^n & \longrightarrow & Y \\ \cup & & \cup \\ S_G^n & \xrightarrow[f]{} & X \end{array} \tag{2}$$

in the category $\mathbf{Top}_0^*$ where f is G-equivariant and where CS_G^n is the cone of S_G^n in (1) in $\mathbf{Top}^*$ with the obvious action of G. The action of G on Y is uniquely determined by the property that all maps in (2) are G-equivariant.

For a G-space X and for a homomorphism $\varphi\colon G \to H$ we obtain a morphism (φ, j) with the following universal property. Consider the diagram

$$\begin{array}{ccc} (G, X) & \xrightarrow{(\varphi, j)} & (H, \varphi_* X) \\ & {\scriptstyle (\varphi, f)} \searrow \quad \swarrow {\scriptstyle (1, \bar{f})} & \\ & (H, Y) & \end{array} \tag{3}$$

in $\mathbf{Gr}\,\mathbf{Top}_0^*$ where f is any φ equivariant map. Then there exists a unique map $(1, \bar{f})$ such that the diagram commutes where 1 denotes the identity of H. One obtains $\varphi_* X$ as a quotient

$$\varphi_* X = \left(\bigvee_H X\right) \Big/ \sim \tag{4}$$

where the equivalence relation is given by $(x, h) \sim (x \cdot g, \varphi(g^{-1})h)$ for $x \in X$, $g \in G$, $h \in H$. Here (x, h) represents an element in the one point union $\bigvee_H X =$

$(X \times H)/(* \times H)$ which is a quotient of the product $X \times H$. The map j carries x to $(x, 1)$ and $\bar{f}$ carries (x, h) to $f(x) \cdot h$. We now say that (φ, f) in (3) is a **cofibration** if $(1, \bar{f})$ in (3) is given by a well-ordered succession of elementary cofibrations as in (2), (which may be transfinite; for this we observe that limits of sequences of elementary cofibrations exist in $\mathbf{Gr\,Top}_0^*$). Next we define a cylinder

$$I(G, X) = (G, I_* X) \tag{5}$$

by use of the cylinder in (I.1.1) which has a G-action by $(t, x) \cdot g = (t, x \cdot g)$ for $t \in I$, $x \in X$, $g \in G$. Moreover we define **homotopies** $H\colon (\varphi, f) \simeq (\psi, g)$ by the condition $\varphi = \psi$ and by the cylinder (5).

(D.3) **Definition.** We say that an object (G, X) in $\mathbf{Gr\,Top}_0^*$ is a **semi free CW-complex** if X is the direct limit $X = \varinjlim X^n$ in $\mathbf{Top}^*$ of a skeletal filtration

$$* = X^0 \subset X^1 \subset X^2 \subset \cdots \subset X^n \subset X^{n+1} \subset \cdots \tag{1}$$

where $X^1 = \bigvee S_G^1$ is a one point union of 1-dimensional spherical objects. Moreover X^{n+1} is obtained from X^n ($n \geq 1$) by a push out diagram in $\mathbf{Top}^*$

$$\begin{array}{ccc} C(\bigvee S_G^n) & \longrightarrow & X^{n+1} \\ \cup & & \cup \\ \bigvee S_G^n & \xrightarrow[f]{} & X^n \end{array} \tag{2}$$

where $\bigvee S_G^n = \bigvee_Z S_G^n$ is a one point union and where f is G-equivariant. In this case we call $Z = Z_{n+1}$ the set of $\boldsymbol{G - (n+1)}$**-cells** of X and $f = f_{n+1}$ the attaching map. Let

$$\mathbf{GCW} \subset \mathbf{Gr\,Top}_0^* \tag{3}$$

be the subcategory of such semi free CW-complexes and of filtration preserving maps.

The semi free CW-complexes are special cofibrant objects in $\mathbf{Gr\,Top}_0^*$; in fact, they form a sufficiently large class of such cofibrant objects since we have the following result which seems to be new:

(D.4) **Theorem.** *The category* $\mathbf{Gr\,Top}_0^*$ *with cofibrations and weak equivalences as in* (D.2) *is a cofibration category in which all objects are fibrant. Moreover there is an equivalence of categories*

$$M\colon \mathrm{Ho}\,\mathbf{Gr\,Top}_0^* \xrightarrow{\sim} \mathbf{GCW}/\simeq$$

Compare (III.4.1) and (III.4.5). This result is the equivariant extension of a classical result on CW-models of spaces, compare (I.5.6) in Baues (AH). The functor M carries a G-space (G, X) to a cofibrant model

$$* \rightarrowtail (G, MX) \xrightarrow{\sim} (G, X)$$

where MX can be chosen to be a semi free CW-complex. The G-homotopy type of MX is well defined by X; moreover MX can be chosen with trivial n-skeleton if X is an n-connected space. We call MX the **semi free CW-model** (or CW-resolution) of the G-space X with fixpoint $*$. A proof of (D.4) can be given along the same lines as in the non equivariant case, for the existence of a factorization as in (C3) see for example (1.4.5) in Baues (OT). The existence of push outs (C2) uses push outs of groups and an inductive argument for elementary cofibrations.

We would like to emphasize that actually most results described in this book have a canonical equivariant extension for the categories in (D.4). For example there is a CW-tower of categories for the categories $\mathbf{GCW}/\overset{0}{\simeq}$ and $\mathbf{GCW}/\simeq$ as in chapter II. The category of homotopy systems of order $(n+1)$ for $\mathbf{GCW}/\overset{0}{\simeq}$, denoted by $\mathbf{GH}^c_{n+1}$, is defined similarly as in (II.2.1) by objects (C, f_{n+1}, X^n) where X^n is an n-dimensional object in **GCW** and where C is a chain complex of $\pi_1 X^n$-modules with a basis of free G-sets. The natural systems $H_n\Gamma_m$ are defined by cohomology groups given by cochains which are equivariant with respect to the actions of π_1 and of the group G. Then one has a linear extension of categories

$$\text{(D.5)} \qquad H^n\Gamma_n \longrightarrow \mathbf{GH}^c_{n+1} \xrightarrow{\lambda} \mathbf{GH}^c_n \longrightarrow H^{n+1}\Gamma_n$$

as in (II.3.3). All results in chapter II hold in a similar way for this exact sequence. (In fact (D.5) can be obtained as a special case of the results in chapter VII of Baues (AH) since Ob(**GCW**) is a very good class of complexes).

(D.6) **Definition.** Let **C** be a category of totally free objects like **H** or **Q** respectively. Then we define the full subcategory

$$\mathbf{GC} \subset \mathbf{Gr\,C}$$

of **totally free group objects** as follows. An object (G, X) in **Gr C** is an object in **GC** if a basis of X can be chosen which is a free G-set and if the action of G on X is induced by the action of G on this basis. In particular we get the categories **GH** and **GQ** by use of the categories **H** in (III.2.7) an **Q** in (3.4). We also shall use the categories $\mathbf{GRQ}_n$ and $\mathbf{GRH}_n$ given by $\mathbf{RQ}_n$ and $\mathbf{RH}_n$ in (C.7), (C.8). Homotopies in such categories **GC** are given by maps $IX \to Y$ in **Gr C** where IX is a cylinder in **C**, that is, homotopies are as well compatible with the group action.

We now are able to generalize the main result (7.7) to the equivariant case.

(D.7) **Theorem.** *One has the commutative diagram of functors*

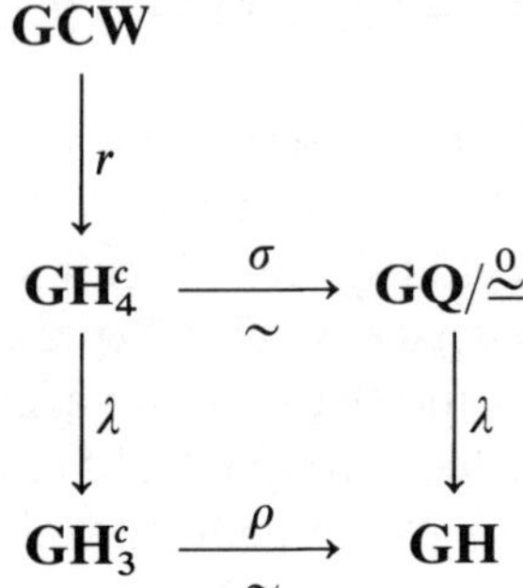

where σ and ρ are equivalences of categories. The functor $\rho = \rho\lambda r = \lambda\sigma r$ is defined in the same way as ρ in (III.2.8) *and the functor $\sigma = \sigma r$ is given by an equivariant choice of $\sigma(X)$ in* (6.8)(4). *Each functor in the diagram induces a functor between homotopy categories with respect to homotopies $\simeq$ and σ and ρ are as well equivalences of categories for these homotopy categories.*

The next result classifies homotopy types of 4-dimensional semi free group actions.

(D.8) **Corollary.** *Homotopy types of* 4*-dimensional semi free* CW*-complexes in* $\mathbf{GCW}/\simeq$ *are in* 1-1 *correspondence with homotopy types of* 4*-dimensional objects in* $\mathbf{GQ}$. *Moreover the full subcategory of* 3*-dimensional objects in* $\mathbf{GCW}/\simeq$ *is equivalent to the full subcategory of* 3*-dimensional objects in* $\mathbf{GQ}$.

A similar generalization as in (D.7) is available for theorem (C.13) concerning reduced CW-complexes with bottom degree $n \geq 2$, compare the notation in (C.12).

(D.9) **Theorem.** *Let $n \geq 2$. Then one has the commutative diagram of functors*

$$\begin{array}{ccc}
R_n\mathbf{GCW} & & \\
\downarrow{\scriptstyle r} & & \\
R_n\,\mathbf{GH}^c_{n+2} & \xrightarrow[\sim]{\sigma} & \mathbf{GRQ}_n/\overset{0}{\simeq} \\
\downarrow{\scriptstyle \lambda} & & \downarrow{\scriptstyle \lambda} \\
R_n\,\mathbf{GH}^c_{n+1} & \xrightarrow[\sim]{\rho} & \mathbf{GRH}_n
\end{array}$$

where σ and ρ are equivalences of categories. The left hand side is given by restriction of the CW*-tower to* **GCW**. *The diagram is compatible with the suspension functor Σ on each category and all functors induce functors between homotopy categories defined by homotopies ≃ and σ and ρ are as well equivalences of categories for these homotopy categories* (*as in* (C.13)).

This theorem implies the equivariant version of corollary (C.14); we leave it to the reader to formulate this corollary explicitly.

We are interested in semi free group actions since such actions arise naturally by considering quotients of the universal covering of a CW-complex. In particular we shall use the **quotient functor** ($n \geq 1$)

(D.10) $$Q = Q_n \colon \mathbf{CW} \to R_n\mathbf{GCW}$$

which carries a CW-complex X to the semi free CW-complex

$$Q_n(X) = (\pi_1(X), \hat{X}/\hat{X}^{n-1}). \tag{1}$$

Using the functor in (I.2.1) we see that Q_n is well defined. Moreover Q_n induces a functor

$$Q_n \colon \mathbf{CW}/\overset{0}{\simeq} \to R_n\mathbf{GCW}/\overset{0}{\simeq} \tag{2}$$

which is actually compatible with the CW-towers approximating these categories. The 0-homotopy type of $Q_n(X)$ thus may be considered as an invariant of the 0-homotopy type of X. We shall see that Steenrod squares with local coefficients for X and also Pontrjagin squares with local coefficients for X, see (I.5.3)(7), depend only on such quotients Q_nX. We finally describe an algebraic functor E for which the diagram

$$\begin{array}{ccc} \mathbf{CW}/\overset{0}{\simeq} & \xrightarrow{Q_2} & R_2\mathbf{GCW}/\overset{0}{\simeq} \\ \downarrow{\scriptstyle\sigma} & & \downarrow{\scriptstyle\sigma} \\ \mathbf{Q}/\overset{0}{\simeq} & \xrightarrow{E} & \mathbf{GRQ}_2/\overset{0}{\simeq} \end{array} \tag{3}$$

commutes. Here σ is given by (D.9) and by (6.8)(4) respectively. Let σ be an object in $\mathbf{Q}$ with basis $Z_n \subset \sigma_n$, $n \geq 1$. We define the free π-nil(2)-group

$$E_2 = \langle Z_2 \times \pi \rangle / \Gamma_3 \langle Z_2 \times \pi \rangle \tag{4}$$

by the free π-set $Z_2 \times \pi$ with $\pi = \pi_1(\sigma)$. There is an obvious action of π on E_2. We now consider E_2 to be the free nil(2)-module $O_E \colon E_2 \to \pi$ with trivial boundary $0_E: E_2 \to 0 \in \pi$ and with basis Z_2. Since $d_2: \sigma_2 \to \sigma_1$ is a free nil(2)-module with basis $f: Z_2 \to \sigma_1$ we obtain a commutative diagram

$$\begin{array}{ccccccccc} \cdots \longrightarrow & \sigma_4 & \xrightarrow{d_4} & \sigma_3 & \xrightarrow{d_3} & \sigma_2 & \xrightarrow{d_2} & \sigma_1 \\ & \| & & \downarrow{\scriptstyle q_3} & & \downarrow{\scriptstyle q_2} & & \downarrow{\scriptstyle q_1} \\ \cdots \longrightarrow & E_4 & \xrightarrow[\partial_4]{} & E_3 & \xrightarrow[\partial_3]{} & E_2 & \xrightarrow[0_E]{} & \pi \end{array} \tag{5}$$

where q_1 is the quotient map with $q_1 d_2 = 0$ and where q_2 is given by the unique map $(q_2, q_1): d_2 \to 0_E$ between nil(2)-modules which is the identity on the basis Z_2. Next let $(\partial_3: E_3 \to E_2, \omega_E)$ be the free reduced quadratic module with basis $Z_3 \times \pi \to E_2$ which carries (x, g) to $q_2(x) \cdot g$. Then $(\omega_E, \partial_3, 0_E)$ is actually a quadratic module so that there is a unique map q_3 as in (5) such that (q_3, q_2, q_1) is a map between quadratic modules which is the identity on C_2 so that $q_3 \omega = \omega_E$. Finally let $E_n = \sigma_n$ for $n \geq 4$. Then the bottom row in (5) gives us the object

$$E(\sigma) = (\cdots \to E_3 \to E_2, \omega_E) \tag{6}$$

in the category $\mathbf{GRQ}_2$ which defines the functor E in (3) by the naturality of diagram (5). We shall use this functor in (V.5.5)(2) and (VI.6.3)(1) below. Using the description of $\sigma(X)$ in terms of tracks as in (VI.§ 6) we see that diagram (3) commutes. The equivariant cellular boundary invariant of $Q_2(X)$ is considered in (V.6.1) below.

Chapter V

Cohomological invariants

Recall that a cohomological invariant α associates with a CW-complex X an element

$$\alpha_X \in \hat{H}^*(X, A\hat{C}_*(X))$$

where A is a functor which carries chain complexes to modules over group rings. The element α_X needs only be defined for objects X in a full subcategory $\mathbf{K}$ of $\mathbf{CW}$. Moreover α_X is natural with respect to maps $f: X \to Y$ in $\mathbf{K}$ in the sense that $f_*\alpha_X = f^*\alpha_Y$. In (I.§ 6) we have seen that cup products, the classical Pontrjagin square and the Steenrod square are determined by such cohomological invariants, see (I.6.3)(1), (I.6.5), (I.6.6).

In this chapter we deduce new cohomological invariants from the cellular boundary invariants β_X of a space X. In (II.§ 4) we have seen that the cellular boundary invariants can be used for the computation of the obstruction operator $\mathcal{O}$ of the CW-tower. Using the functor

$$\sigma: \mathbf{CW} \xrightarrow{\sim} \mathbf{Q}/\overset{0}{\simeq}$$

described in chapter IV we here give an algebraic characterization of the cellular boundary invariant β_X. In particular we describe explicitly the Postnikov chain functor, see (1.5). Using these results we obtain a new classification of all 4-dimensional homotopy types X in terms of the crossed chain complex $\rho(X)$ and a chain map β_X. It seems that this is the simplest algebraic structure which can possibly describe all 4-dimensional homotopy types, see (1.14). We compare this result with Whitehead's classification of simply connected 4-dimensional homotopy types. For this we deduce from β_X three new cohomological invariants which we call the 'strong Pontrjagin element', the 'Pontrjagin element' and the 'weak Pontrjagin element'. These elements, however, are only defined for a space X if certain obstructions (depending on $\hat{C}_*X$) vanish. In the next chapter we shall see that the 'universal obstruction' for the existence of the Pontrjagin element is a non trivial element of order 2 in the second cohomology of the category of free modules. This solves the old problem on the existence of the Pontrjagin squares with local coefficients described in (I.5.3)(7). Such Pontrjagin squares do not exist for all spaces X.

On the other hand we deduce from the cellular boundary invariant a new cohomological invariant $\beta_4 X \in \hat{H}^4(X, C_2 \otimes C_2/\nabla_B)$ which is actually defined for all connected spaces X and which yields the classical cup product element by the formula $t_{4*}\beta_4 X = i_2 \cup i_2$, see (2.6).

§ 1 The classification of 4-dimensional homotopy types

The equivalence of categories

$$\sigma\colon \mathbf{H}_4^c/\simeq \xrightarrow{\sim} \mathbf{Q}/\simeq$$

in (IV.7.7) shows that homotopy types of connected 4-dimensional CW-complexes X are in 1-1 correspondence with the homotopy types of 4-dimensional totally free quadratic chain complexes in $\mathbf{Q}/\simeq$. In this section we describe an algebraic model of the homotopy type of X which is somewhat simpler than the quadratic chain complex $\sigma(X)$. This model is a pair $(\rho(X), \beta_X)$ where $\rho(X)$ is the crossed chain complex of X and where

$$\beta_X\colon C\rho(X) \to P\rho(X)$$

is a chain map.

Here $P\rho(X)$ is given by an algebraic description of the Postnikov chain functor qP in (II.4.28) and β_X is the cellular boundary invariant of X. We show how β_X can be deduced from the quadratic chain complex $\sigma(X)$. This way we obtain an algebraic classification of 4-dimensional homotopy types by cellular boundary invariants. In the following sections we deduce from the cellular boundary invariant β_X various cohomological invariants of X. In particular we shall see how the classical Pontrjagin square is related to β_X.

We consider the following diagram of functors

$$\text{(1.1)}\qquad \begin{array}{ccc} \mathbf{H} & \xrightarrow{P} & \mathbf{Chain}_{\mathbb{Z}}^{\wedge}/\simeq_2 \\ & {\scriptstyle \rho}\nwarrow{\scriptstyle\sim} \quad \nearrow{\scriptstyle qP} & \\ & \mathbf{H}_3^c & \end{array}$$

where qP is the Postnikov chain functor, see (II.4.5) and (II.4.28), and where ρ is the equivalence of categories in (III.2.9). The main purpose of this section is the algebraic construction of a functor P in (1.1) for which qP is naturally weak equivalent to $P\rho$, see (1.5).

We obtain the functor P on $\mathbf{H}$ as follows. Let $\rho = \{\rho_n, d_n\}$ be a totally free crossed chain complex in $\mathbf{H}$ and let $C = C(\rho)$ be the chain complex of ρ

given by (III.2.11). We choose for the free crossed module $d_2: \rho_2 \to \rho_1$ a free nil(2)-module $\partial: \sigma_2 \to \rho_1$ with $\partial^{cr} = d_2$, see (III.1.11)(3). Then we get the following commutative diagram of unbroken arrows:

(1.2)

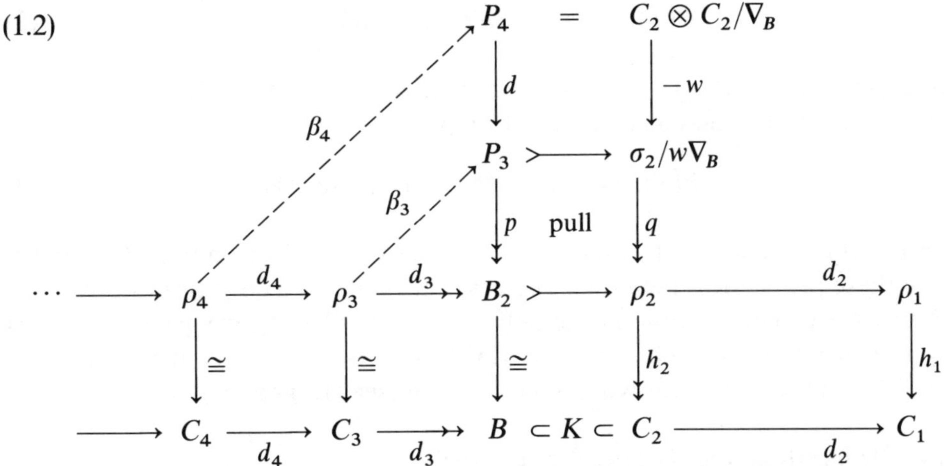

Here $B = d_3 C_3$ and $K = \text{kernel}\,(d_2: C_2 \to C_1)$ are the modules of 2-boundaries and 2-cycles respectively in C. Moreover

$$\nabla_B = B \otimes B + \tau[B, C_2] \subset C_2 \otimes C_2 \tag{1}$$

is defined by the Whitehead product subgroup $[B, C_2] \subset \Gamma(C_2)$. We have the canonical inclusion

$$\tau\Delta_B = \tau(\Gamma(B) + [B, K]) \subset \nabla_B \tag{2}$$

where Δ_B is the subgroup of $\Gamma(K)$ in (IV.2.13). The maps w and p are induced by the corresponding maps in (IV.1.8) so that

$$\text{kernel}(p) = \text{image}(w). \tag{3}$$

We define P_3 by the pull back diagram 'pull' in (1.2), equivalently $P_3 = p^{-1}(B_2)$. This group lies in the kernel of ∂. Therefore (IV.1.6) and the choice of ∇_B in (1) shows that P_3 is actually an abelian group and a π_1-module. We now are ready for the definition of the functor P in (1.1).

(1.3) **Definition.** The **Postnikov chain functor** P in (1.1) is defined for an object ρ in $\mathbf{H}$ by the chain complex $P(\rho) = \{P_i; d: P_i \to P_{i-1}\}$ with 2-skeleton $P(\rho)^2 = C(\rho)^2$ and with $P_i = 0$ for $i \geq 5$. Moreover

$$P_4 = C_2 \otimes C_2/\nabla_B \tag{1}$$

and $P_3 = p^{-1}(B_2)$ are defined as in (1.2). The boundary maps $d: P_4 \to P_3$, resp. $d: P_3 \to C_2 = P_2$ are given by $-w$, resp. $h_2 p$ in (1.2). A map $f: \rho \to \rho'$ in $\mathbf{H}$ induces a well defined map

$$P(f): P(\rho) \to P(\rho') \quad \text{in } \mathbf{Chain}_{\mathbb{Z}}^{\wedge}/\simeq_2 \tag{2}$$

as follows. We set $P(f)^2 = C(f)^2$ on 2-skeleta. Moreover, in degree 4 the map $P(f)_4$ is given by the tensor product map

$$P(f)_4 = \{C(f)_2 \otimes C(f)_2\}, \quad \text{see (1).} \tag{3}$$

Naturality of diagram (1.2) shows that there exists a chain map $P(f)$ as in (2) with these properties. Two such choices differ only by a ($\simeq_2$)-homotopy since $H_3 P(\rho') = 0$. This completes the definition of P. Clearly, we get $H_3 P(\rho) = 0$ by the exactness of the sequence (IV.1.8). Moreover, the injection τ_0 in (IV.2.13) induces the following isomorphism (see (1.7) below):

(1.4) **Proposition.** $\Theta_*: H_4(P(\rho)) \cong \Gamma(\pi_2(\rho))$.

Therefore $P(\rho)$ has the same homology groups as the Postnikov chain functor $P(X)$ with $X \in \mathbf{H}_3^c$, see (II.4.11) and (I.4.11). In fact, we shall prove the following result, see (1.9) below.

(1.5) **Theorem.** *For the functors in* (1.1) *there exists a natural weak equivalence* $\chi: qP \xrightarrow{\sim} P\rho$ *in* $\mathbf{Chain}_{\mathbb{Z}}^{\wedge}/\simeq_2$.

For the proof of (1.4) we need the next lemma.

(1.6) **Lemma.** *Let K be a free abelian group and let B be a subgroup of K. Then we have the following equality of subgroups in $K \otimes K$*

$$\tau\Gamma K \cap (B \otimes B + \tau[B, K]) = \tau\Gamma B + \tau[B, K].$$

The left hand side is the intersection of two subgroups in $K \otimes K$ where $\tau: \Gamma K \to K \otimes K$; the sum $+$ denotes the sum of subgroups, that is $U + V = \{u + v: u \in U, v \in V\}$.

Proof. For $\pi_2 = K/B$ we have the diagram with exact rows

$$\begin{array}{ccccc}
\Gamma B + [B, K] & \subset & \Gamma K & \twoheadrightarrow & \Gamma(\pi_2) \\
\| & & \uparrow & & \cup \\
\Delta_B & \subset & A & \twoheadrightarrow & \Delta(\pi_2)
\end{array} \tag{1}$$

where $A = \tau^{-1}(\tau\Gamma K \cap (B \otimes B + \tau[B, K]))$. The top row is given by (I.4.9) or (IV.2.13). We define $\Delta(\pi_2)$ by the image of A in $\Gamma(\pi_2)$. The proposition of lemma (1.6) is equivalent to the equation

$$\Delta(\pi_2) = 0. \tag{2}$$

For the proof of (2) we first observe that Δ is actually a functor from abelian groups to abelian groups: For a homomorphism $\varphi\colon \pi_2 \to \pi'_2$ we choose a map $F\colon (K, B) \to (K', B')$ between pairs inducing $\varphi\colon \pi_2 = K/B \to \pi'_2 = K'/B'$. The map F induces a map $F_*\colon A \to A'$, see (1), and hence a homomorphism $\Delta(\varphi) = F_*\colon \Delta(\pi_2) \to \Delta(\pi'_2)$. Here $\Delta(\varphi)$ does not depend on the choice of F since $\Delta(\varphi)$ is a restriction of $\Gamma(\varphi)$. Next we prove that Δ satisfies

$$\Delta(\pi_2 \oplus \pi'_2) = \Delta(\pi_2) \oplus \Delta(\pi'_2). \tag{3}$$

Here the isomorphism $\Delta(i) + \Delta(i')$ is given by the inclusions i (i') of π_2 (π'_2) into $\pi_2 \oplus \pi'_2$. For the proof of (3) we use (I.4.8). Hence it is enough to show that in $K \otimes K' \oplus K' \otimes K$ we get

$$\tau[K, K'] \cap X = Y \quad \text{where} \tag{4}$$

$$X = B \otimes B' \oplus B' \otimes B + \tau[B, K'] + \tau[B', K], \quad \text{and}$$

$$Y = \tau[B, B'] + \tau[B, K'] + \tau[B', K].$$

Recall that $[K, K'] = K \otimes K'$ by (I.4.8) and that the composition

$$[K, K'] \xrightarrow{\tau} K \otimes K' \oplus K' \otimes K \xrightarrow{pr} K \otimes K' \tag{5}$$

is an isomorphism; here pr is the projection and τ is given by

$$\tau[x, x'] = x \otimes x' + x' \otimes x, \quad (x \in K, x' \in K').$$

Therefore (4) is equivalent to

$$pr(X) = B \otimes B' + B \otimes K' + K \otimes B' = pr(Y). \tag{6}$$

This completes the proof of (3). One readily checks that for a cyclic group we have

$$\Delta(\mathbb{Z}/n) = 0 \quad (n \geq 0). \tag{7}$$

Therefore (3) shows that (2) is satisfied for all finitely generated abelian groups.

Moreover, a limit argument shows that (2) holds for arbitrary abelian groups π_2. □

(1.7) *Proof of* (1.4). We have

$$H_4P(\rho) = \text{kernel}(w\colon C_2 \otimes C_2/\nabla_B \to \sigma_2/w\nabla_B) \tag{1}$$

$$= (\tau\Gamma K + \nabla_B)/\nabla_B \tag{2}$$

Equation (2) follows from the exactness of (IV.1.8). Since $\Gamma(\pi_2) = \Gamma K/\Delta_B$ we have to show

$$\tau\Gamma K \cap \nabla_B = \tau\Delta_B \quad \text{in } C_2 \otimes C_2. \tag{3}$$

This follows from (1.6) since K, as an abelian group, is a direct summand of C_2. Here we use the fact that C_1 is a free abelian group. Now $C_2 = K \oplus R$ yields $[B, C_2] = [B, K] \oplus [B, R]$. Thus (1.6) and $\tau\Gamma K \cap [B, R] = 0$ prove (1.4). □

For the proof of theorem (1.5) we construct the **cellular boundary invariant** β_σ of a (totally free) quadratic chain complex σ. If $\sigma = \sigma(X)$ is given by an object $X \in \mathbf{H}_4^c$ via the functor σ in (IV.7.7) then the cellular boundary invariant β_σ is 'equivalent' to $q\beta_X$ in (II.4.27), see (1.7). The element

(1.8) $$\beta_\sigma \in [C, I^2, P\rho], \quad \rho = \lambda(\sigma), \quad C = C(\rho),$$

is given by a chain map $\beta = \{\beta_4, \beta_3\}$ as indicated by the broken arrows in diagram (1.2). Here we use the functor λ in (IV.3.2) which gives us the crossed chain complex $\rho = \lambda(\sigma)$. We deduce from (IV.3.3) the following commutative diagram of unbroken arrows in which the columns are exact sequences, here σ is given by $\sigma = (\omega, \partial_n\colon \sigma_n \to \sigma_{n-1}, n \geq 2)$.

$$\begin{array}{ccccccc}
 & & & & \Gamma(\pi_2) & & \\
 & & & & \downarrow & & \\
 & & C_2 \otimes C_2/\nabla_B & = & P_4 & & \\
 & \overset{-\beta_4}{\nearrow} & \downarrow \omega & & \downarrow -d & & \\
\sigma_4 & \xrightarrow{\partial_4} & \sigma_3/\omega\nabla_B & \xrightarrow{\partial_3} & P_3 & \subset & \sigma_2/\omega\nabla_B \\
\| q_4 & & s \uparrow \downarrow q_3 & \overset{\beta_3}{\nearrow} & \downarrow & & \downarrow q_2 \\
\rho_4 & \xrightarrow[d_4]{} & \rho_3 & \xrightarrow[d_3]{} & B_2 & \subset & \rho_2
\end{array} \tag{1}$$

The equations (IV.1.10)(3), (4) show that $\sigma_3/\omega\nabla_B$ is actually a π_1-module. Since $\rho_3 = C_3$ is a free π_1-module there exists a π_1-splitting s of q_3. Now we define the chain map $\{\beta_3, \beta_4\}$ representing β_σ in (1.8) by

$$\left.\begin{aligned} \beta_3 &= \partial_3 s \\ -\beta_4 &= \omega^{-1}(-\partial_4 + sd_4q_4). \end{aligned}\right\} \tag{2}$$

The $(\simeq_2)$-homotopy class β_σ of $\{\beta_3, \beta_4\}$ in (1.8) does not depend on the choice of the splitting s. One readily checks that β_σ is natural with respect to maps $f: \sigma \to \sigma'$ in $\mathbf{Q}/\overset{0}{\simeq}$, that is

$$f_*\beta_\sigma = f^*\beta_{\sigma'}. \tag{3}$$

In addition to (1.5) we now get:

(1.9) **Theorem.** *There is a natural weak equivalence* $\chi: qP \xrightarrow{\sim} P\rho$ *in* $\mathbf{Chain}_{\mathbb{Z}}^{\wedge}/\simeq_2$ *with*

$$\chi_*\beta_X = \beta_{\sigma X}$$

for $X \in \mathbf{H}_4^c$. *Here the induced map*

$$\chi_*: [C, I^2, qP(\lambda X)] \xrightarrow{\approx} [C, I^2, P\rho(\lambda X)]$$

is a bijection by (II.4.16).

Proof of (1.9). For X in $\mathbf{H}_3^c$ let P_2X^3 be given as in (II.4.5)(1). We apply the functor σ to P_2X^3. For $\sigma(P_2X^3)$ we obtain the cellular boundary as above:

$$\beta = \beta_{\sigma P_3X}: \hat{C}_*P_2X^3 \to P\rho P_2X^3 = P\rho' X \tag{1}$$

It can be deduced from (1.4), (IV.3.11) and (II.4.11) that β induces a weak equivalence

$$\chi: qP(X) = (\hat{C}_*P_3X^3)^{(4)} \xrightarrow{\sim} P\rho(X). \tag{2}$$

Moreover, by naturality in (1.8)(3) above we see that χ is a natural transformation in $\mathbf{Chain}_{\mathbb{Z}}^{\wedge}/\simeq_2$. Now let X be an object in $\mathbf{H}_4^c$ and let

$$p_3: X^5 \to P_2X^3 \tag{3}$$

be given as in (II.4.6)(1). The map p_3 induces by (1.8)(3) the following diagram which commutes in $\mathbf{Chain}_{\mathbb{Z}}^{\wedge}/\simeq_2$.

$$\begin{array}{ccccc}
\hat{C}_* P_2 X^3 & \longrightarrow & P\rho P_2 X^3 & = & P\rho'(X) \\
\uparrow \hat{C}_*(p_3) & & & & \uparrow P\rho(p_3) = 1 \\
C_* X^5 & \longrightarrow & P\rho X^5 & = & P\rho'(X).
\end{array} \tag{4}$$

Since $\hat{C}_*(p_3)$ yields β_X one gets

$$\chi_* \beta_X = \beta_{\sigma X} \tag{5}$$

by definition of χ in (2) and (4). □

Next we construct a detecting functor on the category $\mathbf{H}_4^c/\simeq \;\sim \mathbf{Q}/\simeq$ which is defined similarly as the detecting functor β in (II.4.30). As in (II.4.29) we first define the following category.

(1.10) **Definition.** We associate with the functor P in (1.1) the **category of cellular boundary invariants**, denoted by $(\mathbf{H}, P)$. Objects are pairs (ρ, β) where ρ is an object in $\mathbf{H}$ and where β is an element

$$\beta \in [C(\rho), I^2, P(\rho)] \tag{1}$$

see (II.4.15), where I^2 is the identity of $C(\rho)^2$. A morphism $(\rho, \beta) \to (\rho', \beta')$ is a map $\xi\colon \rho \to \rho'$ in $\mathbf{H}$ which satisfies

$$\beta' \xi = P(\xi) \beta \tag{2}$$

in $\mathbf{Chain}_{\mathbb{Z}}^{\wedge}/\simeq_2$. Morphisms are homotopic if they are homotopic in $\mathbf{H}$, let $(\mathbf{H}, P)/\simeq$ be the quotient category. We call an object (ρ, β) in $(\mathbf{H}, P)$ a **cellular boundary invariant**.

Using the cellular boundary invariant $\beta_{\sigma X}$ in (1.8) we obtain functors

$$\text{(1.11)} \qquad \begin{cases} \beta\sigma\colon \mathbf{H}_4^c \xrightarrow{\sim} \mathbf{Q}/\overset{0}{\simeq} \xrightarrow{\beta} (\mathbf{H}, P), \\ \beta\sigma\colon \mathbf{H}_4^c/\simeq \xrightarrow{\sim} \mathbf{Q}/\simeq \xrightarrow{\beta} (\mathbf{H}, P)/\simeq \end{cases}$$

by setting $\beta\sigma(X) = (\rho\lambda X, \beta_\sigma)$ for X in $\mathbf{H}_4^c$ and $\beta(\sigma) = (\lambda(\sigma), \beta_\sigma)$ for σ in $\mathbf{Q}$. Theorem (1.9) and (II.4.31) imply the next result.

(1.12) **Theorem.** *The functors β and $\beta\sigma$ in each row of* (1.11) *are detecting functors.*

The objects in $(\mathbf{H}, P)$ are somewhat simpler than the quadratic chain complexes in $\mathbf{Q}$. Clearly we have by (1.12) and (I.0.5):

(1.13) **Corollary.** *The homotopy types in the categories* $\mathbf{H}_4^c/\simeq$, $\mathbf{Q}/\simeq$, *and* $(\mathbf{H}, P)/\simeq$ *respectively are in* 1-1 *correspondence with each other.*

By (II.2.14)(e) we know that 4-dimensional homotopy types are determined by 4-dimensional homotopy systems of order 4. We therefore derive from (1.13):

(1.14) **Classification theorem.** *The homotopy types of connected* 4*-dimensional* CW*-complexes are in* 1-1 *correspondence with the algebraic homotopy types of* 4*-dimensional totally free quadratic chain complexes* σ *in* $\mathbf{Q}/\simeq$, *and with the homotopy types of cellular boundary invariants* (ρ, β) *in* $(\mathbf{H}, P)/\simeq$ *with* $\dim \rho \leq 4$.

This is our best result on the classification of arbitrary 4-dimensional homotopy types. In section §4 below we describe special 4-dimensional homotopy types which can be classified by cohomological invariants which generalize the classical Pontrjagin square. Arbitrary 4-dimensional homotopy types cannot be classified by such cohomological invariants.

We point out that the cellular boundary invariants yield part of Whitehead's exact sequence by (II.4.11). In particular we get for a chain map

$$\beta\colon C(\rho) \to P(\rho), \quad \rho = \rho(X), \quad X \in \mathbf{CW}$$

representing $\beta_{\sigma X}$ in (1.8) the following commutative diagram where C_β is the algebraic mapping cone of β, see (II.4.10).

(1.15)
$$\begin{array}{ccccccccc}
H_4C(\rho) & \xrightarrow{\beta_*} & H_4P(\rho) & \xrightarrow{r'} & H_4C_\beta & \xrightarrow{\partial} & H_3C(\rho) & \longrightarrow & H_3P(\rho) \\
\| & & \downarrow\cong & & \downarrow\cong & & \| & & \| \\
\hat{H}_4X & \longrightarrow & \Gamma\pi_2X & \longrightarrow & \pi_3X & \longrightarrow & \hat{H}_3X & \longrightarrow & 0
\end{array}$$

This is a commutative diagram of $\pi_1(X)$-modules with exact rows. The bottom row is described in (I.3.7). Clearly (1.15) is an important tool for the computation of π_3X as a $\pi_1(X)$-module. For the computation of $\pi_3(X)$ as a functor on X, however, one needs the isomorphism $\pi_3X = \pi_3\sigma(X)$ in (IV.7.5). As a special application of (1.15) we consider the following case where X is of dimension 3.

(1.16) **Computation of $\pi_3 X$ as a $\pi_1 X$-module for** $\dim(X) \leq 3$. In this case we have the short exact sequence of $\pi_1(X)$-modules

$$\Gamma(\pi_2 X) \rightarrowtail \pi_3 \twoheadrightarrow \hat{H}_3 X \tag{1}$$

Here $\hat{H}_3 X$ is free as an abelian group but needs not to be free as a $\pi_1(X)$-module. Therefore the exact sequence (1) need not be split. Whence the problem arises to compute $\pi_3 X$ as a $\pi_1 X$-module. This problem is solved as follows. We first compute the crossed chain complex $\rho = \rho(X)$ and then we choose any chain map $\beta: C(\rho) \to P(\rho)$ extending the identity in degree ≤ 2. That is we choose any lift β_3 as in diagram (1.2). Then the algebraic mapping cone of β yields the isomorphism

$$\pi_3 X \cong H_4 C_\beta \tag{2}$$

of $\pi_1(X)$-modules as in (1.15).

§ 2 A new cohomological invariant and the cup product

Recall that a cohomological invariant of a space X is a natural element $\alpha_X \in \hat{H}^*(X, A\hat{C}_* X)$ where A is a functor which carries a chain complex (π, C) to a π-module, see (I.6.1). For example the cup product

$$i_2 \cup i_2 \in \hat{H}^4(X, A\hat{C}_* X) \quad \text{with} \quad A(C) = (C_2/dC_3) \otimes_{\mathbb{Z}} (C_2/dC_3)$$

is a cohomological invariant where i_2 is represented by the cocycle $C_2 \to C_2/dC_3$, $C = \hat{C}_* X$, see (I.6.3). In this section we deduce from the cellular boundary invariant (1.8) a new cohomological invariant

$$\beta_4(X) \in \hat{H}^4(X, A'\hat{C}_* X) \quad \text{with} \quad A'(C) = (C_2 \otimes C_2)/\nabla_B.$$

This invariant is non trivial since there is a natural transformation $t_4: A' \to A$ which carries $\beta_4(X)$ to the cup product $i_2 \cup i_2$.

Let X be a CW-complex in **CW** and let $\beta_{\sigma X}$ be the cellular boundary invariant of X. We obtain $\beta_{\sigma X}$ by the quadratic chain complex $\sigma = \sigma X$ as in (1.8). The map

$$-\beta_4 = \omega^{-1}(-\partial_4 + sd_4 q_4): C_4 \to C_2 \otimes C_2/\nabla_B$$

with $\beta_{\sigma X} = \{\beta_4, \beta_3\}$, see (1.8)(1), is a cocycle since β_4 is part of a chain map.

Hence β_4 represents an element

(2.1) $\beta_4(X) = \{\beta_4\} \in \hat{H}^4(X, A'\hat{C}_* X)$ with $A'(C) = C_2 \otimes C_2/\nabla_B$.

(2.2) **Theorem.** *The element $\beta_4(X)$ is a cohomological invariant on* **CW**/$\simeq$ *in the sense of* (I.6.1).

We have to check naturality of the element (2.1). This is clear since the cellular boundary invariant is natural in the sense of (1.10)(2). We now consider the kernel $A'[X]$ of $(i_0)_* - (i_1)_*$ in (I.6.2). For this we introduce the map

(2.3) $\delta = \{1 \otimes d_2 + d_2 \otimes 1\}\colon C_2 \otimes C_2/\nabla_B \to (C_2 \otimes C_1 \oplus C_1 \otimes C_2)/[B, C_1]$

where $B = d_3 C_3$ and where $d_2\colon C_2 \to C_1$ is given by the chain complex C. Moreover, $[B, C_1]$ denotes the submodule of all elements $(b \otimes x, x \otimes b) \in C_2 \otimes C_1 \oplus C_1 \otimes C_2$ with $b \in B$, $x \in C_1$. The proof of the next result is similar to the one in (I.6.4).

(2.4) **Proposition.** *The* kernel $A'[X]$ *of* $(i_0)_* - (i_1)$ *coincides with the kernel of the map* $\delta_* = H_4(X, \delta)$ *where* δ *is defined in* (2.3). *In particular, the element* $\beta_4 X$ *satisfies* $\delta_*(\beta_4 X) = 0$.

Next we consider the canonical quotient map

(2.5) $$t_4\colon C_2 \otimes C_2/\nabla_B \to C_2/B \otimes C_2/B$$

which is induced by the identity on $C_2 \otimes C_2$ and which is a natural transformation $t_4\colon A' \to A$. The following result gives us the connection of the invariant $\beta_4(X)$ with the cup product element $i_2 \cup i_2$.

(2.6) **Theorem.** $$(t_4)_* \beta_4(X) = i_2 \cup i_2.$$

Proof. Since both sides of the equation are cohomological invariants it is enough to prove the theorem in case X is a 0-reduced simplicial set. In this case we have an explicit description of $\sigma(X)$ in (IV.6.2). On the other hand the cup product $i_2 \cup i_2$ is represented by the Alexander Whitney map

$$\Delta\colon C_4 \to C_2 \otimes C_2 \to C_2/B \otimes C_2/B \qquad (1)$$

which carries a generator $\xi \in C_4$ to

$$\Delta\xi = \xi_{012} \otimes \xi_{234}^{\xi - \xi_{02}}.$$

Since the sequence

$$J \longrightarrow C_2 \otimes C_2 \twoheadrightarrow C_2/B \otimes C_2/B \tag{2}$$

is exact, see (IV.3.4)(1), we get the next commutative diagram in which the right hand column is a short exact sequence. We derive the diagram from (IV.3.4)(1) where we divide out J. This shows that v in (IV.3.4)(1) induces a splitting $\overline{v}$ as indicated:

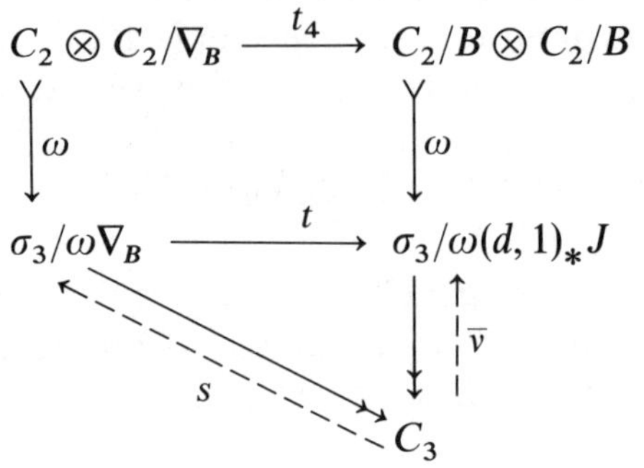

$$\begin{array}{ccc} C_2 \otimes C_2/\nabla_B & \xrightarrow{t_4} & C_2/B \otimes C_2/B \\ \big\downarrow \omega & & \big\downarrow \omega \\ \sigma_3/\omega\nabla_B & \xrightarrow{t} & \sigma_3/\omega(d,1)_* J \\ & \searrow \nwarrow s & \big\downarrow \uparrow \overline{v} \\ & & C_3 \end{array} \tag{3}$$

Here t is the quotient map. We choose a splitting s with $ts = \overline{v}$. Then we get by (2.1) and by definition of β_4 in (1.8)(2)

$$t_4\beta_4(\xi) = -t_4\omega^{-1}(-\partial_4(\xi) + sd_4\xi) \tag{4}$$

$$= -\omega^{-1}(-t\partial_4(\xi) + \overline{v}d_4\xi) \tag{5}$$

$$= -\omega^{-1}(-(tv(a) + \omega\Delta(\xi)) + \overline{v}d_4\xi) \tag{6}$$

$$= +\Delta(\xi) \tag{7}$$

In (6) we use equation (IV.6.4) where $tv(a)$ corresponds to $\overline{v}d_4\xi$ and where $t\omega(b)$ corresponds to $\omega\Delta(\xi)$ in (1). By (7) and (1) the proof of (2.6) is complete. □

§ 3 Obstructions for the existence of certain chain maps and the primary obstruction for the realizability of a chain complex

In this section we consider an obstruction for the existence of certain chain maps. This is a simple and general construction which also will be used in § 6 and § 7 below. As a special case we obtain the primary obstruction for the realizability of a chain complex, $\mathcal{O}_5(C)$, considered in Appendix (III.A). We show that the obstructions are torsion elements if π is a finite group. This yields an easy proof of theorem (III.A.4).

Let π be a group and let C be a chain complex of free π-modules. Moreover let Q be a chain complex of π-modules with $Q^n = C^n$ and $dQ_{n+1} = dC_{n+1} = B_n$ and suppose that $Q_i = 0$ for $i > n+2$ and $H_{n+1}(Q) = 0$. We consider the following diagram of π-equivariant homomorphisms.

(3.1)

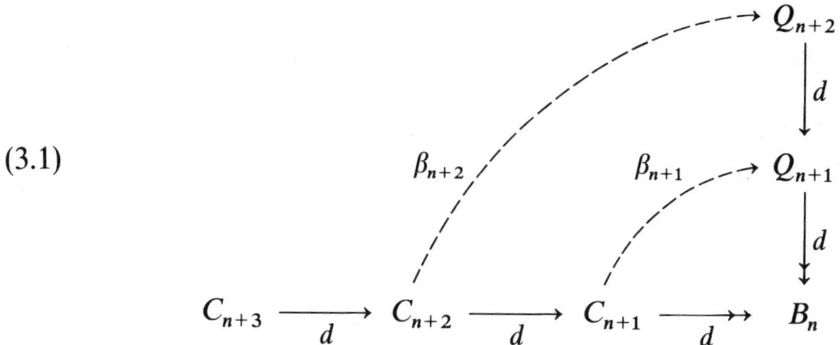

For example, for $n = 2$ we have such a diagram as a part of (1.2) with $Q = P(\rho)$ given by the Postnikov chain functor on **H**. Since C is free and since $H_{n+1}Q = 0$ we can choose maps β_{n+2}, β_{n+1} such that diagram (3.1) commutes. The pair $(\beta_{n+2}, \beta_{n+1})$ is a chain map iff $\beta_{n+2}d = 0$; in this case the pair $(\beta_{n+2}, \beta_{n+1})$ represents an element

$$\beta \in [C, I^n, Q] \tag{1}$$

where I^n is the identity of $C^n = Q^n$, compare (II.4.15). We derive from the pair (C, Q) the following two **obstructions**. The map

$$\beta_{n+2}d\colon C_{n+3} \to H_{n+2}Q \subset Q_{n+2}$$

is a cocycle which represents the obstruction

$$\mathcal{O}(C, I^n, Q) \in \hat{H}^{n+3}(C, H_{n+2}Q) \tag{2}$$

Moreover, the map

$$\beta_{n+1}d\colon C_{n+2} \to dQ_{n+2} \subset Q_{n+1}$$

is a cocycle which represents the obstruction

$$\mathcal{O}'(C, I^n, Q) \in \hat{H}^{n+2}(C; dQ_{n+2}). \tag{3}$$

One readily checks that the cohomology classes (2) and (3) do not depend on the choice of $(\beta_{n+2}, \beta_{n+1})$. Moreover, the short exact sequence $H_{n+2}Q \rightarrowtail Q_{n+2} \twoheadrightarrow dQ_{n+2}$ induces a boundary operator

$$\partial\colon \hat{H}^{n+2}(C, dQ_{n+2}) \to \hat{H}^{n+3}(C, H_{n+2}Q) \tag{4}$$

such that

$$\partial \mathcal{O}'(C, I^n, Q) = \mathcal{O}(C, I^n, Q). \tag{5}$$

(3.2) **Lemma.** *The set $[C, I^n, Q]$ is not empty if and only if $\mathcal{O}(C, I^n, Q) = 0$.*

This is clear by the definitions in (3.1)(2). We now assume that $\mathcal{O}(C, I^n, Q) = 0$; in this case we get a transitive and effective action

(3.3) $$\hat{H}^{n+2}(C, H_{n+2}Q) \xrightarrow[\approx]{+} [C, I^n, Q]$$

compare (II.4.17). Here the bijection $+$ carries $\{\xi\}$ to $\beta_0 + \{\xi\}$ with $\beta_0 \in [C, I^n, Q]$. We have the natural function

$$\varepsilon\colon [C, I^n, Q] \to \hat{H}^{n+2}(C, Q_{n+2}) \tag{1}$$

which carries the $(\simeq_n)$-homotopy class $\beta = \{\beta_{n+2}, \beta_{n+1}\}$ to the cohomology class $\varepsilon(\beta) = \{\beta_{n+2}\}$ represented by the cocycle β_{n+2}. Here we use the assumption that $Q_{n+3} = 0$. Clearly the map ε in (1) is equivariant with respect to the action (3.3), that is

$$\varepsilon(\beta_0 + \{\xi\}) = \varepsilon(\beta_0) + i_*\{\xi\} \tag{2}$$

where $i\colon H_{n+2}Q \subset Q_{n+2}$ is the inclusion. Moreover, $d\colon Q_{n+2} \twoheadrightarrow dQ_{n+2}$ yields the formula

$$d_*\varepsilon(\beta) = \mathcal{O}'(C, I^n, Q). \tag{3}$$

The function ε, however, needs not to be injective nor surjective. Now the obstruction $\mathcal{O}'(C, I^n, Q)$ has the following property.

(3.4) **Lemma.** *Assume $\mathcal{O}'(C, I^n, Q) = 0$. Then there is a natural bijection*

$$\varepsilon'\colon [C, I^n, Q] \xrightarrow{\approx} \hat{H}^{n+2}(C, H_{n+2}Q)$$

which is equivariant with respect to the action (3.3) and which satisfies $i_*\varepsilon' = \varepsilon$.

Proof. We have $\mathcal{O}'(C, I^n, Q) = 0$ if and only if there exists β_{n+1} in (3.1) with $\beta_{n+1}d = 0$. Whence we can choose $\beta_{n+2} = 0$ in this case. Moreover, the $(\simeq_n)$-homotopy class $\beta_0 = \{\beta_{n+1}, \beta_{n+2}\}$ with $\beta_{n+2} = 0$ does not depend on the choice of β_{n+1}. We now define ε' by the inverse of the bijection (3.3).

Equivalently ε' can be defined as follows. Each element $\beta \in [C, I^n, Q]$ can be represented by a chain map $(\beta_{n+2}, \beta_{n+1})$ with $\beta_{n+1} d = d\beta_{n+2} = 0$. Then

$$\beta_{n+2}\colon C_{n+2} \to H_{n+2}Q \subset Q_{n+2}$$

is a cocycle which represents $\varepsilon'(\beta)$. □

(3.5) **Lemma.** *Let π be a finite group. Then the obstructions* (3.1)(2), (3) *satisfy*

$$|\pi| \cdot \mathcal{O}(C, I^n, Q) = 0, \quad |\pi| \cdot \mathcal{O}'(C, I^n, Q) = 0$$

where $|\pi|$ is the number of elements in π.

Proof. By (3.1)(5) it is enough to prove the second equation in (3.5). We consider the extension of π-modules

$$dQ_{n+2} \overset{i}{\rightarrowtail} Q_{n+1} \twoheadrightarrow B_n \tag{1}$$

where $B_n \subset C_n$. Since C_n is a free π-module we see that B_n as a group is free abelian. Therefore the extension (1) admits a homomorphism

$$r\colon Q_{n+1} \to dQ_{n+2} \tag{2}$$

between abelian groups with $ri = 1$. We now define a map

$$\left.\begin{aligned} &R\colon Q_{n+1} \to dQ_{n+2} \\ &R(x) = \sum_{\alpha \in \pi} r(x^{\alpha})^{-\alpha} \end{aligned}\right\} \tag{3}$$

One readily checks that R is π-equivariant and that

$$Ri(y) = |\pi| \cdot y \quad \text{for } y \in Q_{n+2}. \tag{4}$$

Now $|\pi| \mathcal{O}'(C, I^n, Q)$ is represented by $|\pi| \cdot \beta_{n+1} d = (Ri\beta_{n+1})d$ where the right hand side is a coboundary. This proves (3.5). □

(3.6) **Lemma.** *Assume that $dC_{n+1} = B_n$ is a projective $\mathbb{Z}[\pi]$-module or assume $d\colon Q_{n+1} \twoheadrightarrow B_n$ admits a splitting. Then $\mathcal{O}'(C, I^n, Q) = 0$.*

Proof. Take $\beta_{n+1} = sd$ where d is the splitting of $d\colon Q_{n+1} \twoheadrightarrow B$. □

We may apply the general remarks above for an example where we set $n = 2$ and where we consider diagram (3.1) as part of diagram (1.2). That is, let ρ be a totally free crossed chain complex, $\rho \in \mathbf{H}$, and let $C = C(\rho)$ and $Q = P(\rho)$ be given as in diagram (1.2). In this case we get

(3.7) **Proposition.** *The primary obstruction $\mathcal{O}_5(C)$ for the realizability of C in* (III.A.2) *satisfies the equation*

$$\mathcal{O}_5(C) = \mathcal{O}(C, I^2, P(\rho)) \in \hat{H}^5(C, \Gamma(H_2C))$$

where $\mathcal{O}(C, I^2, P(\rho))$ is defined as in (3.1)(2).

Therefore theorem (III.A.4) is an immediate consequence of (3.5) above.

Proof of (3.7). The proposition follows from the commutativity of diagram (II.4.12)(3) where we set $n = 3$. This diagram shows that the restriction of β_4 to Z_4 corresponds via the isomorphism Θ to the restriction of f_4 to Z_4. Therefore the equation (3.7) is obtained by the definition of $\mathcal{O}_5(C)$ in (II.3.13)(3). □

Finally we obtain the following application of the map ε in (3.3)(1) which yields for $n = 2$ the natural function

$$\varepsilon\colon [C, I^2, P(\rho)] \to \hat{H}^4(C, C_2 \otimes C_2/\nabla_B).$$

The definition of the cohomological invariant $\beta_4 X$ in (2.1) shows that for $\rho = \rho(X)$, $C = \hat{C}_* X$, $X \in \mathbf{CW}$,

$$\beta_4 X = \varepsilon(\beta_{\sigma X}) \tag{3.8}$$

where $\beta_{\sigma X}$ is the cellular boundary invariant of the quadratic chain complex σX, see (1.8). We point out that ε is equivariant as in (3.3)(2). If the map

$$\hat{H}^4(C, \Gamma H_2 C) \to \hat{H}^4(C, C_2 \otimes C_2/\nabla_B), \quad \text{see (1.4)},$$

is injective this implies that also ε is injective. In this case the pair $(\hat{C}_* X, \beta_4 X)$ determines the homotopy type of X with $\dim(X) \leq 4$, see (1.14).

§ 4 The classification of special 4-dimensional homotopy types by Pontrjagin squares

Recall that a simply connected 4-dimensional homotopy type $\{X\}$ is determined by the Pontrjagin invariant

$$\mathscr{P}(X) = (C, \xi) \quad \text{with} \quad \xi \in H^4(C, \Gamma H_2 C) \tag{1}$$

where C is the (reduced) cellular chain complex of X and where $\xi = \not{p}_2(X)$ is the Pontrjagin element of X, compare § 8 in chapter I. In this section we consider the question whether 4-dimensional homotopy types $\{X\}$ with non trivial fundamental group $\pi_1 X \neq 0$ can be similarly classified by pairs (C, ξ) as in (1) where now $C = \hat{C}_* X$ is the cellular chain complex of the universal covering of X. We show that this is not possible in general. In fact, such an invariant (C, ξ) for X only exists if a certain obstruction $\mathcal{O}'(X, I^2, P(\rho))$ vanishes. In this case we say that C or X is 'strongly Pontrjagin' and that $\xi = \not{p}_2(X)$ is the 'strong Pontrjagin element' of X. In case X is simply connected this is the classical Pontrjagin element above. There are many CW-complexes which are not strongly Pontrjagin.

This answers a question discussed in the introduction of (II.§ 4) where we considered the first obstruction operator in the CW-tower. This obstruction operator in general is not determined by invariants, but restricted to strongly Pontrjagin objects X, $Y \in \mathbf{H}_4^c$ it satisfies the formula

$$\mathcal{O}_{X,Y}(F) = F_* \not{p}_2(X) - F^* \not{p}_2(Y). \tag{2}$$

Moreover the class of strongly Pontrjagin objects is the largest class of objects in $\mathbf{H}_4^c$ for which the obstruction operator is determined by invariants as in (2). The results in this section generalize Whitehead's results on the homotopy classification of simply connected 4-dimensional CW-complexes described in (I.§ 8).

(4.1) **Definition.** We say that a chain complex C with $C = C(\rho)$, $\rho \in \mathbf{H}$, is **strongly Pontrjagin** if the obstruction

$$\mathcal{O} = \mathcal{O}'(C, I^2, P(\rho)) \in \hat{H}^4(C, C_2 \otimes C_2/\nabla) \tag{1}$$

vanishes. Here $P(\rho)$ is given by the Postnikov chain functor in (1.3) and $C_2 \otimes C_2/\nabla$ is the image of d: $P(\rho)_4 \to P(\rho)_3$ given by the submodule

$$\nabla = B \otimes B + \tau[C_2, B] + \tau\Gamma(K) \subset C_2 \otimes C_2 \tag{2}$$

where $B \subset K \subset C_2$ are the groups of 2-boundaries and 2-cycles in C as in (1.2)(1), compare (1.4). The obstruction (1) is defined as in (3.1)(3) where we set $n = 2$, $Q = P(\rho)$. We call a CW-complex X or a homotopy system X of order 4 strongly Pontrjagin if $C = C\rho$ with $\rho = \rho(X)$ is strongly Pontrjagin as in (1). In this case we define the **strong Pontrjagin element**

$$\not{p}_2(X) = \varepsilon'(\beta_{\sigma X}) \in \hat{H}^4(C, \Gamma H_2 C) \tag{3}$$

by the natural bijection ε' in (3.4). Here $\beta_{\sigma X}$ is the cellular boundary invariant of X in (1.9).

For example a chain complex C is strongly Pontrjagin if the module $B = dC_3$ is projective. Therefore all simply connected CW-complexes are strongly Pontrjagin.

(4.3) **Proposition.** *The strong Pontrjagin element $\wp_2(X)$ is a cohomological invariant on the homotopy category of all strongly Pontrjagin* CW*-complexes X. Moreover, if X is simply connected, this invariant coincides with the classical Pontrjagin element in* (I.6.8).

Proof. The naturality of the strong Pontrjagin element is clear by the naturality of the cellular boundary invariant and by the naturality of ε' in (3.4). For the second proposition we use arguments as in the proof of (I.7.6) and formula (4.4) below. □

If $X \in \mathbf{CW}$ is strongly Pontrjagin we get formulas as in (I.6.12), namely

$$\begin{cases} \mu\wp_2(X) = b_4, \\ -j^*i^*q_*\Delta^{-1}\wp_2(X) = \{\pi_3\}, \end{cases}$$

compare the notation in (II.5.4). Moreover, we have in this case the following property of the cup product element

$$i_2 \cup i_2 = (i \otimes i)_*\tau_*\wp_2(X). \tag{4.4}$$

Here $i\colon \pi_2 X = H_2\hat{X} \subset C_2/B_2$ is the inclusion and $\tau\colon \Gamma\pi_2 \to \pi_2 \otimes \pi_2$ is the map in (I.4.2). Equation (4.4) is an easy consequence of (3.8) and (2.6). Since i induces the operator $\mu\colon \hat{H}^2(X, G) \to \mathrm{Hom}_\pi(H_2\hat{X}, G)$ we get by (4.4) the following result.

(4.5) **Proposition.** *Let X be strongly Pontrjagin and let $a \in \hat{H}^2(X, G)$, $b \in \hat{H}^2(X, G')$. Then the cup product satisfies the formula*

$$a \cup b = (\mu a \otimes \mu b)_*\tau_*\wp_2(X),$$

in particular, $a \cup b$ is determined by the elements μa and μb.

(4.6) **Example.** If a CW-complex X has a non trivial cup product $a \cup b \neq 0$ with $\mu a = 0$, then by (4.5) the CW-complex X is not strongly Pontrjagin. This shows that the real projective 4-space $\mathbb{R}P_4$ or the product $P_n \times P_n$ are not strongly Pontrjagin. Here $P_n = S^1 \cup_n e^2$ is a pseudo projective plane. On the other hand the connected sum

$$M = (S^1 \times S^3)\#(S^2 \times S^2)$$

is a strongly Pontrjagin CW-complex which is not simply connected, (here we actually have $B_2 = 0$), see also (V.§9) in Baues (AH) where we describe the group of homotopy equivalences of M.

Next we generalize the category $\mathbf{P}_2^2$ in (I.8.4).

(4.7) **Definition of the category $\hat{\mathbf{P}}_2^2$.** Objects are pairs (C, ξ) where X is a strongly Pontrjagin 4-dimensional chain complex (that is $C = C(\rho)$ with $\mathcal{O}'(C, I^2, P(\rho)) = 0$ for $\rho \in \mathbf{H}$) and where ξ is an element $\xi \in \hat{H}^4(C, \Gamma(H_2 C))$. Morphisms $F: (C, \xi) \to (C', \xi')$ are homotopy classes of chain maps $F: C \to C'$ in $\mathbf{H}_1/\simeq$ which satisfy $F_*(\xi) = F^*(\xi')$. Here F_* is induced by $\Gamma(H_2 F)$. The category $\mathbf{P}_2^2$ in (I.8.4) is a full subcategory of $\hat{\mathbf{P}}_2^2$.

Let $\mathbf{Pontrjagin}^4$ be the full subcategory of $\mathbf{CW}/\simeq$ consisting of strongly Pontrjagin 4-dimensional CW-complexes. Then the strong Pontrjagin element (4.9) gives us the functor

$$\mathscr{P}: \mathbf{Pontrjagin}^4 \to \hat{\mathbf{P}}_2^2 \tag{4.8}$$

which carries X to $\mathscr{P}(X) = (\hat{C}_* X, \wp_2 X)$. The next result generalizes the classification result of J.H.C. Whitehead in (I.8.7).

(4.9) **Theorem.** *The functor $\mathscr{P}$ in* (4.8) *is a detecting functor.*

Proof. This follows easily from (II.1.16) since we have the following property of the obstruction operator $\mathcal{O}$ in the CW-tower. □

(4.10) **Proposition.** *Let X, Y be homotopy systems of order* 4 *and let $F: \lambda X \to \lambda Y$ be a map in $\mathbf{H}_3^c$. If X and Y are strongly Pontrjagin the obstruction operator of the* CW*-tower satisfies the formula*

$$\mathcal{O}_{X,Y}(F) = F_* \wp_2(X) - F^* \wp_2(Y).$$

This formula generalizes formula (I.8.9)(2) where we consider the simply connected case. The strongly Pontrjagin objects form the largest class of objects in $\mathbf{H}_4^c$ for which the obstruction operator $\mathcal{O}$ is given by invariants, compare the notation in (II.1.16). The proposition follows from (II.4.21), (1.9) and (3.4).

§ 5 Natural quotients of the Postnikov chain functor

In section § 1 of this chapter we defined the Postnikov chain functor

$$P: \mathbf{H} \to \mathbf{Chain}_{\mathbb{Z}}^{\wedge}/\simeq_2.$$

This functor relies mainly on the properties of free nil(2)-modules. In this section we consider functors

$$N, M: \mathbf{H}_1 \to \mathbf{Chain}_{\mathbb{Z}}^{\wedge}/\simeq_2$$

which are similarly defined as P above in terms of the properties of free nil(2)-groups. Moreover, for $\rho \in \mathbf{H}$ there are natural quotient maps

$$P(\rho) \overset{u}{\twoheadrightarrow} N(C\rho) \overset{t}{\twoheadrightarrow} M(C\rho)$$

where $C: \mathbf{H} \to \mathbf{H}_1$ is the chain functor. For many computational purposes it is easier to use the functors N and M instead of P since for the definition of N and M only nil(2)-groups are needed. Their structure is simpler than the structure of nil(2)-modules used in the definition of P. This, for example, is useful in the computation of the primary obstruction for the realizability of a chain complex C, see (5.8). In the next section we shall use N and M for the definition of cohomology operations with local coefficients which generalize the classical Pontrjagin square.

Let C be a free π-module with basis Z. The **free π-nil(2)-group** E generated by Z is given by

(5.1) $$E = \langle Z \times \pi \rangle / \Gamma_3 \langle Z \times \pi \rangle.$$

There is an obvious action of the group π on E by $(x, \alpha)^{\beta} = (x, \alpha + \beta)$ for $x \in Z$, $\alpha, \beta \in \pi$. The abelianization of E is the free π-module E^{ab}. Now let C be a chain complex of free π-modules and let a free π-nil(2)-group E_n with $E_n^{ab} = C_n$ be given as in (5.1). Then we obtain the following commutative diagram of unbroken arrows which is similar to (1.2).

(5.2)

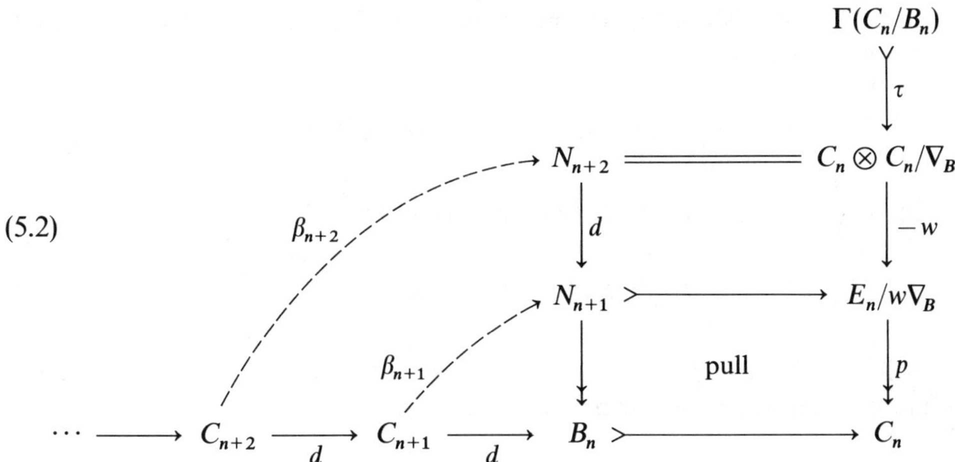

Here we set $B_n = dC_{n+1}$ and the π-submodule

$$\nabla_B = B_n \otimes B_n + \tau[B_n, C_n] \subset C_n \otimes C_n \tag{1}$$

is defined as in (1.2)(1). The right hand column of (5.2) is an exact sequence which is induced by the exact sequence

$$\Gamma(C_n) \rightarrowtail^{\tau} C_n \otimes C_n \xrightarrow{w} E_n \overset{p}{\twoheadrightarrow} C_n \tag{2}$$

given by (IV.1.2). The map τ in (5.2) is injective since we can use (1.6). Moreover, the map w in (5.2) satisfies

$$dN_{n+2} = w(C_n \otimes C_n/\nabla_B) \cong C_n \wedge C_n/B_n \wedge B_n. \tag{3}$$

Since we divide out $w(B_n \otimes B_n) \cong B_n \wedge B_n$ the pull back group $N_{n+1} = p^{-1}(B_n)$ in (5.2) is abelian. All maps in (5.2) are equivariant with respect to the action of π, in particular, N_{n+1} is a π-module.

(5.3) **Definition.** We obtain the **nil(2)-chain functor** ($n \geq 0$)

$$N = N^{(n)}: \mathbf{H}_1 \to \mathbf{Chain}_{\mathbb{Z}}^{\wedge}/\simeq_n$$

as follows. Here $\mathbf{H}_1$ is the category of reduced free chain complexes $C = (\pi, C)$ in (III.2.10). The chain complex $N(C) = \{N_i, d: N_i \to N_{i-1}\}$ with n-skeleton $N(C)^n = C^n$ and with $N_i = 0$ for $i > n + 2$ is given by

$$N_{n+2} = C_n \otimes C_n/\nabla_B, \quad N_{n+1} = p^{-1}(B) \tag{1}$$

as in (5.2) above. Moreover the boundaries in $N(C)$ are given by the maps in (5.2). We clearly have

$$H_{n+1}N(C) = 0, \quad H_{n+2}N(C) = \Gamma(C_n/B_n). \tag{2}$$

A chain map $f\colon C \to C'$ in $\mathbf{H}_1$ induces a well defined map

$$N(f)\colon N(C) \to N(C') \text{ in } \mathbf{Chain}_{\mathbb{Z}}^{\wedge}/\simeq_n \tag{3}$$

as follows. We set $N(f)^n = f^n$ on the n-skeleton. In degree $n + 2$ the map $N(f)$ is given by the tensor product map

$$N(f)_{n+2} = \{f_n \otimes f_n\}, \quad \text{see (1).} \tag{4}$$

Moreover, naturality of (5.2)(2) shows that we can choose a map $N(f)_{n+1}\colon N_{n+1} \to N'_{n+1}$ such that $N(f)$ is a chain map. Two such choices differ only by a $(\simeq)_n$-homotopy since $H_{n+1}N' = 0$ by (2).

We now consider the diagram of functors

(5.4)
$$\begin{array}{ccc} \mathbf{H} & \xrightarrow{P} & \mathbf{Chain}_{\mathbb{Z}}^{\wedge}/\simeq_2 \\ & {\scriptstyle C}\searrow \quad \nearrow{\scriptstyle N = N^{(2)}} & \\ & \mathbf{H}_1 & \end{array}$$

where P is the Postnikov chain functor in (1.1) and where C is the functor in (III.2.10).

(5.5) **Proposition.** *For $\rho \in \mathbf{H}$ there is a canonical quotient map*

$$u\colon P(\rho) \to NC(\rho)$$

which is a natural transformation of the functors in (5.4) *in the category* $\mathbf{Chain}_{\mathbb{Z}}^{\wedge}/\simeq_2$.

Proof. For the chain complexes $P = P(\rho)$ and $N = NC(\rho)$ we obtain a chain map $u\colon P \to N$ for which the following diagram commutes with $B = B_2$.

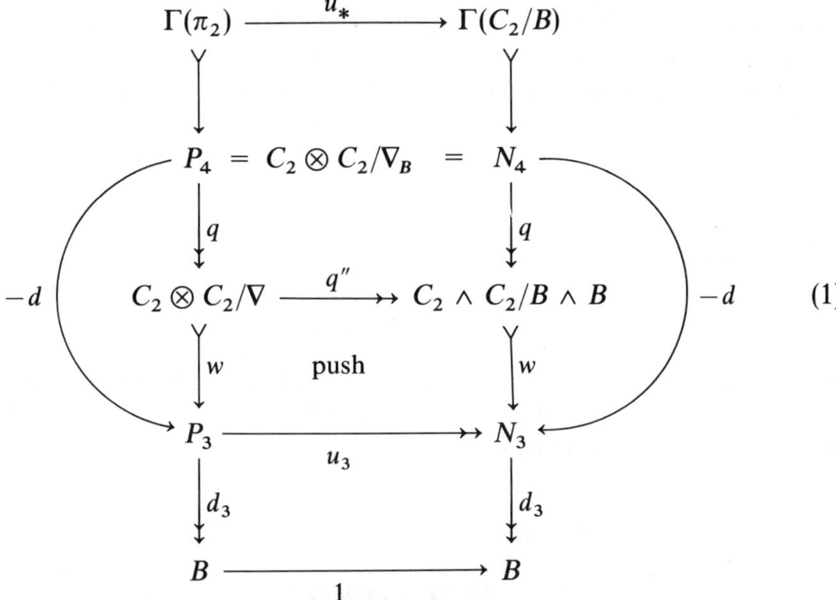

Here the left hand column is determined by the chain complex P in (1.2) and the right hand side is given by N in (5.3). Each vertical sequence $\rightarrowtail \cdot \twoheadrightarrow$ in the diagram is a short exact sequence. Therefore the subdiagram push is a push out of π-modules. The map $d = d_4$ is the boundary map of P, resp. of N. The chain map u is the identity on P_4 so that q'' is the canonical quotient map. Moreover, the map $u_* = H_4(u)$ is given by $u_* = \Gamma(i)$ where $i\colon \pi_2 = H_2(C) \subset C_2/B$ is the inclusion. We now construct the map u_3 in (1).

For the free nil(2)-module $\partial\colon \sigma_2 \to \sigma_1 = \rho_1$ in (1.1) and for the free π-nil(2)-group E_2 in (5.2) we have the following commutative diagram, (compare also (VI.6.3)(1) below).

$$\begin{array}{ccccccc}
\Gamma K & \rightarrowtail & C_2 \otimes C_2 & \xrightarrow{w} & \sigma_2 & \overset{p}{\twoheadrightarrow} & \rho_2 = \sigma_2^{cr} \\
\downarrow{\scriptstyle \Gamma i} & & \| & & \downarrow{\scriptstyle h} & & \downarrow{\scriptstyle h_2} \\
\Gamma C_2 & \overset{\tau}{\rightarrowtail} & C_2 \otimes C_2 & \xrightarrow{w} & E_2 & \overset{p}{\twoheadrightarrow} & C_2 = \rho_2^{ab}
\end{array} \qquad (2)$$

Here the top row is the exact sequence in (IV.1.8) and the bottom row is the exact sequence in (5.2)(2) above. Assume $Z_2 \subset \sigma_2$ is a basis of the free nil(2)-module $\partial\colon \sigma_2 \to \rho_1$. Then Z_2 is a basis of the free π-module C_2 and of the free π-nil(2)-group E_2. The group E_2 can be considered to be the free nil(2)-module $0_E\colon E_2 \to \pi$ with trivial boundary $0_E\colon E_2 \to 0 \in \pi$ and with basis Z_2. Since σ_2 is given by a free nil(2)-module there is a unique map h as in (2) such that $h\colon \partial \to 0_E$ is a map between nil(2)-modules which is the identity on Z_2, and which is $\rho_1 \twoheadrightarrow \pi$ equivariant. Here $\rho_1 \twoheadrightarrow \pi = \operatorname{cok}(\partial)$ is the quotient map. Naturality of the sequence (IV.1.8) shows that diagram (2) commutes. Moreover, diagram (2) is natural with respect to maps $\partial \to \partial'$ between totally free nil(2)-modules. We now obtain the map u_3 in (1) by the restriction

$$\begin{array}{ccc} P_3 & \xrightarrow{u_3} & N_3 \\ \cap & & \cap \\ \sigma_2/\omega\nabla_B & \xrightarrow{h} & E_2/w\nabla_B \end{array} \tag{3}$$

where h is induced by h in (2). Now one readily checks by the definitions of the functors P and N that $u\colon P \to N$ is a well defined surjective chain map which is natural in $\mathbf{Chain}_{\mathbb{Z}}^{\wedge}/\simeq_2$. □

Next we define a functor

(5.6) $$M\colon \mathbf{H}_1 \to \mathbf{Chain}_{\mathbb{Z}}^{\wedge}/\simeq_2$$

together with a natural quotient map

$$t\colon N(C) \to M(C), \quad C \in \mathbf{H}_1, \tag{1}$$

as follows. Consider the following commutative diagram.

$$\begin{array}{ccc} \Gamma(C_2/B) & \xrightarrow{\tau} & \tau\Gamma(C_2/B) \\ \downarrow & & \downarrow \\ N_4 = C_2 \otimes C_2/\nabla_B & \xrightarrow{t_4} & C_2/B \otimes C_2/B = M_4 \\ \downarrow q & & \downarrow q \\ C_2 \wedge C_2/B \wedge B & \xrightarrow{q} & C_2/B \wedge C_2/B \\ \downarrow w & \text{push} & \downarrow w \\ N_3 & \xrightarrow{t_3} & M_3 \\ \downarrow d_3 & & \downarrow d_3 \\ B & \xrightarrow{1} & B \end{array} \tag{2}$$

(with curved arrows $-d_4\colon N_4 \to N_3$ and $-d_4\colon M_4 \to M_3$)

The column at the left hand side is given by the chain complex $N = N^{(2)}(C)$ in (5.2) and the column at the right hand side describes the chain complex $M = M(C)$. The maps q and t_4 are the obvious quotient maps, see also (2.5), and M_3 is given by a push out as indicated. This completes the definition of $M(C)$ and t in (1). The naturality of (2) shows that M in (5.6) is a well defined functor.

We clearly have

$$H_3 M(C) = 0 \tag{3}$$

and in degree 4 we obtain the homology groups as in the commutative diagram

$$\begin{array}{ccc} \Gamma(C_2/B) & \xrightarrow{\tau} & \tau\Gamma(C_2/B) \\ \| & & \| \\ H_4 N(C) & \xrightarrow{\tau_*} & H_4 M(C) \end{array} \tag{4}$$

Here the surjection τ is given by $\tau\colon \Gamma(C_2/B) \to C_2/B \otimes C_2/B$. In fact τ in (2) is an isomorphism provided

$$\tau\colon \Gamma\pi_2 \to \pi_2 \otimes \pi_2, \quad \pi_2 \subset H_2 C, \tag{5}$$

is injective; in this case the map t in (1) is a weak equivalence.

We now consider the primary obstruction for the realizability of the chain complex C, see (3.7). With the notation in (3.1)(2) we clearly get for the functors N and M above:

(5.7)
$$\begin{cases} u_* \mathcal{O}(C, I^2, P(\rho)) = \mathcal{O}(C, I^2, N(C)), \\ t_* \mathcal{O}(C, I^2, N(C)) = \mathcal{O}(C, I^2, M(C)). \end{cases}$$

Similar equations hold for the obstructions $\mathcal{O}'$ in (3.1)(3). By (5.7) we immediately get:

(5.8) **Proposition.** *If $\mathcal{O}(C, I^2, N(C)) \neq 0$ or $\mathcal{O}(C, I^2, M(C)) \neq 0$ then C is not realizable.*

The proposition is useful since the chain complexes $N(C)$ and $M(C)$ are much easier to compute than the Postnikov chain complex $P(\rho)$ in (3.7).

§6 Pontrjagin squares with local coefficients

In this section we solve the following problem which was already considered in (I.5.3)(7). Let X be a connected space and let G be a $\pi_1(X)$-module. Do there exist binatural functions

$$\wp: \hat{H}^2(X, G) \to \hat{H}^4(X, \Gamma G), \tag{1}$$

$$\mathcal{S}q: \hat{H}^n(X, G) \to \hat{H}^{n+2}(X, G \otimes \mathbb{Z}/2), \quad n \geq 2, \tag{2}$$

such that these functions coincide with the classical Pontrjagin square, resp. Steenrod square in case G is a trivial $\pi_1(X)$-module? Moreover these functions should satisfy formulas as in (I.5.3). We show that in general such functions do not exist for all X. The function $\wp$, however, exists provided an obstruction $\mathcal{O}'(C, I^2, N(C)) = 0$ vanishes where $C = \hat{C}_* X$ is the chain complex of the universal covering of X. In this case we call C or X 'Pontrjagin' and $\wp$ the (generalized) Pontrjagin square. A similar result is obtained for the Steenrod square in (2). We give examples of spaces X which are not Pontrjagin and we show that all spaces X with finite fundamental group of odd order are Pontrjagin. Moreover we describe the connection of the Pontrjagin square (1) with the 'strong Pontrjagin element' in §4 and also with a 'weak Pontrjagin element'.

We point out that Gitler introduced Steenrod power operations with local coefficients $\mathbb{Z}/p$, $p = $ prime, where $\mathbb{Z}/p$ is endowed with a $\pi_1(X)$-actions. This clearly makes sense only in connection with odd primes p since then $\mathrm{Aut}(\mathbb{Z}/p) \neq 0$. Moreover, our approach in (2) is different from the one in Gitler since we allow the coefficient module G to be an arbitrary $\pi_1(X)$-module.

Our construction of the Pontrjagin square with local coefficients is based on the cellular boundary invariant $\beta_{\sigma X}$ in (1.9) and on the natural transformations

$$P(\rho) \xrightarrow{u} N(C) \xrightarrow{t} M(C)$$

(with $\rho = \rho(X), C = C(\rho) = \hat{C}_* X$, $X \in \mathbf{CW}$) described in §5 above. Using these maps we obtain the elements

(6.1) $$\beta_X^N = u_* \beta_{\sigma X} \in [C, I^2, NC],$$

(6.2) $$\beta_X^M = t_* u_* \beta_{\sigma X} \in [C, I^2, MC].$$

These elements are natural with respect to maps in **CW** in the same sense as $\beta_{\sigma X}$, compare (1.11). The element β_X^N is the same as the **equivariant cellular boundary invariant** of the quotient space

$$Q_2(X) = (\pi_2 X, \hat{X}/\hat{X}^1),$$

compare (IV.D.10)(3).

The element β_X^M is related to the cup product element $i_2 \cup i_2$ by the formula

(6.3) $$\varepsilon\beta_X^M = i_1 \cup i_2$$

where ε is the function in (3.3)(1). This formula follows from (3.8) and (2.6). The cup product element $i_2 \cup i_2$ and the element $\beta_4(X)$ in § 2 are cohomological invariants for all connected spaces X. We now describe new cohomological invariants which are, however, only defined for restricted classes of spaces X. The following definition is similar to the definition in (4.1) above.

(6.4) **Definition.** We call a chain complex C in $\mathbf{H}_1$ **Pontrjagin** if the obstruction

$$0 = \mathcal{O}'(C, I^2, N(C)) \in \hat{H}^4(C, C_2 \wedge C_2/B_2 \wedge B_2) \tag{1}$$

vanishes. Here $N(C)$ is given by the nil(2)-chain functor $N = N^{(2)}$ in (5.3) and we use the boundary group $dN_4 \cong C_2 \wedge C_2/B_2 \wedge B_2$ in (5.5)(1). The obstruction (1) is defined as in (3.1)(3) where we set $n = 2$, $Q = N(C)$. We call a CW-complex X or a homotopy system X of order 4 Pontrjagin if $C = \hat{C}_* X$ is Pontrjagin as in (1). In this case the **Pontrjagin element**

$$\wp_2^N(X) = \varepsilon' \beta_X^N \in \hat{H}^4(C, \Gamma(C_2/B_2)) \tag{2}$$

is defined by β_X^N in (6.1) and by the natural bijection ε' in (3.4). The Pontrjagin element determines the **Pontrjagin square**

$$\begin{aligned} &\wp\colon \hat{H}^2(X, G) \to \hat{H}^4(X, \Gamma G), \\ &\wp\{a\} = \Gamma(a)_* \wp_2^N(X) \end{aligned} \tag{3}$$

where $a\colon C_2/B \to G$ is a cocycle representing the cohomology class $\{a\}$. Here G may be any $\pi_1(X)$-module.

The definition of the Pontrjagin square relies on the nil(2)-chain functor N which is a quotient of the Postnikov chain functor P, see (5.5). We now use in the same way the chain functor M which is a quotient of N, see (5.6).

(6.5) **Definition.** We call a chain complex X in $\mathbf{H}_1$ **weakly Pontrjagin** if the obstruction

$$0 = \mathcal{O}'(C, I^2, M(C)) \in \hat{H}^4(C, C_2/B_2 \wedge C_2/B_2) \tag{1}$$

vanishes. Here we use the boundary group $dM_4 = C_2/B_2 \wedge C_2/B_2$ in (5.6)(2). We call a CW-complex X or a homotopy system X of order 4 weakly Pontrjagin if $C = \hat{C}_* X$ satisfies (1). In this case the **weak Pontrjagin element**

$$\mathscr{P}_2^M(X) = \varepsilon' \beta_X^M \in \hat{H}^4(C; \tau\Gamma(C_2/B_2)) \tag{2}$$

is defined by β_X^M in (6.2) and by ε' in (3.4). This element determines the **weak Pontrjagin square**

$$\begin{aligned} \mathscr{P}'\colon \hat{H}^2(X, G) &\to \hat{H}^4(X, \tau\Gamma(G)), \\ \mathscr{P}'\{a\} &= \tau\Gamma(a)_* \mathscr{P}_2^M(X) \end{aligned} \tag{3}$$

where $a\colon C_2/B_2 \to G$ is a cocycle representing the cohomology class $\{a\}$. Again G may be any $\pi_1(X)$-module.

We obtain the cohomology operations $\mathscr{P}$ and $\mathscr{P}'$ from the cohomological invariants $\mathscr{P}_2^N(X)$ and $\mathscr{P}_2^M(X)$ respectively in the same way as the cup product from the cup product element, see (I.6.3)(1). We now describe the connections between the various Pontrjagin elements. Using the natural quotient maps u and t in (5.5) and (5.6) we clearly get the following implications for a CW-complex X

(6.6) strongly Pontrjagin $\Rightarrow$ Pontrjagin $\Rightarrow$ weakly Pontrjagin.

Compare (4.1), (6.4) and (6.5). This follows since the $\mathcal{O}'$-obstructions satisfy

$$u_* \mathcal{O}'(C, I^2, P(\rho)) = \mathcal{O}'(C, I^2, N(C)), \tag{1}$$

$$t_* \mathcal{O}'(C, I^2, N(C)) = \mathcal{O}'(C, I^2, M(C)). \tag{2}$$

Similarly one gets the following equations for the various Pontrjagin elements provided these elements are defined

$$\begin{aligned} \Gamma(i)_* \mathscr{P}_2(X) &= \mathscr{P}_2^N(X), \\ \tau_* \mathscr{P}_2^N(X) &= \mathscr{P}_2^M(X). \end{aligned} \tag{3}$$

Here $i\colon \pi_2 X = H_2 \hat{X} \subset C_2/B_2$ is the inclusion and $\tau\colon \Gamma(C_2/B_2) \twoheadrightarrow \tau\Gamma(C_2/B_2)$ is the surjection, compare the top row of (5.6)(2). By (3) we also get

$$\tau_* \wp = \wp' \tag{4}$$

for the cohomology operations in (6.4)(3) and (6.5)(3) respectively. Moreover if G is a trivial $\pi_1(X)$-module then $\wp$ in (6.4)(3) coincides by (4.2) with the classical Pontrjagin square in (I.5.3). As in (I.5.3) the Pontrjagin square with local coefficients satisfies for $x, y \in \hat{H}^2(X, G)$ the equations

(6.7)
$$\begin{cases} \wp(-x) = \wp(x) \\ \wp(x+y) - \wp(x) - \wp(y) = [1,1]_*(x \cup y), \\ \tau_* \wp(x) = x \cup x \end{cases}$$

provided X is Pontrjagin. Similar equations are available for $\wp'$ if X is weakly Pontrjagin. The equations in (6.7) follow from (2.6) and from the binaturality of the cohomology operations $\wp, \wp'$. In fact, using the naturality of the cellular boundary invariant we get as in (4.2) the following proposition.

(6.8) **Proposition.** *The Pontrjagin element is a cohomological invariant on the full category of Pontrjagin* CW*-complexes in* **CW**. *Moreover the Pontrjagin square $\wp$ is binatural on this category.*

In the same way we get the next result for the larger category of weakly Pontrjagin CW-complexes.

(6.9) **Proposition.** *The weak Pontrjagin element is a cohomological invariant on the full category of weakly Pontrjagin* CW*-complexes in* **CW**. *Moreover the weak Pontrjagin square is binatural on this category.*

We next describe some properties of the obstructions $\mathcal{O}'(C, I^2, N(C))$ and $\mathcal{O}'(C, I^2, M(C))$ which are used to define Pontrjagin and weakly Pontrjagin chain complexes.

(6.10) **Theorem.** *Let X be a* CW*-complex with $X^0 = *$ and let $C = \hat{C}_* X$. Then* $2 \cdot \mathcal{O}'(C, I^2, M(C)) = 0$.

Proof. For the quotient map

$$q = d\colon C_2/B \otimes C_2/B \to C_2/B \wedge C_2/B \tag{1}$$

given by the boundary d_4 in (5.6)(2) we have by (6.3) and (3.3)(3), $M = M(C)$,

$$\mathcal{O}'(X, I^2, M) = d_* \varepsilon(\beta_X^M) = d_*(i_2 \cup i_2). \tag{2}$$

Moreover, the interchange map T in (I.5.2) satisfies

$$T_*(i_2 \cup i_2) = i_2 \cup i_2. \tag{3}$$

Since $d(1 + T) = 0$ for d in (1) we get by (3)

$$2d_*(i_2 \cup i_2) = d_*(1 + T)_*(i_2 \cup i_2) = 0. \tag{4}$$

This proves the proposition by (2). □

We have seen in (3.5) that for a finite group π one has

(6.11)
$$\begin{cases} |\pi| \cdot \mathcal{O}'(C, I^2, N(C)) = 0, \\ |\pi| \cdot \mathcal{O}'(C, I^2, M(C)) = 0. \end{cases}$$

Therefore (6.10) implies the following corollary.

(6.12) **Corollary.** *Let X be a* CW*-complex in* **CW** *with finite fundamental group $\pi = \pi_1 X$ of odd order. Then X is weakly Pontrjagin.*

Theorem (6.10) as well is a consequence of the following more delicate result.

(6.13) **Theorem.** *Let C be a chain complex in $\mathbf{H}_1$. Then $2\mathcal{O}'(C, I^2, N(C)) = 0$.*

We prove this theorem via a long calculation in (VI.2.10) below. Now (6.11) implies as in (6.12) the

(6.14) **Corollary.** *Let X be a* CW*-complex in* **CW** *with finite fundamental group $\pi = \pi_1(X)$ of odd order. Then X is Pontrjagin and the Pontrjagin square is defined for X.*

Moreover, we define the **Steenrod square with local coefficients**

(6.15)
$$\mathcal{S}q\colon \hat{H}^2(X, G) \to \hat{H}^4(X, G \otimes \mathbb{Z}/2)$$

by the formula $\sigma_* \wp(x) = \mathcal{S}q(x)$, see (I.5.3)(6). Again $\mathcal{S}q$ is defined for any right $\pi_1(X)$-module G provided X is Pontrjagin.

(6.16) **Remark.** We say that a CW-complex X with $X^0 = *$ is **Pontrjagin in degree** n if the obstruction $\mathcal{O}'(X, I^n, N^{(n)}C) = 0$ is trivial, $C = \hat{C}_* X$, see (3.1)(3) and (5.3). For such complexes X there is a binatural **Steenrod square with local coefficients**

$$\mathcal{S}q\colon \hat{H}^n(X,G) \to \hat{H}^{n+2}(X, G \otimes \mathbb{Z}/2)$$

where G is any right $\pi_1(X)$-module. For $n = 2$ this is the operation in (6.15) above and for trivial coefficients this is the classical Steenrod square in (I.5.3)(5). As in (6.13) one has $2 \cdot \mathcal{O}'(X, I^n, N^{(n)}C) = 0$ since $N^{(n)}$ differs from $N^{(2)}$ only by a shift of degree. Moreover, for a finite group π we have as in (6.11) the equation $|\pi| \cdot \mathcal{O}'(C, I^n, N^{(n)}X) = 0$. This shows that a CW-complex X with a finite fundamental group of odd order is Pontrjagin in degree $n \geq 2$. In particular, the Steenrod square above is defined for such X.

In the following examples we describe CW-complexes X, Y with finite fundamental group of even order where X is still Pontrjagin and where Y is not weakly Pontrjagin.

(6.17) **Example.** Let $C = \hat{C}_*(\mathbb{R}P_\infty)$ be the chain complex of **real projective space**. Whence C_n is the free $\mathbb{Z}[\mathbb{Z}/2]$-module generated by e_n and $de_n = e_{n-1} + (-1)^n e^e_{n-1}$ where $e \in \mathbb{Z}/2$, $e \neq 0$, $n \geq 2$. Hence the group

$$C_n \wedge C_n \cong \mathbb{Z} \tag{1}$$

is generated by $e_n \wedge e^e_n$ and the group $\mathbb{Z}/2$ acts on $\mathbb{Z}$ in (1) by -1 since $(e_n \wedge e^e_n)^e = e^e_n \wedge e_n = -e_n \wedge e^e_n$. The subgroup $B_n \wedge B_n$ of (1) is trivial since $(de_{n+1})^e = (-1)^{n+1} de_{n+1}$. The obstruction

$$\mathcal{O}'(C, I^n, N^{(n)}C) \in \hat{H}^{n+2}(C, C_n \wedge C_n) \tag{2}$$

defined by (3.1)(3) and (5.3), see (6.4), satisfies

$$\mathcal{O}'(C, I^n, N^{(n)}C) \begin{cases} = 0, & n \text{ even}, \\ \neq 0, & n \text{ odd}. \end{cases} \tag{3}$$

We only prove (3) for n odd. For this we consider the diagram in the proof of (VI.2.10) below where we set $C = \hat{C}_*(\mathbb{R}P_\infty)$. We define $\delta\colon G_n \to G_{n-1}$ by $\delta(e_n) = e_{n-1} + (-1)^n e^e_{n-1}$. For n odd we thus get

$$\begin{aligned} \delta\delta(e_{n+2}) &= \delta(e_{n+1} - e^e_{n+1}) = \delta e_{n+1} - (\delta e_{n+1})^e \\ &= e_n + e^e_n - (e_n + e^e_n)^e = e_n + e^e_n - e_n - e^e_n. \end{aligned} \tag{4}$$

This shows that $\delta\delta(e_{n+2}) = w((-e_n) \wedge (-e^e_n)) = w(e_n \wedge e^e_n)$. Hence the obstruction (3) is represented by the cocycle $\mathcal{O}(\delta)\colon C_{n+2} \to C_n \wedge C_n$ which carries e_{n+2} to $e_n \wedge e^e_n$. This cocycle represents the generator in the cohomology (3), n odd. Hence we get (3) by (VI.2.10)(4).

By (3) we know that $X = \mathbb{R}P_\infty$ is Pontrjagin for all even degrees n, so that the Pontrjagin square (6.4) and the Steenrod square (6.16), in even degrees n, exist for all $\pi_1(X)$-modules G. In odd degrees n, however, the Steenrod square (6.16) is not defined for all $\pi_1(X)$-modules G with $X = \mathbb{R}P_\infty$.

(6.18) **Example.** Let $C = \hat{C}_*(\mathbb{R}P_2 \times \mathbb{R}P_2) = \hat{C}_* \mathbb{R}P_2 \otimes_{\mathbb{Z}} \hat{C}_* \mathbb{R}P_2$ be the chain complex of a product of two projective planes. Using a computer my student A. Hohmann showed that the obstruction

$$0 \neq \mathcal{O}'(C, I^2, M(C)) \in \hat{H}^4(C, C_2/B_2 \wedge C_2/B_2)$$

is non trivial. This implies by (6.6)(2) that also $\mathcal{O}'(C, I^2, N(C)) \neq 0$ is non trivial. Whence the Pontrjagin element and the weak Pontrjagin element are not defined for $X = \mathbb{R}P_2 \times \mathbb{R}P_2$.

The stable equivalence classes of finite 4-dimensional complexes

We say that two complexes X, Y are **stably equivalent** if there exists a homotopy equivalence of their N-fold suspensions

$$\Sigma^N X \simeq \Sigma^N Y$$

for some $N \gg 0$. If $\dim X$, $\dim Y \leq 4$ we may assume that $N = 3$. We now classify completely all stable equivalence classes of finite 4-dimensional complexes, this result is due to M. Hennes.

(A.1) **Definition.** An **elementary Moore space** is a Moore space $M(A,n)$ where $A=\mathbb{Z}$ or where $A=\mathbb{Z}/p^i$ (p=prime, $i\geq 1$) is a cyclic group of prime power order. Clearly $M(\mathbb{Z},n)=S^n$ is a sphere and $M(\mathbb{Z}/q,n)=\Sigma^{n-1}P_q$ is a suspended pseudo projective plane $P_q=S^1 \cup_q e^2$, $n\geq 1$. We have $\dim(M(\mathbb{Z}/q,n))=n+1$.

(A.2) **Definition.** An **elementary cup square space** is a connected finite 4-dimensional complex X for which the cup square function

$$H^2(X,\mathbb{Z}/2) \xrightarrow{\cong} H^4(X,\mathbb{Z}/2), \quad x \longmapsto x \cup x,$$

is an isomorphism and for which X has homology groups $H_*(X,\mathbb{Z}) = (H_1, H_2, H_3, H_4)$ as in one of the rows of the following list.

X	H_1	H_2	H_3	H_4	
$X({}_p\xi^q)$	$\mathbb{Z}/2^p$	0	$\mathbb{Z}/2^q$	0	$p \geq q \geq 1$
$X(t\xi^q)$	0	$\mathbb{Z}/2^t$	$\mathbb{Z}/2^q$	0	$t \geq 1$, $q \geq 1$
$X(t\xi)$	0	$\mathbb{Z}/2^t$	0	$\mathbb{Z}$	$t \geq 1$
$X(\xi^q)$	0	$\mathbb{Z}$	$\mathbb{Z}/2^q$	0	$q \geq 1$
$X(\xi)$	0	$\mathbb{Z}$	0	$\mathbb{Z}$	

(A.3) **Theorem.** *Elementary cup square spaces exist and two elementary cup square spaces X, Y are stably equivalent if and only if there is an isomorphism of homology groups $H_*(X,\mathbb{Z}) \cong H_*(Y,\mathbb{Z})$.*

Whence the rows of diagram (A.2) describe exactly all stable equivalence classes of elementary cup square spaces.

(A.4) **Theorem.** *Let X be a connected finite 4-dimensional* CW*-complex. Then X is stably equivalent to a one point union $X_1 \vee \cdots \vee X_n$ where the complexes X_i for $i = 1, \ldots, n$ are elementary Moore spaces of dimension ≤ 4 or elementary cup square spaces. Moreover the one point union is unique up to permutation.*

The theorem gives us a complete list of all stable equivalence classes of connected finite 4-dimensional CW-complexes; in fact, the set of all such equivalence classes is the free abelian monoid generated by the elementary Moore spaces of dimension ≤ 4 and the elementary cup square spaces. Addition in the monoid is given by the one point union of complexes.

Some of the elementary cup square spaces are well known, for example

$$X(\xi) = \mathbb{C}P_2 \quad \text{and} \quad X({}_1\xi^1) = \mathbb{R}P_4 \tag{A.5}$$

are the complex projective plane and the real projective 4-space respectively. Moreover $X({}_p\xi^p)$ can be chosen to be the 4-skeleton of an Eilenberg-Mac Lane space $K(\mathbb{Z}/2^p, 1)$. The space $X(\xi^1)$ can be chosen to be the quotient $\mathbb{R}P_4/S^1$, compare (IV.A.9). Moreover we obtain $X(w) = C_f$ for $w = \xi$, ξ^α, $t\xi$, $t\xi^\alpha$ by the mapping cone C_f of a map $f = f(w)$. Here $f(\xi) = \eta\colon S^3 \to S^2$ is the Hopf map and

$$f(\xi^q) = i_2\eta + 2^q i_3\colon S^3 \to S^3 \vee S^2,$$

$$f(t\xi) = (\eta, 2^t i_2)\colon S^3 \vee S^2 \to S^2,$$

$$f(t\xi^q) = (i_2\eta + 2^q i_3, 2^t i_2)\colon S^3 \vee S^2 \to S^3 \vee S^2.$$

The maps i_2 and i_3 denote the inclusions of the S^2 and S^3 respectively. The main difficulty lies in the construction of the elementary cup square space $X({}_p\xi^q)$ which actually only exists for $p \geq q \geq 1$; for this the theory of quadratic chain complexes and theorem (2.6) on the cup square of quadratic chain complexes are essential.

(A.6) **Definition.** We obtain the elementary cup square space $X({}_p\xi^q)$ for $p \geq q \geq 1$ by a 4-dimensional CW-complex

$$X({}_p\xi^q) = * \cup e \cup e_2 \cup e_3 \cup e_4 \tag{1}$$

with exactly one n-cell e_n in each dimension $n \leq 4$; we set $e = e_1$. The homotopy type of $X({}_p\xi^q)$ is determined by the quadratic chain complex $\sigma(X({}_p\xi^q))$

which is defined by the following boundary formulas:

$$d(e_2) = 2^p e = e + \cdots + e \quad (2^p\text{-fold sum}) \tag{2}$$

$$d(e_3) = e_2 - e_2^{2^{p-q}e} \tag{3}$$

$$d(e_4) = \omega(\{e_2\} \otimes \{e_2\}) + e_3 + \sum_{j=1}^{2^q-1} e_3^{j2^{p-q}e}. \tag{4}$$

We leave it to the reader to show that $dd = 0$ is satisfied and that the universal covering of $X({}_p\xi^q)$ is homotopy equivalent to a one point union of 2-spheres and 4-spheres. Using theorem (2.6) one can check that $X({}_p\xi^q)$ is an elementary cup square space. In fact one gets immediately by definition of β_4 in (2.1) the formula

$$\beta_4(e_4) = \{e_2 \otimes e_2\}. \tag{5}$$

Hence (2.6) and (I.6.3)(1) show that the cup square in (A.2) carries a generator to a generator. The condition $p \geq q \geq 1$ for $X = X({}_p\xi^q)$ in (A.2) is necessary since the diagram

$$\begin{array}{ccc} H^2(X,\mathbb{Z}) \otimes H^2(X,\mathbb{Z}) & \xrightarrow{\cup} & H^4(X,\mathbb{Z}) \\ \big\downarrow \chi_* \otimes \chi_* & & \big\downarrow \chi_* \\ H^2(X,\mathbb{Z}/2) \otimes H^2(X,\mathbb{Z}/2) & \xrightarrow{\cup} & H^4(X,\mathbb{Z}/2) \end{array} \tag{6}$$

commutes. Here $\chi\colon \mathbb{Z} \to \mathbb{Z}/2$ is the quotient map which induces a surjection χ_*. For $p < q$, however, the composition $\chi_*\cup$ is trivial since $H^4(X,\mathbb{Z}) = \mathbb{Z}/2$ and $H^2(X,\mathbb{Z}) = \mathbb{Z}/2^p$.

The proof of (A.3) and (A.4) is based on the construction of the elementary cup square spaces above and on the main result of Baues-Hennes.

Chapter VI

The cohomology of categories and the calculus of tracks

The cohomology of a category **C** with coefficients in **C**-modules and **C**-bimodules was studied by Grothendieck and Mitchell respectively. In this chapter we use the more general notion of cohomology groups of **C** with coefficients in a natural system D on **C**. As we shall see, this notion is most appropriate for the purposes of homotopy theory. We used already the cohomology group $H^2(\mathbf{C}, D)$ for the classification of linear extensions of the category **C** by D, see (II.1.3). We discuss in § 2 and § 6 three examples of linear extensions given by the categories **nil**, **nil**^ and $\mathbf{Q}^2$ respectively. Here **nil** is the category of free nil(2)-groups and **nil**^ is the category of free nil(2)-groups with operators in groups, moreover $\mathbf{Q}^2$ is the category of totally free nil(2)-modules. One has canonical functors

$$\mathbf{nil} \rightarrowtail \mathbf{Q}^2 \longrightarrow \mathbf{nil}^{\wedge}. \tag{1}$$

In section § 2 we prove that the corresponding cohomology classes $\{\mathbf{nil}\}$ and $\{\mathbf{nil}^{\wedge}\}$ both are non trivial elements of order 2, see (2.7). The proof involves laborious calculations on explicit cocycles which we use in § 4, § 5 for a new identification of Igusa's associativity classes $\chi(1)$ and $\chi(\pi)$. We are interested in the class $\{\mathbf{nil}^{\wedge}\}$ since this class can be considered to be the universal obstruction for the existence of Pontrjagin squares with local coefficients, compare (2.10).

The cohomology group $H^3(\mathbf{C}, D)$ has a deep connection with homotopy theory since this group classifies all linear track extensions of **C** by D, see (3.15). Such a track extension is given by a track category $T\mathbf{C}$ which is the same as a groupoid enriched category. For example each full subcategory $E\mathbf{C} \subset \mathbf{Top}^*$ is such a track category with tracks (also called 2-morphisms or 2-cells) given by homotopy classes of homotopies. The homotopy category $E\mathbf{C}/\simeq\, = \mathbf{C}$ determines the characteristic cohomology class

$$\langle \mathbf{C} \rangle \in H^3(\mathbf{C}, D_\Sigma). \tag{2}$$

As special cases we consider the homotopy categories $\mathbf{S}(2)$, $\hat{\mathbf{S}}(2)$ and $\mathbf{CW}^2/\overset{0}{\simeq}$ which are connected by functors

$$\mathbf{S}(2) \rightarrowtail^{i} \mathbf{CW}^2/\overset{0}{\simeq} \xrightarrow{Q} \hat{\mathbf{S}}(2) \tag{3}$$

Here $\mathbf{S}(n)$ is the homotopy category of one point unions of n-spheres and $\hat{\mathbf{S}}(n)$ is the homotopy category of one point unions of n-spheres with operators in groups. Moreover $\mathbf{CW}^2$ is the category of 2-dimensional CW-complexes and of cellular maps and $\overset{0}{\simeq}$ denotes the relation given by 0-homotopies. There is an obvious inclusion i in (3); the functor Q is given by a quotient of the universal covering, that is $Q(X) = \hat{X}/\hat{X}^1$. The track categories associated to the homotopy categories in (2) are highly related to the categories in (1) since we prove that the characteristic cohomology classes are given by the equations

$$\langle \mathbf{S}(n) \rangle = \beta_n\{\mathbf{nil}\}, \tag{4}$$

$$\langle \hat{\mathbf{S}}(n) \rangle = \beta_n\{\mathbf{nil}^\wedge\}, \tag{5}$$

$$\langle \mathbf{CW}^2/\overset{0}{\simeq} \rangle = \beta\{\mathbf{Q}^2\}. \tag{6}$$

Here β_n and β are appropriate Bockstein operators. Moreover these equations are compatible with the functors in (1) and (3) respectively. The equations are obtained by describing explicit equivalences between topological and algebraic track categories, see (4.5), (5.6) and (6.4). These equivalences determine algebraically the "calculus of tracks" in the corresponding topological track categories. This is used in section § 7 for the characterization of 4-dimensional CW-complexes by 'track-models' in the track category given by $\mathbf{Q}^2$ in (6). We then show that the track model of X determines the quadratic chain complex $\sigma(X)$. This way we obtain a good geometric interpretation of the algebraic ingredients of a quadratic chain complex.

Finally we point out that there is a connection of the characteristic cohomology classes $\langle \mathbf{S}(n) \rangle$ and $\langle \hat{\mathbf{S}}(n) \rangle$ above with algebraic K-theory. For this recall that Igusa introduced 'associativity classes' $\chi(1)$ and $\chi(\pi)$ for the construction of invariants on $K_3(\mathbb{Z})$ and $K_3(\mathbb{Z}[\pi])$ respectively. Indeed the stable 3-stem $\pi_3^s = \mathbb{Z}/24$ of homotopy groups of spheres is a subgroup of $K_3(\mathbb{Z}) = \mathbb{Z}/48$ so that

$$\pi_3^s \rightarrowtail K_3(\mathbb{Z}) \xrightarrow{\chi(1)_*} \mathbb{Z}/2 \tag{7}$$

is exact where the surjection $\chi(1)_*$ is induced by the cohomology class $\chi(1)$. We show that $\chi(1)$ and $\chi(\pi)$ are actually restrictions of the cohomology classes $\langle \mathbf{S}(n) \rangle$ and $\langle \hat{\mathbf{S}}(n) \rangle$ respectively ($n \geq 3$), see (4.6)(6) and (5.7)(5). Igusa defined the cohomology class $\chi(\pi)$ only by describing an explicit and very complicated cocycle representing the class $\chi(\pi)$. By use of the equation (5) one now has a simple and elegant definition of $\chi(\pi)$ by $\beta_n\{\mathbf{nil}^\wedge\}$ and also a new topological interpretation of $\chi(\pi)$ by the characteristic cohomology class $\langle\hat{\mathbf{S}}(n)\rangle$. For the trivial group $\pi = 1$ this shows by (7) that the cohomology class $\langle\mathbf{S}(n)\rangle$ actually detects the exotic element in $K_3(\mathbb{Z})$. On the other hand we see by the track models in § 7 that the cohomology classes $\langle\mathbf{S}(2)\rangle$, $\langle\hat{\mathbf{S}}(2)\rangle$ and $\langle\mathbf{CW}^2/\overset{0}{\simeq}\rangle$ are also responsible for the miraculous nature of 4-dimensional homotopy types.

§ 1 The cohomology of categories

In this section we define the cohomology of a category $\mathbf{C}$ with coefficients in a natural system D on $\mathbf{C}$ and we describe some basic facts concerning such cohomology groups. Further information can be found in (IV.§ 5) of Baues (AH), Baues-Wirsching and Baues-Dreckmann.

Recall that the **category of factorizations** in $\mathbf{C}$, denoted by $F\mathbf{C}$, is given as follows, see (II.1.1). Objects are the morphisms $f, g, \dots,$ in $\mathbf{C}$ and morphisms $f \to g$ are pairs (α, β) for which

$$\begin{array}{ccc} B & \xrightarrow{\alpha} & B' \\ {\scriptstyle f}\uparrow & & \uparrow{\scriptstyle g} \\ A & \xleftarrow[\beta]{} & A' \end{array}$$

commutes in $\mathbf{C}$. Composition is defined by $(\alpha', \beta')(\alpha, \beta) = (\alpha'\alpha, \beta\beta')$ so that $(\alpha, \beta) = (\alpha, 1)(1, \beta) = (1, \beta)(\alpha, 1)$. A **natural system** (of abelian groups) on $\mathbf{C}$ is a functor,

$$D\colon F\mathbf{C} \to \mathbf{Ab}, \tag{1.1}$$

from $F\mathbf{C}$ to the category of abelian groups. The functor D carries the object f to $D_f = D(f)$ and carries the morphism (α, β) to

$$D(\alpha, \beta) = \alpha_*\beta^*\colon D_f \to D_{\alpha f \beta} = D_g \tag{1}$$

where $D(\alpha, 1) = \alpha_*$ and $D(1, \beta) = \beta^*$. We have obvious functors

$$F\mathbf{C} \xrightarrow{\pi} \mathbf{C}^{op} \times \mathbf{C} \xrightarrow{p} \mathbf{C} \tag{2}$$

which show that a **C-module** $F\colon \mathbf{C} \to \mathbf{Ab}$ and a **C-bimodule** $G\colon \mathbf{C}^{op} \times \mathbf{C} \to \mathbf{Ab}$

yield in a canonical way natural systems $(p\pi)^*F$, π^*G as well denoted by F and G respectively. Let M, M': $\mathbf{Ab} \to \mathbf{Ab}$ be functors then we get as an example the **Ab**-bimodule

$$\operatorname{Hom}(M', M)\colon \mathbf{Ab}^{op} \times \mathbf{Ab} \to \mathbf{Ab} \tag{3}$$

which carries the object (A, B) to the group $\operatorname{Hom}(M'A, MB)$; in case M' is the identical functor we write $\operatorname{Hom}(-, M)$ for (3). Next we define the cohomology of a category $\mathbf{C}$ with coefficients in a natural system D on $\mathbf{C}$. In order to get cohomology groups which are actually sets we have to assume that $\mathbf{C}$ is a small category; in principle it would also be possible to define the cohomology in case $\mathbf{C}$ is not small.

(1.2) **Definition.** Let $\mathbf{C}$ be a small category and let $N_n(\mathbf{C})$ be the set of sequences $(\lambda_1, \ldots, \lambda_n)$ of n composable morphisms in $\mathbf{C}$ (which are the n-simplices of the **nerve** of $\mathbf{C}$). For $n = 0$ let $N_0(C) = \operatorname{Ob}(\mathbf{C})$ be the set of objects in $\mathbf{C}$. The n-th cochain group $F^n = F^n(\mathbf{C}, D)$ is the abelian group of all functions

$$c\colon N_n(\mathbf{C}) \to \dot{\bigcup_{g \in \operatorname{Mor}(\mathbf{C})}} D_g = D^{\cdot} \tag{1}$$

with $c(\lambda_1, \ldots, \lambda_n) \in D_{\lambda_1 \circ \cdots \circ \lambda_n}$. Addition in F^n is given by adding pointwise in the abelian groups D_g. The coboundary $\delta\colon F^{n+1} \to F^n$ is defined by the formula

$$\begin{aligned}(\delta c)(\lambda_1, \ldots, \lambda_n) &= (\lambda_1)_* c(\lambda_2, \ldots, \lambda_n)\\ &\quad + \sum_{i=1}^{n-1} (-1)^i c(\lambda_1, \ldots, \lambda_i \lambda_{i+1}, \ldots, \lambda_n) \\ &\quad + (-1)^n \lambda_n^* c(\lambda_1, \ldots, \lambda_{n-1})\end{aligned} \tag{2}$$

For $n = 1$ we have $(\delta c)(\lambda) = \lambda_* c(A) - \lambda^* c(B)$ for $\lambda\colon A \to B \in N_1(\mathbf{C})$. One can check that $\delta c \in F^n$ for $c \in F^{n-1}$ and that $\delta\delta = 0$. Whence the **cohomology groups**

$$H^n(\mathbf{C}, D) = H^n(F^*(\mathbf{C}, D), \delta) \tag{3}$$

are defined, $n \geq 0$. These groups are discussed in Baues (AH); in particular they coincide for $D = (p\pi)^*F$ and $D = \pi^*G$, see (1.1)(2), with the cohomology groups introduced in Grothendieck and in Mitchell respectively. In case $\mathbf{C}$ is not small we may denote by $H^n(\mathbf{C}, D)$ the directed system of all cohomology groups $H^n(\mathbf{C}', D)$ where $\mathbf{C}'$ is a small subcategory of $\mathbf{C}$. A functor $\phi\colon \mathbf{C}' \to \mathbf{C}$ induces the homomorphism

(1.3) $$\phi^*: H^n(\mathbf{C}, D) \to H^n(\mathbf{C}', \phi^* D)$$

where $\phi^* D$ is the natural system given by $(\phi^* D)_f = D_{\phi(f)}$. On cochains the map ϕ^* is given by the formula

$$(\phi^* f)(\lambda'_1, \ldots, \lambda'_n) = f(\phi\lambda'_1, \ldots, \phi\lambda'_n)$$

where $(\lambda'_1, \ldots, \lambda'_n) \in N_n(\mathbf{C}')$. In (IV.5.8) of Baues (AH) we show:

(1.4) **Proposition.** *Let $\phi: \mathbf{C} \to \mathbf{C}'$ be an equivalence of categories. Then ϕ^* in* (1.3) *is an isomorphism of groups.*

A natural transformation $\tau: D \to D'$ between natural systems induces a homomorphism

(1.5) $$\tau_*: H^n(\mathbf{C}, D) \to H^n(\mathbf{C}, D')$$

by $(\tau_* f)(\lambda_1, \ldots, \lambda_n) = \tau_\lambda f(\lambda_1, \ldots, \lambda_n)$ where $\tau_\lambda: D_\lambda \to D'_\lambda$ with $\lambda = \lambda_1 \circ \cdots \circ \lambda_n$ is given by the transformation τ. Now let

$$D'' \overset{\iota}{\rightarrowtail} D \overset{\tau}{\twoheadrightarrow} D'$$

be a short exact sequence of natural systems on $\mathbf{C}$. Then we obtain as usual the natural long exact sequence

(1.6) $$\longrightarrow H^n(\mathbf{C}, D') \xrightarrow{\iota_*} H^n(\mathbf{C}, D) \xrightarrow{\tau_*} H^n(\mathbf{C}, D'') \xrightarrow{\beta} H^{n+1}(\mathbf{C}, D') \longrightarrow$$

where β is the Bockstein homomorphism. For a cocycle c'' representing a class $\{c''\}$ in $H^n(\mathbf{C}, D'')$ we obtain $\beta\{c''\}$ by choosing a cochain c as in (1.2)(1) with $\tau c = c''$. This is possible since τ is surjective. Then $\iota^{-1}\delta c$ is a cocycle which represents $\beta\{c''\}$.

(1.7) **Remark.** The cohomology (1.2) generalizes the **cohomology of a group**. In fact, let G be a group and let $\mathbf{G}$ be the corresponding category with a single object and with morphisms given by the elements in G. A G-module D yields a natural system $\bar{D}: F\mathbf{G} \to \mathbf{Ab}$ by $\bar{D}g = D$ for $g \in G$. The induced maps are given by $f^*(x) = x^f$ and $h_*(y) = y$, $f, h \in G$. Then the classical definition of the cohomology $H^n(G, D)$ coincides with the definition of

$$H^n(\mathbf{G}, \bar{D}) = H^n(G, D)$$

given by (1.2). Using an Eilenberg-Mac Lane space $K(G, 1)$ this is as well the cohomology in (I.4.14) and (III.4.16).

§ 2 The category of free nil(2)-groups and the existence of Pontrjagin squares

We have already seen in (II.1.3) that linear extensions of a category **C** represent cohomology classes in the second cohomology of **C**. In this section we consider the category of free nil(2)-groups, **nil**, which is a linear extension of the category of free abelian groups. This extension represents a cohomology class which is an element of order 2. More generally the category, **nil**^, of free nil(2)-groups with operators in groups represents as well such a cohomology class of order 2. This is the deeper reason for the equation $2\mathcal{O}'(C, I^2, NC) = 0$ in theorem (V.6.13) where $\mathcal{O}'(C, I^2, NC)$ is the obstruction for the existence of Pontrjagin squares. The laborious computations in this section use explicit descriptions of certain cocycles which shall be important later for a new identification of Igusa's associativity class.

We first recall some facts on linear extensions of categories already described in (II.1.3). Let D be a natural system on the category **C** and let $M(\mathbf{C}, D)$ be the set of equivalence classes of linear extensions of **C** by D. Then one has the canonical bijection

(2.1) $$\psi\colon M(\mathbf{C}, D) \cong H^2(\mathbf{C}, D)$$

which maps the split extensions to the zero element. Here we use the cohomology defined in § 1 above. In (2.1) we assume that **C** is a small category so that ψ is actually a bijection of sets; using, however, 'classes' instead of 'sets' one finds readily appropriate modifications of our definitions which allow **C** to be any category. Let

$$D + \rightarrowtail \mathbf{E} \overset{p}{\twoheadrightarrow} \mathbf{C} \tag{1}$$

be a linear extension representing an element in $M(\mathbf{C}, D)$. We obtain the element $\{\mathbf{E}\} = \psi\mathbf{E} \in H^2(\mathbf{C}, D)$ as follows. First we choose a "splitting" function t for p which associates with each morphism $f\colon A \to B$ in **C** a morphism $f_0 = t(f)$ in **E** with $pf_0 = f$. Then t yields a cocycle $c = \Delta_t$ by the formula

$$t(gf) = t(g)t(f) + \Delta_t(g, f) \tag{2}$$

with $\Delta_t(g, f) \in D_{gf}$. One readily checks that Δ_t is a well defined cocycle. This cocycle represents the cohomology class $\{\mathbf{E}\} = \{\Delta_t\}$. We now describe various examples of linear extensions for which we choose splitting functions t and for which we compute Δ_t explicitly.

Let **nil** be the full subcategory of the category of groups consisting of **free**

nil(2)-groups. Moreover, let $\mathbf{Ab}_c$ be the full subcategory of $\mathbf{Ab}$ consisting of free abelian groups. For a morphism $g: G' \to G$ in **nil** we have the following commutative diagram

$$\text{(2.2)}\qquad \begin{array}{ccc} C' \wedge C' & \xrightarrow{x \wedge x} & C \wedge C \\ \downarrow{\scriptstyle w} & & \downarrow{\scriptstyle w} \\ G' & \xrightarrow{g} & G \\ \downarrow{\scriptstyle p} & & \downarrow{\scriptstyle p} \\ C' & \xrightarrow{x} & C \end{array}$$

Here C, C' are the abelianization of G, G' and x is the abelianization of g, $x = g^{ab}$. The columns in (2.2) are given by the short exact sequence (IV.1.2). We define an action $+$ of

$$\alpha \in \mathrm{Hom}(C', \Lambda^2 C) = \mathrm{Hom}_{\mathbb{Z}}(C', C \wedge C) \tag{1}$$

on the set $\mathbf{nil}(G', G)$ of morphisms $G' \to G$ by

$$g + \alpha = g + w\alpha p. \tag{2}$$

One readily checks that $g^{ab} = f^{ab}$ iff there is α with $f = g + \alpha$. Moreover the abelianization functor $p: \mathbf{nil} \to \mathbf{Ab}_c$ yields an equivalence of categories

$$i: p\mathbf{E} \xrightarrow{\sim} \mathbf{Ab}_c \tag{3}$$

where $p\mathbf{E}$ is the image category. Let $\mathrm{Hom}(-, \Lambda^2 -)$ be the bifunctor on $\mathbf{Ab}_c$ given by (1). Then (2), (3) gives us the weak linear extension of categories

$$\text{(2.3)}\qquad \mathrm{Hom}(-, \Lambda^2) \overset{+}{\rightarrowtail} \mathbf{nil} \overset{p}{\twoheadrightarrow} \mathbf{Ab}_c.$$

Compare the notation in (II.1.7). The category **nil** thus yields the element

$$\{\mathbf{nil}\} \in H^2(\mathbf{Ab}_c, \mathrm{Hom}(-, \Lambda^2)) \tag{1}$$

in the second cohomology of the category $\mathbf{Ab}_c$, compare also (II.1.19)(14). Now let $\mathbf{Ab}'_c \subset \mathbf{Ab}_c$ be the full subcategory of finitely generated free abelian groups which is a small category. Then one actually has an isomorphism

$$H^2(\mathbf{Ab}'_c, \mathrm{Hom}(-, \Lambda^2)) = \mathbb{Z}/2 \tag{2}$$

and the restriction of the class $\{\mathbf{nil}\}$ above is the generator. This is proved in Baues-Dreckmann.

We now consider the equivariant analogue of the linear extension (2.3). Let $\mathbf{nil}^\wedge$ be the following category of **free nil(2)-groups with operators**. Objects are pairs (π, G) where π is a group and where G is a free π-nil(2)-group, see (V.5.1). Morphisms $g = (\varphi, g)\colon (\pi', G') \to (\pi, G)$ are pairs (φ, g) where $\varphi\colon \pi' \to \pi$ is a homomorphism and where $g\colon G' \to G$ is a φ-equivariant homomorphism. Moreover let $(\mathbf{Mod}_{\mathbb{Z}}^{\wedge})_c$ be the full subcategory of the category $\mathbf{Mod}_{\mathbb{Z}}^{\wedge}$ in (I.1.7) consisting of all (π, C) where C is a free π-module. We have the abelianization functor $p\colon \mathbf{nil}^\wedge \to (\mathbf{Mod}_{\mathbb{Z}}^{\wedge})_c$ which yields an equivalence of categories $i\colon p\hat{\mathbf{E}} \xrightarrow{\sim} (\mathbf{Mod}_{\mathbb{Z}}^{\wedge})$ as in (2.2)(3) above. Now assume g in (2.2) is a morphism $g = (\varphi, g)$ in the category $\mathbf{nil}^\wedge$. Then $x = (\varphi, x)$ is a φ-equivariant homomorphism from the π'-module C' to the π-module C. We define the action of π on $C \wedge C$ by $(a \wedge b)^\xi = a^\xi \wedge b^\xi$, $(a, b \in C, \xi \in \pi)$. Then all maps in (2.2) are equivariant homomorphisms. The group

(2.4) $$\mathrm{Hom}(-, \Lambda^2 -)_\varphi = \mathrm{Hom}_\varphi(C', C \wedge C)$$

acts on the set $\mathbf{nil}^\wedge(G', G)$ of morphisms $G' \to G$ by the same formula as in (2.2)(2). Therefore we obtain the weak linear extension of categories

(2.5) $$\mathrm{Hom}(-, \Lambda^2) + \rightarrowtail \mathbf{nil}^\wedge \xrightarrow{p} (\mathbf{Mod}_{\mathbb{Z}}^{\wedge})_c$$

which generalizes the one in (2.3). Here $\mathrm{Hom}(-, \Lambda^2)$ denotes the natural system on the category $(\mathbf{Mod}_{\mathbb{Z}}^{\wedge})_c$ given by (2.4), compare the notation in (II.1.1). As in (2.3)(1) we now get the element

(2.6) $$\{\mathbf{nil}^\wedge\} \in H^2((\mathbf{Mod}_{\mathbb{Z}}^{\wedge})_c, \mathrm{Hom}(-, \Lambda^2))$$

which is represented by the linear extension (2.5). The main result in this section is the

(2.7) **Theorem.** *The element $\{\mathbf{nil}^\wedge\}$ is an element of order* 2.

We have the full inclusion $i\colon \mathbf{Ab}_c \to (\mathbf{Mod}_{\mathbb{Z}}^{\wedge})_c$ which carries C to (π, C) with $\pi = 0$. The extension $\mathbf{nil}$ in (2.3) is the restriction $i^*\mathbf{nil}^\wedge$ of the extension (2.5). This shows by (2.7) that also $i^*\{\mathbf{nil}^\wedge\} = \{\mathbf{nil}\}$ is an element of order 2, in fact, in the proof of (2.7) we first show $2 \cdot \{\mathbf{nil}\} = 0$; this equation is proved by a new explicit coboundary formula and does not rely on (2.3)(2).

We now show that the element $\{\mathbf{nil}^\wedge\}$ is a kind of 'universal' obstruction for the existence of Pontrjagin squares. Let C be a chain complex of free π-modules. We associate with C the commutative square, $n \in \mathbb{Z}$,

(2.8)
$$\mathbf{Q}_n = \left\{ \begin{array}{ccc} & C_{n+1} & \\ {\scriptstyle d}\nearrow & & \searrow{\scriptstyle d} \\ C_{n+2} & \xrightarrow[0]{\quad} & C_n \\ \searrow & & \nearrow \\ & 0 & \end{array} \right\}$$

which is a subcategory of $(\mathbf{Mod}_{\hat{\mathbb{Z}}})_c$. The inclusion functor $j\colon \mathbf{Q}_n \subset (\mathbf{Mod}_{\hat{\mathbb{Z}}})_c$ induces a homomorphism j^* as in the diagram

(2.9)
$$\begin{array}{ccc} H^2((\mathbf{Mod}_{\hat{\mathbb{Z}}})_c, \mathrm{Hom}(-,\Lambda^2)) & \xrightarrow{j^*} & H^2(\mathbf{Q}_n, j^*\mathrm{Hom}(-,\Lambda^2)) \\ & & \| \varphi \\ \hat{H}^{n+2}(C, dN_{n+2}) & \underset{i^*}{\rightarrowtail} & \hat{H}^{n+2}(C^{n+2}, C_n \wedge C_n/B_n \wedge B_n) \end{array}$$

Here we use the chain complex $N = N^{(n)}(C)$ in (V.5.3) and we use (V.5.2)(3) for the equation $dN_{n+2} = C_n \wedge C_n/B_n \wedge B_n$. The inclusion $i\colon C^{n+2} \subset C$ of the $(n+2)$-skeleton induces the injection i^* in (2.9). Moreover, the canonical isomorphism φ of abelian groups is given by (7.10) in Baues-Wirsching. Now the obstruction (V.6.4) and (V.6.16) satisfies the formula

(2.10) **Theorem.** $i^*\mathcal{O}'(C, I^n, N^{(n)}C) = \varphi j^*\{\mathbf{nil}^\wedge\}$.

The equation (V.6.13) is an obvious consequence of (2.10) and (2.7). Using (V.6.4) and (V.6.16) we see that via the formula in the theorem the class $\{\mathbf{nil}^\wedge\}$ can be considered as a universal obstruction for the existence of Pontrjagin squares and Steenrod squares with local coefficients.

Proof of (2.10). Consider the diagram of groups and of π-equivariant maps.

$$\begin{array}{ccccc} C_{n+2} \wedge C_{n+2} & \longrightarrow & C_{n+1} \wedge C_{n+1} & \longrightarrow & C_n \wedge C_n \\ \downarrow w & & \downarrow w & & \downarrow w \\ G_{n+2} & \overset{\delta}{\dashrightarrow} & G_{n+1} & \overset{\delta}{\dashrightarrow} & G_n \\ \downarrow p & & \downarrow p & & \downarrow p \\ C_{n+2} & \underset{d}{\longrightarrow} & C_{n+1} & \underset{d}{\longrightarrow} & C_n \end{array} \tag{1}$$

Here the columns are short exact sequences as in (2.2). By (2.1) the element $j^*\{\mathbf{nil}^\wedge\}$ is the obstruction for finding π-equivariant homomorphisms δ such that the diagram commutes and such that $\delta\delta = 0$. We can describe this obstruction as follows. Since G_m in (1) is a free π-nil(2)-group we can choose π-equivariant homorphisms δ in (1) with $p\delta = dp$. Such δ also satisfy $\delta w = w(d \wedge d)$. The composition $\delta\delta$ in (1), however, needs not to be trivial though $dd = 0$. Since $p\delta\delta = 0$ and since $\delta\delta w = 0$ there is a unique π-equivariant homorphism

$$\mathcal{O}(\delta)\colon C_{n+2} \to C_n \wedge C_n \tag{2}$$

with $w\mathcal{O}(\delta)p = \delta\delta$. Now the composition

$$q\mathcal{O}(\delta)\colon C_{n+2} \to C_n \wedge C_n \to C_n \wedge C_n/B_n \wedge B_n \tag{3}$$

is easily seen to be a cocycle which represents the obstruction

$$\mathcal{O}'(C, I^n, N^{(n)}C) = \{q\mathcal{O}(\delta)\} \in \hat{H}^{n+2}(C, C_n \wedge C_n/B_n \wedge B_n) \tag{4}$$

On the other hand $q\mathcal{O}(\delta)$ as well represents the obstruction $\varphi j^*\{\mathbf{nil}^\wedge\}$; this proves (2.10). An example for (4) is computed in (V.6.17) above. □

The following proof of (2.7) is long; it is, however, a good exercise for the explicit computations with cochains as in (1.2).

(2.11) *Proof of theorem* (2.7). We first describe the following properties of a nil(2)-group G. Let $f\colon M \to G$ be a function and let $<$ and $\ll$ be two total orderings on the set M. Then we have in G the formula

$$\sum_{m\in M}^{\ll} f(m) = \sum_{m\in M}^{<} f(m) + w \sum_{\substack{m \ll m' \\ m' < m}} fm \wedge fm' \tag{1}$$

Here we assume that only finitely many elements $f(m)$, $m \in M$, are non trivial. The element $w(a \wedge b) = (a, b)$ denotes the commutator $-a - b + a + b$. For $a \in G$ and $n \in \mathbb{Z}$ let $na = a + \cdots + a$ be the n-fold sum in G in case $n \geq 0$, and let $na = -|n|a$ for $n < 0$. As in (1) one gets inductively the formula

$$n \sum_{m\in M}^{<} f(m) = \sum_{m\in M}^{<} nf(m) - w\binom{n}{2} \sum_{m<m'} fm \wedge fm' \tag{2}$$

where $\binom{n}{2} = n(n-1)/2$.

We now are ready for the proof of (2.7). We first show that $2\{\mathbf{nil}\} = 0$ where $\{\mathbf{nil}\}$ is defined in (2.3). For this we compute explicitely a cocycle Δ_t representing $\{\mathbf{nil}\}$ and we show that $2\Delta_t$ is a coboundary. We obtain Δ_t by a splitting function t, compare (2.1) (2), which we construct as follows. We choose for each object C in $(\mathbf{Ab})_c$ a basis Z of the free abelian group C and we choose a well ordering $<$ for the set Z. Now consider the diagram

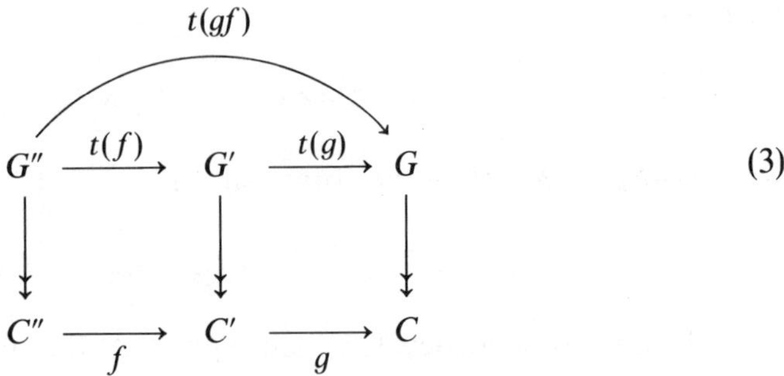

(3)

where f and g are maps in $\mathbf{Ab}_c$ and where G is the free nil(2)-group generated by Z, that is $G = \langle Z \rangle / \Gamma_3 \langle Z \rangle$. Similarly we have the free nil(2)-groups G', resp. G'', by the basis $Z' \subset C'$, resp. $Z'' \subset C''$. The homomorphism g is completely determined by its values on the basis Z', namely

$$g(x') = \sum_{x \in Z} g(x', x)x, \quad x' \in Z', \tag{4}$$

where $g(x', x) \in \mathbb{Z}$. We now define the splitting function t. For g in (4) let $t(g)$ in (3) be the unique homomorphism with

$$t(g)(x') = \sum_{x \in Z}^{<} g(x', x)x, \quad x' \in Z'. \tag{5}$$

Here the sum is defined by the ordering $<$ of the basis Z. Below we compute the cocycle Δ_t with

$$t(gf) = t(g)t(f) + \Delta_t(g, f). \tag{6}$$

Moreover, we show that $-2\Delta_t$ is a coboundary; that is

$$-2\Delta_t(g, f) = g_* \delta_f - \delta_{gf} + f^* \delta_g \tag{7}$$

where δ is a function which carries g to an element $\delta_g \in \mathrm{Hom}(C', C \wedge C)$. The coboundary formula (7) corresponds to (1.2)(2) above. For the definition of δ we consider the following diagram of homomorphisms between free abelian groups

$$\begin{array}{ccccc} C'' & \xrightarrow{f} & C' & \xrightarrow{g} & C \\ \downarrow{\scriptstyle i} & & \downarrow{\scriptstyle i} & & \downarrow{\scriptstyle i} \\ C'' \otimes C'' & \xrightarrow{f \otimes f} & C' \otimes C' & \xrightarrow{g \otimes g} & C \otimes C \\ \downarrow{\scriptstyle p} & & \downarrow{\scriptstyle p} & & \downarrow{\scriptstyle p} \\ C'' \wedge C'' & \xrightarrow[f \wedge f]{} & C' \wedge C' & \xrightarrow[g \wedge g]{} & C \wedge C \end{array} \tag{8}$$

In each column the maps i and p are given by the choice of the ordered basis. We define i on C by

$$i(x) = x \otimes x \quad \text{for} \quad x \in Z, \tag{9}$$

and we define p on $C \otimes C$ by

$$p(x \otimes y) = \begin{cases} x \wedge y & x < y, \\ 0 & \text{otherwise} \end{cases} \tag{10}$$

for $x, y \in Z$. Now the function δ associates with g the composition

$$\delta_g = p(g \otimes g)i \tag{11}$$

which is given by the maps in (8). Similarly the maps on the right hand side of (7) are all given by composition of maps in diagram (8). We have to show that formula (7) is actually satisfied. For this we first compute Δ_t in (6). The left hand side $t(gf)$ in (6) is given as in (5) by

$$t(gf)(x'') = \sum_{x \in Z}^{<} (gf)(x'', x)x, \quad x'' \in Z'', \tag{12}$$

$$= \sum_{x \in Z}^{<} \sum_{x' \in Z'}^{\ll} f(x'', x')g(x', x)x = \mathcal{U} \tag{13}$$

Here we denote the ordering of Z' by $\ll$. On the other hand we get

$$t(g)t(f)(x'') = t(g)\left(\sum_{x' \in Z'}^{\ll} f(x'', x')x'\right), \tag{14}$$

$$= \sum_{x' \in Z'}^{\ll} f(x'', x')t(g)(x'), \tag{15}$$

$$= \sum_{x' \in Z'}^{\ll} f(x'', x') \left(\sum_{x \in Z}^{<} g(x', x) x \right). \tag{16}$$

One can compare (13) and (16) by the rules (1) and (2) above. For this we apply first (2) and we get (16) $= \mathscr{U}' - w(R_1)$ with

$$\mathscr{U}' = \sum_{x' \in Z'}^{\ll} \sum_{x \in \mathbb{Z}}^{<} f(x'', x') g(x', x) x, \tag{17}$$

$$R_1 = \sum_{x' \in Z'} \binom{f(x'', x')}{2} \sum_{x<y} g(x', x) g(x', y)(x \wedge y). \tag{18}$$

Moreover we compare (13) and (17) by (1) and we get $\mathscr{U} = \mathscr{U}' + w(R_2)$ where

$$R_2 = \sum_{\substack{x<y \\ y' \ll x'}} f(x'', x') g(x', x) f(x'', y') g(y', y) x \wedge y. \tag{19}$$

Thus (6), (12) and (14) yield the equation

$$\Delta_t(g, f)(x'') = R_2 + R_1. \tag{20}$$

Using (20) we can check (7). For $N(x', x) = f(x'', x') g(x', x)$ we get

$$-2\Delta_t(g, f)(x'') = -2R_2 - 2R_1 \tag{21}$$

$$= -\sum_{x<y} \left(2 \sum_{y' \ll x'} N(x', x) N(y', y) + \sum_{x'} N(x', x) N(x', y) \right) x \wedge y \tag{22}$$

$$+ \sum_{x<y} \left(\sum_{x'} f(x'', x') g(x', x) g(x', y) \right) x \wedge y. \tag{23}$$

Here the summand (23) coincides with $f^* \delta_g$ in (7). Moreover one readily checks that (22) coincides with $g_* \delta_f - \delta_{gf}$ in (7) since

$$g_* \delta_f = \sum_{x' \ll y'} \sum_{x, y} N(x', x) N(y', y) x \wedge y, \tag{24}$$

$$-\delta_{gf} = -\sum_{x<y} \sum_{x', y'} N(x', x) N(y', y) x \wedge y. \tag{24}$$

This completes the proof of (7).

Next we show that $2\{\mathbf{nil}^\wedge\} = 0$ where $\{\mathbf{nil}^\wedge\}$ is defined in (2.6). For this we assume that the bottom row of (3) is given by morphisms in $(\mathbf{Mod}_{\mathbb{Z}}^{\wedge})_c$. Moreover, we assume that

$$Z = Z_0 \times \pi, \qquad Z' = Z'_0 \times \pi', \quad Z'' = Z''_0 \times \pi'' \tag{25}$$

where π, π', π'' are groups. The homomorphism f is φ-equivariant and g is ψ-equivariant where $\varphi\colon \pi'' \to \pi'$, $\psi\colon \pi' \to \pi$ are homomorphisms between groups. Let C_0 be the free abelian group generated by Z_0 and let $i_0\colon C_0 \subset C$ be given by the inclusion $Z_0 \subset Z_0 \times \pi$, $x \mapsto (x, 0)$. The homomorphism $t(g)$ in (3) is not ψ-equivariant; therefore we define a splitting function $\hat{t}$ by

$$\hat{t}(g) = t(gi_0) \odot \psi. \tag{26}$$

This is the ψ-equivariant extension of $t(gi_0)$ or equivalently $\hat{t}(g)$ is the unique ψ-equivariant homomorphisms with

$$\hat{t}(g)(x') = \sum_{x \in Z}^{<} g(x', x)x, \quad x' \in Z'_0. \tag{27}$$

Moreover, we define a function $\hat{\delta}$ which carries g to $\hat{\delta}_g \in \mathrm{Hom}_\psi(C', C \wedge C)$ by

$$\hat{\delta}_g = (\delta_g \circ i_0) \odot \psi. \tag{28}$$

This is the ψ-equivariant extension of the composition $\delta_{gi_0} = \delta_g i_0\colon C'_0 \subset C' \to C \wedge C$. We claim that $-2\Delta_{\hat{t}}$ is a coboundary, namely

$$-2\Delta_{\hat{t}}(g, f) = g_* \hat{\delta}_f - \hat{\delta}_{gf} + f^* \hat{\delta}_g. \tag{29}$$

This proves the proposition in (2.7). We derive (29) from (7). It is enough to check

$$-2\Delta_{\hat{t}}(g, f)i_0 = g_* \delta_{fi_0} - \delta_{gfi_0} + (fi_0)^* \hat{\delta}_g \tag{30}$$

since both sides of (29) are $\psi\varphi$-equivariant. Let $i_0\colon G''_0 = \langle Z''_0 \rangle / \Gamma_3 \subset G''$ be defined by $Z''_0 \subset Z''$. Then $\Delta_{\hat{t}}(g, f)i_0$ is given by

$$t(gfi_0) = \hat{t}(gf)i_0 = \hat{t}(g)\hat{t}(f)i_0 + w\Delta_{\hat{t}}(g, f)i_0 \tag{31}$$

where $\hat{t}(f)i_0 = t(fi_0)$. Now let

$$\nabla_g\colon C' \to C \wedge C \tag{32}$$

be the unique homorphism between abelian groups with $\hat{t}(g) = t(g) + w(\nabla_g)$. Then (31) and (6) show

$$\Delta_{\hat{t}}(g, f)i_0 = \Delta_t(g, fi_0) - (fi_0)^*\nabla_g. \tag{33}$$

Now we apply (7) and we get

$$-2\Delta_{\hat{t}}(g, f)i_0 = g_*\delta_{fi_0} - \delta_{gfi_0} + (fi_0)^*\delta_g + 2(fi_0)^*\nabla_g. \tag{34}$$

Whence (30) is a consequence of the equation

$$\hat{\delta}_g - \delta_g = 2\nabla_g. \tag{35}$$

Therefore the proof of the theorem is complete when we check (35). We first compute ∇_g. For this we remark that the ψ-equivariance of g implies

$$g((x')^\alpha, x) = g(x', x^{-\psi\alpha}), \quad \text{see (4).} \tag{36}$$

For $x' \in Z_0$, $\alpha \in \pi'$ we have

$$\begin{aligned} \hat{t}(g)(x'^\alpha) &= \sum_{x \in Z}^{<} g(x', x)x^{\psi\alpha}, \\ &= \sum_{y \in Z}^{<\alpha} g(x', y^{-\psi\alpha})y. \end{aligned} \tag{37}$$

Here we set $x <^\alpha y \Leftrightarrow x^{-\psi\alpha} < y^{-\psi\alpha}$. Using (1) and (36) we get (37) $= t(g)(x'^\alpha) + w(\nabla_g(x'^\alpha))$ where

$$\nabla_g(x'^\alpha) = \sum_{\substack{x <^\alpha y \\ y < x}} M(x, y)x \wedge y \tag{38}$$

with $M(x, y) = g(x'^\alpha, x) \cdot g(x'^\alpha, y)$. On the other hand we get by (28) and (11)

$$\begin{aligned} (\hat{\delta}_g - \delta_g)(x'^\alpha) &= (p(g \otimes g)ix')^{\psi\alpha} - p(g \otimes g)i(x'^\alpha) \\ &= (p(gx' \otimes gx'))^{\psi\alpha} - p(g(x'^\alpha) \otimes g(x'^\alpha)). \end{aligned} \tag{39}$$

The definition of p yields the formulas

$$\begin{aligned} (p(gx' \otimes gx'))^{\psi\alpha} &= \sum_{x < y} g(x', x)g(x', y)x^{\psi\alpha} \wedge y^{\psi\alpha}, \\ &= \sum_{x <^\alpha y} M(x, y)x \wedge y, \end{aligned} \tag{40}$$

$$p(g(x'^\alpha) \otimes g(x'^\alpha)) = \sum_{x < y} M(x, y)x \wedge y. \tag{41}$$

Thus we get

$$(\hat{\delta}_g - \delta_g)(x'^{\alpha}) = \sum_{x <^{\alpha} y} M(x,y) x \wedge y - \sum_{x<y} M(x,y) x \wedge y$$
$$= \sum_{\substack{x <^{\alpha} y \\ y<x}} M(x,y) x \wedge y - \sum_{\substack{x<y \\ y <^{\alpha} x}} M(x,y) x \wedge y. \qquad (42)$$

This shows (35) since $M(x,y) = M(y,x)$ and $-x \wedge y = y \wedge x$. □

§ 3 Linear track extensions of categories

In homotopy theory 'tracks' are homotopy classes of homotopies between maps. Such tracks yield the structure of a groupoid on the set of all maps from X to Y in **Top***. This example leads to the notion of a 'track category' which is essentially the same as a groupoid enriched category. Given a category **C** and a natural system D on **C** we introduce the notion of a linear track extension

$$D + \rightarrowtail T \rightrightarrows \mathbf{E} \twoheadrightarrow \mathbf{C} \qquad (1)$$

of **C** by D. Here $(\mathbf{E}, T)$ is a track category with homotopy category **C** and with the number of tracks measured by the natural system D. Equivalence classes of such linear track extensions are in 1-1 correspondence with the elements in the cohomology group $H^3(\mathbf{C}, D)$. As an example we consider track categories **E** in **Top*** consisting of suspensions. Such track categories have in a canonical way the structure of a linear track extension as in (1) and hence yield a characteristic class

$$\langle \mathbf{C} \rangle \in H^3(\mathbf{C}, D) \qquad (2)$$

with $\mathbf{C} \subset \mathbf{Top}^*/\simeq$. This class determines all Toda brackets in **C**.

Recall that $\mathbf{K}(A,B)$ denotes the set of all morphisms $A \to B$ in a category **K**. Assume for all objects A, B in **K** we have an equivalence relation $\simeq$ on $\mathbf{K}(A,B)$. Then $\simeq$ is said to be a **natural equivalence relation** on **K** if $f \simeq g$ and $x \simeq y$ implies $xf \simeq yg$ for $f, g \in \mathrm{Mor}(A,B)$ and $x, y \in \mathrm{Mor}(B,C)$. In this case we obtain the **quotient category** $\mathbf{K}/\simeq$ which has the same objects as **K** and for which the morphisms $A \to B$ are the equivalence classes $\{f\}$ in $\mathbf{K}(A,B)/\simeq$. The identical morphism is $\{i_A\}$ where $i = i_A$ is the identity on A.

(3.1) **Definition.** A **track category** $T\mathbf{K}$ or $T \Longrightarrow \mathbf{K}$ is a category $\mathbf{K}$ with the following additional structure T of tracks.

(i) For $f, g \in \mathbf{K}(A, B)$ a set $T(f, g)$ is given. We write $f \simeq g$ if $T(f, g)$ is non empty and we write $H: f \simeq g$ if $H \in T(f, g)$. We call H a **track** from f to g and we indicate H also by the diagram

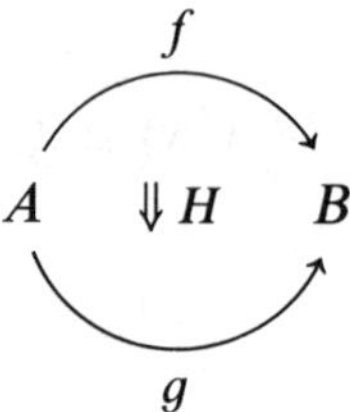

(ii) An element $0 = 0_f \in T(f, f)$ and functions

$$+: T(f, g) \times T(g, h) \to T(f, h),$$

$$-: T(f, g) \to T(g, f)$$

are given. We call 0 the **trivial track** and + is the **addition of tracks** H and G denoted by $H + G: f \simeq h$. The function $-$ maps $H: f \simeq g$ to the **negative** $-H: g \simeq f$ of H.

(iii) **Induced functions**

$$b_*: T(f, g) \to T(bf, bg),$$

$$a^*: T(f, g) \to T(fa, ga)$$

are given for $b: B \to B'$ and $a: A' \to A$ where $A, A', B, B' \in \mathrm{Ob}(\mathbf{K})$.

(iv) This structure satisfies the following conditions

$$H + (G + F) = (H + G) + F,$$

$$H + 0 = 0 + H = H,$$

$$H + (-H) = 0, \quad (-H) + H = 0,$$

$$a^*(H + G) = (a^*H) + (a^*G),$$

$$a^*(-H) = -a^*(H),$$

$$b_*(H + G) = (b_*H) + (b_*G),$$

$$b_*(-H) = -(b_*H),$$

$$(a'a)^* = a^*(a')^*, \quad 1^* = \text{identity},$$

$$(b'b)_* = (b')_*b_*, \quad 1_* = \text{identity},$$

$$b_*a^* = a^*b_*, \quad \text{and}$$

$$f^*D + y_*H = x_*H + g^*D \quad \text{for}$$

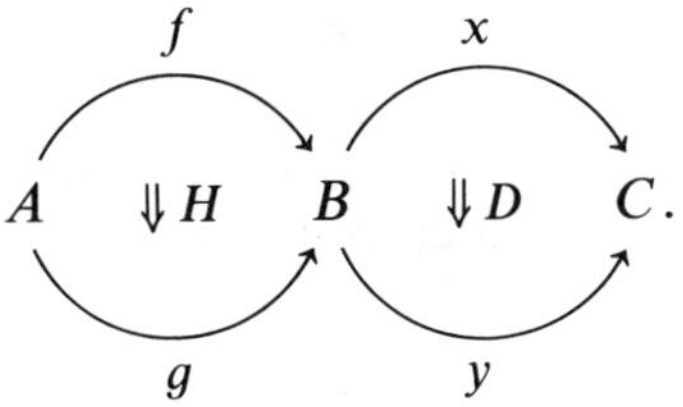

$$\begin{array}{ccccc} & f & & x & \\ A & \Downarrow H & B & \Downarrow D & C. \\ & g & & y & \end{array}$$

One readily checks that the relation $\simeq$ in (3.1)(i) is a natural equivalence relation on **K**. We call $\mathbf{K}/\!\simeq$ the **homotopy category** of the track category and we write

(3.2) $$T \Rrightarrow \mathbf{K} \xrightarrow{p} \mathbf{C}$$

if p is a full functor which is the identity on objects and which satisfies $p(f) = p(g)$ iff $f \simeq g$. Hence p induces the isomorphism $\mathbf{K}/\!\simeq\ \cong \mathbf{C}$. More generally we use the notation in (3.2) if p induces an equivalence of categories $\mathbf{K}/\!\simeq\ \xrightarrow{\sim} \mathbf{C}$. A track category $T\mathbf{K}$ as above has the following additional properties. Recall that a **groupoid** is a small category whose morphisms are invertible. We define a groupoid $\mathbf{T}(A, B)$ for objects $A, B \in \mathbf{K}$ as follows. The set of objects in $\mathbf{T}(A, B)$ is the set $\mathbf{K}(A, B)$ and the elements $H \in T(f, g)$ are the morphisms

(3.3) $$H\colon g \to f \quad \text{in } \mathbf{T}(A, B).$$

The composition law is given by the operation $+$ of tracks above and the identical morphism of f is 0_f. We obtain a bifunctor

$$\begin{aligned} *\colon \mathbf{T}(B, C) \times \mathbf{T}(A, B) &\to \mathbf{T}(A, C) \\ D * H = f^*D + y_*H &= x_*H + g^*D \end{aligned} \tag{3.4}$$

by the final equation in (3.1)(iv). This bifunctor satisfies $0_y * H = y_*H$ and $D * 0_f = f^*D$. Moreover the $*$-operation (3.4) is associative. This shows that, up to the convention in (3.3), a track category is the same as a *groupoid enriched category* or equivalently a category based on the monoidal category of groupoids, compare Fantham-Moore.

The operations on tracks can be combined to give the more general operation of **pasting**. For example

(3.5)
f u
$\Downarrow H$ h $\Downarrow G$
g v

is meant to indicate the track $u_* H + f^* G\colon uf \simeq uhg \simeq vg$. One can generalize this, so as to give meaning to such multiple composites as indicated by the diagram

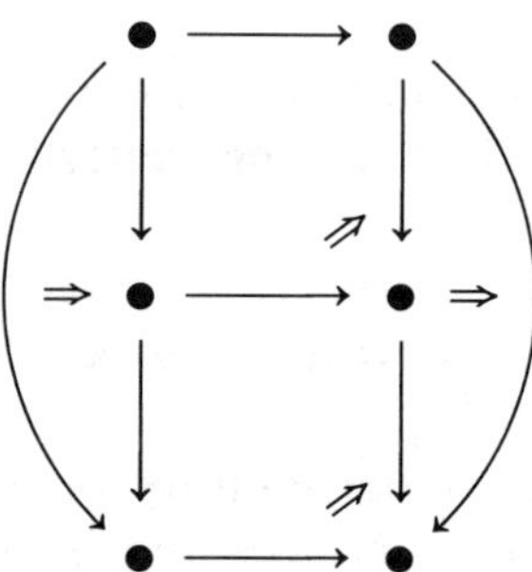

This diagram will play a role in the context of algebraic models of 4-dimensional CW-complexes.

(3.6) **Definition.** We define a **functor** $t\colon T\mathbf{K} \to T'\mathbf{K}'$ **between track categories** by a functor $t\colon \mathbf{K} \to \mathbf{K}'$ and by functions

$$t = t_{f,g}\colon T(f,g) \to T'(tf, tg)$$

which are compatible with the structure (ii) and (iii) in (3.1), that is $t(0) = 0$, $t(H + G) = (tH) + (tG)$, $t(-H) = -(tH)$, $t(b_* H) = (tb)_*(tH)$, $t(a^* H) = (ta)^*(tH)$. We say that t is **injective, surjective**, and **bijective on tracks** if the function $t_{f,g}$ is injective, surjective, and bijective respectively for all f, g. Clearly a functor $t\colon T\mathbf{K} \to T'\mathbf{K}'$ induces a functor $t\colon \mathbf{K}/\simeq \to \mathbf{K}'/\simeq$ between homotopy categories.

Our standard example of a track category is given by tracks in the category **Top***. Such tracks are homotopy classes of homotopies which are elements in the set

(3.7) $$T(f,g) = [I_*X, Y]^{(f,g)}$$

where we use the reduced cylinder I_*X and where the right hand side is the set of homotopy classes relative $(f,g)\colon X \vee X \to Y$ as defined in (III.4.6)(2). Addition of tracks is given by addition of homotopies. More precisely there are homeomorphisms

$$m\colon I_*X \approx I_*X \cup_X I'_*X, \quad n\colon I_*X \approx I_*X$$

given by $n(t,x) = (1-t,x)$ and by $m(t,x) = (2t,x) \in I_*X$ for $t \leq 1/2$ and $m(t,x) = (2t-1,x) \in I'_*X = I_*X$ for $t \geq 1/2$. Now we define the structure (ii), (iii) in (3.1) by

$$\begin{cases} 0 = \{fp\} \in T(f,f) \\ \{H\} + \{G\} = \{(H,G)m\} \\ -\{H\} = \{Hn\} \\ b_*\{H\} = \{bH\} \\ a^*\{H\} = \{H(I_*a)\} \end{cases}$$

Here $I_*a\colon I_*A' \to I_*A$ carries (t,x) to $(t,a(x))$. The following lemma is classical, see for example Brown.

(3.8) **Lemma.** *The category* **Top*** *with the structure* T *in* (3.7) *is a track category.*

(3.9) **Remark.** The lemma has the following generalization. Let **C** be a cofibration category with an initial object $*$. Then the full subcategory $\mathbf{C}_{cf}$ (consisting of cofibrant and fibrant objects in **C**) is a track category. This is proved in (II.5.6) of Baues (AH), compare (III.§4) above.

In a track category the set of tracks $T(f,f)$ is a group by use of $(+,-,0)$. In particular, for (3.7) the group

(3.10) $$T(f,f) = [I_*X, Y]^{(f,f)} = \pi_1(Y^{*X}, f)$$

is the fundamental group of the function space Y^{*X} of basepoint preserving maps $X \to Y$. The group acts transitively and effectively on the set $T(g,f)$ provided $T(g,f)$ is non empty. The action is given by addition of tracks. We now consider a related kind of action on track categories which is part of the following definition of a linear track extension.

(3.11) **Definition.** Let **C** be a category and let D be a natural system on **C**. A **linear track extension** $\mathscr{E}$ of **C** by D, denoted by

$$D \xrightarrow{+} T \rightrightarrows \mathbf{K} \xrightarrow[p]{} \mathbf{C}, \tag{1}$$

is defined by a track category as in (3.1), a functor p and an action of D on T as follows. The functor p is the identity on objects and is full, moreover p satisfies

$$p(f) = p(g) \Leftrightarrow f \simeq g \tag{2}$$

so that p induces an isomorphism $\mathbf{K}/\simeq\ \cong \mathbf{C}$. The **action of** D **on** T is given by isomorphisms of groups

$$\sigma = \sigma_f \colon D_{pf} \cong T(f, f), \quad f \in \operatorname{Mor} \mathbf{K}, \tag{3}$$

such that (4) and (5) hold:

$$\sigma_f(\alpha) + H = H + \sigma_h(\alpha) \quad \text{for} \quad H \in T(f, h), \tag{4}$$

$$\left.\begin{aligned} g^*\sigma_f(\alpha) &= \sigma_{fg}(g^*\alpha), \quad \alpha \in D_{pf}, \\ f_*\sigma_g(\beta) &= \sigma_{fg}(f_*\beta), \quad \beta \in D_{pg}. \end{aligned}\right\} \tag{5}$$

The standard example of an action as in (3.11) arises by the following lemma.

(3.12) **Lemma.** *Assume the set $T(f, g)$ in (3.7) is non empty and assume that $X = \Sigma X'$ is a suspension (or a co-h-group). Then the abelian group $D(X, Y) = [\Sigma X, Y]$ acts transitively and effectively on $T(f, g)$.*

For $\alpha \in [\Sigma X, Y]$ we denote the action in (3.12) by $H + \alpha$.

Proof of (3.12). Since X is a suspension we have the map $v \colon X \to X \vee X$ with $v = -i_1 + i_2$ where i_1, i_2 are the inclusions of X into $X \vee X$. Clearly $(1, 1)v \simeq 0$. Therefore we have a map p for which the composition

$$X \vee X \rightarrowtail C_v \xrightarrow{p} X \tag{1}$$

is the folding map $(1, 1)$. Here C_v is the mapping cone of v. Since p is a homotopy equivalence (under $*$) we see that (1) is a cylinder in the sense of (III.4.6)(1). By uniqueness of cylinders we have a canonical homotopy equivalence $h \colon C_v \simeq I_*X$ under $X \vee X$, (here we can use the lifting lemma (III.4.6)). Using h we identify

$$[C_v, Y]^{(f,g)} = [I_*X, Y]^{(f,g)}. \tag{2}$$

Now the cooperation $\mu\colon C_v \to C_v \vee \Sigma X$ yields the action in (3.12), compare (II.8.9) in Baues (AH). We point out that the map

$$h\colon \Sigma X = C_v/(X \vee X) \to I_*X/(X \vee X) = \Sigma X \tag{3}$$

(induced by h above) is the map $-a \in [\Sigma X, \Sigma X]$ since we define C_v by the cone $CX = I_*X/i_1X$. □

(3.13) **Corollary.** *Let* **K** *be a full subcategory of* **Top*** *such that the objects of* **K** *are suspensions. Then the bifunctor* D_Σ *on* $\mathbf{K}/\simeq$ *with* $D_\Sigma(X, Y) = [\Sigma X, Y]$ *is part of a linear track extension*

$$D_\Sigma + \rightarrowtail T \rightrightarrows \mathbf{K} \twoheadrightarrow \mathbf{K}/\simeq$$

where the track structure T *on* **K** *is defined by* (3.7).

Proof. For $f\colon X \to Y$ in **K** we obtain the isomorphism

$$\sigma_f\colon D_\Sigma(X, Y) \cong T(f,f) \tag{1}$$

by $\sigma_f(\alpha) = 0_f + \alpha$ where we use the action in (3.12). This shows that the action in (3.12) satisfies

$$H + \alpha = H + \sigma_g(\alpha), \quad H \in T(f,g), \tag{2}$$

where the right hand side is given by the addition of tracks. □

We leave it to the reader to generalize the corollary for any cofibration category, see (3.9). We now consider maps between linear track extensions.

(3.14) **Definition.** Let $T\mathbf{K}$ and $T'\mathbf{K}'$ be both linear track extensions of **C** by D as in (3.11). A **D-equivariant map over C**, $t\colon T\mathbf{K} \to T'\mathbf{K}'$, is a functor t as in (3.6) which satisfies

$$pt = p \quad \text{and} \quad t_{f,f}\sigma_f = \sigma_{tf} \tag{1}$$

for $f \in \operatorname{Mor}\mathbf{K}$. Whence all linear track extensions of **C** by D and D-equivariant maps over **C** form a category which we denote by **Track**(**C**, D); here we assume that **C** is small. Two objects in this category are **equivalent**, and we write $T\mathbf{K} \sim T'\mathbf{K}'$, if there exist maps

$$T\mathbf{K} \leftarrow T''\mathbf{K}'' \rightarrow T'\mathbf{K}' \tag{2}$$

in **Track**$(\mathbf{C}, D)$. Then

$$\pi_0 \, \mathbf{Track}(\mathbf{C}, D) = \mathrm{Ob}(\mathbf{Track}(\mathbf{C}, D))/\!\sim \tag{3}$$

is the **class of connected components** of the category **Track**$(\mathbf{C}, D)$. We define the **trivial track extension** in **Track**$(\mathbf{C}, D)$ by $\mathbf{K} = \mathbf{C}$, $p = 1_{\mathbf{C}}$, and

$$T(f, g) = \begin{cases} D(f) & \text{if } f = g \\ \phi & \text{otherwise,} \end{cases} \tag{4}$$

with $\sigma_f = 1_{D(f)}$ and with zero tracks given by zero elements in $D(f)$, $f \in \mathrm{Mor}(\mathbf{C})$.

The next result is a fundamental property of linear track extensions which is of similar nature as the classification of linear extensions in (2.1), it is proved in Baues-Dreckmann.

(3.15) **Theorem.** *There is a canonical bijection*

$$\psi\colon \pi_0 \, \mathbf{Track}(\mathbf{C}, D) \cong H^3(\mathbf{C}, D)$$

which carries the trivial track extension to the zero element of the cohomology group $H^3(\mathbf{C}, D)$ *defined in* (1.2).

(3.16) **Definition.** We define the bijection ψ in (3.15) as follows. Let $T\mathbf{K}$ be a linear track extension of $\mathbf{C}$ by D as in (3.11). We choose functions

$$\left.\begin{aligned} &t\colon \mathrm{Mor}\,\mathbf{C} \rightarrow \mathrm{Mor}\,\mathbf{K} \\ &H\colon N_2\mathbf{C} \rightarrow \bigcup_{f, g \in \mathrm{Mor}(\mathbf{K})} T(f, g) \end{aligned}\right\} \tag{1}$$

with $pt = 1$ and

$$H(f, g) \in T(tf \circ tg, t(fg)). \tag{2}$$

Using such choices of t and H we obtain the cochain

$$c(t, H)\colon N_3(\mathbf{C}) \rightarrow \bigcup_{f \in \mathrm{Mor}\,\mathbf{C}} D(f) \tag{3}$$

by the element $c(t, H)(f, g, h) \in D(fgh)$. This element is obtained by the "operation of pasting" in the following diagram

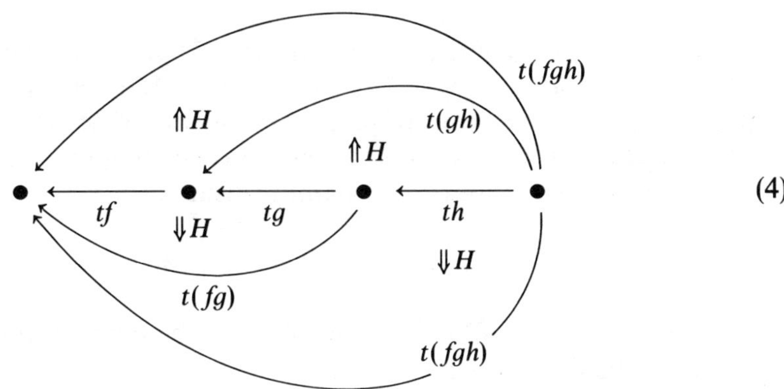

(4)

that is

$$c(t,H)(f,g,h) = \sigma_{t(fgh)}^{-1}(\Delta) \quad \text{with}$$
$$\Delta = -H(f,gh) - (tf)_* H(g,h) + (th)^* H(f,g) + H(fg,h). \tag{5}$$

One can check that $c(t,H)$ is a cocycle which represents the **characteristic cohomology class**

$$\psi\{T\mathbf{K}\} = \{c(t,H)\} \in H^3(\mathbf{C}, D). \tag{6}$$

This cohomology class depends only on the equivalence class $\{T\mathbf{K}\}$ of $T\mathbf{K}$ in $\pi_0\,\mathbf{Track}(\mathbf{C}, D)$ so that ψ in (3.15) is well defined.

We consider two important examples of characteristic cohomology classes as in (3.13). Further examples are described in the following sections.

(3.17) **Example.** Let $\mathbf{C}$ be a full subcategory of the homotopy category $\mathbf{Top}^*/\simeq$ and suppose that the objects of $\mathbf{C}$ are suspensions. Moreover let $E\mathbf{C}$ be the corresponding full subcategory of maps in $\mathbf{Top}^*$ with objects in $\mathbf{C}$ so that $E\mathbf{C}/\simeq\, = \mathbf{C}$. Then we get by (3.13) the linear track extension

$$D_\Sigma + \rightarrowtail T \rightrightarrows E\mathbf{C} \twoheadrightarrow \mathbf{C}$$

which represents the characteristic cohomology class

$$\langle \mathbf{C} \rangle = \psi\{TE\mathbf{C}\} \in H^3(\mathbf{C}, D_\Sigma)$$

which we call the **bracket** or the **characteristic cohomology class** of the homotopy category **C**. In Baues-Dreckmann we show that all classical *Toda brackets* $\langle f, g, h\rangle$ with $f, g, h \in \mathbf{C}$ are determined by the bracket $\langle \mathbf{C}\rangle$. □

(3.18) **Example.** We identify a monoid $(M, \cdot, 1)$ and the category **M** with a single object $*$ and with $M = \mathbf{M}(*, *)$. Now let M be a **topological monoid** and let $\pi_0 M$ be the monoid of path components of M with the projection $p\colon M \to \pi_0 M$ which carries $m \in M$ to the component $p(m) = M_m$ with $m \in M_m$. We assume that the Hurewicz homomorphism

$$\sigma\colon \pi_1(M_m, m) \xrightarrow{\cong} H_1(M_m) \tag{1}$$

is an isomorphism for all $m \in M$. In this case the **fundamental groupoid** πM of M yields the linear track extension

$$\pi_1 M + \rightarrowtail \pi M \rightrightarrows M \twoheadrightarrow \pi_0 M. \tag{2}$$

Here $\pi_1 M$ is the natural system on $\pi_0 M$ which carries $M_m \in \pi_0 M$ to the homology $H_1(M_m)$; induced functions are $(p(m))_* = H_1(l_m)$ and $(p(n))^* = H_1(r_n)$ with $m \cdot n = l_m(n) = r_n(m)$. Let

$$\psi\{\pi M\} \in H^3(\pi_0 M; \pi_1 M) \tag{3}$$

be the characteristic class associated to the track extension (2), see (3.16). In case $\pi_0 M$ is a group we have as well the first k-invariant

$$k_2 \in H^3(\pi_1 BM, \pi_2 BM) \cong H^3(\pi_0 M, \pi_1 M) \tag{4}$$

of the classifying space BM of M. Here the isomorphism is induced by $\pi_n BM \cong \pi_{n-1} M, n \geq 1$. In Baues-Dreckmann we show that actually the cohomology classes (3) and (4) coincide (in case $\pi_0 M$ is a group), that is

$$k_2 = \psi\{\pi M\} \tag{5}$$

where we use the isomorphism in (4) as an identification. Now let $M = \mathscr{E}(X)$ be the **topological monoid of all homotopy equivalences** $X \to X$ **in Top***. If X is a suspension one can check that condition (1) is satisfied. It then follows from (3.10) that the linear track extension (2) for $M = \mathscr{E}(X)$ is the restriction of the linear track extension (3.17) to the group of homotopy equivalences $i\colon \pi_0 \mathscr{E}(X) \subset \mathbf{C}$ with $X \in \mathbf{C}$. Whence we get by (5) the equation

$$k_2 = i^* \langle \mathbf{C} \rangle. \tag{6}$$

Thus the bracket $\langle \mathbf{C}\rangle$ in (3.17) determines all k-invariants k_2 as in (4) with $M = \mathscr{E}(X)$, $X \in \mathbf{C}$.

§ 4 Free nil(2)-groups and tracks for one point unions of n-spheres

Using free nil(2)-groups we define an algebraic track category which is equivalent to the topological track category of one point unions of n-spheres. As we shall see this equivalence is a deeper reason why nilpotency degree 2 (as used in the definition of quadratic modules in chapter IV) is sufficient for the construction of algebraic models of 4-dimensional CW-complexes. The equivalence is also used for the computation of the characteristic class of the homotopy category $\mathbf{S}(n)$ of one point unions of n-spheres. This class satisfies the equation $\langle \mathbf{S}(n) \rangle = \beta_n\{\mathbf{nil}\}$. Here $\{\mathbf{nil}\}$ is the cohomology class given by the category **nil** of free nil(2)-groups and β_n is a Bockstein operator. Moreover a restriction of the cohomology class $\langle \mathbf{S}(n) \rangle$ yields Igusa's associativity class $\chi(1)$ which was used by Igusa to detect the exotic element of $K_3(\mathbb{Z})$.

We first describe a method which derives from a linear extension of categories a linear track extension. Let $\mathbf{C}$ be a category and let

(4.1) $$\mathscr{D}\colon D'' \overset{\iota}{\rightarrowtail} D' \overset{\tau}{\twoheadrightarrow} D$$

be a short exact sequence of natural systems on $\mathbf{C}$. Then we obtain for each linear extension

$$D + \rightarrowtail \mathbf{E} \overset{p}{\twoheadrightarrow} \mathbf{C} \tag{1}$$

of $\mathbf{C}$ by D, see (II.1.2), a **linear track extension $T_{\mathscr{D}}\mathbf{E}$ associated to $\mathscr{D}$**

$$D'' + \rightarrowtail T_{\mathscr{D}} \Rrightarrow \mathbf{E} \overset{p}{\twoheadrightarrow} \mathbf{C}, \tag{2}$$

as follows. For $f, g\colon A \to B$ in $\mathbf{E}$ with $pf = pg$ there is a unique $\Delta \in D(pf)$ with $f + \Delta = g$. Now the track structure $T_{\mathscr{D}}$ in (2) is given by the set

$$T_{\mathscr{D}}(f, g) = \{H \in D'(pf); \tau H = \Delta\}. \tag{3}$$

By exactness of the sequence $\mathscr{D}$ we obtain the isomorphism

$$\sigma = \sigma_f\colon D''(pf) \cong T_{\mathscr{D}}(f, f) \tag{4}$$

by setting $\sigma(\alpha) = \iota(\alpha)$ when ι is the inclusion in (4.1). Addition of tracks in (3) is defined by addition in $D'(pf)$ and the induced functions for the set of tracks in (3) are defined by the induced functions for the natural system D'. One readily checks that (2) is a well defined linear track extension. In case $\mathbf{C}$ is small we have the Bockstein operator

$$\beta_{\mathscr{D}}\colon H^2(\mathbf{C},D)\to H^3(\mathbf{C},D'') \tag{5}$$

defined by (4.1), see (1.6).

(4.2) **Proposition.** *The characteristic cohomology class of the linear track extension* (4.1)(2) *satisfies*

$$\psi\{T_{\mathscr{D}}\mathbf{E}\} = \beta_{\mathscr{D}}\{\mathbf{E}\}.$$

Compare (2.1) and (3.15), we leave the proof as an exercise.

We now consider examples for the linear track extension in (4.1)(2). Recall that the category **nil** of free nil(2)-groups is embedded in the linear extension

(4.3) $$\mathrm{Hom}(-,\Lambda^2-)+\rightarrowtail \mathbf{nil}\twoheadrightarrow \mathbf{Ab}_c$$

as in (4.1)(2) where $\mathbf{Ab}_c$ is the category of free abelian groups. For a group A in $\mathbf{Ab}_c$ we have by (I.4.2) the natural short exact sequence ($n\geq 2$)

$$\Gamma_n^1 A \overset{\tau}{\rightarrowtail} T_n' A \overset{q}{\twoheadrightarrow} \Lambda^2 A \tag{1}$$

where we set $\Gamma_2^1 A=\Gamma A$, $T_2' A=\otimes^2 A$ and ($n\geq 3$) $\Gamma_n^1 A = A\otimes\mathbb{Z}/2$, $T_n' A=\hat{\otimes}^2 A$. This sequence yields the short exact sequence

$$\mathrm{Hom}(-,\Gamma_n^1)\overset{\tau_*}{\rightarrowtail}\mathrm{Hom}(-,T_n')\overset{q_*}{\twoheadrightarrow}\mathrm{Hom}(-,\Lambda^2) \tag{2}$$

of natural systems which are bimodules on $\mathbf{Ab}_c$. Hence we get by (4.1)(2) above the linear track extension

$$\mathrm{Hom}(-,\Gamma_n^1)+\rightarrowtail T_n \Rrightarrow \mathbf{nil}\twoheadrightarrow \mathbf{Ab}_c \tag{3}$$

where T_n is the track structure associated to the exact sequence (2).

On the other hand let $\mathbf{S}(n)$ be the full subcategory of $\mathbf{Top}^*/\simeq$ consisting of one point unions of n-spheres ($n\geq 2$),

(4.4) $$A=\bigvee_Z S^n = \Sigma^{n-1}X \quad\text{with}\quad X=\bigvee_Z S^1$$

where Z is a discrete set. The homotopy category $\mathbf{S}(n)$ is part of the topological linear track extension

$$D_\Sigma + \rightarrowtail T \Rrightarrow E\mathbf{S}(n)\twoheadrightarrow \mathbf{S}(n) \tag{1}$$

described in (3.17). We have the homology functor H_n and the suspension functor Σ

$$H_n\colon \mathbf{S}(n) \xrightarrow[\sim]{\Sigma} \mathbf{S}(n+1) \xrightarrow[\sim]{H_{n+1}} \mathbf{Ab}_c \tag{2}$$

which both are equivalences of categories for $n \geq 2$. Moreover we have the natural isomorphism

$$D_\Sigma(A, B) = [\Sigma A, B] \cong \mathrm{Hom}(H_n A, \Gamma_n^1 H_n B) \tag{3}$$

for $A, B \in \mathbf{S}(n)$, compare (I.4.11) and (I.4.12). We use (2) and (3) as identifications so that (1) corresponds to a linear track extension of $\mathbf{Ab}_c$ by $\mathrm{Hom}(-, \Gamma_n^1)$. Such a track extension is already described purely algebraicly in (4.3)(3). In fact, we show

(4.5) **Theorem.** *The topological linear track extension for* $\mathbf{S}(n)$ *in* (4.4)(1) *is equivalent to the algebraic linear track extension* T_n **nil** *in* (4.3)(3), $n \geq 2$.

Here we use the notion of equivalence of linear track extensions in (3.14). The theorem shows that the track category T_n **nil** is an algebraic model of the topological track category $TE\mathbf{S}(n)$ of one point unions of n-spheres. It is an interesting fact that even for $n \geq 3$ the algebraic model is built by non-abelian nil(2)-groups though the topological category $\mathbf{S}(n)$ is in the stable range for $n \geq 3$. Using (4.2), (4.5) and (3.15) we get the following result on characteristic cohomology classes, see (3.16).

(4.6) **Corollary.** $\beta_n\{\mathbf{nil}\} = \langle S(\mathbf{n})\rangle$.

Here β_n is the Bockstein operator

$$\beta_n\colon H^2(\mathbf{Ab}_c, \mathrm{Hom}(-, \Lambda^2)) \to H^3(\mathbf{Ab}_c, \mathrm{Hom}(-, \Gamma_n^1)) \tag{1}$$

associated to the exact sequence (4.3)(2). The commutative diagram (I.4.2) shows that β_n satisfies the equation

$$\beta_n = \sigma_* \beta_2 \quad \text{for} \quad n \geq 3. \tag{2}$$

Here σ_* is induced by the natural transformation $\mathrm{Hom}(1, \sigma)$ where $\sigma\colon \Gamma A \to A \otimes \mathbb{Z}/2$ is the suspension map in (I.4.12). Using (4.6) one has correspondingly the equations ($n \geq 3$)

$$\langle \mathbf{S}(n)\rangle = \beta_n\{\mathbf{nil}\} = \sigma_* \beta_2\{\mathbf{nil}\} = \Sigma_*^{n-2}\langle \mathbf{S}(2)\rangle \tag{3}$$

where Σ is the suspension operator on the natural system D_Σ and where we use identifications as in (4.4)(2). One gets equation (3) as well directly by considering the linear track extensions (4.4)(1).

In Baues-Dreckmann we show that β_n in (1), restricted to the full subcategory $\mathbf{Ab}_c$ of finitely generated free abelian groups, is an isomorphism. Moreover the restriction of $\langle \mathbf{S}(n) \rangle$ in (4.6) yields by (2.3)(2) a generator of the group

$$H^3(\mathbf{S}'(n), D_\Sigma) \cong H^3(\mathbf{Ab}'_c, \mathrm{Hom}(-, \Gamma_n^1)) = \mathbb{Z}/2. \tag{4}$$

Here $\mathbf{S}'(n)$ is the full subcategory of finite one point unions of spheres in $\mathbf{S}(n)$. Next we describe a connection of the class $\langle \mathbf{S}(n) \rangle$ with the *associativity class* $\chi(1)$ *of Igusa* and with $K_3(\mathbb{Z})$. Let $\mathbf{End}(\bigvee^N S^n)$ with

$$j\colon M_N(\mathbb{Z}) \cong \mathbf{End}\left(\bigvee^N S^n\right) \subset \mathbf{S}(n) \tag{5}$$

be the full subcategory of $\mathbf{S}(n)$ consisting of the single object $\bigvee^N S^n$ which is the N-fold one point union of the n-sphere. Using the equivalence (4.4)(2) this is the monoid $M_N(\mathbb{Z})$ of integral $N \times N$-matrices. In Baues-Dreckmann we show that Igusa's class $\chi(1)$ satisfies for $n \geq 3$ the equation

$$\chi(1) = \left\langle \mathbf{End}\left(\bigvee^N S^n\right)\right\rangle = j^*\langle \mathbf{S}(n) \rangle = \beta_n j^*\{\mathbf{nil}\}. \tag{6}$$

For the tedious proof of this formula one needs the explicit cocycle Δ_t computed in (2.11)(20), compare the proof of theorem 7.9 in Baues-Dreckmann. We point out that Igusa defined the class $\chi(1)$ only by describing an explicit and complicated cocycle representing $\chi(1)$. By (4) one has now a simple and elegant algebraic definition of $\chi(1)$ by $\beta_n\{\mathbf{nil}\}$ and also a new topological interpretation by the characteristic cohomology class $\langle \mathbf{S}(n) \rangle$. This shows that the class $\langle \mathbf{S}(n) \rangle$ is related to algebraic K-theory as follows. It is known that the algebraic K-group $K_3(\mathbb{Z})$ is the cyclic group of order 48 and that the stable 3-stem $\pi_3^s = \mathbb{Z}/24$ of homotopy groups of spheres is a canonical subgroup of $K_3(\mathbb{Z})$. The element $\chi(1)$ was used by Igusa to define a surjective homomorphism $\chi(1)_*$ such that

$$\pi_3^S \rightarrowtail K_3(\mathbb{Z}) \xrightarrow{\chi(1)_*} \mathbb{Z}/2 \tag{7}$$

is exact. In this sense the cohomology class $\chi(1)$ in (6) and hence the class $\langle \mathbf{S}(n) \rangle$ detect the exotic element in $K_3(\mathbb{Z})$.

(4.7) *Proof of* (4.5). For the proof of (4.5) it is enough to construct maps between linear track extensions

$$TES(n) \xleftarrow{\ i\ } T\Sigma(n) \xrightarrow{\ t\ } T_n\,\mathbf{nil}. \tag{1}$$

Here $\Sigma(n)$ is the subcategory of $ES(n)$ consisting of all maps which are (n-1)-fold suspensions $\Sigma^{n-1}f$ with f in $ES(1)$. Moreover the map i is the inclusion of the corresponding track category. Clearly this is a map as in (3.14) since each homotopy class in $\mathbf{S}(n)$ can be represented by a map in $\Sigma(n)$. The map t in (1) is more complicated. The functor

$$t\colon \Sigma(n) \to \mathbf{nil} \tag{2}$$

carries the object $A = \Sigma^{n-1}X$ in (4.4) to the free nil(2)-group

$$t(A) = G_X = \pi_1(X)/\Gamma_3\pi_1(X) \tag{3}$$

where X is a one point union of 1-spheres as in (4.4). Moreover t in (2) carries $\Sigma^{n-1}f$ to the map induced by $\pi_1(f)$. We identify in the obvious way

$$t(A)^{ab} = \pi_1(X)^{ab} = H_nA. \tag{4}$$

We now get the map t in (1) on tracks by the function

$$t = t_{f,g}\colon T(\Sigma^{n-1}f, \Sigma^{n-1}g) \xrightarrow{\ \approx\ } T_n(tf, tg). \tag{5}$$

We define this function first for $n = 2$ as follows. Let $f, g\colon X \to Y$ be maps in **Top*** where X and Y are one point unions of 1-spheres. Then we get $t_{f,g}$ in (5) by the composition of the following bijections

$$T(\Sigma f, \Sigma g) = [I_*\Sigma X, \Sigma Y]^{\Sigma f, \Sigma g} \tag{6}$$

$$\cong [I_*X, \Omega\Sigma Y]^{if, ig} \tag{7}$$

$$\cong [I_*X, JY]^{jf, jg} \tag{8}$$

$$\cong [\rho I_*X, \rho JY]^{f_*, g_*} \tag{9}$$

$$\cong [I_*\rho X, \rho_J(Y)]^{hf_*, hg_*} \tag{10}$$

$$\cong T_2(tf, tg). \tag{11}$$

Naturality of these bijections shows that t yields a well defined map as in (1) which is equivariant with respect to the action of the natural system $\mathrm{Hom}(-, \Gamma-)$. In (7) we use the adjunction of the loop space $\Omega\Sigma Y$ and the map $i\colon Y \to \Omega\Sigma Y$ adjoint to the identity of ΣY. For the infinite reduced product $J(Y)$ of James we have the homotopy equivalence $J(Y) \simeq \Omega\Sigma Y$ which carries the inclusion $j\colon Y \subset J(Y)$ to i and which induces the bijection (8). Next we use the functor ρ in (III.§2) which yields the bijection (9) since $\dim(I_* X) \leq 2$. Here we set $f_* = \rho(if)$ and $g_* = \rho(ig)$ and $\rho I_* X$ is the cylinder in (III.§3). Now we get the bijection (10) by a projection

$$h\colon \rho JY \to \rho_J(Y) \quad \text{in } \mathbf{crosschain} \tag{12}$$

which induces an isomorphism $\pi_1(h)$ for $i = 1, 2$. Here $\rho_J(Y)$ is the crossed module

$$\partial = w : \rho_J(Y)_2 = \otimes^2 H_1 Y \to \rho_J(Y)_1 = G_Y \tag{13}$$

where w is the commutator map of G_Y, see (3). The action of G_Y on $\rho_J(Y)_2$ is trivial so that $\partial = w$ is a well defined crossed module since w is central. We set $\rho_J(Y)_i = 0$ for $i \geq 3$. The map h in (12) is obtained by (q_2, q_1) in (III.C.9). The crossed module $\rho_J(Y)$ is a simple model of the 2-type of $\rho J(Y)$. For $n \geq 3$ we obtain t in (5) by suspending the bijection (6)...(1), compare (4.6)(3). □

§ 5 Tracks for one point unions of n-spheres with operators in groups

In this section we describe the equivariant version of the result in § 4 above. We consider the homotopy category $\hat{\mathbf{S}}(n)$ consisting of pairs (π, A) where A is a one point union of free π-spheres $S^n_\pi = \bigvee_\pi S^n$ of dimension n where π is a group. The morphisms of $\hat{\mathbf{S}}(n)$ are homotopy classes of equivariant maps. Using the category $\mathbf{nil}^\wedge$ of free nil(2)-groups with operators in groups, see (2.5), we define an algebraic track category which is equivalent to the topological track category given by $\hat{\mathbf{S}}(n)$. The equivalence is used for the computation of the characteristic class $\langle \hat{\mathbf{S}}(n) \rangle$ which satisfies the formula $\langle \hat{\mathbf{S}}(n) \rangle = \beta_n \{\mathbf{nil}^\wedge\}$ where β_n is a Bockstein operator. A restriction of the cohomology class $\langle \hat{\mathbf{S}}(n) \rangle$ yields Igusa's associativity class $\chi(\pi)$.

Let π be a group. A **free π-sphere** S^n_π of dimension n is a one point union

$$S^n_\pi = \bigvee_\pi S^n \tag{5.1}$$

of n-spheres with index set π. There is an obvious action of π on S^n_π by permuting the spheres S^n in S^n_π. The homology of the free π-sphere,

$$H_n(S^n_\pi) = \mathbb{Z}[\pi], \tag{1}$$

is thus the group ring of π. We now consider one point unions of free π-spheres (compare (4.4))

$$A = \bigvee_Z S^n_\pi = \Sigma^{n-1}\left(\bigvee_Z S^1_\pi\right) = \Sigma^{n-1}X \tag{2}$$

with arbitrary index set Z. The inclusion $i\colon S^n \subset S^n_\pi$ corresponding to $0 \in \pi$ yields the inclusion

$$i\colon A_0 = \bigvee_Z S^n \subset A = \bigvee_Z S^n_\pi. \tag{3}$$

This inclusion has the following universal property. Let U be a space with a (right) action of a group π' for which $* \in U$ is a fixpoint and let $\varphi\colon \pi \to \pi'$ be a homomorphism. Then any basepoint preserving map $u\colon A_0 \to U$ in **Top*** yields a unique φ-equivariant map

$$\bar{u} = u \odot \varphi\colon A \to U \tag{4}$$

with $\bar{u}i = u$. Moreover, any φ-equivariant map $\bar{u}\colon A \to U$ is of this form, that is $\bar{u} = u \odot \varphi$ for $u = \bar{u}i$. Similar properties hold for homotopy classes of φ-equivariant maps $A \to U$. We now are ready for the definition of the 'equivariant analogue' of the category $\mathbf{S}(n)$ in (4.4).

(5.2) **Definition.** We define the homotopy category $\hat{\mathbf{S}}(n)$ of one point unions of n-spheres with operators in groups. Objects are pairs (π, A) where A is a one point union of free π-spheres as in (5.1)(2). Morphisms are pairs $(\varphi, \{f\})\colon (\pi, A) \to (\pi', B)$ where $\varphi\colon \pi \to \pi'$ is a homomorphism and where $\{f\}$ is a homotopy class of φ-equivariant maps $f\colon A \to B$.

The homotopy category $\hat{\mathbf{S}}(n)$ is part of the topological linear track extension

(5.3) $$\hat{D}_\Sigma \rightarrowtail T \rightrightarrows E\hat{\mathbf{S}}(n) \twoheadrightarrow \hat{\mathbf{S}}(n)$$

which is the equivariant analogue of the linear track extension in (4.4)(1). The category $E\hat{\mathbf{S}}(n)$ has the same objects as $\hat{\mathbf{S}}(n)$ but the morphisms are all pairs (φ, f) where $f\colon A \to B$ is a φ-equivariant map. The set of tracks $(\varphi, f) \simeq (\psi, g)$ is non empty only if $\varphi = \psi$ and this set is defined by

$$T((\varphi, f), (\varphi, g)) = T(f, g)_\varphi = [I_* A, B]^{f,g}_\varphi. \tag{1}$$

Here the right hand side denotes the subset of φ-equivariant tracks in $T(f, g)$, see (3.7) and (4.4)(1). The natural system $\hat{D}_\Sigma$ in (5.3) is defined for $(\varphi, \{f\})$ by the group

$$\hat{D}_\Sigma(\varphi, \{f\}) = [\Sigma A, B]_\varphi \tag{2}$$

of φ-equivariant homotopy classes $\Sigma A \to B$. We obtain the isomorphism σ (see (3.11)(3)) for the linear track extension (5.3) by the composition

$$\sigma_f\colon [\Sigma A, B]_\varphi = [\Sigma A_0, B] = T(fi, fi) = T(f, f)_\varphi \tag{3}$$

where we use the inclusion $i\colon A_0 \subset A$ in (5.1)(3) and the isomorphism σ in (3.13). Moreover we use in (3) the universal property of i described in (5.1)(4). It is easy to check that (5.3) is a well defined linear track extension of the category $\hat{\mathbf{S}}(n)$, see (3.13). We now describe the algebraic equivalent of this track extension. For this we first observe that the homology functor H_n and the suspension functor Σ yield equivalences of categories

$$H_n\colon \hat{\mathbf{S}}(n) \xrightarrow[\sim]{\Sigma} \mathbf{S}(n+1) \xrightarrow[\sim]{H_{n+1}} (\mathbf{Mod}_{\mathbb{Z}}^{\wedge})_c \tag{4}$$

for $n \geq 2$. Recall that $(\mathbf{Mod}_{\mathbb{Z}}^{\wedge})_c$ is the category consisting of pairs (π, C) where C is a free π-module. Therefore we obtain the equivalence (4) by (5.1)(1). Moreover we have the natural isomorphism

$$\hat{D}_\Sigma(\varphi, \{f\}) = [\Sigma A, B]_\varphi \cong \operatorname{Hom}_\varphi(H_n A, \Gamma_n^1 H_n B) \tag{5}$$

for $(\pi, A), (\pi', B) \in \hat{\mathbf{S}}(n)$; compare (4.4)(1). We use (4) and (5) as identifications so that (5.3) corresponds to a linear track extension of $(\mathbf{Mod}_{\mathbb{Z}}^{\wedge})_c$ by the natural system $\operatorname{Hom}(-, \Gamma_n^1 -)$. Such a linear track extension is obtained algebraically as follows. Recall that the category $\mathbf{nil}^\wedge$ of free π-nil(2)-groups is embedded in the linear extension

(5.5) $$\operatorname{Hom}(-, \Lambda^2) + \rightarrowtail \mathbf{nil}^\wedge \twoheadrightarrow (\mathbf{Mod}_{\mathbb{Z}}^{\wedge})_c$$

as in (2.5). The short exact sequence (4.3)(1) yields the short exact sequence

$$\operatorname{Hom}(A, \Gamma_n^1 B)_\varphi \xrightarrow{\tau_*} \operatorname{Hom}(A, T_n' B)_\varphi \xrightarrow{q_*} \operatorname{Hom}(A, \Lambda^2 B)_\varphi \tag{1}$$

of natural systems with $A, B \in (\mathbf{Mod}_{\mathbb{Z}}^{\wedge})_c$. Whence we get by (4.1)(2) the linear track extension

$$\operatorname{Hom}(-, \Gamma_n^1) + \rightarrowtail T_n \Rrightarrow \mathbf{nil}^\wedge \twoheadrightarrow (\mathbf{Mod}_{\mathbb{Z}}^{\wedge})_c \tag{2}$$

where T_n is the track structure associated to the exact sequence (1).

(5.6) **Theorem.** *The topological linear track extension for $\hat{\mathbf{S}}(n)$ in (5.3) is equivalent to the algebraic linear track extension $T_n\,\mathbf{nil}^\wedge$ in (5.5)(2), $n \geq 2$.*

Proof. We construct maps between linear track extensions

$$TE\hat{\mathbf{S}}(n) \xleftarrow{i} T\hat{\mathbf{\Sigma}}(n) \xrightarrow{\hat{t}} T_n\,\mathbf{nil}^\wedge \tag{1}$$

Here $\hat{\mathbf{\Sigma}}(n)$ is the subcategory of $E\hat{\mathbf{S}}(n)$ consisting of all maps which are $(n-1)$-fold suspensions $\Sigma^{n-1}(f)$ with f in $E\hat{\mathbf{S}}(1)$. Moreover the map i is the inclusion of track categories. We define the map $\hat{t}$ similarly as in (4.7)(2). The functor

$$\hat{t}\colon \hat{\mathbf{\Sigma}}(n) \to \mathbf{nil}^\wedge \tag{2}$$

carries the object $A = \Sigma^{n-1}X$ in (5.1)(2) to the free π-nil(2)-group

$$\hat{t}(A) = G_X = \pi_1(X)/\Gamma_3\pi_1(X). \tag{3}$$

Moreover $\hat{t}$ carries $\Sigma^{n-1}f$ to the map induced by $\pi_1(f)$. We then get the map $\hat{t}$ on tracks by t in (4.7)(5) and by the universal property (5.1)(4). That is, $\hat{t}$ is the composition

$$\begin{aligned} T(\Sigma^{n-1}f, \Sigma^{n-1}g)_\varphi &\cong T(\Sigma^{n-1}(fi), \Sigma^{n-1}(gi)) \\ &\cong T_n(t(fi), t(gi)) \\ &\cong T_n(\hat{t}f, \hat{t}g)_\varphi \end{aligned} \tag{4}$$

where the second bijection is given by t in (4.7)(5). □

Using (4.2), (5.6) and (3.16) we get the following result on the characteristic cohomology class $\langle\hat{\mathbf{S}}(n)\rangle$ of the linear track extension (5.3).

(5.7) **Corollary.** $\beta_n\{\mathbf{nil}^\wedge\} = \langle\hat{\mathbf{S}}(n)\rangle$.

Here β_n is the Bockstein operator

$$\beta_n\colon H^2(\mathbf{C}, \mathrm{Hom}(-, \Lambda^2)) \to H^3(\mathbf{C}, \mathrm{Hom}(-, \Gamma_n^1)) \tag{1}$$

associated to the exact sequence of natural systems (5.5)(1) on $\mathbf{C} = (\mathbf{Mod}_{\mathbb{Z}}^{\wedge})_c$. Moreover $\{\mathbf{nil}^\wedge\}$ is the cohomology class in (2.6) which, as we have seen in (2.7), is an element of order 2. By (5.7) also the class $\langle\hat{\mathbf{S}}(n)\rangle$ is an element of order 2. The Bockstein operator (1) has similar properties as in (4.6)(2), (3), that is

$$\beta_n = \sigma_*\beta_2 \quad \text{for} \quad n \geq 3 \tag{2}$$

where σ_* is induced by $\mathrm{Hom}(1, \sigma)$. Whence

$$\beta_n\{\mathbf{nil}^\wedge\} = \sigma_*\beta_2\{\mathbf{nil}^\wedge\} = \Sigma_*^{n-2}\langle\hat{\mathbf{S}}(2)\rangle = \langle\hat{\mathbf{S}}(n)\rangle. \tag{3}$$

Here Σ is the suspension operator on the natural system $\hat{D}_\Sigma$ and in the right hand equation of (3) we use identifications as in (5.3)(4). Next we describe a connection of the class $\langle\hat{\mathbf{S}}(n)\rangle$ with the algebraic K-theory of group rings. Let

$$j\colon M_N(\mathbb{Z}[\pi]) \cong \mathbf{End}\left(\bigvee^N S_\pi^n\right) \subset \hat{\mathbf{S}}(n) \tag{4}$$

be the full subcategory of $\hat{\mathbf{S}}(n)$ consisting of the single object $\bigvee^N S_\pi^n$ which is the N-fold one point union of the free π-sphere S_π^n. Using the equivalence (5.3)(4) this is the monoid $M_N(\mathbb{Z}[\pi])$ of $N \times N$-matrices over the group ring $\mathbb{Z}[\pi]$. One can check that the class $\chi(\pi)$ defined in Igusa actually satisfies for $n \geq 3$ the equation

$$\chi(\pi) = \left\langle \mathbf{End}\left(\bigvee^N S_\pi^n\right)\right\rangle = j^*\langle\hat{\mathbf{S}}(n)\rangle = \beta_n j^*\{\mathbf{nil}^\wedge\}. \tag{5}$$

As in (4.6)(6) we point out that Igusa defined the class $\chi(\pi)$ only by describing an explicit and very complicated cocycle representing $\chi(\pi)$. By (5) one has now a simple and elegant algebraic definition of $\chi(\pi)$ by $\beta_n\{\mathbf{nil}^\wedge\}$ and also a new topological interpretation by the characteristic cohomology class $\langle\hat{\mathbf{S}}(n)\rangle$. This yields by Igusa's work the connection of this class with algebraic K-theory. The tedious proof of (5) consists in the comparison of Igusa's cocycle with a cocycle representing $\beta_n\{\mathbf{nil}^\wedge\}$ which can be derived from the explicit cocycle $\Delta_{\hat{\imath}}$ representing $\{\mathbf{nil}^\wedge\}$ in (2.11)(26).

We also remark that the restriction of the class $\langle\hat{\mathbf{S}}(n)\rangle$ to the group of automorphisms

$$GL_N(\mathbb{Z}[\pi]) = \mathbf{Aut}\left(\bigvee^N S_\pi^n\right) \subset \hat{\mathbf{S}}(n) \tag{6}$$

yields the first k-invariant k_2 of the classifying space $B(\mathscr{E}_N)$ where $\mathscr{E}_N$ is the topological monoid of equivariant homotopy equivalences of $\bigvee^N S_\pi^n$, compare (3.18)(4). This holds for $n \geq 2$.

§ 6 Free nil(2)-modules and tracks for 2-dimensional CW-complexes

In this section we describe the connection of free nil(2)-modules and 2-dimensional CW-complexes. Using free nil(2)-modules we obtain an algebraic track category which is equivalent to the topological track category of 0-tracks on 2-dimensional CW-complexes. This fact is actually a reason why nil(2)-modules play a crucial role in the definition of quadratic chain complexes in chapter IV. The equivalence of track categories here is related to the corresponding equivalences of track-categories in (4.5) and (5.6) above.

Recall that a nil(2)-module is a pre-crossed module of Peiffer nilpotency degree 2. For a totally free nil(2)-module $\partial: \sigma_2 \to \sigma_1$ we have the following natural commutative diagram.

(6.1)

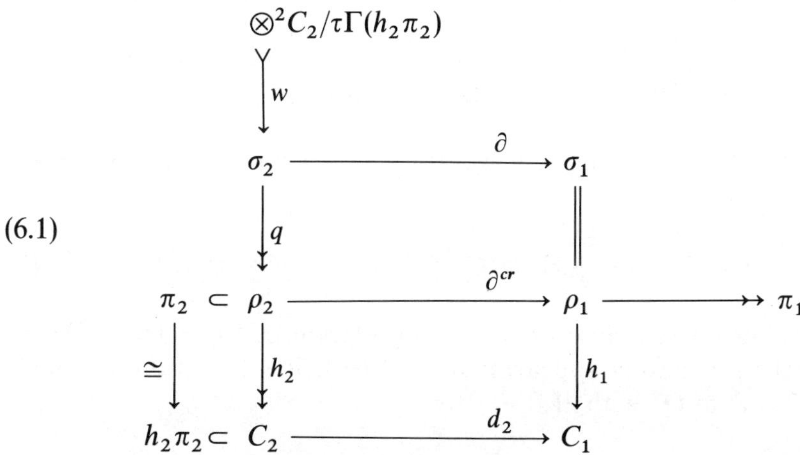

Here $\rho_2 = \sigma_2^{cr} = \mathrm{cok}(w)$ is the cokernel of the Peiffer commutator map w given by the exact sequence (IV.1.8). Moreover $\pi_2 = \ker(\partial^{cr})$ is the kernel of the associated crossed module ∂^{cr} and $C_2 = \rho_2^{ab}$ is the abelianization. The maps q and h_2 are the quotient maps and h_1 and d_2 are given as in (III.2.11) so that $h_2\pi_2 = \ker(d_2)$. Clearly $\pi_1 = \mathrm{cok}\, \partial^{cr}$.

Let $\mathbf{H}^2$ be the category of totally free crossed modules and let $\mathbf{Q}^2$ be the category of totally free nil(2)-modules. These are the subcategories of 2-dimensional objects in $\mathbf{H}$ and $\mathbf{Q}$ respectively, see (IV.1.14). Using diagram (6.1) we obtain a linear extension of categories

(6.2) $$\mathrm{Hom}(C_2, \Delta) + \rightarrowtail \mathbf{Q}^2 \xrightarrow{\lambda} \mathbf{H}^2.$$

Here λ carries ∂ to ∂^{cr} as in (IV.3.2). For the natural system $\mathrm{Hom}(C_2, \Delta)$ we need the functors

$$C_2, \pi_2, \Delta: \mathbf{H}^2 \to \mathbf{Mod}_{\mathbb{Z}}^{\wedge}. \tag{1}$$

For an object $\delta: \rho_2 \to \rho_1$ in $\mathbf{H}^2$ we obtain these functors as in (6.1) by $C_2 = \rho_2^{ab}$,

$\pi_2 = \ker(\delta)$ and

$$\Delta(\delta) = \otimes^2 C_2/\tau\Gamma(h_2\pi_2). \tag{2}$$

This yields the short exact sequence

$$\Gamma\pi_2 \rightarrowtail \otimes^2 C_2 \twoheadrightarrow \Delta \tag{3}$$

of functors $\mathbf{H} \to \mathbf{Mod}_{\mathbb{Z}}^{\wedge}$. The natural system $\mathrm{Hom}(C_2, \Delta)$ in (6.2) carries a map $f: \delta \to \delta'$, which induces $\varphi = \pi_1(f)$ on fundamental groups, to the abelian group

$$\alpha \in \mathrm{Hom}(C_2, \Delta)_f = \mathrm{Hom}_\varphi(C_2(\delta), \Delta(\delta')). \tag{4}$$

Since $C_2(\delta)$ is a free $\pi_1(\delta)$-module we see that (3) induces a short exact sequence

$$\mathrm{Hom}(C_2, \Gamma\pi_2) \rightarrowtail \mathrm{Hom}(C_2, \otimes^2 C_2) \twoheadrightarrow \mathrm{Hom}(C_2, \Delta) \tag{5}$$

of natural systems on $\mathbf{H}^2$. Here we set as in (4) $\mathrm{Hom}(M, N)_f = \mathrm{Hom}_\varphi(M(\delta), N(\delta'))$ for functors M and N appearing in (5). We define the action of α in (4) on a map $g: \partial \to \partial'$ in $\mathbf{Q}^2$ with $\lambda(g) = f$ by

$$g + \alpha = (g_2 + w\alpha h_2 q, g_1) \tag{6}$$

where $g = (g_2, g_1)$. Since w is central and since $\ker(q) = \mathrm{image}(w)$, see (6.1), we see that the linear extension (6.2) is well defined. We now get by (4.1)(2) the linear track extension

$$\mathrm{Hom}(C_2, \Gamma\pi_2) + \rightarrowtail T_0 \Rrightarrow \mathbf{Q}^2 \overset{\lambda}{\twoheadrightarrow} \mathbf{H}^2 \tag{7}$$

where T_0 is the track structure associated to the exact sequence (5). We point out that for maps f, g in $\mathbf{Q}^2$ the set $T_0(f, g)$ in (7) coincides with the set of 0-homotopies $\alpha_2: f \overset{0}{\simeq} g$ defined in (IV.4.8). Therefore we have as in (7) the isomorphism

$$\mathbf{Q}^2/\overset{0}{\simeq} \;\cong \mathbf{H}^2. \tag{8}$$

We now describe a topological linear track extension

(6.3) $$\mathrm{Hom}(C_2, \Gamma\pi_2) + \rightarrowtail T_0 \Rrightarrow \mathbf{CW}^2 \longrightarrow \mathbf{CW}^2/\overset{0}{\simeq}$$

which is equivalent to the algebraic linear track extension in (6.2)(7) above.

Here $\mathbf{CW}^2$ is the category of 2-dimensional CW-complexes X with $X^0 = *$ and of cellular maps $f: X \to Y$. A 0-homotopy $H: f \overset{0}{\simeq} g$ is a homotopy running through cellular maps H_t, $t \in I$, see (I.2.4). A 0-**track** $\{H\}: f \overset{0}{\simeq} g$ is a homotopy class rel $X \vee X$ of such 0-homotopies. Let

$$T_0(f,g) = [(I_* X, I_* X^1), (Y, Y^1)]^{(f,g)} \tag{1}$$

be the set of such 0-tracks. Here the right hand side is a set of homotopy classes under (f,g) in the cofibration category $\mathbf{C} = \mathbf{Pair(Top^*)}$, see (II.1.5) in Baues (AH). We know by (III.7.1) that we have an equivalence of categories

$$\rho: \mathbf{CW}^2/\overset{0}{\simeq} \xrightarrow{\sim} \mathbf{H}^2 \tag{2}$$

which we use as an identification. Using this equivalence the natural system $\mathrm{Hom}(C_2, \Gamma\pi_2)$ in (6.3) corresponds to the natural system in (6.2)(7), that is

$$\mathrm{Hom}(C_2, \Gamma\pi_2)(f) = \mathrm{Hom}_\varphi(\hat{C}_2 X, \Gamma\pi_2 Y) \tag{3}$$

with $\varphi = \pi_1(f)$. For the linear track extension in (6.3) we need the isomorphism (see (3.11))

$$\sigma: \mathrm{Hom}_\varphi(\hat{C}_2 X, \Gamma\pi_2 Y) \cong T_0(f,f) \tag{4}$$

which we obtain as follows. For this we first observe that $T_0(f,f)$ satisfies

(i) $$T_0(f,f) \overset{\cong}{\longleftarrow} [I_* X, Y]^F$$

where $F: (X \vee X) \cup I_* X^1 \to Y$ is given by (f,f) on $X \vee X$ and by $f^1 p$ on $I_* X^1$. Here $p: I_* X^1 \to X^1$ is the projection. There is a canonical map as in (i) which is a bijection since $\pi_2 Y^1 = 0$ and $\pi_3 Y^1 = 0$ are trivial groups. We now choose an attaching map of 2-cells in X,

$$f_2: \bigvee_Z S^1 = A \to X^1,$$

as in (I.3.10). Then we know that $I_* X = C_w$ is the mapping cone of a map

$$w: \Sigma A \to (X \vee X) \cup I^* X^1,$$

compare (II.8.12) Baues (AH). Whence we get by (II.8.9) Baues (AH) and by (I.4.11) the isomorphism:

$$\begin{aligned} T_0(f,f) &\cong [I_*X, Y]^F \\ &\cong [\Sigma^2 A, Y] \\ \text{(ii)} \qquad &\cong \underset{Z}{\times}\, \Gamma(\pi_2 Y) \\ &\cong \mathrm{Hom}_\varphi(\hat{C}_2 X, \Gamma\pi_2 Y) \end{aligned}$$

which carries the trivial track to the trivial homomorphism. For the isomorphism (ii) we use the basis $Z \subset \hat{C}_2 X$ which is given by f_2 as in (I.3.12). The composition of isomorphisms in (ii) yields σ in (4). We leave it to the reader to check that σ has the naturality properties described in (3.11), so that (6.3) is a well defined linear track extension. Using the identification of $\mathbf{H}^2$ and $\mathbf{CW}^2/\overset{0}{\simeq}$ in (6.3)(2) we now can state the following result:

(6.4) **Theorem.** *The topological linear track extension for* $\mathbf{CW}^2/\overset{0}{\simeq}$ *in* (6.3) *and the algebraic linear track extension for* $\mathbf{H}^2$ *in* (6.2)(7) *are equivalent.*

Proof. We construct maps between linear track extensions

$$T_0\mathbf{CW}^2 \xleftarrow{\;i\;} T_0\mathbf{W} \xrightarrow{\;j\;} T_0\mathbf{Q}^2. \tag{1}$$

Here $\mathbf{W}$ is the following category obtained by the construction of $\sigma(X)$ in (IV.6.8)(5). Objects in $\mathbf{W}$ are triples $(X, \sigma X, b)$ with $X \in \mathbf{CW}^2$, see (IV.6.8)(2). Morphisms in $\mathbf{W}$ from $(X, \sigma X, b^X)$ to $(Y, \sigma Y, b^Y)$ are triples $f' = (f, \sigma f, H_f)$ where $f\colon X \to Y$ is a map in $\mathbf{CW}^2$ and where

$$\begin{array}{ccc} \sigma_S(iX) & \overset{b^X}{\underset{\sim}{\longleftarrow}} & j\sigma X \\ \Big\downarrow f_* & \overset{H_f}{\Leftarrow} & \Big\downarrow \sigma f \\ \sigma_S(iY) & \overset{b^Y}{\underset{\sim}{\longleftarrow}} & j\sigma Y \end{array} \tag{2}$$

is a diagram in the track category $\mathbf{Fil(Q)}$, see (3.9). There is an obvious composition law for such morphisms. A track H in $T_0\mathbf{W}$ from f' to g' is the same as a track $H \in T_0(f,g)$, see (6.3)(1). The functor i in (1) is the forgetful functor which is the identity on tracks. Moreover the functor j in (1) carries f' to σf and carries the track H to the unique track jH with

$$b^Y_* jH = -H_f + H_*\psi + H_g. \tag{3}$$

Here $\psi\colon I_*\sigma_S(iX) \xrightarrow{\sim} \sigma_S I_* iX$ is an equivalence of cylinders in $\mathbf{Fil(Q)}$. The functors i and j both are equivariant with respect to the action of the natural system $\mathrm{Hom}(C_2, \Gamma\pi_2)$. □

Using (4.2), (6.4) and (3.16) we get the following result on the characteristic cohomology class $\langle \mathbf{CW}^2/\overset{0}{\simeq} \rangle$ of the linear track extension (6.3)

(6.5) Corollary. $\beta\{\mathbf{Q}^2\} = \langle \mathbf{CW}^2/\overset{0}{\simeq} \rangle$

Here β is the Bockstein operator

$$\beta\colon H^2(\mathbf{H}^2, \mathrm{Hom}(C_2, \Delta)) \to H^3(\mathbf{H}^2, \mathrm{Hom}(C_2, \Gamma\pi_2))$$

associated to the exact sequence of natural systems on $\mathbf{H}^2$ in (6.2)(5). Moreover $\{\mathbf{Q}^2\}$ is the cohomology class given by the category $\mathbf{Q}^2$ of totally free nil(2)-modules via the linear extension (6.2), see (2.1). The equation (6.5) corresponds to the equations for $\langle \mathbf{S}(2) \rangle$ and $\langle \hat{\mathbf{S}}(2) \rangle$ in (4.6) and (5.7) respectively. In fact the cohomology class in (6.5) is non trivial since $\mathbf{S}(2) \subset \mathbf{CW}^2/\overset{0}{\simeq}$ is a full subcategory. We know that $2\langle \mathbf{S}(2) \rangle = 0$ and $2\langle \hat{\mathbf{S}}(2) \rangle = 0$, however, we do not know whether $2\langle \mathbf{CW}^2/\overset{0}{\simeq} \rangle$ is trivial or not. There might be a possibility by use of the method in (2.7) to show that $2\{\mathbf{Q}^2\}$ is trivial.

Next we compare the class $\langle \mathbf{CW}^2/\overset{0}{\simeq} \rangle$ with the corresponding classes $\langle \mathbf{S}(2) \rangle$ and $\langle \hat{\mathbf{S}}(2) \rangle$. For this we compare the linear extension for $\mathbf{Q}^2$ in (6.2) with the linear extensions for **nil** and **nil**^ respectively, see (4.3) and (5.5). One has the following commutative diagram of linear extensions.

$$(6.6)\qquad \begin{array}{ccccc}
\mathrm{Hom}(-, \Lambda^2) & \overset{+}{\rightarrowtail} & \mathbf{nil} & \twoheadrightarrow & \mathbf{Ab}_c \\
\downarrow 1 & & \downarrow i & & \downarrow i \\
\mathrm{Hom}(C_2, \Delta) & \overset{+}{\rightarrowtail} & \mathbf{Q}^2 & \twoheadrightarrow & \mathbf{H}^2 \\
\downarrow p_* & & \downarrow E_2 & & \downarrow C_2 \\
\mathrm{Hom}(-, \Lambda^2) & \overset{+}{\rightarrowtail} & \mathbf{nil}^\wedge & \twoheadrightarrow & (\mathbf{Mod}_{\mathbb{Z}}^{\wedge})_c
\end{array}$$

Here the top row is the restriction of the extension $\mathbf{Q}^2 \twoheadrightarrow \mathbf{H}^2$ to all objects $\rho_2 \to \rho_1$ in $\mathbf{H}^2$ with $\rho_1 = 0$; in this case ρ_2 is just a free abelian group. The functor i is the inclusion. The functor C_2 is already defined in (6.2)(1). We now obtain the functor E_2 in (6.6) by the following natural commutative diagram.

$$\begin{array}{ccccc} \Delta(\partial) & \rightarrowtail & \sigma_2 & \xrightarrow{q} \!\!\!\!\twoheadrightarrow & \rho_2 \\ \downarrow p & & \downarrow h & & \downarrow h_2 \\ \Lambda^2 C_2 & \rightarrowtail & E_2 & \longrightarrow & C_2 \end{array} \tag{1}$$

Here $\partial: \sigma_2 \to \sigma_1$ is a totally free nil(2)-module in $\mathbf{Q}^2$ with basis $Z_2 \to \sigma_1$ and E_2 is the free $\pi_1(\partial)$-nil(2)-group with basis Z_2. The map h is given as follows. We may consider the trivial map

$$0_E: E_2 \to \pi_1(\partial) \tag{2}$$

to be the free nil(2)-module with trivial basis $0(x) = 0$ for $x \in Z_2$. Since $\partial: \sigma_2 \to \sigma_1$ is free there is a unique map $h: \partial \to 0_E$ between nil(2)-modules which is the identity on Z_2 and which is the quotient map $\sigma_1 \twoheadrightarrow \pi_1(\partial)$ in degree 1. The map h yields the natural surjection h in (1). The map h commutes with the quotient map p in (1) which is given by $\otimes^2 C_2 \to \Lambda^2 C_2$. Now it is clear that diagram (1) yields a functor E_2 as in (6.6) which carries ∂ to $E_2(\partial) = E_2$; (we used diagram (1) already in the proof (V.5.5)(2)). The maps between linear extensions in (6.6) yield as well obvious functors between track categories

(6.7) $$T_2\,\mathbf{nil} \xrightarrow{i} T_0\mathbf{Q}^2 \xrightarrow{E_2} T_2\,\mathbf{nil}^\wedge$$

which are given by (4.3)(3), (6.2)(7) and (5.5)(2) respectively. Here i is the inclusion and E_2 is injective on tracks by $\Gamma(\pi_2) \rightarrowtail \Gamma(C_2)$. There are functors between topological track categories

(6.8) $$T\mathbf{ES}(2) \xrightarrow{i} T_0\mathbf{CW}^2 \xrightarrow{Q} T\mathbf{E}\hat{\mathbf{S}}(2)$$

such that these functors are "equivalent" to the corresponding algebraic functors in (6.7). The equivalences of track categories are already described in (4.5), (5.6) and (6.4) respectively. The functor i in (6.8) is the inclusion which carries a one point union $\bigvee S^2$ of 2-spheres to the CW-complex $X = \bigvee S^2$ with $X^1 = *$. The **quotient functor** Q in (6.8) carries $X \in \mathbf{CW}^2$ to the quotient space

$$Q(X) = \hat{X}/\hat{X}^1 = \bigvee_Z S^2_\pi \tag{1}$$

where $\hat{X}$ is the universal covering of X with 1-skeleton $\hat{X}^1$. The space $Q(X)$ is a one point union of free π-spheres S^2_π where $\pi = \pi_1 X$ and where Z in (1) is the set of 2-cells in X. The inclusion (5.1)(3) corresponds to the inclusion $Z \subset \hat{C}_2 X = H_2 Q(X)$ given as in (I.3.11). The functor Q carries a cellular map

$f: X \to Y$ to the map $Q(f)$ induced by $\hat{f}$, see (I.2.1). Moreover Q carries a homotopy H to the homotopy QH with $(QH)_t = Q(H_t)$ for $t \in I$. On tracks the functor Q is equivariant with respect to the injective homomorphism

$$\begin{array}{ccc} \mathrm{Hom}(\hat{C}_2 X, \Gamma\pi_2 Y)_\varphi & \xrightarrow{(\Gamma i)_*} & \mathrm{Hom}(\hat{C}_2 X, \Gamma\hat{C}_2 Y)_\varphi \\ \wr\| & & \wr\| \\ T_0(f,f) & \xrightarrow{Q} & T(Qf, Qf)_\varphi \end{array} \qquad (2)$$

where $\varphi = \pi_1(f)$, see (5.3)(3), (5). The map $(\Gamma i)_*$ is induced by the inclusion $i: \pi_2 Y \cong H_2 \hat{C}_* Y \subset \hat{C}_2 Y$. This shows that the functor Q in (6.8) is injective on tracks. The functor Q in (6.8) is equivalent to the functor E_2 in (6.7) in the sense that there is a commutative diagram of functors between track categories:

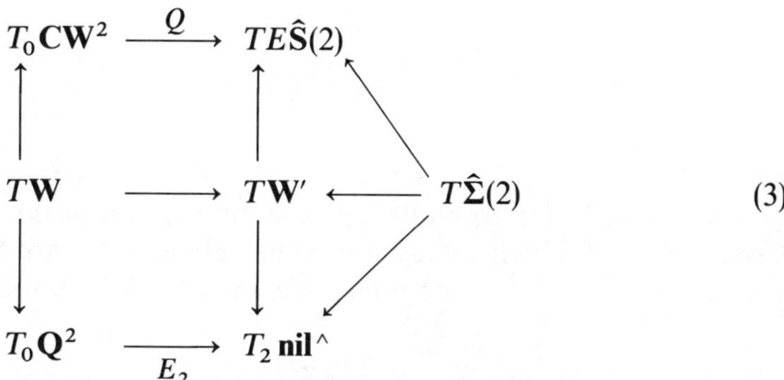

(3)

All vertical arrows are bijective on tracks. Here we use the equivalences constructed in the proof of (6.4) and (5.6) respectively. We leave it to the reader to find an appropriate definition for the track category $T\mathbf{W}'$ in (3). In a similar way the inclusion functor i in (6.8) is equivalent to the inclusion functor i in (6.7).

§7 Track-models for 4-dimensional CW-complexes

We have seen in the last section §6 that the category of 0-tracks on 2-dimensional CW-complexes is equivalent to the algebraic track category $T_0\mathbf{Q}^2$ of 0-tracks on totally free nil(2)-modules. In this section we use this equivalence for the construction of the quadratic chain complex $\sigma(X)$ of a 4-dimensional CW-complex X. For this we first describe X by use of a 0-track H_X on 2-dimensional CW-complexes. This 0-track H_X corresponds by the equivalence in §6 to a 0-track in $T_0\mathbf{Q}^2$ which in turn yields the boundary map

d_4 in $\sigma(X)$ by a certain algebraic construction. This way one gets a further geometric interpretation of the quadratic chain complex $\sigma(X)$.

Let $X = X^4$ be a CW-complex of dimension 4 with trivial 0-skeleton $X^0 = *$. We assume that X has basepoint preserving attaching maps

$$f_n\colon \Sigma^{n-2}V_n \to X^{n-1} \quad \text{with} \quad V_n = \bigvee_{Z_n} S^1. \tag{7.1}$$

Here Z_n is the discrete set of n-cells of X and the mapping cone Cf_n of f_n is the n-skeleton X^n. Thus we obtain X by the iterated mapping cone.

$$\begin{array}{rcl} & & X^4 = Cf_4 \\ & & \cup \\ \Sigma^2 V_4 & \xrightarrow{f_4} & X^3 = Cf_3 \\ & & \cup \\ \Sigma V_3 & \xrightarrow{f_3} & X^2 = Cf_2 \\ & & \cup \\ V_2 & \xrightarrow{f_2} & X^1 = V_1 \end{array} \tag{7.2}$$

where f_2 corresponds to a presentation $f_2\colon Z_2 \to \langle Z_1 \rangle$ of the fundamental group $\pi = \pi_1 X$. Up to homotopy the map f_4 can be constructed as follows. Consider the following diagram where H'_X and G_0 are tracks and where 0 denotes the trivial map, compare also (VII.1.21) in Baues (AH).

$$\begin{array}{ccccc} & \Sigma V_4 & \xrightarrow{\quad 0 \quad} & & \\ & \downarrow \delta_4^V & \nearrow G_0 & & \searrow \\ 0 \;\; \overset{\Rightarrow}{H'_X} & \Sigma V_3 \vee X^1 & \subset & C\Sigma V_3 \vee X^1 & \\ & \downarrow (f_3, i) & \text{push} & \downarrow \pi & \\ & X^2 & \underset{i}{\subset} & X^3 & \end{array} \tag{7.3}$$

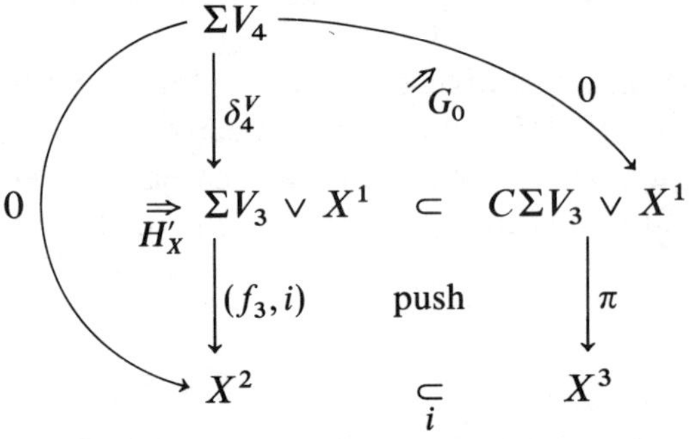

Here $C\Sigma V_3$ is the cone of ΣV_3 and the subdiagram "push" is a push out diagram by definition of the mapping cone $X^3 = Cf_3$. By addition of tracks we get the track $iH'_X + \pi G_0\colon 0 \simeq 0$. This track defines the homotopy class of a map f_4, namely

$$f_4 = iH'_X + \pi G_0\colon \Sigma^2 V_4 \to X^3. \tag{7.4}$$

Since $\pi_3 X^1 = 0$ there is only a unique track G_0. Whence the homotopy class of f_4 is completely determined by the track H'_X. For δ_4^V in (7.3) the diagram

(7.5)
$$\begin{array}{ccccc} Z_4 & \xrightarrow{\delta_4^V} & \pi_2(\Sigma V_3 \vee X^1) & = & \mathbb{Z}[Z_3 \times \sigma_1] \\ \cap & & & & \downarrow q \\ C_4 & \xrightarrow{d_4} & C_3 & = & \mathbb{Z}[Z_3 \times \pi] \end{array}$$

commutes. Here $q\colon \sigma_1 = \langle Z_1 \rangle \twoheadrightarrow \pi = \pi_1 X$ is the projection and d_4 is the differential of $\hat{C}_* X$.

(7.6) **Lemma.** *Let f_4 and d_4 be given by X in* (7.2) *and choose δ_4^V such that* (7.5) *commutes. Then there exists a track $H'_X\colon 0 \simeq (f_3, i)\delta_4^V$ for which equation* (7.4) *is satisfied.*

Proof. We know by (III.5.11) and (V.2.6) in Baues (AH) that f_4 is a twisted maps associated to an appropriate map

$$\delta_4''\colon \Sigma V_4 \to \Sigma V_3 \vee X^2 \tag{1}$$

which is trivial on X^2. Equivalently

$$f_4 \in E_{f_3}(\delta_4'') \tag{2}$$

is a functional suspension. The map δ_4'' admits up to homotopy a factorization

$$\Sigma V_4 \xrightarrow{\delta_4^V} \Sigma V_3 \vee X^1 \overset{i}{\subset} \Sigma V_3 \vee X^2 \tag{3}$$

where i is given by the inclusion $X^1 \subset X^2$, see (III.5.11)(c) Baues (AH). The construction of a functional suspension f_4 as in (2) exactly corresponds to equation (7.4). Moreover, δ_4'' in (1) induces d_4 in (7.5), see (III.5.10)(1) Baues (AH). □

We shall use the following notation. The one point union $A \vee D$ in **Top*** is a space under and over D by

(7.7)
$$D \xrightarrow{i_2} A \vee D \xrightarrow{(0,1)} D$$

where i_2 is the inclusion and where $p = (0, 1)$ is the projection which maps A to the basepoint. The cylinder of $A \vee D$ satisfies $I_*(A \vee D) = I_*A \vee I_*D$. Whence for maps $(f, g), (f', g')\colon A \vee D \to U$ a track $H\colon (f, g) \simeq (f', g')$ is of the form $H = (H_1, H_2)$ where $H_1\colon f \simeq f'$ and $H_2\colon g \simeq g'$ are tracks. We say that H is **constant** on D if $g = g'$ and if $H_2 = 0$ is the trivial track. We now associate with diagram (7.3) the following diagram which we call the **track-model** of X.

(7.8)

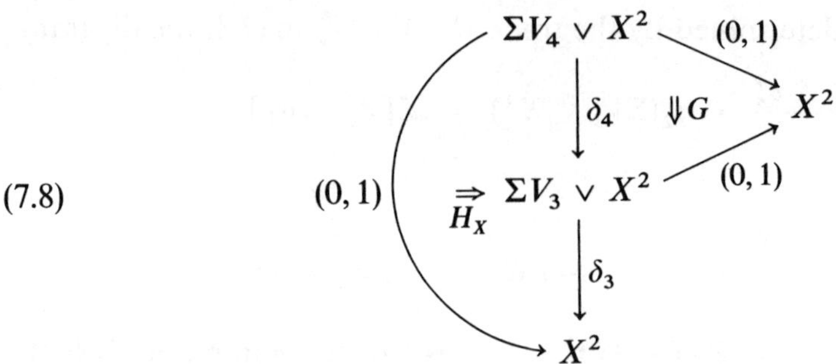

Here δ_4 and δ_3 are maps under X_2 and H_X and G are tracks which are constant on X^2. Moreover, δ_4 and G are **good** in the following sense. The map $\delta_4 i_1 = i\delta_4^V$ admits a factorization δ_4^V as in the following diagram

(7.9)

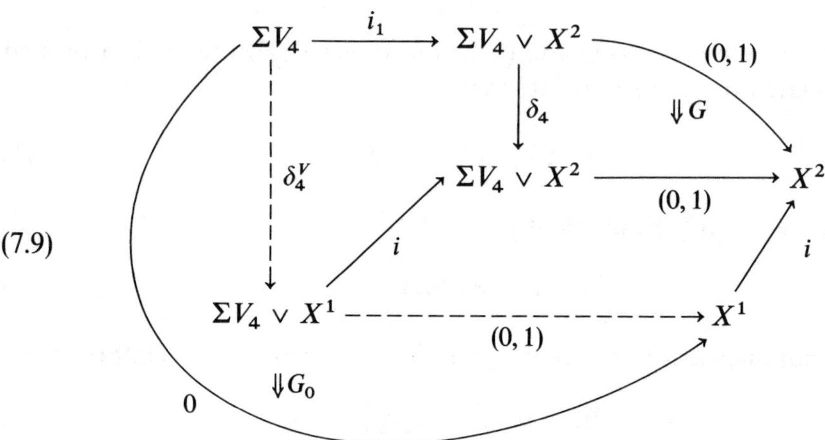

and the addition of tracks in (7.9) yields the trivial track from $0 = (0, 1)i_1$ to $0 = i0$. We point out that there is a unique track G_0, so that there is a unique good track G in (7.8).

We derive a diagram as in (7.9) from (7.3) by setting $\delta_4 i_1 = i\delta_4^V$, $\delta_3 = (f_3, 1)$ and $H_X = (H'_X, 0)$, see (7.7). The space $\Sigma V_4 \vee X^2$ is a 2-dimensional CW-complex with 1-skeleton X^1 since ΣV_4 is a one point union of 2-spheres which has a trivial 1-skeleton. This shows that the tracks in the track-model of X are actually 0-tracks in $T_0\mathbf{CW}^2$, see (6.3)(1). Therefore we can apply the equivalence in theorem (6.4) to diagram (7.8). This will give us an 'algebraic track model of the 4-dimensional CW-complex X'.

To this end we introduce the following totally free nil(2)-modules which are objects in the algebraic track category $T_0\mathbf{Q}^2$, see (6.2)(7). Let

(7.10) $$\partial: \sigma_2 \to \sigma_1 = \langle Z_1 \rangle$$

be the free nil(2)-module with basis $f_2: Z_2 \to \langle Z_1 \rangle$ given by f_2 in (7.1). Moreover let

$$0: E_n \to \sigma_1 \tag{7.11}$$

be the free nil(2)-module with bases $Z_n \to 0 \in \sigma_1$, $n = 3, 4$. Here Z_n is the set of n-cells of X in (7.2). As in (IV.2.2) we have the **sum under** σ_1

(7.12)

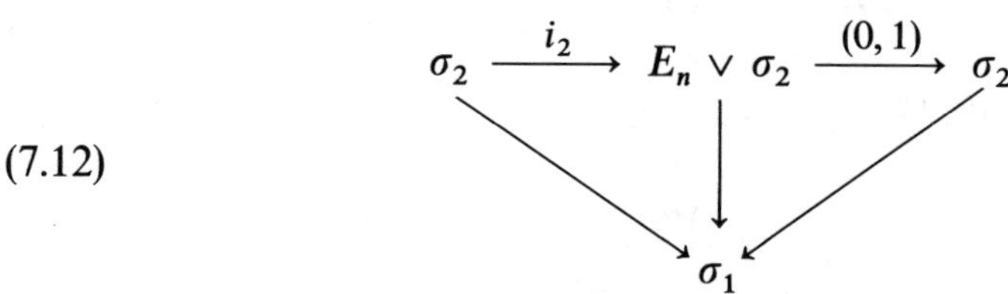

which is an object under and over σ_2 in the category $\mathbf{Q}^2$. The nil(2)-module $(0, \partial): E_n \vee \sigma_2 \to \sigma_1$ is actually the free nil(2)-module with basis

$$(0, f): Z_n \cup Z_2 \to \langle Z_1 \rangle \tag{1}$$

where $Z_n \cup Z_2$ is the disjoint union of sets. Here $(0, f)$ corresponds to the attaching map of the 2-dimensional CW-complex

$$\Sigma V_n \vee X^2 = C(0, f) \tag{2}$$

with $(0, f): V_n \vee V_2 \to X^1$, see (7.2).

We now apply the equivalence in theorem (6.4) to the track model in (7.8). This leads to the following diagram in the algebraic track category $T_0\mathbf{Q}^2$ which we call the **algebraic track model** of X.

(7.13)

$$\begin{array}{ccccc} & & E_4 \vee \sigma_2 & \xrightarrow{(0,1)} & \\ & & \downarrow \delta_4 & \Downarrow G & \sigma_2 \\ (0,1) & \overset{\Rightarrow}{H_X} & E_3 \vee \sigma_2 & \xrightarrow{(0,1)} & \\ & & \downarrow \delta_3 & & \\ & & \sigma_2 & & \end{array}$$

This diagram is obtained by choosing a diagram $\mathbf{D}$ in $T_0\mathbf{W}$ which projects via i to (7.8), see (6.4)(1), and by mapping $\mathbf{D}$ to $T_0\mathbf{Q}^2$ via j in (6.4)(1). Since G in (7.8) is good we see that G in (7.13) is actually the trivial track 0 and that $(0, 1)\delta_4 = (0, 1)$. Whence δ_4 in (7.13) is a map under and over σ_2. Moreover δ_3

is a map under σ_2 and H_X is constant on σ_2. Therefore (7.13) is an example for the following definition.

(7.14) **Definition.** An algebraic track model $(H_X, \delta_4, \delta_3)$ is a diagram

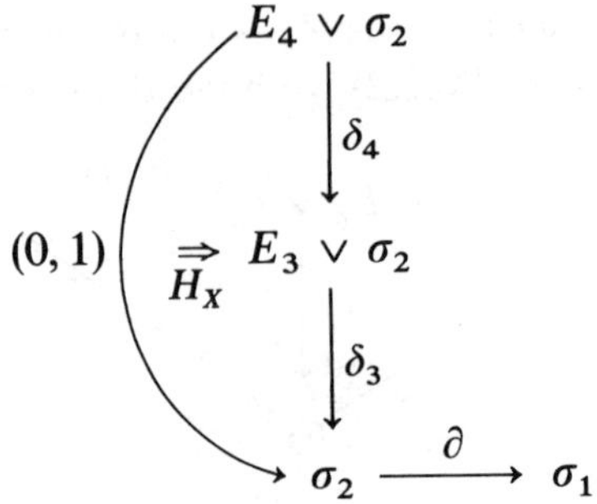

in the track category $T_0\mathbf{Q}^2$ where we use free nil(2)-modules as in (7.12). The map δ_4 is a map under and over σ_2 and δ_3 is a map under σ_2. Moreover the track H_X is **constant on** σ_2, this means that the homotopy

$$H_X: C_4 \oplus C_2 \to C_2 \otimes C_2 \tag{1}$$

is trivial if restricted to C_2, that is H_X is of the form $(H_X, 0)$ where $H_X: C_4 \to C_2 \otimes C_2$. Here we use the free π-modules

$$C_2 = \sigma_2^{cr\,ab} \quad \text{and} \quad C_n = E_n^{ab} \quad \text{for } n = 3, 4 \tag{2}$$

where $\pi = \operatorname{cok}(\partial)$ is the fundamental group. The homotopy H_X is a homorphism of π-modules as in (1) by the definition of 0-tracks in $T_0\mathbf{Q}^2$, see (6.2)(7) and (IV.2.8)(4).

Using (7.13) we associate with each 4-dimensional CW-complex X an algebraic track model of X as in (7.14). The equivalences in (6.4) and (7.4) show that each algebraic track model as in (7.13) arises as the algebraic track model of an appropriate 4-dimensional CW-complex.

(7.15) **Remark.** An algebraic track model (7.14) is uniquely determined by the functions

$$\left.\begin{aligned} &f_2: Z_2 \to \langle Z_1 \rangle = \sigma_1 \\ &\delta_3: Z_3 \to \sigma_2 \\ &\delta_4: Z_4 \to (E_3 \vee \sigma_2)_2 \\ &H_X: Z_4 \to C_2 \otimes C_2 \end{aligned}\right\} \tag{1}$$

which are given by the restrictions of the corresponding functions in (7.14). We call these functions the **basis** of the algebraic track model. Any set of functions as in (1) arises as a basis provided the following conditions are satisfied. The group σ_2 is given by the free nil(2)-module $\partial: \sigma_2 \to \sigma_1$ with basis f_2. The equation

$$\partial\delta_3(x) = 0 \tag{2}$$

holds for $x \in Z_3$; this implies that $\delta_3: E_3 \vee \sigma_2 \to \sigma_2$ is well defined by δ_3 in (1), see (IV.2.6). Finally, the equation

$$\delta_3\delta_4(y) = wH_X(y) \tag{3}$$

holds for $y \in Z_4$. Here $w: C_2 \otimes C_2 \to \sigma_2$ is the Peiffer commutator map for $\partial: \sigma_2 \to \sigma_1$.

Recall that $(E_n \vee \sigma_2)_2$ denotes the kernel of the projection $(0, 1): E_n \vee \sigma_2 \to \sigma_2$. We have the exact sequence

(7.16) $$\begin{cases} J_n \xrightarrow{w_n} (E_n \vee \sigma_2)_2 \overset{p}{\twoheadrightarrow} C_n \\ J_n = C_n \otimes C_n \oplus C_n \otimes C_2 \oplus C_2 \otimes C_n, \end{cases}$$

compare (IV.2.7). Using (7.16) we associate with an algebraic track model (7.14) the following *commutative* diagram of σ_1-groups.

(7.17)

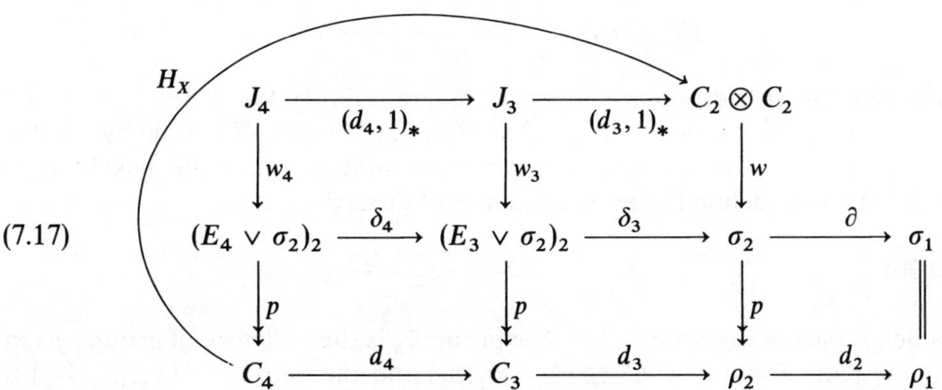

The columns of this diagram are exact sequences. The maps δ_4, δ_3, H_X are restrictions of the corresponding maps in (7.14). The maps in the bottom row are induced by δ_4, δ_3 and ∂ respectively. Recall that $\rho_2 = \sigma_2^{cr}$ and that $C_2 = \rho_2^{ab}$. One can check that the bottom row of (7.17) is a well defined totally free crossed chain complex $\rho(X)$ if (7.17) is given by the algebraic track model (7.13) of a 4-dimensional CW-complex X.

(7.18) **Lemma.** *Let ρ be a 4-dimensional totally free crossed chain complex. Then there is an algebraic track model $(H_X, \delta_4, \delta_3)$ which induces ρ.*

Proof. Let $f_2\colon Z_2 \to \rho_1$ be a basis of $d_2\colon \rho_2 \to \rho_1$. Then we choose ∂ by the free nil(2)-module with basis f_2. Let Z_n be a basis of the free $\pi_1(\rho)$-module C_n, $(n = 3, 4)$. We define E_n in (7.17) by Z_n, see (7.12). Moreover, we choose $\delta_3'\colon Z_3 \to \sigma_2$ with $\partial\delta_3' = 0$ and $p\delta_3' = d_3$. This is possible since p in (7.17) is surjective. Similarly we choose $\delta_4'\colon Z_4 \to (E_3 \vee \sigma_2)_2$ with $p\delta_4' = d_4$. The maps δ_3', δ_4' determine maps δ_3 and δ_4 as in (7.17) and (7.14). The composition $\delta_3\delta_4$, however, needs not to be trivial. Since

$$p(\delta_3\delta_4) = d_3 d_4 p = 0$$

we can choose $H_X\colon Z_4 \to C_2 \otimes C_2$ with $\delta_3\delta_4' = wH_X$. This map determines H_X in (7.17). □

Next we derive from an algebraic track model $X = (H_X, \delta_4, \delta_3)$ a totally free quadratic chain complex $\sigma = \sigma(X)$ as follows. We first consider the diagram

(7.19)

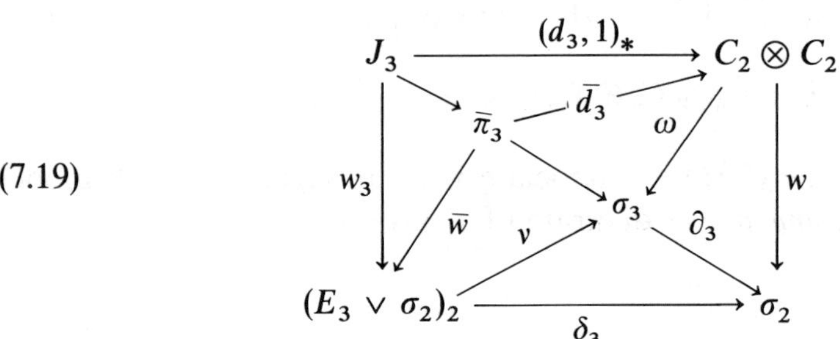

in which the outside square is part of diagram (7.17). We define (σ_3, v, ω) by the *central push out* of $(w_3, (d_3, 1)_*)$ so that $\partial_3 = (\delta_3, w)$. We know by (IV.2.9) that $(\omega, \partial_3, \partial_2 = \partial)$ is a totally free quadratic module; this is the 3-skeleton of $\sigma(X)$. We now define the exact sequence of groups

(7.20) $$J_3 \xrightarrow{d^J} \bar{\pi}_3 \xrightarrow{\nabla} \sigma_3 \xrightarrow{\partial_3} \sigma_2$$

which is part of diagram (7.19). The group $\bar{\pi}_3$ is the *pull back* of groups given by the pair of maps $(w\colon C_2 \otimes C_2 \to \sigma_2, \delta_3)$ and the map $d^J = (w_3, (d_3, 1)_*)$ is induced by the maps in (7.19). We now define ∇ by the difference

(7.21) $$\nabla = v\bar{w} - \omega\bar{d}_3$$

This is a well defined homomorphism since $\omega(C_2 \otimes C_2)$ is central in σ_3.

(7.22) **Lemma.** *The sequence* (7.20) *is exact.*

Proof. Clearly we have

$$\partial_3 \nabla = \partial_3(v\bar{w} - \omega\bar{d_3}) = \delta_3 \bar{w} - w_3 \bar{d_3} = 0,$$

$$v d^J = (v\bar{w} - \omega\bar{d_3}) d^J = v w_3 - \omega(d_3, 1)_* = 0.$$

We obtain the pull back $\bar{\pi}_3$ by the subgroup

$$i: \bar{\pi}_3 \rightarrowtail (E_3 \vee \sigma_2)_2 \times (C_2 \otimes C_2) \tag{1}$$

consisting of all pairs (x_1, x_2) with $\delta_3(x_1) = w(x_2)$. The maps $\bar{w}, \bar{d_3}$ are given by

$$i(z) = (\bar{w}z, \bar{d_3}z), z \in \bar{\pi}_3. \tag{2}$$

On the other hand the central push out σ_3 is the quotient group (see (I.4.19))

$$q: (E_3 \vee \sigma_2)_2 \times (C_2 \otimes C_2) \twoheadrightarrow \sigma_3 \tag{3}$$

for which the kernel of q is given by all pairs $(w_3 y, (d_3, 1)_*(-y))$, $y \in J_3$. Here we use the fact that $w_3 y$ is central. The maps v, ω are given by

$$q(x_1, x_2) = v(x_1) + \omega(x_2). \tag{4}$$

Now it is easy to show that (7.20) is exact. In fact, let $y = vx_1 + \omega x_2 \in \sigma_3$ with $\partial_3 y = 0$. Then we have $\partial_3(vx_1 + \omega x_2) = \delta_3 x_1 + \omega x_2 = 0$ and hence $(x_1, -x_2) \in \bar{\pi}_3$ with

$$\nabla(x_1, -x_2) = (v\bar{\omega} - \omega\bar{d_3})(x_1, -x_2)$$

$$= vx_1 + \omega x_2 = y$$

Next let $(x_1, x_2) \in \bar{\pi}_3$ with $\nabla(x_1, x_2) = 0$, that is $vx_1 + \omega(-x_2) = q(x_1, -x_2) = 0$. Whence, there is y with $(x_1, -x_2) = (w_3 y, (d_3, 1)_*(-y))$ by (3). Equivalently we get $(x_1, x_2) = d^J(y)$. □

We now proceed in the construction of the quadratic chain complex $\sigma = \sigma(X)$ given by $X = (H_X, \delta_4, \delta_3)$. Since the 3-skeleton of $\sigma(X)$ is already defined in (7.19) it remains to define the boundary map $\partial_4: \sigma_4 = C_4 \to \sigma_3$. For this we consider the following commutative diagram

$$\begin{array}{ccc} (E_4 \vee \sigma_2)_2 & \xrightarrow{\partial'_4} & \bar{\pi}_3 \\ \downarrow p & & \downarrow \nabla \\ C_4 & \xrightarrow[\partial_4]{} & \sigma_3 \end{array} \tag{7.23}$$

where p is part of (7.17) and where ∇ is defined in (7.21). The map ∂'_4 is induced by maps in the commutative diagram (7.17), namely $\partial'_4 = (\delta_4, H_X p)$ where we use the pull back property of $\bar{\pi}_3$. The map ∂'_4 is well defined since $wH_X p = \delta_3 \delta_4$ by (7.17).

(7.24) **Lemma.** *The map ∂'_4 induces a homomorphism ∂_4 such that* (7.23) *commutes.*

Proof. We consider the map $w_4\colon J_4 \to (E_4 \vee \sigma_2)_2$ in (7.17) which maps onto the kernel of p in (7.23) and we show $\nabla \partial'_4 w_4 = 0$. This implies (7.24). In fact, we have

$$\nabla \partial'_4 w_4 = \nabla d^J (d_4, 1)_* = 0$$

by definition of ∂'_4 and by the commutativity of diagram (7.17). Clearly $(d_3, 1)_*(d_4, 1)_* = 0$ since $d_3 d_4 = 0$. □

The next result describes the connection of an algebraic track model $(H_X, \delta_4, \delta_3)$ of a 4-dimensional CW-model X and of the quadratic chain complex of X. This result yields via diagram (7.19) the geometric meaning of the free quadratic module (constructed in (IV.2.9)) in terms of tracks.

(7.25) **Theorem.** *Let X be a 4-dimensional* CW-*complex with $X^0 = *$ and let $(H_X, \delta_4, \delta_3)$ be an algebraic track model of X as constructed in* (7.13). *Then $(H_X, \delta_4, \delta_3)$ yields the quadratic chain complex $\sigma = \sigma(X)$ of X in* (IV.6.8)(4). *In fact σ is given by*

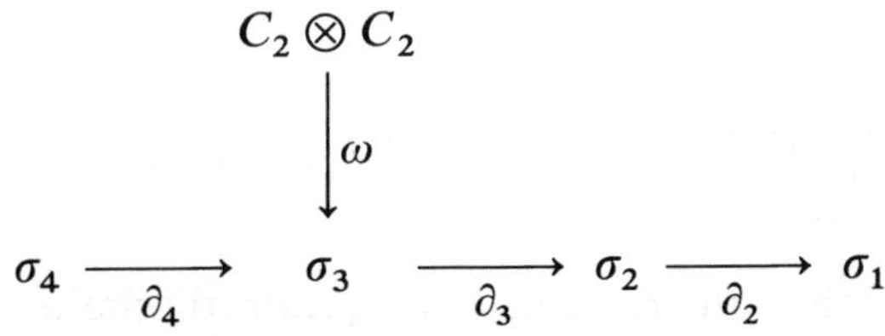

where $(\omega, \partial_3, \partial_2)$ with $\partial_2 = \partial$ is defined by (7.19) *and where ∂_4 is defined by* (7.23).

On the other hand we obtain an algebraic track model $(H_X, \delta_4, \delta_3)$ for X from $\sigma(X)$ by choosing ∂'_4 in (7.23) and by choosing δ_3 in (7.17), here we use the exactness of (7.20). The algebraic track models (7.14) are objects of a category which we define as follows. For this we introduce first the following

(7.26) **Definition.** Consider a track

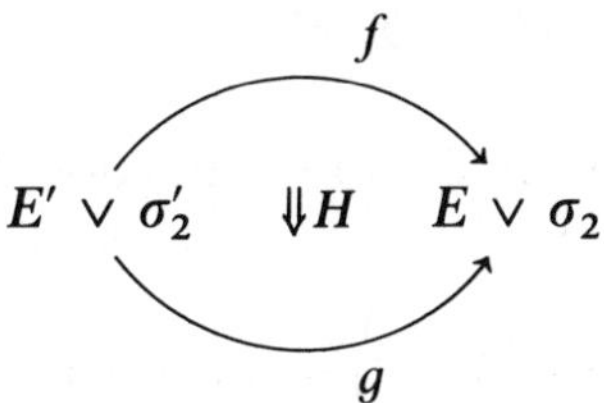

in the category $T_0\mathbf{Q}^2$. Here $(0, \partial')$: $E' \vee \sigma_2' \to \sigma_1'$ and $(0, \partial)$: $E \vee \sigma_2 \to \sigma_1$ are free nil(2)-modules as in (7.12). The track H is a homomorphism

$$H: C' \oplus C_2' \to (C \oplus C_2) \otimes (C \oplus C_2)$$

by (6.2)(7) and (IV.2.8)(4) with $C' = (E')^{ab}$, $C = E^{ab}$. We say that H is **constant on** σ_2' if $Hi_2 = 0$ where i_2: $C_2' \subset C' \oplus C_2'$ is the inclusion. Moreover, we say that H is **good** if $(p_2 \oplus p_2)H = 0$ where p_2: $C \oplus C_2 \to C_2$ is the projection.

With the notation in (7.26) we define the **category of algebraic track models** as follows. Consider the diagram in $T_0\mathbf{Q}^2$:

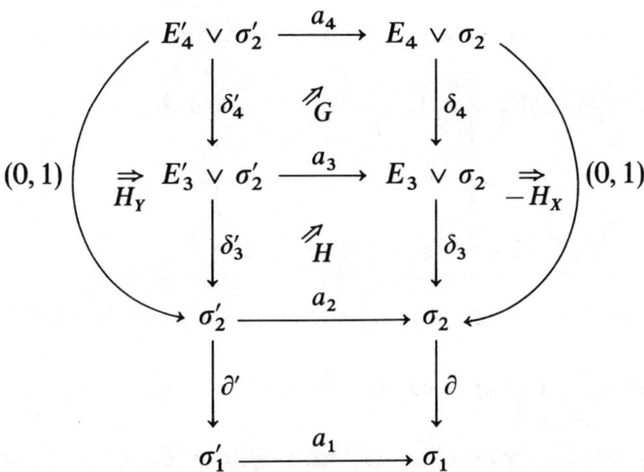

where $Y = (H_Y, \delta_4', \delta_3')$ and $X = (H_X, \delta_4, \delta_3)$ are algebraic track models. We say that

(7.27) $$(a, G, H): Y \to X$$

with G, H and $a = (a_1, a_2, a_3, a_4)$ as in the diagram above is a **map between algebraic track models** if the following conditions hold. The maps (a_n, a_1), $n = 1, 2, 3$, are maps between nil(2)-modules and the maps a_3 and a_4 are maps under and over a_2, see (7.12). Moreover, track addition in the diagram yields the trivial track, that is

$$0 = a_{2*}H_Y + \delta_4'^*H + \delta_{3*}G - a_4^*H_X. \tag{1}$$

Here H and G are constant on σ_2' and G is good, see (7.26). Composition of such maps is defined by

$$(a, G, H)(b, G', H') = (ab, G * G', H * H') \tag{2}$$

where $G * G'$ and $H * H'$ are defined as in (3.4). One readily checks that algebraic track models (7.14) with maps as in (7.27) form a well defined category which we denote by **trackmodels**. We obtain a functor

(7.28) $$\sigma\colon \mathbf{trackmodels} \to \mathbf{Q}^4$$

as follows. On objects this functor is defined as in (7.25) by the construction in (7.19) and (7.23). On morphisms (7.27) this functor is given by the quadratic chain map

$$f = \sigma(a, G, H)\colon \sigma(Y) \to \sigma(X) \tag{1}$$

with $f_1 = a_1, f_2 = a_2$. Moreover f_4 is the map which makes the diagram

$$\begin{array}{ccc} (E_4' \vee \sigma_2')_2 & \xrightarrow{a_4} & (E_4 \vee \sigma_2)_2 \\ {\scriptstyle p}\downarrow & & \downarrow{\scriptstyle p} \\ C_4' = \sigma_4' & \xrightarrow[f_4]{} & \sigma_4 = C_4 \end{array} \tag{2}$$

commutative. Finally we define f_3 by the formula (see (IV.3.4)(1))

$$f_3(v'x_1 + \omega'x_2) = va_3x_1 - \omega Hpx_1 + \omega(\bar{a}_2 \otimes \bar{a}_2)x_2 \tag{3}$$

where $x_1 \in (E_3' \vee \sigma_2')_2$, $x_2 \in C_2' \otimes C_2'$ and where $\bar{a}_2\colon C_2' \to C_2$ is the map induced by a_2. The homotopy $H\colon C_2' \to C_2 \otimes C_2$ is given by $H = (H, 0)$ in (7.27). This completes the definition of f in (1). In fact, the map f does not depend on G; but the existence of G and equation (7.27)(1) imply that $\partial_4 f_4 = f_3\partial_4'$ in (1). Moreover, since H is given by a homotopy as in (7.27) we get $\partial_3 f_3 = f_2\partial_3'$. Also by (3) we have $f_3\omega' = \omega(\bar{a}_2 \otimes \bar{a}_2)$. Therefore f in (1) is a well defined map between quadratic chain complexes. Moreover, one readily checks that (1) is compatible with composition so that σ in (7.28) is a well defined functor.

(7.29) **Lemma.** *The functor σ in (7.28) satisfies the realizability condition for objects and maps.*

Compare the remark following theorem (7.25) above.

(7.30) **Remark.** The maps between the track models in (7.27) are motivated by the construction in (VII.1.21) Baues (AH) where we set $n = 2$ and where we interchange X and Y. By applying the equivalence (6.4) to the diagram in Baues (AH) we get the diagram for (7.27). This shows that similar methods as discussed in this section are useful in many different cofibration categories. This also shows that the obstruction operator in the CW-tower can be calculated by the computation of certain track categories.

Bibliography

Barratt, M.G., (HR) Homotopy ringoids and homotopy groups, Q. J. Math. Ox (2) 5 (1954), 271–290

—, (T) Track groups (II), Proc. London Math. Soc. (3) 5 (1955), 285–329

Bauer, St., The homotopy type of a 4-manifold with finite fundamental group, Lecture Notes in Math. 1361 (1988), 1–6

Baues, H.J., (AH) *Algebraic Homotopy*, Cambridge Studies in Advanced Mathematics 15, Cambridge University Press (1988), 450 pages

—, (CC) *Commutator calculus and groups of homotopy classes*, London Math. Soc. LNS 50 (1981)

—, (M) On the homotopy category of 1-connected 4-dimensional manifolds, In preparation

—, (OT) *Obstruction Theory*, Lecture Notes in Math. 628 (1977), Springer-Verlag

—, (QF) Quadratic functors and metastable homotopy I, II. Max-Planck-Institut für Mathematik, preprint 89-24, 89-31

Baues, H.J. and Conduché, D., The central series for Peiffer commutators in groups with operators, J. of Algebra, 133 (1990), 1–34

Baues, H.J. and Dreckmann, W., The cohomology of homotopy categories and the general linear group. *K*-Theory, 3 (1989), 307–338

Baues, H.J. and Hennes, M., The homotopy classification of $(n-1)$-connected $(n+3)$-dimensional polyhedra, Preprint Max-Planck-Institut für Mathematik, Bonn (1989), MPI/89-21, to appear in Topology

Baues, H.J. and Wirsching, G., The cohomology of small categories, J. Pure and Appl. Alg. 38 (1985), 187–211

Brown, K.S., *Cohomology of groups*, GTM 87 Springer-Verlag N.Y. (1982)

Brown, R., *Topology*, Ellis Horwood Series in Mathematics and its Applications, Ellis Horwood Limited Publishers Chichester (1988), 460 pages

Brown, R. and Gilbert, N.D., Algebraic models of 3-types and automorphism structures for crossed modules, Proc. London Math. Soc. (3) 59 (1989), 51–73

Brown, R. and Golasinski, M., A model structure for the homotopy theory of crossed complexes, Cah. Top. Geom. Diff. Cat. 30 (1989), 22 pp

Brown, R. and Higgins, P.J. (AC) On the algebra of cubes, Journal of Pure and Applied Algebra 21 (1981), 233–260

—, (CC) Crossed complexes and non-abelian extensions, Proc. Int. Conf. on Category Theory, Gummersbach 1981 Lecture Notes in Math (Springer)

—, (TP) Tensor products and homotopies for ω-groupoids and crossed complexes, Journal of Pure and Applied Algebra 47 (1987), 1–33

Brown, R. and Huebschmann, J., Identities among relations, in "Low dimensional topology", London Math. Soc. Lecture Note 48, Cambridge University Press 1982

Brown, R. and Loday, J.-L., Van Kampen theorems for diagrams of spaces, Topology 26 (1987), 311–335

Conduché, D., Modules croisés généralisés de longueur 2, Journal of Pure and Applied Algebra 34 (1984), 155–178

Curtis, E.B., (SH) Simplicial homotopy theory, Advances in Math. 6 (1971), 107–209

—, (SR) Some relations between homotopy and homology, Ann. of Math. 82 (1965), 386–413

Eilenberg, S. and Mac Lane, S., On the groups $H(\pi, n)$, II. Ann. of Math. (2) 60 (1954), 49–139

Fantham, P.H.H. and Moore, E.J., Groupoid enriched categories and homotopy theory, Can. J. Math. 35 (1983), 385–416

Gabriel, P. and Zisman, M., *Calculus of fractions and homotopy theory*. Ergebnisse der Math. und ihrer Grenzgebiete 35, Springer Verlag 1967

Gitler, S., Cohomology operations with local coefficients, American J. Math. 85 (1963), 156–188

Grothendieck, A., Sur quelques point d'algèbre homologique, Tohoku Math. J. 9 (1957), 119–221

Hambleton, I. and Kreck, M., On the classification of topological 4-manifolds with finite fundamental group, Math. Ann. 280, 85–104 (1988)

Hartl, M., The second cohomology of the category of finitely generated abelian groups, Preprint Bonn (1985), Diplomarbeit Math. Inst. der Universität Bonn

Hendricks, H., Obstruction theory in 3-dimensional topology: classification theorems, Bull. AMS 83 (1977), 737–738

Hilton, P., *Homotopy theory and duality*, Nelson (1965) Gordon Breach

Hilton, P. and Stammbach, U., A course in homological algebra, Springer GTM 4. New York 1971

Hohmann, A., Das Tensorprodukt quadratischer Kettenkomplexe, Diplomarbeit Math. Inst. der Universität Bonn (1989)

Igusa, K., On the algebraic K-theory of A_∞-ring spaces, in Algebraic K-Theory, Proc. Oberwolfach 1980 part II, Springer Lecture Notes in Math. 967 (1982), 146–194

James, I.M., Reduced product spaces, Ann. of Math. 62 (1955), 170–197

Jibladze, M.A. and Pirashvili, T.I., Some linear extensions of the category of finitely generated free modules, Bulletin of the Academy of Sciences of the Georgian SSR. 123 No. 3, 1986 (Translated from Russian)

Johansson, J., Zu den zweidimensionalen Homotopiegruppen, Norsk Mat. Forenings Skr., (2) Nr. 1/12 (1933), 55–59

Kampen, E.H. Van, On the connection between the fundamental group of some related spaces, Amer. Journ. Math. 55 (1933), 261–267

Kan, D.M., A relation between *CW*-complexes and free C.s.s. groups, Am. J. of Math. 81 (1959), 512–528

Kreck, M. and Schafer, J.A., Classification and stable classification of manifolds: some examples, Comment. Math. Helv. 59 (1984), 12–38

Loday, J.-L., Spaces with finitely many homotopy groups, Journal of Pure and Applied Algebra 34 (1984), 155–178

Mac Lane, S. and Whitehead, J.H.C., On the 3-type of a complex, Proc. Nat. Acad. Sci 36 (1950), 41–48

Magnus, W. and Karrass, A. and Solitar, D., *Combinatorial group theory*, Interscience, New York (1966)

Massey, W.S., *Algebraic Topology*: An Introduction. Harcourt, Brace and World, Inc. New York (1967) Springer Verlag (1977)

Metzler, W., Über den Homotopietyp zweidimensionaler *CW*-Komplexe und Elementartransformationen bei Darstellungen von Gruppen durch Erzeugende und definierende Relation, J. reine u. angew. Math. 285 (1976), 7–23

Mitchell, B., Rings with several objects, Advances in Math. 8 (1972), 1–161

Olum, P., Self-equivalences of pseudo-projective planes, Topology 4 (1965), 109–127

Reidemeister, K., Homotopiegruppen von Komplexen, Abh. math. Sem. Hamburg Univ. (1934)

Rutter, J.W., Homotopy classification of maps between pseudo-projective planes, Quaestiones Math. 11 (1988), 409–422

Seifert, H. and Threlfall, W., *Lehrbuch der Topologie*, Chelsea Publishing Company New York

Sieradski, A., A semigroup of simple homotopy types, Math. Zeit. 153 (1977), 135–148

—, Stabilization of self equivalences of the pseudo projective spaces. Michigan Math. J. 19 (1972) 109–119

Sieradski, A. and Dyer, M.L., Distinguishing arithmetic for certain stably isomorphic modules, J. of Pure and Appl. A. 15 (1979), 199–217

Smith, J.R., Topological realizations of chain complexes I. The general theory, Topology and its Appl. 22 (1986), 301–313

Steenrod, N.E., Homology with local coefficients, Ann. of Math. 44 (1943), 610–627

Thomas, E., The generalized Pontrjagin cohomology operations and rings with divided powers, Memoirs of the AMS 27 (1957), 82 pages

Toda, H., *Composition methods in homotopy groups of spheres*, Princeton University Press (1962)

Unsöld, H.M., On the classification of spaces with free homology, Dissertation Math. Inst. der Freien Universität Berlin 87

Wall, C.T.C., Finiteness conditions for *CW*-complexes II, Proc. Roy. Soc. 295 (1966), 129–139

—, (P) Poincaré complexes, Ann. Math. 86 (1962), 213–245

—, (S) *Surgery on compact manifolds*, Academic Press London New York (1970)

Weick, M., Diplomarbeit Bonn (1988)

Whitehead, G.W., *Elements of homotopy theory*, Graduate Texts in Math. 61, Springer (1978)

Whitehead, J.H.C., (C) Combinatorial Homotopy I, Bull. Amer. Math. Soc. 55 (1949), 213–245

—, (CE) A certain exact sequence, Ann. of Math. 52 (1950), 51–110

—, (CH) Combinatorial homotopy II, Bull. AMS 55 (1949), 213–245

—, (HT) The homotopy type of a special kind of polyhedron, Annales de la Soc. Polonaise de Math. 21 (1948), 176–186

—, (SB) The secondary boundary operator, Proc. Nat. Acad. Sci. 36 (1950), 55–60

—, (SC) On simply connected 4-dimensional polyhedra, Comment. Math. Helv. 22 (1949), 48–92

—, (SH) Simple homotopy types, Amer. J. Math. 72 (1950), 1–57

List of Symbols

(i) Notation for categories

Boldface letters like $\mathbf{C}$, $\mathbf{K}$, ... denote categories (page numbers)

(ii) Topological and algebraic notation

Index

UNIV DUNELM

DURHAM UNIVERSITY LIBRARY
3 0104 00639674 5